Sustainable Consumption

Alberto do Amaral Junior • Lucila de Almeida • Luciane Klein Vieira
Editors

Sustainable Consumption

The Right to a Healthy Environment

Springer

Editors
Alberto do Amaral Junior
School of Law
University of São Paulo
São Paulo, Brazil

Lucila de Almeida
Florence School of Regulation, Robert Schuman Centre for Advanced Studies, European University Institute
Florence, Italy

Luciane Klein Vieira
Department of Law
UNISINOS University of the Rio dos Sinos Valley and University of Buenos Aires
São Leopoldo, Brazil

ISBN 978-3-030-16987-9 ISBN 978-3-030-16985-5 (eBook)
https://doi.org/10.1007/978-3-030-16985-5

This Springer imprint is published by the registered company Springer Nature Switzerland AG.
The registered company address is: Gewerbestrasse 11, 6330 Cham, Switzerland

Foreword

The rise of consumerism in the second half of the twentieth century brought the need to protect consumers, a phenomenon reflected in most legal traditions predominantly through private law approaches. Yet, by the end of the century, consumers had themselves become a threat because of the environmental footprint of their combined choices. Now legal systems struggled to find solutions, largely because curtailing consumption is politically delicate. Over time, however, a wide range of techniques has been developed to bring consumption to more sustainable levels. Until now, these techniques have remained scattered across different areas of law, with the resulting challenge for anyone interested in their forms and operation. This volume addresses this challenge by providing a comprehensive survey of the legal techniques developed to pursue more sustainable levels of consumption. The editors and contributors should be praised for offering such a comprehensive and timely account of this strategically important legal development.

University of Cambridge,
Cambridge, UK

Jorge E. Viñuales

Foreword

Sustainable Consumption has been on the agenda for more than 20 years. After the initial hype in the aftermath of the greening of the United Nations Guidelines on Consumer Protection in the 1990s, sustainable consumption has been broken down into the various disciplines involved. In law, sustainability has never ranked high on the agenda. Setting Environmental Law aside, it has played a role at the margins but not at the core of the law that governs economic transactions: Trade Law, Competition Law, state aid, Labour Law and Consumer Law. Sustainability has led its own life, in Environmental Law and policy, and in consumer policy, but not so much in Consumer Law.

The two fields have remained neatly separated, despite forceful voices calling for a holistic perspective, and despite the growing interest in ethical consumption. The 'greening' of Competition Law or of Trade Law is another bumpy road towards sustainable consumption. In this context, it is apt to consider Trentmann's reconstruction of the history of consumption since the fifteenth century.[1] At the final chapter of the book, his outlook is not very optimistic. Private consumption, *we* the consumers are standing in the way for a world where sustainability is taken seriously. Consumer law and the way it has developed over the last 50 years rests on the ideals of the consumer society, more, better, and always newer products. The remedies of Consumer Law, with the European Union being at the forefront of the development, is built around replacement and not about repair, about fast substitution and not long lasting endurance. Taking sustainability seriously requires a different economy, a circular economy, to use another catchword of our times. The 'scream'[2] of sustainable consumption cannot be achieved without engaging

[1]Trentmann (2016).

[2]'We can only start from where we are, from where we are but do not want to be, from where we scream'. Holloway (2010), p. 4; Lindahl (2018), p. 177, with further examples from the opponents of globalisation.

into the political economy of the law. It seems that the Global South is far ahead of the Global North.

This book, with its broad scope that reaches beyond the core of Consumer Law and Environmental Law that engages into the political and economic dimension of sustainability, fills a huge gap in the academic literature. Those who are writing blurbs tend to use a kind of standard formula 'this is a must read for everybody working in the field'. Sustainable consumption: the right to a health environment goes beyond. The book formulates a political agenda on the law we need to realise sustainable consumption. In 1962, John F. Kennedy famously coined the formula 'consumers we are all'. The consumer rights he advocated and which form still the core of consumer policy programmes around the world, must be complemented by consumer responsibilities. There is no unlimited right to consumption at the costs of the planet. It is not by coincidence that the initiative to the current book stems from the Global South. The contributions provide for a broad array of proposals of what could be done here and now, and on what kind of law we need to achieve a different economy and a different society.

European University Institute, Hans-W. Micklitz
Florence, Italy

References

Holloway J (2010) Change the world without taking power: the meaning of revolution today, 2nd edn. Pluto Press, London

Lindahl H (2018) Authority and the globalisation of inclusion and exclusion. Cambridge University Press, Cambridge

Trentmann F (2016) The empire of things: how we became a world of consumers, from the fifteenth century to the twenty-first. Allen Lane, London

Contents

Abbreviations and Acronyms

AAP	Asociación Argentina de Publicidad
AB	Appellate Body
AEI	Agro-ecological intensification
ANEEL	Brazilian National Energy Agency
AMF	Autorité des marchés financiers
APEC	Asia-Pacific Economic Cooperation
APMS	Advances in Production Management Systems
ASEAN	Association of Southeast Asian Nations
BCBS	Basel Committee for Banking Supervision
BCH	Biosafety Clearing-House
BG	Base Capacity Generation
BRA	Base Residual Auction
CAA	Cámara Argentina de Anunciantes
CBD	Convention on Biological Diversity
CBDR	Common but differentiated responsibilities
CCC	Civil and Commercial Code
CCLC	Civil Code of Lower Canada
CDP	Carbon Disclosure Project
CETA	Comprehensive Economic and Trade Agreement
CETESB	Environmental Company of the State of São Paulo
CGF	Consumer Goods Forum
CIDIP	Conferencias Especializadas Interamericanas sobre Derecho Internacional Privado
CMHC	Canada Mortgage and Housing Corporation
COO	Country-of-Origin
COOL	Country of Origin Labelling
CPATTP	Comprehensive and Progressive Agreement for Trans-Pacific Partnership
CPC	Consumer Protection Code
CRA	Consumption Reduction Agents
CSA	Climate smart agriculture

CSR	Corporate Social Responsibility
CTE	Committee on Trade and Environment
DER	Distributed energy resource
DiD	Difference-in-Difference
DISCO	Electricity distribution company
DR	Demand Resources
EC	European Commission
EE	Energy Efficiency
EEE	Electrical and electronical equipment
EEM	Energy-efficiency measures
EESC	European Economic and Social Committee
EGA	Environmental Goods Agreement
EGS	Environmental goods and services
EMIT	Environmental Measures and International Trade
EP	European Parliament
EPP	Environmentally preferable products
EPR	Extended Producer Responsibility
ERSE	Entidade Reguladora dos Serviços Energéticos
ESC	Electricity Savings Commitments
ESCO	Energy Service Companies
ETOSS	Entre Tripartito de Obras y Servicios Sanitarios
EU	European Union
EWS	Efficient World Scenario
FAO	Food and Agriculture Organization
FCC	Federal Communications Commission
FCTC	Framework Convention on Tobacco Control
FGTS	Fundo de Garantia por Tempo de Serviço
FiT	Feed-in tariff
FRFI	Federally-regulated financial institutions
FSB	Financial Stability Board
GAP	Good Agricultural Practice
GATS	General Agreement on Trade in Services
GATT	General Agreement on Tariffs and Trade
GDP	Gross Domestic Product
GFSI	Global Food Safety Initiative
GHG	Greenhouse gases
GMO	Genetically Modified Organism
GPSD	General Product Safety Directive
GSA	General Services Administration
GSC	Global Services Coalition
GSCP	Global Social Compliance Program
HRW	Human Rights Watch
IAIS	International Association of Insurance Supervisors
IBRD	International Bank for Reconstruction and Development

ICSID	International Centre for Settlement of Investment Disputes
IEA	International Energy Agency
IMF	International Monetary Fund
IPPC	International Plant Protection Convention
ITU	International Telecommunication Union
IWRM	Integrated water resources management
JPOI	Johannesburg Plan of Implementation
M&V	Measurement and verification
MBI	Market-based instruments
MDG	Millennium Development Goals
MEA	Multilateral environmental agreements
NAPE	Brazilian National Action Plan for Energy Efficiency
NGO	Non-governmental organizations
NOI	Net Operating Income
NPV	Net Present Value
NSWP	Brazilian National Solid Waste Policy
OAS	Organization of American States
OECD	Organization for Economic Co-operation and Development
OPC	Office de la protection du consommateur
OSFI	Office of the Superintendent of Financial Institutions
PEE	Policy for Energy Efficiency
PES	Payment for environmental services
PNEf	Brazilian National Plan for Energy Efficiency
PPEC	Plan to Promote Efficiency in Electricity Consumption
PPM	Processes and production methods
PSIRU	Public Services International Research Unit
PV	Photovoltaic (solar)
RCEP	Regional Comprehensive Economic Partnership
RCT	Rational choice theories
RMUP	Residential Mortgage Underwriting Policy
RoO	Rules on origin
RPM	Reliability Pricing Model
RRI	Responsible Research and Innovation
SABESP	Sanitation Company of the State of São Paulo
SAD	Single Administrative Document
SAI	Sustainable agricultural intensification
SCP	Sustainable consumption and production
SDG	Sustainable Development Goals
SFP	Sustainable Fisheries Partnership
SNIS	System for Information on Basic Sanitation
SPC	Sustainable Production and Consumption
SPS	Sanitary and Phytosanitary
TBT	Technical Barriers to Trade
TEU	Treaty on European Union

TiSA	Trade in Services Agreement
TPP	Trans-Pacific Partnership
TTIP	Transatlantic Trade and Investment Partnership
TWC	Tradable White Certificate
UN	United Nations
UNCED	United Nations Conference on Environment and Development
UNCESCR	United Nations Covenant on Economic, Social and Cultural Rights
UNCTAD	United Nations Conference on Trade and Development
UNSPSC	United Nations Standard Products and Services Code
UNTRAD	United Nations Conference on Trade and Development
UNDESA	United Nations Department for Economic and Social Affairs
UNEP	United Nations Environmental Program
UNESCO	United Nations Educational, Scientific and Cultural Organization
UNFCCC	United Nations Framework Convention on Climate Change
UNFPA	United Nations Fund for Population Activities
UNGCP	United Nations Guidelines for Consumer Protection
UNIC	Unione Nazionale Industria Conciaria
VOC	Volatile organic compounds
WBCSD	World Business Council for Sustainable Development
WEE	Waste electrical & electronic equipment
WEEE	Waste management of electrical and electronic equipment
WHO	World Health Organization
WPN	Working Party on Nanotechnology
WSSD	World Summit on Sustainable Development
WTO	World Trade Organization

List of Figures

List of Tables

Chapter 1
An Introduction to Sustainable Consumption and the Law

Alberto do Amaral Junior, Lucila de Almeida, and Luciane Klein Vieira

Contents

Abstract This chapter introduces this edited book by shedding light on its normative and conceptual framework, as well as its contribution to the field as collection of interconnected works on how the law does and/or should play a role in influencing patterns of sustainable consumption.

Barely a quarter-century old, sustainable consumption is now emerging as a leading field of research and practice. The social sciences, including economics, sociology, anthropology, political science and public policy, have been at the forefront of these developments. Sustainable consumption—alongside its siblings, sustainable production and sustainable development—is now recognized as an important field of interdisciplinary research linking virtually all social scientific disciplines.[1]

[1]Reisch and Thøgersen (2015), pp. 1, 13.

A. d. Amaral Junior
University of São Paulo (USP), São Paulo, Brazil

L. de Almeida (✉)
Florence School of Regulation, Robert Schuman Centre for Advanced Studies, European University Institute, Florence, Italy
e-mail: lucila.dealmeida@eui.eu

L. Klein Vieira
Universidade do Vale do Rio dos Sinos (UNISINOS), São Leopoldo, Brazil

University of Buenos Aires (UBA), São Leopoldo, Brazil
e-mail: lucianevieira@unisinos.br

A. d. Amaral Junior et al. (eds.), *Sustainable Consumption*,
https://doi.org/10.1007/978-3-030-16985-5_1

Legal scholarship, in contrast, has neglected the importance of the field of sustainable consumption, although the normative legal implications of sustainable consumption became clear decades ago. Sustainable consumption as a legal concept was born out of an international declaration endorsed by 178 states: the 1992 Rio Declaration. Emerging from the United Nations Earth Summit of the same year, the Declaration brought to the world's attention the two sides of the coin regarding the influential factors underpinning risks to the global environment: unsustainable harmful overproduction and unsustainable consumption.[2] The silence of legal scholars on this question is all the more puzzling when we consider that sustainable consumption is at the forefront of the UN's approach to human development, a commitment enshrined in both its Guidelines for Consumer Protection[3] and the 2030 Agenda for Sustainable Development.[4]

This book seeks to remedy this by bringing the voice of legal scholarship to the table. It aims to provide a broad understanding of how the law *does* or *should* play a role in influencing patterns of sustainable consumption. Bringing together legal scholars from the Global South and the Global North, this edited volume project examines the issue of sustainable consumption from the perspective of different legal disciplines, legal systems, and legal cultures. The first two sections of the book consider the notion of sustainable consumption from the perspective of traditional legal disciplines such as Consumer Law, Environmental Law, Constitutional Law, Tort Law, International Law and Economic Law. The other three sections, in contrast, embody an understanding of fragmentation[5] and pluralism[6] in law. The chapters in these sections shed much-needed light on sector-specific engagements with sustainable consumption within national, transnational, and international legal orders. The collection of chapters here address crucial questions on sustainable consumption and the law, and aim to open a pathway to future work in the field.

1.1 Contribution to the Field

We believe that, as a field of research, sustainable consumption and the law has remained mired in narrow debates as an exclusive matter of either environmental sustainability or consumer empowerment. The first contribution this edited volume therefore makes is to bring other fields of legal scholarship up to speed on the conceptualization of sustainable consumption adopted widely—and to great effect—throughout the social sciences.

---

[2]UNCED (1992), para 4.3 and 4.4.

[3]UNGCP (2003), para. 4 and 42.

[4]UN General Assembly (2015), Goal 12.

[5]Teubner and Fischer-Lescano (2004); Koskenniemi and Leino (2002).

[6]Teubner (1997), Michaels (2005), Berman (2007) and Krisch (2010).

The field of research on sustainable consumption has flourished in the quarter-century since the Rio Declaration. States, organizations, communities, and individuals have grown increasingly aware of the unsustainability of the current trajectory of human consumption patterns. An impressive corpus of empirical research has provided irrefutable evidence for the claim that sustainable development can only be achieved if consumption patterns undergo fundamental change.[7] Frankly, humanity consumes too many natural resources, produces too much toxic material and generates too much hazardous waste. The impact of consumption over time has degraded basic 'ecosystem services', such as the provision of fresh water, fertile soil, and a protective ozone layer for future generations.

As mentioned, the non-legal fields have already shed much-needed light on the causes and effects of—and potential remedies for—unsustainable consumption. For their part, economists downplay the potential for efficiency gains in production to address the problem, arguing that such advances simply provide more resources to be consumed.[8] Sociologists have shown how individual consumption behavior—sustainable or otherwise —is fundamentally embedded in wider social structures.[9] Psychologists and behavioral economists have rapidly advanced our understanding of consumer preferences and biases.[10] Political science enhances the view of individuals as active 'citizen–consumers' in the 'political marketplace for a more sustainable way of life for themselves and their communities.'[11] For all these crucial insights, the absence of broad-based *legal* perspectives on sustainable consumption leaves our understanding of the issue incomplete at best, and radically limited at worst.

Certainly, legal researchers have not been entirely absent from the discussion. In this sense, the edited volume did not start from scratch. However, the existing literature has a distinctly narrow focus; the only legal disciplines having anything to say on the matter being Consumer Law and Environmental Law. Furthermore, existing scholarship has tended to question how the discourse on sustainable consumption affects the legal system, instead of formulating questions the other way around: how the law influences sustainable consumption.

Environmental lawyers acknowledge that sustainable consumption is *inter alia* a legal concept, but the phenomenon is still widely perceived as a criticism of the effectiveness of existing environmental rules to underpin and enhance a sustainable environment.[12] Here we are speaking of those rules addressed to producers, mainly

---

[7]The multiregional input-output database EXIOBASE 2.2, which describes the world economy based on date from 43 countries, 5 world regions, and 200 productions sectors, shows that household consumption for 2006 contributes more than 60% of global greenhouse gas emission and 80% of total land, material and water use. See the detailed research of Ivanova et al. (2016), pp. 526–536. See also prior works published by UNEP (2010), Sachs (2015), Haas et al. (2005), Hertwich (2011) and Sachs (2015).

[8]Geels et al. (2015), Herring and Sorrell (2009) and Barbir and Ulgiati (2008). See also the classic and pioneering analysis by Jevons (1965).

[9]Goldsmith (2015) and Matthies and Wallis (2015).

[10]Sunstein (2015), Young et al. (2010) and Witt (2011).

[11]Stolle and Micheletti (2015) and Shah et al. (2012).

[12]See, for e.g., Salzman (1997).

to limit carbon dioxide emissions, or procedures for the disposal of waste or licences granted in the interests of environmental protection. Consumer lawyers, instead, picture sustainable consumption as a novel complementary tool in the field of Consumer Law that on occasion may (or may not) conflict with consumer protection. Government regulation on eco-labelling, energy efficiency, and packaging are often framed in this regard as a form of market regulation that steers (or empowers) consumers towards sustainable consumption patterns, albeit to the benefit of society at large.[13]

Notwithstanding these important contributions from environmental and consumer lawyers, legal scholarship, broadly defined, has had virtually nothing to say on the impact of other fields of law on sustainable consumption. A number of questions arise in this context. Is it the case that World Trade Organization (WTO) rules preclude or foster sustainable consumption and, if so, do they conflict with other international or bilateral treaties? Do states (or supranational entities like the EU), while discussing policies for addressing over-indebtedness of consumers, distinguish liabilities accumulated debits by households from excessive consumption of non-essential goods and services for life? Is it possible to envisage a rationalization or harmonization of eco-labelling regulations that will make 'eco-marks' an effective signal of sustainability for consumers? These are but a sample of the research questions raised by the contributors in this edited volume.

1.2 Conceptual and Normative Frameworks

Social scientists have long recognized that universal agreement on a definition of sustainable consumption is beyond reach. At the same time, they generally concur that the term implies far more than the sum of individual consumer choices in the marketplace. On the contrary, all mainstream treatments of sustainable consumption agree that it covers the environmental impact of the entire product lifecycle, from production-related investment decisions to logistics and marketing, and from retailing to waste management.

This edited volume defines sustainability in terms of consumption patterns that address basic needs and improve quality of life with either *fewer* or *better* goods and services produced in such a way as to minimize the depletion of natural resources and the production of toxic material, waste, and pollution. For example, a local government that imposes on retailers the obligation to allocate the most prominent retail space to local products can nudge consumers to purchase products with less environmental impact through lower freight carbon emissions. Similarly, sustainability is advanced when states provide economic incentives for banks to lend money to householders for solar panels to expand renewable energy consumption.

[13]See, for e.g., Wilhelmsson (1998).

With these broad concepts in mind, we advance the claim that the study of sustainable consumption and the law demands conceptual and normative frameworks that go well beyond the existing corpus of legal research. In extending the normative and empirical coverage beyond existing policy documents, soft law, and statutes, we seek to contribute significantly to the understanding of the current legal discourse, as well to have a hand in reforming it. In so doing, we seek to establish a range of ambitions for sustainable consumption and the law as a field of research.

The law being principles and rules—and thus a way to shape individual behavior—legal scholarship will inevitably add to the success or failure of accomplishing sustainable consumption. At the most basic level, therefore, the field must explore whether the State has an obligation to impose or 'nudge' certain market behaviors towards more sustainable consumption patterns. But it must also concern itself with whether sustainable consumption should be pursued because it is a good public policy in itself, or because it affects rules targeting other unwanted market behavior, such as unfair or anticompetitive commercial practices. Moreover, a principle line of enquiry must be to ascertain the extent to which sustainability is being referenced in the normative justification of law, for example in the preparatory documents of legislators or in court decisions. Above all, legal scholars should question whether sustainable consumption entails for us—producers, consumers, citizens—duties and rights to a healthy environment.

1.3 Structure of the Book

This edited volume has the ambition to guide legal scholars and practitioners, judges, law students, and social scientists through the infant and promising field of sustainable consumption and the law. With this aim in mind, we regard the clarity and simplicity of the structure of the proposed edited book as strength.

All the chapters of the edited volume are identified according to an established field of legal research. The chapters pinpoint how traditional legal disciplines, sector-related regulation, and legal rules in specific issue-areas engage with the concept of sustainable consumption. Some contributions describe this relationship by isolating a national legal system. Others come from the vantage point of legal pluralism, exploring the conflicts and convergences of rules between multiple international treaties (or guidelines) and those between the rules of international and transnational law (or both) vis-à-vis national legal systems.

Part I covers the various approaches of Consumer Law to sustainable consumption. We understand that the remit of Consumer Law is no longer limited to consumer protection, with the flourishing of subfields since the 1980s. Rather than selecting randomly from among them, we propose to cover them all by dedicating a chapter to each. Mateja Durovic and Franciszek Lech (Chap. 2) provide an introduction to the internationalization and transnationalization of Consumer Law on sustainable consumption. The chapter describes the growing insertion of the notion of sustainable consumption on binding and non-binding normative documents, namely the UN Guidelines, EU law, OECD Guidelines and Specialized Inter-American Conferences on International Private Law. Thierry Bourgoignie (Chap. 3) reconciles the notion of consumer protections and sustainable consumption. The contribution provides a view on areas where consumer protection tools and sustainable development interact and, in doing so, focuses on the conflict between the social concern for sustainability and the obsolescence of consumer products. Patrícia Galindo da Fonseca (Chap. 4) describes the shift from consumer protection to consumer empowerment, where the latter is argued to be a new consumer law. The contribution argues that the new consumer law aims to change consumer's behavior, making them aware of the ecosystems fragility through the adoption of sustainable consumption patterns. DiógenesFaria de Carvalho and Vitor Hugo Do Amaral Ferreira (Chap. 5) shed lights on the most recent developments in the field of Consumer Law, namely consumer behavior. The contribution addresses sustainable consumption as a State responsibility to nudge consumers towards a more sustainable purchase's practices.

Part II challenges the view that sustainable consumption is an exclusive matter of Consumer Law, as the notion has been incorporated into the epistemological discourses of other fields of law besides Consumer Law. Délton Winter de Carvalho and Fernanda D. L. Damacena (Chap. 6) analyze the role of sustainable consumption and Environmental Law by focusing on the Brazilian legal framework for disaster risk and management. Teresa Villac and Maria Cecília Loschiavo dos Santos (Chap. 7) introduce the concept of sustainable public procurement in Brazilian administrative law. The chapter sheds lights on cases where sustainable criteria in public procurement converge with the notion of sustainable consumption. Claudia Lujan Oviedo (Chap. 8) explains the relationship between sustainable consumption and the principle of prevention within tort law. The contribution argues that the principle of prevention, which was recently introduced into the new Argentinean Civil and Consumer Code, ensures consumers the right to challenge the risks of harm to their healthy environment. Sergio Sebastián Barocelli (Chap. 9) claims that unsustainable business practices are ultimately unfair business practices. The contribution argues that unsustainable practices infringe the principle of sustainability introduced in the Argentinean Civil and Consumer Code. Ana Cândida Muniz Cipriano (Chap. 10) provides a historical account of the endorsement of the sustainable consumption goals by an international organization, namely the United Nations. The contribution concludes by recommending future actions in the field of consumer protection and sustainability, and of the work of the United Nations. Finally, Guilherme José Pastana de Figueiredo (Chap. 11) analyzes the treatment of sustainable consumption into international trade agreements. The contribution does so by identifying different definitions of environmental goods and services, environmentally preferable goods and services and their interrelationship with sustainable consumption.

Part III zooms in on a major issue common to the matters of sustainable consumption and the law: eco-labelling and packaging. While sustainable consumption conveys the idea that consumers need to consume less and better, the law is the means to ensure that consumers have access to information about sustainable methods of production of goods and services or to punish misleading or fraudulent advertising. Contributions, therefore, analyzes the efficacy and compliance with eco-labelling and packaging schemes. James P. Nehf (Chap. 12) challenges the assumption that environmental-friendly marketing—namely eco-labelling—has been at all effective in the United States. The chapter argues that green marketing can be overstated, confusing, misleading, or outright false advertising, and there is a vast room for creating a specific regulatory framework besides the general rules of the Federal Trade Commission Act and state laws. Suvi Sankari (Chap. 13) assesses whether the concept of sustainable consumption would support a reinterpretation of relevant trade law, namely EU and WTO rules, allowing for robust and harmonious country-of-origin (COO) labelling. Fabiana D'Andrea Ramos and Tatiana Cardoso Squeff (Chap. 14) discuss the correlations among food security, consumption and the right of information, addressing, in particular, the labelling of food items and its role in helping consumers choose their groceries properly. Cristiane G. F. Vianna and Luís Renato Vedovato (Chap. 15) argue that national laws on plain packaging of tobacco products violate neither commercial freedom of expression nor WTO rules.

Parts IV address the development of sustainable consumption in sector-related regulation pursued by international and regional organizations. The promotion of sustainable consumption often requires coordination between states. Consequently, states agree on a common regulatory framework to enhance sustainable consumption by addressing sector-specific regulation on the matter. Alberto do Amaral Junior and Luciane Klein Vieira (Chap. 16) focuses on the contradictory regulation of living modified organisms in contemporary international law. The chapter explores the reasons why states support scientific evidence in the Agreement on Sanitary and Phytosanitary Measures yet backed the precautionary principle in the Biosafety Protocol. Kleverton M. Carvalho and Clara Minaverry (Chap. 17) compare the regulatory framework of water security programs in the EU and the MERCOSUR. The chapter concludes that treaties are effective when there are economic coercion clauses such as those in the Basel and Stockholm Conventions. Andreia Costa Vieira (Chap. 18) draws a picture of the entrance of the private sector through foreign direct investment in the water business in parallel with a world-wide claim for sustainable water consumption and a human right to water. Rodrigo C. A. Lima and Josiane Godoy Lima (Chap. 19) address the inherent challenges in characterizing sustainable consumption and production (SCP) as a tangible reality when it comes to the food and apparel sectors. Finally, Tiago Matsuoka Megale (Chap. 20) analyzes the extent to which private standards of sustainability imposed by supermarkets protect common values, such as human life and health, rather than constituting disguised restrictions to trade.

Part V, in contrast to the previous section, focuses on the development of national sector-specific regulatory frameworks that either enhance or preclude sustainable consumption. Marc Lacoursière (Chap. 21) presents an analysis of the legislative and regulatory approach to real estate credit in Canada and Quebec, designed to encourage a responsible approach to borrowing. Maria Paula Costa Bertran Muñoz (Chap. 22) sheds light on the potential for well-meaning but poorly thought-out state interventions in Brazilian consumption indebtedness in the last 15 years. The chapter focus on the Brazilian government's practice of releasing severance pay fund assets, the so-called FGTS. Tiago de Barros Correia, Gabriel Moreira Pinto and Vitor Hugo da Silva Oliveira (Chap. 23) examine the design of auctions to procure energy efficiency measures as a distributed energy resource, which aims at the reduction of energy consumption to end users. The chapter evaluates the pilot auction model implemented by the Brazilian Electricity Agency (ANEEL) in Roraima and compares it with international experience from Germany, Portugal, Switzerland, and the United States. Claudia Lima Marques and Matheus Linck Bassani (Chap. 24) analyze the legal framework provided by ANEEL Resolution No. 482/2012 for the current mini- and microgrid distributed generation in Brazil. The chapter tests the advantages and disadvantages of this regulatory framework in light with sustainable consumption principles. Cláudio José Franzolin (Chap. 25) articulates the business practice of planned obsolescence with the Law on the National Policy on Solid Waste (Law No. 12305/2010) under the perspective of systematic interpretation, as well as the inclusion, in the judicial speech, of the sensitivity of the institute regarding ecological values. Wilson Engelmann, Haide Maria Hupffer and Raquel Von Hohendorff (Chap. 26) propose a new self-regulatory model to manage potential environmental risks that could be generated by nanoparticles. The ISO TC229 (on nanotechnology) and the comprehensive principles regarding nanotechnologies and materials advanced by NanoAction (a project of the International Centre for Technology Assessment) are the inspiration for this proposition.

We conclude this edited volume (Chap. 27) by proposing a future research agenda for the field of research on sustainable consumption and the law. While this edited volume develops the topic of sustainable consumption and the law using a variety of research methods from the perspective of multiple legal disciplines, our conclusion envisions a step forward that embodies a closer tie between the concept of sustainable consumption and economic development.

Lastly, we would like to thank Hans-W. Micklitz and Jorge E. Viñuales for the forewords to this edited volume, Pia Letto-Vanamo, Dean of the Faculty of Law at the University of Helsinki, for providing funds from the Academy of Finland under the Finnish Distinguished Professor Programme (FiDiPro) [283002], and Simon P. Watmough, chairman of The Manuscript Whisperer, for carefully editing the language and formatting all contributions.

References

Barbir F, Ulgiati S (eds) (2008) Sustainable energy production and consumption: benefits, strategies and environmental costing, NATO science for peace and security series C: environmental security, 1st edn. Springer, Berlin

Berman PS (2007) Global legal pluralism. South Calif Rev 80:1155

Fischer-Lescano A, Teubner G (2004) Regime-collisions: the vain search for legal unity in the fragmentation of global law. Mich J Int'l L 25(4):999–1046

Geels FW, McMeekin A, Mylan J, Southerton D (2015) A critical appraisal of sustainable consumption and production research: the reformist, revolutionary and reconfiguration positions. Glob Environ Change 34:1–12

Goldsmith EB (2015) Social influence and sustainable consumption. International series on consumer science. Springer International Publishing, Cham

Haas W, Hertwich RG, Hubaceck K, Korytarova K, Ornetzeder M, Weisz H (2005) The enviromental impact of consumption: research methods and driving forces. NTNU program for industriell Økologi, Working Paper no. 3/2005

Herring H, Sorrell S (eds) (2009) Energy efficiency and sustainable consumption: the rebound effect (Energy, Climate and the Environment). Palgrave Macmillan, London

Hertwich EG (2011) The life cycle environmental impacts of consumption. Econ Syst Res 23 (1):27–47

Ivanova D, Stadler K, Steen-Olsen K, Wood R, Vita G, Tukker A, Hertwich EG (2016) Environmental impact assessment of household consumption. J Ind Ecol 20(3):526–536

Jevons S (1965) The coal questions: an inquiry considering the progress of the nation, and probable exhaustion of coalmines. A. M. Kelley, New York

Koskenniemi M, Leino P (2002) Fragmentation of international law? Postmodern Anxieties. Leiden J Int Law 15(3):553–579

Krisch N (2010) Beyond constitutionalism: the pluralist structure of postnational law. Oxford University Press, Oxford

Matthies E, Wallis H (2015) Family socialization and sustainable consumption. In: Reisch LA, Thørgersen J (eds) Handbook of research on sustainable consumption. Edward Elgar Publisher, Cheltenham

Michaels R (2005) The Re-State-Ment of non-state law: the state, choice of law, and challenge from global legal pluralism. Wayne Law Rev 51:1209

Reisch LA, Thørgersen J (2015) Research on sustainable consumption: introduction and overview. In: Reisch LA, Thørgersen J (eds) Handbook of research on sustainable consumption. Edward Elgar Publisher, Cheltenham, p 1, 13

Sachs JD (2015) The age of sustainable development. Columbia University Press, Berlin

Salzman J (1997) Sustainable consumption and the law. Environ Law 27:1244–1293

Shah DV, Lewis A, Friedland CW, Kim YM, Rojas H, Micheletti M, Stolle D (2012) Sustainable citizenship and the new politics of consumption. Ann Am Acad Polit Soc Sci 644(1):88–120

Stolle D, Micheletti M (2015) Political consumerism: global responsibility in action. Reprint edition. Cambridge University Press, New York

Sunstein CR (2015) Behavioural economics, consumption and environmental protection. In: Reisch LA, Thørgersen J (eds) Handbook of research on sustainable consumption. Edward Elgar Publisher, Cheltenham

Teubner G (1997) Global Bukowina: legal pluralism in the World Society. In: Teubner G (ed) Global law without a State. Dartmout, Aldershot

UN General Assembly (2015) Transforming our World: the 2030 agenda for sustainable development, 2015, A/RES//70/1, Goal 12: ensure sustainable consumption and production patterns

UNCED (1992) The Rio Declaration on environment and development, 1992, para 4.3 and 4.4

UNEP (2010) Assessing the environmental impact of consumption and production: priority products and materials

UNGCP (2003) The United Nations guidelines for consumer protection, 2003, para. 4, and 42

Wilhelmsson T (1998) Consumer law and the environment: from consumer to citizen. J Consum Policy 21(1):45–70

Witt U (2011) The dynamics of consumer behavior and the transition to sustainable consumption patterns. Environ Innov Soc Trans 1(1):109–114

Young W, Kumju H, McDonald S, Oates SJ (2010) Sustainable consumption: green consumer behaviour when purchasing products. Sustain Dev 18(1):20–31

Part I
Consumer Law and Sustainable Consumption

Chapter 2
International and Transnational Consumer Law on Sustainable Consumption

Mateja Durovic and Franciszek Lech

Contents

Abstract This chapter examines the little-known area of the International and Transnational Consumer Law on sustainable consumption. Sustainable consumption, as such, has not attracted much academic consideration, despite the increasing prominence that environmental awareness and protection plays in international legislative and regulatory efforts. Nonetheless, most of the international efforts fail to effect an actual, direct encouragement of sustainable development. The fatal flaws of these efforts are mainly their non-binding nature and lack of sufficient consensus between the countries to compromise and impose more radical reform. Therefore, international consumer law on sustainable consumption is presently confined to the UN Guidelines on Consumer Protection and its precatory clauses.

M. Durovic (✉) · F. Lech
King's College London, London, UK
e-mail: mateja.durovic@kcl.ac.uk; franciszek.lech@kcl.ac.uk

A. d. Amaral Junior et al. (eds.), *Sustainable Consumption*,
https://doi.org/10.1007/978-3-030-16985-5_2

2.1 Introduction

The present chapter aims to carve a path through the *terra incognita* of International and Transnational Consumer Law on sustainable consumption. Curiously, sustainable consumption has not attracted much academic consideration, despite the increasing prominence that environmental awareness and protection plays in international legislative and regulatory efforts. Nonetheless, virtually all of the international efforts, both at the global level by the UN or the OECD, at the regional level by the EU and the Organization of American States (OAS)—not to mention the various NGO proposals—have failed to effect an actual, direct encouragement of sustainable development. The fatal flaws of these efforts are mainly their non-binding nature and lack of sufficient consensus between the Member States to compromise and impose more radical reform. Therefore, international consumer law on sustainable consumption is presently confined to the UN Guidelines and precatory clauses.

In our chapter, we begin by providing some background to the concept of sustainable consumption, before discussing the major international documents in the area. These include an examination of the global-scale efforts (UN and OECD), regional (EU and OAS) and NGO (Consumers International) guidelines to analyse the level and mode of their impact on sustainable consumption.

2.2 Sustainable Consumption

The United Nations understands the term 'sustainable consumption' to include 'meeting the needs of present and future generations for goods and services in ways that are economically, socially and environmentally sustainable'.[1] Beneath this phrase hides the desire to protect 'the planet and the people who live on it'.[2] The concept and the underlying ideology is a response to the environmental challenge posed by unsustainable consumption patterns. It has been recognised—indeed some claim that it is 'well-established'—that the 'consumption and production patterns of the *industrialized* world are environmentally unsustainable'.[3] This claim gives rise to a paradox: Schroeder and Anantharaman (and most writers, for that matter) recognise that the origin, and (one could say) responsibility, for unsustainable consumption patterns, and the associated environmental damage, lies with the industrialised world. The solution and the key to reverse that damage, however, lies on the side of the *developing* countries, with their 'leap' to sustainable lifestyles being envisaged as the key for the future sustainability of the planet. It seems that the previous

[1]UN ECOSOC Res E/1999/INF/2 (26 July 1999), Guidelines on Consumer Protection, (1999 Guidelines), Cl 42.

[2]Consumers International (2016), p. 10.

[3]Schroeder and Anantharaman (2017), p. 4 (emphasis added).

conduct of the developed economies will force the developing economies to adopt radical reforms, which could be controversial.

The abovementioned authors also differentiate between weak and strong sustainable consumption. The weak version focuses on 'improving the eco-efficiency of consumption activities' while the strong version includes 'absolute reductions in consumption levels' and 'fundamental shifts in consumption patterns'.[4]

2.3 The UN Guidelines: The First Global Document on Consumer Protection

The United Nations has traditionally played a very important role in the area of consumer protection.[5] The initial guidelines published by the UN in 1985 were contained in Annex I to UN General Assembly Resolution 39/248.[6] It has a specific focus on the needs of consumers in 'developing countries'.[7] Notably, the document also recognises the 'imbalance in economic terms, educational levels and bargaining power' of the consumer.[8] The document also introduced a number of objectives for the guidelines, a key one among them (and the one most frequently criticised) being 'to encourage high levels of ethical conduct for those engaged in the production and distribution of goods and services to consumers'.[9]

Already in 1985 (without addressing it expressly), the UN made a step towards encouraging sustainable consumption, *inter alia* via encouraging informed consumer choice[10] and by calling for consumer education to also cover the environmental impacts of consumption.[11] We contend that, especially bearing in mind the subsequent adjustments made to the Guidelines, it would be an exaggeration to argue that suitable consumption, as an objective, was one of the main aims of the 1985 Guidelines. Rather, the 1985 Guidelines marked the significant baby steps sustainable consumption was making in the transnational regulatory consciousness; after all, acknowledging the problem is the first step to solving it. Instead, the focal point of the 1985 Guidelines was establishing consumer protection across all Member States and control of restrictive and abusive trading practices.[12]

---

[4] Schroeder and Anantharaman (2017), p. 6.

[5] See: Durovic and Micklitz (2017).

[6] UNGA Res 39/248 (16 April 1985) A/RES/39/248, (1985 Guidelines), available at: http://www.un.org/documents/ga/res/39/a39r248.htm. Accessed 10 Feb 2019.

[7] 1985 Guidelines, Annex, cl 1.

[8] 1985 Guidelines, cl 1.

[9] 1985 Guidelines cl 1(c).

[10] See 1985 Guidelines, cl 3(c).

[11] See 1985 Guidelines, cl 33(f).

[12] See for example 1985 Guidelines cl 4, 15 or 19.

The Guidelines were amended for the first time in 1999. Compared with the previous version, one can notice the growing awareness of the environmental impact of consumption. Another 'legitimate need' was added: consumer education about 'environmental, social and economic aspects of consumer choice'. Clearly building on Clause 33(f) from 1985, it lays out expressly the nexus between consumer choice and the environment. More importantly, 'sustainable consumption' as a genuinely pursued objective is more present than in 1985—as one should perhaps expect. Clause 4 is a juxtaposition to clause 1 from 1985, but instead of focusing on the emerging economies, the Guidelines emphasise that '[u]nsustainable patterns of production, particularly in industrialized countries, are the major cause of the continued deterioration of the global environment. All countries should strive to promote sustainable consumption patterns; developed countries should take lead in achieving sustainable consumption patterns'.[13] This shifts the focus from emerging economies to the industrialised countries, although in future, as evident from the Schroeder and Anantharaman's article, the spotlight shall return to the emerging economies.

Clause 5 of the 1999 amended version of the Guidelines stresses that in pursuing sustainable consumption, the broader economic goals 'should' be considered, perhaps, placing achieving sustainable patterns of consumption as secondary to the more directly pressing objectives of eradicating poverty, satisfying the basic human needs of all members of society, and reducing inequality within, and between, countries. It should be kept in mind, though, that attaining these broader goals can get us, the global population, closer to sustainable consumption.[14]

Clause 24 of the 1999 Guidelines reiterates the importance of informed consumer choice in effecting sustainable consumption, by inviting the Member States to 'encourage' 'consumer access to accurate information about the environmental impact of products and services'.[15]

Crucially, 'Promotion of Sustainable Consumption' is recognised as its own heading within the guidelines, illustrating its rising importance, and, usefully, defining the concept of 'sustainable consumption'.[16] The document goes on to assign responsibility for sustainable consumption to 'all members and organizations of society', although special responsibility is attributed to informed consumers, governments, business, labour, consumer and environmental organisations.[17] Clause 44 states that governments, in concord with business and 'relevant organizations [...] *should* develop and implement *strategies*' that would promote sustainable consumption through a proposed mix of policies which would include for example

[13] 1999 Guidelines, cl 4.

[14] 1999 Guidelines, cl 5.

[15] 1999 Guidelines, cl 24, cf 1985 Guidelines, cl 33(f) where information about environmental impacts of consumption where only considered important when 'appropriate', and not absolutely.

[16] 1999 Guidelines, cl 42.

[17] 1999 Guidelines, cl 43.

sectoral policies for 'land use, transport, housing, energy', information programmes, withdrawal of subsidisation of unsustainable production patterns, and other 'economic and social instruments'. The words 'should' and 'strategies' are emphasised to illustrate the general and non-imposing manner of the Guidelines; they do not lay down binding obligations—which could be seen as their weakness. Nevertheless, it must be remembered that the Guidelines represent a (supposedly) hard-fought for and fragile consensus built in a committee of the most populous intergovernmental organisation, which could justify why the document does not contain any absolute obligations. Clause 50, like clause 42, is addressed at governments and asks that they, together with the private sector, work to 'encourage the transformation of unsustainable consumption patterns through the development and use of new environmentally sound products and services and new technologies, including information and communication technologies, that can meet consumer needs while reducing pollution and depletion of natural resources.'

The change between the 1985 and the 1999 versions, in terms of the prominence of sustainable consumption within them, is very noticeable. This adaptation is praiseworthy, as the UN saw a global problem and acted to remedy it. Perhaps the environmental depletion from unsustainable production and consumption patterns was not yet seen in 1985, or perhaps the first steps made in 1985 resulted in the changes made to the Guidelines in 1999. In 2015–2016, the UN yet again updated the Guidelines.[18] One interesting observation can be made by comparing clause 6 from 2015 with Clause 4 from 1999. In 2015 the UN observes that unsustainable production and consumption patterns,

> particularly in industrialized countries, are the major cause of the continued deterioration of our environment. All Member States should strive to take the lead in achieving sustainable consumption patterns; developing countries should seek to achieve sustainable consumption patterns in their development process, having due regard for the principle of common but differentiated responsibilities.[19]

The emphasis on the practices in industrialised countries, and the different situation of the developing countries is more stressed, and a 'principle of common but differentiated responsibilities' is introduced. Unlike the previous two versions, the 2015 version introduces Principles for Good Business Practices, which are the 'benchmarks' for good business practice.[20] Within the Guidelines themselves, there is a longer then before section devoted to promotion of sustainable consumption.[21] Some sections are transposed from the 1999 version, and some are new.

We pause to make one caveat to the proceeding analysis: 'little commentary is available on the UN Guidelines'.[22] Thus, the present work is an attempt to explore

---

[18]UNGA Res 70/186 on Consumer Protection (22 December 2015) A/RES/70/186 (2015 Guidelines).

[19]2015 Guidelines, cl 6.

[20]2015 Guidelines, cl 11.

[21]2015 Guidelines, cl 49–62.

[22]Gawith (2010), p. 484.

this uncharted ground, notwithstanding the drought of academic commentary. Accordingly, our exploration is necessarily confined.

The UN's effort has not been immune to criticism. One of the most vociferous critics, Weidenbaum, has claimed that the UN should not even engage in economic regulation, let alone protection of consumers, and specifically, promoting sustainable consumption, arguing that:

> [R]egulation of the production and distribution of goods and services via these Guidelines is a far cry from the central role of the United Nations which is, according to the UN charter, 'to maintain international peace and security'. That role deserves the greatest amount of emphasis in the dangerous world in which we live. It is sad to see the UN diverting its resources to large-scale forms of economic regulation when it is doing such an inadequate job of carrying out the basic tasks for which it was established.[23]

Weidenbaum continues that even if one could surmount the fact that it is not UN's job to engage in such action, 'no convincing case has been made for the participation of international agencies in such a basic domestic consideration as the protection of consumers'.[24] That criticism, while perhaps valid in 1987, is hard to accept today. With the growth of transborder trade, e-commerce, and the like, the need for transnational consumer protection is self-evident.

Moreover, Weidenbaum argues that 'the UN Guidelines for "Consumer Protection" are an exemplary model of vagueness and overblown phraseology. Lofty goals are set forth in language that is reminiscent of the first draft of a blueprint for a centrally directed society',[25] as they call for greater government interference within the market.

The Guidelines have their proponents too. Peterson has responded to the criticism levelled by Weidenbaum. She refutes the accusations cast by Weidenbaum, convincingly arguing that condemning the Guidelines' language for being 'vague' or 'lofty' is analogous to criticising the US Constitution for its postulates to 'establish justice', and other ideals couched in similarly broad terms.[26] Moreover, the UN Guidelines, unlike the US Constitution, is not binding on any State.[27] The Guidelines, in Peterson's view, express 'goals and objectives for all nations to aspire to in protecting their own people, their consumers'.[28]

According to Peterson, if Weidenbaum's main contention—namely, that the Guidelines should become binding—were implemented, this would result in the UN forcing governments to intervene in markets. She notes that this is untenable, given that

> for anyone inclined to take Weidenbaum's strictures seriously, not one provision of the UN Consumer Guidelines can be imposed on any country by the UN or by any other outside

[23]Weidenbaum (1987), p. 430.

[24]Weidenbaum (1987), p. 425.

[25]Weidenbaum (1987), p. 425.

[26]Peterson (1987), pp. 434–435.

[27]Peterson (1987), p. 434.

[28]Peterson (1987), p. 434.

> force or agency. Thus, the Guidelines are not just a 'pretense' of being voluntary; they have no 'bite' whatsoever, except to the extent that civilized nations and civilized people elect to adopt or adapt them to their own needs.[29]

For her, the Guidelines simply envision 'guidance, support, technical assistance, international expertise extended to nations which, with limited economic and professional resources, cannot yet possibly match the resources of the industrialised nations in protecting the health and safety of their citizens from the hazards of the marketplace'.[30]

Writing in 2010, Gawith seemed to agree with Weidenbaum that the language present in the 1985 and 1999 versions is 'vague' and 'overblown', citing by way of example the 'objectives to "promote just, equitable and sustainable economic and social development"' or 'to "encourage high levels of ethical conduct".'[31] Gawith went as far as to say that one could 'sympathise with Peterson *and* Weidenbaum—there being little practical effect in the UN Guidelines'.[32]

It is evident that the cardinal criticism of the UN efforts should not be the 'lofty', vague language, but precisely that 'little practical effect' of the UN Guidelines. While Gawith noticed there being 'little' effect of the UN Guidelines, there was not a complete lack thereof. In response to the document,

> [C]ountries such as Australia, Canada, Hong Kong, New Zealand, Singapore and the United States established various mechanisms for consumers to obtain redress with respect to national consumer transactions, including the introduction of special procedures into previously existing lower courts, and the creation of new courts and tribunals which specialise in small consumer claims.[33]

One could question if the introduction of such largely procedural and formal reform had an impact in promoting sustainable consumption, but Gawith argues that 'the value of the UN Guidelines, as guidelines, should still be acknowledged. That is to say, while remaining abstract, it is more beneficial to have commonly accepted principles than to not have them at all'.[34] Such a summary of the impact of the Guidelines seems to contradict the claim that the UN Guidelines had a profound impact on promoting sustainable consumption globally.

---

[29] Peterson (1987), p. 436.
[30] Peterson (1987), p. 437.
[31] Gawith (2010), p. 484.
[32] Gawith (2010), p. 485.
[33] Gawith (2010), p. 485.
[34] Gawith (2010), p. 486.

2.4 Other Transnational Regulation Attempts: The OECD Guidelines

The OECD Guidelines from 2000 focus on '[c]onsumer laws, policies, and practices [that] limit fraudulent, misleading and unfair commercial conduct. Such protections are indispensable in building consumer confidence and establishing a more balanced relationship between businesses and consumers in commercial transactions'.[35] Notably, the OECD acknowledges that:

> [t]he global network environment challenges the abilities of each country or jurisdiction to adequately address issues related to consumer protection in the context of electronic commerce. Disparate national policies may impede the growth of electronic commerce, and as such, these consumer protection issues may be addressed most effectively through international consultation and co-operation.[36]

This, in a way, is a response to the comment made by Weidenbaum in relation to the UN Guidelines that transnational regulation is not needed. However, because of the focus on e-commerce, the OECD Guidelines do not import anything more into the quest to estimate the impact of international consumer law on sustainable consumption, although it must be admitted that the OECD document was not studied as extensively as the UN one.

The OECD Guidelines were designed to 'help ensure that consumers are no less protected shopping on-line than they are when they buy from their local store or order from a catalogue',[37] but fatally, much like their UN equivalent, the guidelines are non-binding and lack the bite. Gawith sums up that '[w]hile these are commendable principle, in practice little is being achieved through the OECD Guidelines, at an international level.[38] The criticism here is not of the OECD's Guidelines as such, but rather that 'those purported to be guided by [them] are persisting with parochial mind-sets: the initatives are all regional and in some cases even legally insignificant.'[39]

2.5 Regional Consumer Law: The European Union Legal Framework

We have already seen the scepticism expressed by Gawith towards the regional regulation of consumer protection, but the efforts by the EU should still be analysed to unearth their possible impact on effecting sustainable consumption. For the

---

[35]OECD (2000), p. 9.

[36]OECD (2000), p. 9.

[37]OECD (2000), p. 3.

[38]Gawith (2010), p. 488.

[39]Gawith (2010), p. 489.

present analysis, the following legislation is crucial: 'first, the directive on unfair terms in consumer contracts [Dir 93/13], secondly, the directive on protection of consumers in respect of distance contracts [Dir 97/7] and finally the recent directive on certain aspects of the sale of consumer goods and associated guarantees [Dir 99/44]'.[40] Gawith added the E-Commerce Directive (Dir 2000/31). However, it is important to note that both Gawith and Pinna analyse those directives from the perspective of successful consumer protection and that they do not comment upon the impact of the EU law on the promotion of sustainable consumption. Still, Pinna and Gawith agree that the EU directives have some advantages over the UN Guidelines. Importantly for the promotion of sustainable consumption: 'under the [Brussels] Regulation,[41] the consumer is sovereign in disputes and may choose to sue in their country or in the country of the vendor'.[42]

Consequently, the:

> Rome Convention and the Regulation establish and perpetuate a situation where website businesses have to comply with the consumer protection laws of 27 or more different Member States, i.e. they may have to comply with the laws of any EU country from which their sites are accessible. No doubt this will be so confusing and such a threat to some businesses that many will actually stay out of e-commerce altogether.[43]

Thus, European businesses, especially SMEs, could be disadvantaged by this de facto restriction on market access, causing lower revenues and, in turn, limited investment possibilities, including in sustainable technologies. Additionally, as noted before, consumer choice is crucial for consumers being able to select environmentally sustainable products, and thus by limiting competition and supply that conduit to sustainable consumption is being impeded.

2.6 Regional Consumer Law: Las Conferencias Especializadas Interamericanas Sobre Derecho Internacional Privado (CIDIP)

CIDIP is a conference on private international law convened by the Organization of American States (OAS). The acronym is taken from the Spanish name: las Conferencias Especializadas Interamericanas sobre Derecho Internacional Privado [Specialized Inter-American Conferences on International Private Law]. The first conference was held in 1975, although it was apparently a continuation of a process

[40] Pinna (2002), p. 325.

[41] Regulation 1215/12 on jurisdiction and the recognition and enforcement of judgements in civil and commercial matters [2012] OJ L 351/1 (Brussels Regulation).

[42] Gawith (2010), p. 496.

[43] Gawith (2010), p. 496.

that has its roots at the end of the nineteenth century.[44] The most recent CIDIP conference—CIDIP VII, held in 2003—focused on 'building a practical framework for Consumer Protection'. The CIDIP VII proposals that ensued have been inspired by the successes and failures of the EU—notably the perceived failure of the Rome I Convention. Brazil, Canada and the United States have each proposed a solution, interestingly with each proponent taking a different approach.

Brazil proposed the most radical document, a Draft Convention on the Law Applicable to International Consumer Contracts and Transaction, a sort of treaty solution. Canada proposed a Model Law on Jurisdiction and Choice of Law to Consumer Contracts, while the United States has submitted a Draft of Legislative Guidelines on Availability of Consumer Dispute Resolution and Redress for Consumers.[45] Arroyo argues that a soft law approach (such as the one proposed by the United States) 'is likely to be a faster way to achieve harmonization of consumer law', and that 'model laws and legislative guides offer a degree of flexibility and adaptability in accordance with the needs and wishes of each State that facilitates negotiation'.[46] The main issue, however, remains the jurisdiction question:

> [a]s consumer law contains a lot of mandatory provisions their application depends heavily on the jurisdiction seized. Nonetheless, the negotiation of fora and other jurisdictional issues are the most difficult to compromise as jurisdiction is intrinsically linked with national legal orders and often seen as a question of sovereignty.[47]

However, in Arroyo's view 'the success of the enterprise depends largely on the policies adopted by the States and by the OAS', but beyond that 'consumer protection needs more than a good system of rules and proceedings'; it needs education, advertising or promotion of ethical and fair conduct.[48] No matter which of the proposals is adopted, one can speculate that the effect on promotion of sustainable consumption, in the absence of an express provision dealing with it, is likely to be at most indirect (i.e. via market mechanisms). In such a schema, we would predict that, as the market opens and both consumer and seller confidence is boosted by new and clear rules, competition will not only enhance consumer choice, allowing environmentally-conscious consumers to decide accordingly, but also perhaps force producers to adopt a more sustainable production pattern. All that being said, given the scale of climate change and exponentially increasing need for adequate consumer laws, gradual and incremental change, such as that under contemplation by CIDP, is too slow in proportion to the scale of the problem and a more radical reform is needed. Perhaps the next CIDIP conference will yield such a reform in the

[44]Organization of American States (n.d.), 'Historia del Proceso de las CIDIP's' (unknown date of publication), accessed July 15, available at: https://www.oas.org/dil/esp/derecho_internacional_privado_desarrollo.htm. See also http://www.oas.org/en/sla/dil/private_international_law_history_cidip_process.asp for an English version.

[45]Fernández Arroyo (2009), p. 411.

[46]Ibid, p. 418.

[47]Ibid, p. 419.

[48]Ibid, p. 422.

Western Hemisphere—the proposed topics focus on e-Commerce and the protection of consumers.[49] Even if the most optimal outcome is reached at such a conference, the OAS's efforts are still 'doomed to irrelevance where international co-operation is required'[50] such as is the case with the need for sustainable consumption. Regional consumer law, when isolated and ill coordinated, are just not the appropriate weapon to tackle the global environmental problem.

2.7 Initiatives by NGOs: Consumers International

Previously known as 'ICPEN', the goal of this organisation is to 'gather and share information concerning consumer protection issues, addressing cross-border violation of consumer protection laws, facilitating cross-border remedies, promoting enforcement of consumer protection laws, and encouraging wider cooperation among consumer protection enforcement organ'.[51] CI themselves say that they act to 'ensure that the voice of consumers is as powerful as the governments and corporations they interact with. Achieving this will ensure everyone can realise the benefits of our increasingly globalized and digital world'.[52]

In the field of sustainable consumption, CI works to 'provide fair social conditions for current and future generations' and to 'reduce the confusion around sustainability by ensuring that producers, retailers and policymakers do all they can to make products safe, durable and resource-efficient; before providing clear, reliable information to guide consumer choice'.[53]

2.8 Conclusion

In conclusion, it is possible to observe that the international consumer law fails to launch an effective initiative to promote sustainable consumption, considering the 'lifestyle-leapfrogging paradox'. The main international and transnational efforts have been discussed and shown to be ineffective. We have argued that international consumer law has failed to shift consumption patterns in favour of a more sustainable solution. Nevertheless, it has not been disputed that these efforts are the crucial

[49]Permanent Council of the OAS and Comission of Judicial and Political Affairs (2011).

[50]Gawith (2010), pp. 488–489.

[51]Delisle and Trujillo (2010), p.160.

[52]Consumers International, 'Who We Are': https://www.consumersinternational.org/who-we-are/, accessed on 9 February 2019.

[53]CI 'Consumer Protection' at: https://www.consumersinternational.org/what-we-do/consumer-protection/sustainable-consumption/, accessed on 9 February 2019.

first step—the acknowledgement of the problem—and that they also evidence a growing global awareness of a need to promote sustainable development.

Gawith has argued that 'while there are plenty of examples of effective national consumer protection laws worldwide—at least within the territorial reach of such laws—there are, as yet, no truly international consumer laws'.[54] Delisle has argued that 'much international cooperation [in international consumer protection] is more informal or aspirational'.[55] Despite this, the question is whether there can and should be a comprehensive international document for the promotion of sustainable consumption. Prothero, et al. have suggested that 'a sustainable future cannot be achieved by relying solely on initiating "bottom-up" changes in individual consumer behavior' and thus some international cooperation is required.[56] Gawith agrees that 'a solution to this problem in the form of a single international consumer law made out of or deeply consistent with the best parts of a range of existing national consumer laws is probably not realistic owing to the extreme relative inefficiency of working with so many frequently incompatible elements'.[57]

Ultimately, it can be concluded that neither of the initiatives is perfect. Meanwhile, the impact of the UN's guidelines on the promotion of sustainable consumption is debatable. Despite the growing presence of the concept in the updated versions of the Guidelines, it is fair to say that the actual impact of them is limited owing to the general and non-binding way they are structured.

References

Consejo Permanente De La Organización De Los Estados Americanos, Comisión De Asuntos Jurídicos y Politicos (2011) Selección De Temas Para Una Posible Octava Conferencia Especializada Interamericana Sobre Derecho Internacional Privado', (29 November 2011) available at: http://www.oas.org/consejo/sp/cajp/dip.asp#CIDIP. Accessed 10 Feb 2019. (Permanent Council of OAS and Comission of Judicial and Political Affairs, 'Selection of topics for a possible eighth specialized inter-American conference on International Private Law')

Consumers International (2016) Practical guide to the UN guidelines for consumer protection available at: https://www.consumersinternational.org/news-resources/publications/id/3053. Accessed 10 Feb 2019

Delisle J, Trujillo E (2010) Consumer protection in transnational contexts. Am J Comp Law 58:135

Durovic M, Micklitz H-W (2017) Internationalisation of consumer law. Springer, Berlin

Fernández Arroyo DP (2009) Current approaches towards harmonization of consumer private international law in the Americas. Int Comp Law Q 58(2):411

Gawith D (2010) Model laws relevant to preparation for the international regulation of international consumer transactions. J Bus Law (6):474

[54] Gawith (2010), p. 479.

[55] Delisle and Trujillo (2010), p. 159.

[56] Protehro et al. (2011), p. 35.

[57] Gawith (2010), p. 476.

OECD (2000) Guidelines for consumer protection in the context of electronic commerce. Available at: http://www.oecd.org/sti/consumer/oecdguidelinesforconsumerprotectioninthecontextofele ctroniccommerce1999.htm. Accessed 10 Feb 2019

Organization of American States (n.d.) Historia del Proceso de las CIDIP's' (unknown date of publication), Accessed July 15. Available at: https://www.oas.org/dil/esp/derecho_ internacional_privado_desarrollo.htm, see also http://www.oas.org/en/sla/dil/private_interna tional_law_history_cidip_process.asp for an English version. Accessed 10 Feb 2019

Peterson E (1987) The case against 'The Case Against the UN Guidelines for Consumer Protection'. J Consumer Policy 10:433

Pinna A (2002) The protection of European Consumer: some aspects of private international law. Tilburg Foreign Law Rev 10:325

Protehro SD, Freund J, Kilbourne WE, Luchs MG, Ozanne LK, Thøgersen J (2011) Sustainable consumption: opportunities for consumer research and public policy. J Public Policy Market 30 (1):31–38

Schroeder P, Anantharaman M (2017) Lifestyle leapfrogging in emerging economies: enabling systemic shifts to sustainable consumption. J Consumer Policy 40:3

UN Guidelines on Consumer Protection reform in 1999 (UN Department for Economic and Social Affairs, at http://unctad.org/en/PublicationsLibrary/UN-DESA_GCP1999_en.pdf) Accessed 10 Feb 2019

Weidenbaum M (1987) The case against the UN guidelines for consumer protection. J Consumer Policy 10:425

Chapter 3
Sustainable Consumption and Obsolescence of Consumer Products

Thierry Bourgoignie

Contents

Abstract Consumer protection tools and sustainable development interact in many areas. The present contribution focuses on the conflict between sustainability and the obsolescence of consumer products. Working for the longer lifespan of consumer products and combating product obsolescence contribute both to the protection of consumer economic interests and to sustainable development. Tools likely to contribute to the increased sustainability of consumer products are identified and proposals made for controlling the obsolescence of consumer products.

T. Bourgoignie (✉)
Faculté de science politique et de droit, Université du Québec à Montréal, Montreal, QC, Canada
e-mail: bourgoignie.thierry@uqam.ca

A. d. Amaral Junior et al. (eds.), *Sustainable Consumption*,
https://doi.org/10.1007/978-3-030-16985-5_3

Recommendations include the prohibition of planned obsolescence, the formulation of a general obligation of sustainability associated with the placing of products on the market, setting objective deadlines for the sustainability of consumer products and defining repairability requirements for consumer products made available on the market.

3.1 Introduction

Contemporary development models are principally influenced by the paradigms of capitalism, productivity, materialism and individualism. These generate a large ecological footprint and a global ecological deficit.[1] At the same time, the dominant models have been joined by some alternative developmental approaches in recent decades: ecodevelopment, social ecology, decline, sustainable development, eco-humanism, the circular economy, the share economy, and so on. These, however, have not profoundly questioned the underlying foundations of existing models.[2] Instead, the new alternative models interrogate the sustainability of traditional modes of consumption and call for novel approaches to consumption, including collaborative, sustainable and responsible consumption.

Sustainable development is arguably the new reference model, which has had the strongest reception in national and international spheres. It is likely to profoundly influence so-called *unsustainable* consumption and production patterns, as one of the goals of sustainable development is to transform these patterns within industrial and post-industrial societies so that they become sustainable. In this context, there is a clear convergence between sustainable development policy and consumer protection policy. Actually, the sustainable development paradigm helps to foreground the social and collective dimension of consumer laws and policy. Consumer law, therefore, finds in sustainable development not a constraint or a brake on its development but the opportunity for a 'salutary mutation'.[3]

Adopting a critical discourse of consumerism and distancing itself from the traditional 'normes sociales de consummation',[4] consumer protection policy would seek to ensure long-term values, such as sustainable development, prevail over short-term quantitative interests, such as the benefit of a lower price. Thus, free consumer choice is no longer established as an absolute principle, as sustainable development will impose constraints and derogations in several cases.[5] It is also important to note

[1]On the concept of ecological footprint, see the work of the Global Footprint Network online. See also Roch (2014).

[2]See Roch (2013).

[3]Bourgoignie (2008a), pp. 845–866.

[4]On this concept, see Bourgoignie (1988).

[5]Bourgoignie (2017), pp. 165–177. See also Friant-Perrot (2008), pp. 567–588.

that the approach to sustainable consumption closely and directly links consumption to production: the search for new consumption patterns goes together with the definition of new production patterns. Indeed, what is the point of requiring the consumer to show responsibility if the production system proves to be irresponsible?[6]

Consumer protection tools and sustainable development ones interact in many areas. The present contribution focuses on one only: the conflict between the social concern for sustainability and the obsolescence of consumer products. Working for the longer lifespan of consumer products and combating product obsolescence contribute both to the protection of consumer economic interests and to sustainable development. At the individual level, the concern is to protect the purchasing power of the consumer against repeated, successive and often useless purchases of products that are likely to be still fit for use even after they are thrown away. At the collective level, the stakes are to reduce waste and to prevent or reduce the negative impacts of overconsumption on the environment. For example, the massive production of cell phones, which requires the extraction of rare metals, means clearing and destroying fertile land. It is also well known that major pollution and health problems are caused by the impossibility or the difficulty of recycling certain products, especially those containing dangerous chemical substances.

Five main tools have been identified as likely to contribute to the increased sustainability of consumer products:

1. Prohibiting planned obsolescence;
2. Introducing a general obligation of sustainability associated with the placing of products on the market;
3. Setting objective deadlines for the sustainability of consumer products;
4. Introducing repairability requirements for consumer products made available on the market;
5. Enhancing the role of standardization and certification on this matter.

3.2 Prohibiting Planned Obsolescence

3.2.1 A Typology of Obsolescence

Several authors have developed a typology of obsolescence. However, as pointed out by Claudia Déméné, obsolescence categories are very close to each other.[7] For Serge Latouche, there is a symbiosis between psychological, economic and technical obsolescence.[8]

[6]See also Duguay (2008), pp. 239–261.

[7]Déméné (2015).

[8]Latouche (2012), p. 49.

We may speak of technical obsolescence or economic obsolescence. Technical obsolescence will include functional obsolescence (for example, the occurrence of a defect); functional improvement of product features through additions, changes or updates; obsolescence by incompatibility (for example, software). Economic obsolescence includes loss in quality or performance, design making repair or maintenance difficult or impossible (for example, replacement of batteries), indirect obsolescence (when the product's accessories make the product obsolete, although the product is still functional, for example, the ink cartridge of a printer or the charger for a cell phone). Obviously, there are also psychological or social factors that make the product obsolete in the consumer's mind. The decline in the use value of the product or consumer's satisfaction with the product is because of subjective considerations, such as aesthetics, fashion or social status. Marketing techniques can be shrewdly used to convince consumers of the obsolete nature of the product they use and to replace it with a product presented as more fashionable, trendier, prettier or just newer.

3.2.2 Definition of Planned Obsolescence

Planned obsolescence is defined as techniques or processes set up at the design or production stage of the product, with the purpose of reducing the lifespan of the product and make consumers replace it with a purchase of another product.[9] It is about deliberately reducing the duration of product usage. Currently, France is the only country whose legislation contains a definition of planned obsolescence. Article L 213-4-1 of the Consumer Code, recently introduced by Article 99 of Law 2015–92 of 17 August 2015 on energy transition for growth, defines planned obsolescence as a set of techniques by which the economic operator who places the product on the market intends to reduce the lifespan of the product and hence forces or induces consumers to replace it.

In 2009, *The Economist* defined planned obsolescence as:

> A business strategy in which the obsolescence (the process of becoming obsolete—that is, unfashionable or no longer usable) of a product is planned and built into it from its conception.[10]

In a major report on the lifespan of electrical and electronic equipment published in July 2012,[11] the French Environment and Energy Management Agency (ADEME) defines planned obsolescence as a scheme by which a good would see its standard duration consciously reduced from the time of its conception. The distinguished feature of planned obsolescence is the deliberate programming of a

[9]Slades (2007).

[10]Planned Obsolescence, *The Economist* (23 March 2009), online: www.economist.com.

[11]Agence de l'environnement et de la maitrise de l'énergie (2012).

reduced or limited lifespan of the product. A mechanism, a tool or a technique is conceived and inserted into the design of the product so that this will have a limited lifespan.

As Serge Latouche says:

> L'idéal ... [est] que l'objet tombe en panne juste à l'expiration du délai de garantie. Une sorte de guérilla s'ouvre alors entre producteur et consommateur, avec la possibilité pour ce dernier de souscrire une assurance complémentaire payante afin de prolonger la garantie.[12]

According to Latouche, implied warranties associated with the sales of consumer products are a double-edged sword, because, if they offer to the consumer the assurance of a minimum lifespan of the product, they 'also authorize the manufacturer to make this minimum the maximum lifespan of the product'.[13]

3.2.3 *Current Legal Standards*

There are no international documents that directly refer to planned obsolescence. For Lydie Tollemer,[14] this can be explained by the dependence of the capitalist economic model on systematic growth; a ban on planned obsolescence would cause a significant economic slowdown. There is no provision or legislative project dealing specifically with planned obsolescence that exists in Quebec or elsewhere in Canada. Except in France, no national law on consumer protection addresses the issue.

The aforementioned 2015 French law on energy transition for growth certainly brought changes to both the Environment Protection Code and the Consumer Code, which introduce some regulation of planned obsolescence. Article 70.v of the law amends Article L. 541–1 of the Environment Protection Code by adding the following objective to the national policy for waste prevention and management:

> Lutter contre l'obsolescence programmée des produits manufacturés grâce à l'information des consommateurs. Des expérimentations peuvent être lancées, sur la base du volontariat, sur l'affichage de la durée de vie des produits afin de favoriser l'allongement de la durée d'usage des produits manufacturés grâce à l'information des consommateurs. Elles contribuent à la mise en place de normes partagées par les acteurs économiques des filières concernées sur la notion de durée de vie. La liste des catégories de produits concernés ainsi que le délai de mise en œuvre sont fixés en tenant compte des temps de transition technique et économique des entreprises de production.

Moreover, Article 99 of the law of 2015 introduces into the Consumer Code a new Article L213-4-1 that reads as follows:

---

[12] Latouche (2012), p. 114.

[13] Ibid. p. 93.

[14] Tollemer (2011–2012), p. 168.

I. L'obsolescence programmée se définit par l'ensemble des techniques par lesquelles un metteur sur le marché vise à réduire délibérément la durée de vie d'un produit pour en augmenter le taux de remplacement.
II. L'obsolescence programmée est punie d'une peine de deux ans d'emprisonnement et de 300 000 € d'amende.
III. Le montant de l'amende peut être porté, de manière proportionnée aux avantages tirés du manquement, à 5 % du chiffre d'affaires moyen annuel, calculé sur les trois derniers chiffres d'affaires annuels connus à la date des faits.

Planned obsolescence is erected as a crime punishable by law. This criminal dimension has its drawbacks. It will justify a most restrictive interpretation of the text of the law, which is likely to bump into new and imaginative practices of manufacturers and product designers, as well as marketing and advertising managers. Moreover, bringing evidence of the deliberate intent existing at the time when the product was placed on the market will prove to be extremely difficult.

3.3 General Obligation of Product Sustainability

In all jurisdictions worldwide, warranty law requires that a product sold to the consumer can be used for a 'reasonable period of time'. The review of the prevailing standards in force in Canada and in the European Union will help us in concluding whether this formulation meets our goal of promoting sustainability.

3.3.1 *Quebec*

In Quebec, the Consumer Protection Act provides for the legal or implied warranty of the ordinary usage of the good, which means that a good can be used for the usage that the consumer can normally expect, regardless of the presence of a defect. Article 37 of the LPC specifies that: 'Goods forming the object of a contract must be fit for the purposes for which goods of that kind are ordinarily used.' Article 38 specifies the duration of such a normal usage: 'Goods forming the object of a contract must be durable in normal use for a reasonable length of time, having regard to their price, the terms of the contract and the conditions of use.'

The reasonableness of lifespan is assessed by the court in each situation and for each product. It is hardly possible to draw clear conclusions from case-law decisions that would establish a clear reasonable lifespan for a given product or type of product. On the contrary, there is great diversity in the case law. For example, in the automotive sector, a period of 18 months has been held as unreasonable in case of destruction by fire, 24 months for the appearance of traces of rust, 24 months for a defective motor, 30 months for the breaking of a connecting rod, 36 months for the deterioration of leatherette or fabric seats, and 40 months for break in the transmission system.[15]

Still, the duration most commonly applied by the courts delay for everyday consumer goods is one (1) year. Moreover, both authors and court cases confirm that it is incumbent on the applicant, and therefore on the consumer, to establish the unreasonableness of the period of time in relation to the price of the product, the terms of the contract and the conditions of the product usage.[16] If such evidence is not established, Section 38 of the CPA will not apply. Although the courts' interpretation of Sections 37 and 38 of the CPA is generally favourable to the consumer,[17] there are numerous cases that illustrate the difficulties in determining the 'reasonable duration' of the goods. For example, one judge refused to retain as evidence documentation drawn from the internet,[18] while another refused to decide in the absence of clear provision in the law.[19]

Other provisions complement the CPA. For example, Article 1726 para. 1 of the Civil Code of Quebec defines the warranty of quality:

> The seller is bound to warrant the buyer that the property and its accessories are, at the time of the sale, free of latent defects which render it unfit for the use for which it was intended or which so diminish its usefulness that the buyer would not have bought it or paid so high a price if he had been aware of them.

Under Article 1729 of the Civil Code:

> [I]n a sale by a professional seller, a defect is presumed to have existed at the time of the sale if the property malfunctions or deteriorates prematurely in comparison with identical property or property of the same type; such a presumption is rebutted if the defect is due to improper use of the property by the buyer.

Contracting parties may not derogate from the implied warranties provided for in Articles 37 and 38 of the PLC (Articles 261 and 262 of the PLC). Article 1733 of the Civil Code calls for a similar prohibition of warranty disclaimers.[20]

3.3.2 *New Brunswick*

Article 12 (1) of the New Brunswick Consumer Product Warranty and Liability Act[21] provides that: 'In every contract for the sale or supply of a consumer product there is an implied warranty given by the seller to the buyer that the product and any

---

[15]Bourgoignie (2008b), p. 61 and quoted references.

[16]Lafond (2015), para 414.

[17]Lafond (2015), quoted in footnote 17, para 416.

[18]Hamel c Toshiba du Canada ltée, 2015 QCCQ 117, paras 9–12.

[19]*Paris c Pirelli Pneus inc.*, 2014 QCCQ 1812 para 13.

[20]Bourgoignie (2008b), p. 67.

[21]LN-B 1978, c C-18.1.

components thereof will be durable for a reasonable period of time.' The next paragraph of the same article specifies how to evaluate the reasonable lifespan of a product:

> In determining a reasonable period of time for the purposes of subsection (1), regard shall be had to all relevant circumstances, including the nature of the product, whether it was new or used, its use as contemplated by the seller and buyer at the time of the contract, its actual use and whether it was properly maintained.

In consumer contracts, these provisions are binding on the parties. Under Article 24: 'Where there is a contract for the sale or supply of a consumer product, the parties cannot agree to exclude or restrict any warranty or remedy provided by this Act.'

Case law remains rare and, as in Quebec, the burden of proof of the reasonableness or unreasonableness of the lifespan of the product lies with the consumer.[22]

3.3.3 Saskatchewan

Under Article 19 of the Saskatchewan Consumer Protection and Business Practices Act[23] of 2014, the consumer is entitled to expect a product of acceptable quality, taking into consideration all the circumstances surrounding the sale of the product:

> [A]cceptable quality means the characteristics and the quality of a consumer product that consumers can reasonably expect the product to have, having regard to all the relevant circumstances of the sale of the product, including: (i) the description of the product; (ii) its purchase price; and (iii) the express warranties of the retail seller or manufacturer of the product; and includes merchantable quality within the meaning of The Sale of Goods Act (Article 19).

Sustainability is expressly mentioned. Under Article 19 (g), the consumer has the right to expect

> that the consumer product and all its components are to be durable for a reasonable period, having regard to all the relevant circumstances of the sale, including (i) the description and nature of the consumer product; (ii) the purchase price; (iii) the express warranties of the retail seller or manufacturer; and (iv) the necessary maintenance the consumer product normally requires and the manner in which it has been used.

Unfortunately, there is no jurisprudence that clarifies the concept of a reasonable period under the current legislation. Moreover, if one refers to decisions made under the old Consumer Protection Act, the burden of establishing the reasonable period seems to rest with the consumer.[24]

---

[22] Lagacé c. Bison (Restigouche) Ltd., 2003 NBBR 98 (CanLII), Savoie et Savoie c. René's Service & Trailer Sales Ltd., 2007 NBBR 229 (CanLII).

[23] The Consumer Protection and Business Practices Act, SS 2014, c C-30.2.

[24] Among others, *Rittinger v Concrete Guy Inc*, 2014 SKPC 68 (CanLII).

3.3.4 Countries of the European Union

Directive 1999/44/EC[25] of 25 May 1999, which has been transposed in all Member States, introduces the concept of legal warranty of product conformity to the contract. In particular, it states that:

> [C]onsumer goods are presumed to be in conformity with the contract if they:
>
> (a) comply with the description given by the seller and possess the qualities of the goods which the seller has held out to the consumer as a sample or model;
> (b) are fit for any particular purpose for which the consumer requires them and which he made known to the seller at the time of conclusion of the contract and which the seller has accepted;
> (c) are fit for the purposes for which goods of the same type are normally used;
> (d) show the quality and performance which are normal in goods of the same type and which the consumer can reasonably expect, given the nature of the goods and taking into account any public statements on the specific characteristics of the goods made about them by the seller, the producer or his representative, particularly in advertising or on labelling. The seller shall be held liable where the lack of conformity becomes apparent within two years as from delivery of the goods (art.5, para 1).

The text of the Directive does not impose a single, uniform duration of 2 years for all products; it allows the consumer to benefit from the warranty as long as the defect appears within 2 years. For products with a long lifespan, the warranty provided is therefore limited to 2 years; for products with a normal lifespan of less than 2 years, the guarantee applies if the defect occurs in an unreasonable or too short period compared to the normal lifespan of this product type.

The proof of the lack of conformity lies on the consumer, but the directive establishes a presumption of default when this appears in the first 6 months following the purchase:

> Unless proved otherwise, any lack of conformity which becomes apparent within six months of delivery of the goods shall be presumed to have existed at the time of delivery unless this presumption is incompatible with the nature of the goods or the nature of the lack of conformity.[26]

For example, wear of the product that occurs before the end of the normal lifespan of the product constitutes a breach of the warranty of conformity. However, the normal wear of the product or misuse do not constitute defects of conformity.

The nature, use and the price of the product must also be considered in the assessment of conformity. It should be noted that the text of the Directive does hold any obligation to provide for after-sales services or availability of spare parts. The 1999 Directive is a so-called minimum directive, meaning that in their transposition laws, Member States are still allowed to maintain or adopt provisions that are more protective of the consumer than those of the Directive. Member States may not

---

[25]Directive CE1999/44/CE of the European Parliament and Council of 25 May 1999 on some aspects of sale and guarantee of consumer goods [1999] OJ, L 171/12.

[26]Article 5, para 3.

foresee a warranty period of less than 2 years but implied warranties of more than 2 years are still allowed.

France In France, the provisions of the Consumer Code concerning the legal warranty of conformity of a product result from the transposition of Directive 1999/44/EC mentioned above. The legal warranty of conformity is stated in Article L217–4 of the French Consumer Code. Article L217–5 specifies that the good is in conformity with the contract if it is fit for the purposes for which goods of the same type are normally used. In 2014, Article 15 of Law No. 2014–344[27] introduced an important amendment to Article L217–7 of the Consumer Code by lengthening the duration of the presumption of the defect from 6 months to 2 years (it remains 6 months for used goods).

The Netherlands In The Netherlands, the duration of the legal warranty of conformity is not limited to 2 years. As will be shown in Sect. 3.4 below, longer warranty periods are made applicable to products that have a longer expected lifespan.

Finland In Finland, the Consumer Ombudsman provides guidelines in relation to the Consumer Protection Act.[28] Under these guidelines, the economic operator's liability for breach of the guarantee of conformity should not be limited in time; rather, it should be determined according to the lifespan of that particular product. The consumer might reasonably consider this to be an *expected* or *legitimate period of sustainability*: 'the goods fail, as regards sustainability or otherwise, to conform to what the consumer reasonably expects in purchasing goods of the same type.' The main criterion becomes the legitimate or reasonable expectation of the consumer and the warranty may apply beyond the 2 years prescribed by the European directive as long as the defect occurs within a period corresponding to the expected lifespan of the product. However, 'in respect of durability, the establishment of lack of conformity is based on a general standard and not on the expectations of individual consumers'.

3.4 Setting Fixed Time Limits for Sustainability

As seen, the obligation of duration for a reasonable time enshrined in the legislative texts does not impose one or more objective time limits for the sustainability of consumer products. To avoid problems caused by diversity in the construction of the

---

[27] *Loi relative à la consommation,* Official Journal of 18 March 2014, 5400 text no 1.

[28] Finish Competition and Consumer Authority (2002, modified in 2011), The Consumer Ombudsman's Guidelines. Statutory liability for lack of conformity and guarantee in the sale of consumer goods. On line: http://www.kkv.fi/en/decisions-and-publications/publications/consumer-ombudsmans-guidelines/by-subject/statutory-liability-for-lack-of-conformity-and-guarantee-in-the-sale-of-consumer-goods/.

notions of 'reasonable time' or 'reasonably expected duration', the feasibility and relevance of introducing one or more fixed guarantee periods should be explored. Making concrete the guarantee period(s) in this way would certainly have the merit of simplicity; it would also help to clarify and thus make more effective the obligation of after-sales service and provision of spare parts for products which, in several pieces of legislation, accompanies the obligation of sustainability. Attempts made to this end remain rare. The following relevant developments have been observed in The Netherlands, France and the European Union.

3.4.1 The Netherlands

Work on the lifespan expectancy of products has been carried out in The Netherlands by two professional organizations (UNETO and VNI) in the field of electrical and electronic products. These organizations have published 'guidelines' that make recommendations on the 'normal' lifespan of products. UNETO and VNI have prepared a table of expected product lifespans in years and months.[29] The expectations differ depending on the product categories and the sales price of the product. The ranges of prices suggested are obviously not transferable as such in all jurisdictions (Table 3.1). Adjustments will have to be made according to the situation and prices of local consumer markets.

3.4.2 France

In France, several institutions have studied the question of fixed lifespans for products. A report published in 2012 by ADEME on the lifespan of electrical and electronic equipment[30] offers interesting ways to determine a fixed period of sustainability for certain categories of products:

- **TVs**. Total usage time is estimated to be around 2000 h (e.g., OLED TV) and 80,000 h (e.g., LCOS TV and CRT) or more than 30 years, with a duration of 20,000 to 60,000 h (in optimal conditions) for plasma TVs and TV PHOLEDs.
- **Laptops**. Total usage time of laptops is estimated between 3 and 12 years. Practice reveals a duration of 2.5 and 5 years, depending on professional or domestic use. The depreciation of computers, estimated by insurers, reaches 100% after 6 years.
- **Cell phones**. The potential operating time of a cell phone is estimated at more than 10 years (excluding batteries), but because of functional and social

[29]UNETO and VNI, Tabel met gemiddelde gebruiksduurverwachtingen. See: www.uneto-vni.nl/stream/richtlijnen-afschrijvingsmethoden.

[30]Agence de l'environnement et de la maitrise de l'énergie (2012), pp. 42–45.

Table 3.1 Average product lifespan expectancy, UNETO and VNI

Product category			Expected lifespan: In years	Expected lifespan: In months
TV sets	Sales price (in Euro)	0–299	3	36
		300–499	4	48
		500–999	5	60
		More than 1000	6	72
Devices (TV)	Sales price (in Euro)	0–199	2	24
		More than 200	3	36
Audio	Sales price (in Euro)	0–199	2	24
		200–399	3	36
		400–999	4	48
		More than 1000	5	60
Home appliances (washing machines, dryers)	Sales price (in Euro)	0–199	2	24
		200–299	3	36
		300–399	4	48
		400–499	5	60
		500–599	6	72
		600–699	7	84
		More than 700	8	96
Small appliances	Sales price (in Euro)	0–199	2	24
		More than 200	3	36
Coffee machines	Sales price (in Euro)	0–199	2	24
		200–499	3	36
		More than 500	4	48
Computers (including tablets, laptops, game consoles)	Sales price (in Euro)	0–299	2	24
		More than 300	3	36
Cameras and camera film	Sales price (in Euro)	0–199	2	24
		200–399	3	36
		More than 400	4	48
Telephones (smartphones, smartwatches, GPS, etc.)		All items	2	24
Accessories (e.g. chargers, batteries, adapters, etc.)		All items	2	24

obsolescence, users change on average more than four times during this period. The actual usage time is then estimated between 20 months and 6 years.

- **Printers**. Total usage time of printers, domestic or professional, is estimated between 3 and 6.5 years, depending on the type of the printer (inkjet or laser); 5 years is already considered to be an extended lifespan.
- **Lamps**. Total lamp usage times are varied, depending on the type of the lamp: about 1000 h for conventional incandescent lamps, 6000–20,000 h for compact fluorescent lamps, 6000–30,000 h for LED lamps and 12,000–60,000 h for linear fluorescent lamps.
- **Vacuum cleaners**. Total usage time of vacuum cleaners is estimated between 6 and 8 years on average (or 500 h).
- **Washing machines**. The operating lifespan of washing machines can be estimated in a number of cycles or in years; between 1000 and 5000 cycles of laundry or 9.9 to 22 years depending on the study. The total duration of use would be even more variable, ranging from 2 to 30.8 years, with an average of about 10.1 years in France. For insurers, a washing machine has no value after 8 years.

For ADEME, the measure of setting a fixed product lifespan is nothing but a first step, as it also requires standardized evaluation methods that integrate not only the robustness of the product, but also its repairability, its compatibility with external systems and its ability to evolve.[31] The agency continues to reflect on this issue. In another report on product lifespan cycle extensions published in 2016,[32] it highlights the methodological difficulties of such an approach:

> S'il semble souhaitable d'accroître la durée de vie de la majorité des produits, une approche particulière doit être adoptée pour chaque catégorie de biens. Par ailleurs, la vigilance s'impose pour éviter des pièges tels que la sur-qualité ou les transferts de pollutions. De plus, prendre des décisions pertinentes requiert la poursuite des travaux sur l'analyse environnementale et son interprétation ainsi que la mise en place de méthodologies pour estimer les durées de vie puis la durabilité. Ensuite, des propositions de mesures concrètes à destination des metteurs sur le marché et des fabricants peuvent être discutées.[33]

Among the courses of action proposed by ADEME are the imposition of a minimum level of longevity for a launching product. Based on the evaluation of the normative duration, minimum longevity could limit entry to the market[34] of products that are too unsustainable. Specifically, in each sector, the products should fulfil a number of sustainability criteria (strength, dismountability, repairability, etc.), defined by a regulatory framework. Once this selection is made, the products could be differentiated through a display of their environmental impacts, and more necessarily of their lifespan or sustainability. This system would be able to guard against products featured for their sustainability, despite a bad environmental record overall.

---

[31] Agence de l'environnement et de la maitrise de l'énergie (2016), p. 2.

[32] Ibid, p. 6.

[33] Ibid.

[34] Ibid, p. 46.

In a May 2015 opinion on product lifespan, France's National Consumer Council concluded that there is no one-size-fits-all answer to the question of product lifespan: expectations are different for each consumer and vary according to the products, the industrial stakes are different according to the economic sectors, and the conditions of usage and maintenance of the goods play a determining role on the lifespan of the products.[35] The Council is in favour of carrying out additional studies and, in the shorter term, adopting measures to better inform consumers and to act in favour of repairing goods launched on the market.

Finally, WWF France has proposed a guide on eco-responsible information systems.[36] As minimum computer hardware requirements are doubling every 2 years, the organization advocates for the longest possible use of the same software version and to update software only if it is really necessary. The report notes that: 'Buyers may also require publishers to support their software for longer (more than 5 years) and to commit to the proper functioning of their software, for a given duration, on a given hardware configuration'.[37] Hence, the lifespan of the products would be closely linked to the lifespan of software versions.

3.4.3 The European Union

In 2014, the European Economic and Social Committee (EESC) drafted an important opinion on product lifespan and consumer information,[38] which considers it necessary to launch a system guaranteeing a minimum lifespan of the purchased products. It advocates for the definition of European standards to measure the sustainability of products. The EESC also proposes to extend the guarantee scheme linked to purchases of consumer goods. However, the EESC advocates a nuanced approach that would not increase the lifespan of all products uniformly. It rather proposes to reason in terms of product use and optimization of these uses, which does not automatically imply a longer lifespan expectancy.

3.5 The Obligation to Repair and After-Sales Services

One way to prolong the lifespan of products or to give these a second life is to promote their repair. This is not always the option chosen in case of defect. For ease or because the consumer lost confidence in the good concerned, she may prefer either to replace the defective product with a new one or ask for a refund of the price

[35]Conseil national de la consommation (2015), p. 5.

[36]WWF France (2011).

[37]Ibid, p. 8.

[38]Libaert and Haber (2014).

paid and acquire an alternative. The seller will prefer to avoid all the problems and costs linked to the repair of the product and it may well be that she does not have the necessary spare parts or adequate after-sales service. For commercial purposes, the consumer is satisfied if the seller does not contest the complaint lodged by the former and refunds the price paid or replaces the product with a new or similar product. As known, it is also common practice that the buyer benefits from an express guarantee of absolute satisfaction for a few days or weeks, allowing her to return the purchased goods without having to invoke the presence of any defect.

The cost of repair services is often a decisive obstacle. The ratio between the cost of repair and the purchase of a new product is shrinking more and more, which has the effect of encouraging buying rather than repairing. A study by ADEME shows that, according to this same logic, consumers would be more inclined to have the product repaired when the price of the repair is less than one-third of the price of the newly purchased product.[39] The obligation to repair the defective product and to ensure effective access to the needed spare parts could, therefore, play an important role in our efforts to enhance product sustainability. Such an obligation, however, remains barely developed in Canada and in Europe.

3.5.1 Quebec

In Quebec, Article 39, paragraph 1 of the LPC provides for the obligation to guarantee the availability of spare parts and repair services for a reasonable period of time:

> Where goods being the object of a contract are of a nature that requires maintenance, replacement parts and repair service must be available for a reasonable time after the making of the contract.

However, this provision is not mandatory; the trader or the manufacturer retains the possibility of exemption from the obligation to provide after-sales services. As stated in paragraph 2 of Article 39:

> The merchant or the manufacturer may release himself from this obligation by warning the consumer in writing, before the contract is entered into, that he does not supply replacement parts or repair service.

The text also does not specify what reasonable duration means, or the criteria for assessing the reasonableness of the advertised duration. According to Claude Masse, the reasonable duration provided for in the first paragraph of Article 39 corresponds to the reasonable period provided for in Article 38 of the LPC.[40] This is, in our opinion, too short a period. While it is clear that the duration provided for in Article 39 cannot be less than the period of the legal warranty, a period extending well

---

[39] Agence de l'environnement et de la maitrise de l'énergie (2007), p. 9.

[40] Masse (1999), p. 316.

beyond the legal warranty time limit should apply if sustainability is to be promoted. The duration of the obligation to repair or to have spare parts available should at least correspond to the duration of the commercial, express, or extended guarantees granted to the consumer, and even extend beyond it.[41]

3.5.2 Saskatchewan

In Saskatchewan, Article 19 (h) of the Consumer Protection and Business Practices Act (2013)[42] calls for 'reasonable availability' of spare parts, maintenance and repair services for a period, also defined as 'reasonable', after the sale of the product:

> (h) if the consumer product normally requires repairs (the guarantee is) that spare parts and repair facilities will be reasonably available for a reasonable period after the date of sale of the consumer product.

Unlike in Quebec, disclaimers are not allowed.

3.5.3 Europe

Directive 2006/66/EC of 6 September 2006 on batteries and accumulators, and waste of batteries and accumulators, requires manufacturers of electronic equipment to provide devices, allowing to remove the batteries for easier collection and recycling.

Article 6 of the already quoted Law No. 2014–344 of 2014 creates the obligation to inform the consumer about the guarantee period of the spare parts availability and amends Article L. 111–3 of the Consumer Code accordingly:

> Le fabricant ou l'importateur de biens meubles informe le vendeur professionnel de la période pendant laquelle ou de la date jusqu'à laquelle les pièces détachées indispensables à l'utilisation des biens sont disponibles sur le marché. Cette information est délivrée obligatoirement au consommateur par le vendeur de manière lisible avant la conclusion du contrat et confirmée par écrit lors de l'achat du bien.
>
> Dès lors qu'il a indiqué la période ou la date mentionnée au premier alinéa, le fabricant ou l'importateur fournit obligatoirement, dans un délai de deux mois, aux vendeurs professionnels ou aux réparateurs, agréés ou non, qui le demandent les pièces détachées indispensables à l'utilisation des biens vendus. Un décret précise les modalités et conditions d'application du présent article.

This new provision opens the path towards the recognition of an after-sales service obligation but does not formally introduce such an obligation. What is

[41]Bourgoignie (2008b), p. 62.

[42]Quoted in footnote 24.

actually required is information about the availability of spare parts, but the text falls short of imposing such availability.

In Ireland, the Sale of Goods and Supply of Services Act[43] of 1980, states that spare parts and adequate after-sales service must be available. In Greece and Romania, spare parts must be available for the entire lifespan of the product. The Portuguese Consumer Code[44] formally recognizes the right to spare parts but only for the duration of the legal warranty, which is 2 years. In Italy, there is an obligation to inform the consumer whether the good is still in production or not. In the Republic of Malta, spare parts must be available for a reasonable period of time.[45]

3.6 Standardization and Certification

If the goal is to increase the durability of consumer products from the moment they are designed, product standardization can be a most relevant avenue. Setting product standards calling for sustainability or increased durability will contribute to enhance sustainable production patterns. Although standards remain voluntary, they can be made mandatory when referred to or introduced in technical regulations. They may also be referred to when assessing the conformity of the product with the contract and consumer legitimate expectations.

While 'sustainability' has become a common concern for standardization bodies in Canada and Europe, the issue of 'durability' or product lifespan has much less so. Standards adopted or in preparation in Canada in the area of sustainability focus mostly on the environmental impact or performance of products (energy consumption, materials used, end of lifespan) but not on the product lifespan in terms of years.[46] In Europe, the two following standards are worth noting: the *Nordic Swan/ Nordic Ecolabelling* (Norway and Sweden) and the European *Ecolabel* (European Union).

Nordic Swan/Nordic Ecolabelling evaluates the environmental impact of products through the study of their complete lifespan cycle while taking into consideration the quality, the health impact, as well as the performance and functionality of the product.[47] For example, for televisions and projectors, the standard requires a minimum of 2 years of product operation and 7 years after the end of product production for the availability of compatible replacement parts. The scope of the

[43]Sale of Goods and Supply of Services Act (1980).

[44]Lei De Defensa Do Consumidor, No. 24/96.

[45]ECC-Net Legal guarantees and commercial warranties on consumer goods in the EU, Iceland and Norway.

[46]For example, see: http://shop.csa.ca/fr/canada/normes-et-codes/performance-en-matiere-de-durabilite-des-produits/icat/productsustainabilityperf.

[47]Nordic Ecolabelling. See: http://www.nordic-ecolabel.org/.

standard remains limited; it does not cover computers and coffee machines, among other things.[48]

Created in 1992, the European *Ecolabel* is the only European official eco-label that can be used in all EU Member States.[49] The European *Ecolabel* establishes a required lifespan for certain product categories, excluding, among others, household appliances. For example, the standard prescribed for laptops calls for the availability of spare parts during a period of at least 5 years after production has stopped.[50] For televisions, the period is at least 7 years. Another interesting label is the German label *Blue Angel*, which indeed concern the reference to 'durable' on a number of individual products but there are no details regarding the consideration of product lifespan in the award of certification.[51]

3.7 Conclusions and Way Forward

In its 2014 report, entitled 'Pour une consommation plus durable: la durée de vie des produits de l'industrie et l'information du consommateur au service d'une confiance retrouvée', the Economic and Social Committee of the European Union recommends several avenues towards greater sustainability of consumer products: (a) a ban on products of which defects have been programmed; (b) the introduction of a guarantee of availability of spare parts during the five (5) years following the acquisition of the good; (c) encouragement of voluntary certification procedures; (d) a commitment by manufacturers to publish figures on the most frequently occurred defects and to produce spare parts; (e) the display of the estimated lifespan of the product and; (f) the establishment of a system guaranteeing a minimum lifespan of the purchased products.[52]

Along the same lines but not necessarily in the same order of priority, the following five main measures or tools seem the most relevant to remedy the obsolescence of consumer products. The following measures are all needed and at the same time complementary:

1. **Development of standardized evaluation methods for measuring the lifespan of products** and setting sustainability criteria (robustness, strength, repairability, recyclability, dismountability, etc.) and the promotion of certification mechanisms to attest the conformity of the product with existing standards (labels attesting to the sustainability of the product). In this respect, the development

[48]Ibid.

[49]L'Écolabel européen: qu'est-ce que c'est? See: http://www.ecolabels.fr/fr/l-ecolabel-europeen-qu-est-ce-que-c-est.

[50]The European Ecolabel for Portable Computers. The official EU mark for greener products, on line: http://ec.europa.eu/environment/ecolabel/documents/portable.pdf.

[51]The Blue Angel. See: https://www.blauer-engel.de/en/products.

[52]Opinion quoted in footnote 39.

of a nomenclature of consumer products will be necessary. Among the existing nomenclatures, the one proposed by the United Nations seems the one that could best serve as source of inspiration.[53] For example, the category '05–Furniture, household goods and household maintenance' includes a subcategory: 'Household appliances', which is subdivided into several classes: Major appliances, electric or not (including, for example, refrigerators, stoves, air conditioners, washing machines, vacuum cleaners, heaters, water heaters, microwave ovens, etc.), small household appliances (including, for example, coffee machines, kettles, toasters).

2. **Definition of a general obligation of product durability imposed upon the manufacturer** or his representative or importer, at the time they place the product on the market. The obligation is to place on the market products whose expected lifespan will have to be reached under normal conditions of use and maintenance. Placing on the market means the first entry into circulation of the product. Implementing measures will determine the expected lifespans of products. In the absence of a fixed duration, the requirement of durability will be 3 years. The occurrence of a defect or failure within the period fixed or, failing that, the period of 3 years, will create a presumption of nonconformity of the product with the obligation of durability. It will then be up to the economic operator to prove the abnormal nature of the use or improper maintenance of the product by the user.
3. Confirmation of an obligation to guarantee the repairability of the product for a number of years, fixed by implementing legislation or, failing that, a period of 2 years beyond the expiry of the expected lifespan of the product. This requirement will result in ensuring the availability of essential spare parts, access to repair services or after-sales service. The adequacy of the planned services will be assessed according to the nature of the product, the duration of use and the cost of the services. Accompanying tools will be to reduce the cost of repair services (in particular via a reduction in the taxation of the service) and to promote self-repair.[54]
4. **Adequate consumer information**. Information shall be required at least on the following three (3) elements:

   (a) the expected lifespan of the product;
   (b) the respective start dates, durations, scopes and terms of the implied or legal warranty associated with the product and of the additional and extended commercial guarantees, if any, that will apply within the expected lifespan of the product; and
   (c) the arrangements for having access to spare parts or to repair and after-sales services.

5. **Sanction of planned obsolescence**. There is a consensus that planned obsolescence of products must be sanctioned. Planned obsolescence could be confirmed as a penal offence like in France. An alternative, and probably more effective,

[53]United Nations Standard Products and Services Code (UNSPSC): www.unspsc.org.

[54]Conseil national de la consommation (May 2015), p. 16.

would be to hold planned obsolescence as an unfair commercial practice against consumers, punishable by administrative (including an order to end the practice), civil (reimbursement of the price paid by the consumer, injunction for cessation) and criminal sanctions in the most serious cases or in case of repeated practice.

References

Agence de l'environnement et de la maitrise de l'énergie (2007) Panorama de l'offre de réparation en France, Paris. On line: http://formation.ducretet.net/ftp/fodipeg/Actualites/marche/fev2008/Gouv_reparation_juillet07.pdf

Agence de l'environnement et de la maitrise de l'énergie (2012) Étude sur la durée de vie des équipements électriques et électroniques. Rapport final, Paris. On line: www.ademe.fr/sites/default/files/assets/documents/84636_duree_de_vie_des_eee.pdf

Agence de l'environnement et de la maitrise de l'énergie (2016) L'allongement de la durée de vie des produits, Paris. On line: www.ademe.fr/sites/default/files/assets/documents/avisademe_allongement-duree-vie-produits_201604.pdf

Bourgoignie T (1988) Éléments pour une théorie du droit de la consommation. Cabay-Bruylant, Brussels

Bourgoignie T (2008a) Droit de la consommation: une mutation salutaire. In: Liber Amicorum Jean-Luc Fagnart. Anthémis, Louvain-la-Neuve

Bourgoignie T (2008b) Garanties légales et commerciales liées à la vente de produits de consommation: la confusion des genres. In: Bourgoignie T (ed) Propos autour de l'effectivité du droit de la consommation. Éditions Yvon Blais, Cowansville

Bourgoignie T (2017) Protection du consommateur et développement durable: consommateur souverain, pollueur, responsable ou victime? In: Brovelli G, Sancy M (eds) Environnement et développement durable dans les politiques de l'Union européenne, actualités et défis. Presses Universitaires de Rennes, Rennes

Conseil national de la consommation (2015) Avis du Conseil national de la consommation sur la durée de vie des produits. Paris. On line: www.economie.gouv.fr/files/files/directions_services/cnc/avis/2015/Avis_sur_la_durabilite_des_produits.pdf

Déméné C (2015) Étude exploratoire de la phase d'usage des produits électroniques en vue de minimiser les impacts environnementaux: le cas du téléviseur. Dissertation, Université de Montréal, Montréal. Online: https://papyrus.bib.umontreal.ca/xmlui/handle/1866/11776

Duguay B (2008) Modes de consommation irresponsables: comment en est-on arrivé là? In: Bourgoignie T (ed) Propos autour de l'effectivité du droit de la consommation. Éditions Yvon Blais, Cowansville

Friant-Perrot M (2008) La consommation durable et la protection des consommateurs: réflexions sur les nouveaux rapports entre le droit de la consommation et le concept de développement durable. In: Parent G (ed) Production et consommation durables: de la gouvernance au consommateur-citoyen. Éditions Yvon Blais, Cowansville

Lafond D (2015) L'approche contractuelle: l'imposition d'une obligation légale de garantie au commerçant. In: Lafond P-C, Élie P (eds) Droit de la protection du consommateur: Théorie et pratique. Éditions Yvon Blais, Cowansville

Latouche D (2012) Bon pour la casse, les déraisons de l'obsolescence programmée. Les liens qui libèrent, Paris

Libaert T, Haber J-P (2014) Pour une consommation plus durable: la durée de vie des produits de l'industrie et l'information du consom mateur au service d'une confiance retrouvée (avis d'initiative). European Economic and Social Committee CCMI/112 EESC-2013–1904, Brussels. www.eesc.europa.eu/?i=portal.fr.ccmi-opinions.27387

Masse C (1999) Loi sur la protection du consommateur. Analyse et commentaires. Éditions Yvon Blais, Cowansville

Roch F (2013) Vers un nouveau paradigme en matière de développement? Presses académiques francophones, Saarbrücken

Roch F (2014) Déficit écologique et crise des dettes souveraines: réflexions sur les échecs et les défis des modèles contemporains de développement. In: Pavot D, Dufour G (eds) La crise des dettes souveraines. Lexis/Nexis, Montréal/Toronto

Slades G (2007) Made to break. Technology and obsolescence in America. Harvard University Press, Cambridge

Tollemer L (2011–2012) L'obsolescence programmée. Dissertation, Université Montpellier 1

WWF France (2011) Guide pour un système d'information éco-responsable. On line: http://awsassets.wwfffr.panda.org/downloads/guidentic27042011_110801093454_phpapp01.pdf

Chapter 4
The Shift from Consumer Protection to Consumer Empowerment and the Consequences for Sustainable Consumption

Patrícia Galindo da Fonseca

Contents

Abstract This chapter discusses the evolution of Consumer Law, and the transformation of modes of consumption. Originally, the objective of Consumer Law was to protect and promote the interests of consumers, to inform them about their rights and to ensure their representation in the bodies where decisions concerning them are made. The new Consumer Law now also aims to change consumer behaviour, making them aware of the ecosystem's fragility through the adoption of sustainable consumption patterns. The goal of this paper is to discuss the integration of environmental concerns into Consumer Law, which is far from complete, through the insertion of the concept of collectiveness that is present on both domains.

P. G. da Fonseca (✉)
Universidade Federal Fluminense (UFF), Rio de Janeiro, Brazil

A. d. Amaral Junior et al. (eds.), *Sustainable Consumption*,
https://doi.org/10.1007/978-3-030-16985-5_4

4.1 Introduction

The current consumption model depletes the planet's natural reserves and aggravates social disparities. The model reflects the needs of contemporary consumer society. As pointed out below, consumer society is a society of deep inequalities. It was only after the United Nations Conference on Environment and Development (UNCED) in 1992 in Rio de Janeiro that it was recognised how essential it is to change the present pattern of consumption to protect ecosystems. Gradually, this kind of concern became part of Consumer Law, which led to the emergence of the concept of sustainable consumption.[1]

The United Nations General Assembly adopted the Guidelines on Consumer Protection in its Resolution 39/248 of 9 April 1985. The Guidelines establish a set of consumer law principles that set out a policy framework for national governments so they can promote consumer protection in different areas such as safety, quality and distribution of goods and services, the economic interests of consumers, information and consumer education and redress. The UN General Assembly revised the Guidelines on Consumer Protection to include sustainable consumption in the objectives and general principles, as well as in a specific new section (Section H of the 2015 version) in Resolution 1999/7 of 26 July 1999. The UN Guidelines were revised one more time by the General Assembly in Resolution 70/186 of 22 December 2015 because of the consultation process that began in 2012.

In fact, sustainable consumption derives from the concept of sustainable development, which itself emerged from the landmark 1987 Brundtland Report.[2] It was then defined at the Rio 92 meeting as: 'development that meets the needs of the current generation without compromising the ability of future generations to meet their own needs'. The central idea then rests on the theme of development, which has three different dimensions—economic, social and environmental—all inseparable and interdependent. The formula reflects the search for a point of convergence between economic development, protection of the environment and social solidarity.

Thus, sustainable consumption is the consumption of goods and services that has minimal impact upon the environment and is socially equitable. It implies giving responsibility to the consumer in his or her choices when consuming a particular product, in addition to the responsibility attributed to the producer of the good.[3] Therefore, it implies a broad collaboration among actors operating in the supply chain because it concerns the product life cycle from cradle to grave. Goods and services must be economically viable, as well as meeting basic needs of humans worldwide.

[1]See Friant-Perrot (2008).

[2]United Nations (1987).

[3]Pieratti and Prat (2000).

4.2 Consumer Law and Market Liberal Theory

The original purpose of consumer protection legislation was to protect and promote the interests of consumers and to inform consumers of their rights. In fact, consumer policy has traditionally focused on the economic interests of consumers: price, quality, choice, and redress. Consequently, in most of cases, national consumer laws seek to combat against market irregularities. In some jurisdictions, a much broader view of the consumer relationship prevails, going beyond the contractual framework alone. Through our comparative law analysis of the legislative and institutional building of two different national consumer systems, the Brazilian consumer protection seems to be one of these exceptions.[4] In all cases, it aims to empower consumers to make choices and take legal action.

Like many legal disciplines, Consumer Law cannot be precisely defined and is known as a multidisciplinary field of law. According to Bourgoignie, Consumer Law has a transversal dimension in the very nature of its function and its objectives. The author affirms that Consumer Law is traversed by a diverse range of legal fields, such as Civil Law, Commercial Law, Administrative Law, Banking Law, Competition Law, Judicial Law, Fiscal Law, Criminal Law and Labour Law, and also by various other domains, such as the economy.[5] Consumer Law must continue to adapt and establish new objectives, without losing sight of its larger objective of combating the imbalance between the parties.[6]

According to classical liberal economic theory—championed by Adam Smith, David Ricardo, John Stuart Mill, Leon Walras and Friedrich Hayek—competition is the principle means to achieve an optimal allocation of resources and also to improve consumer satisfaction and welfare. In short, the natural confrontation of individuals prompts competition, which leads people to produce what individuals demand and, in aggregate, the satisfaction of social needs. Far from being a purely economic concept, the market has been an ideological and legal tool to protect individual autonomy and personal liberty. Under this thesis, all contractors are deemed to be lucid, reasonable and capable of assuming their interests. Personal autonomy and the freedom to contract became the basis of the legal system established in the nineteenth century under the Code Napoleon, a system that reigned victorious until the end of the twentieth century in civil law jurisdictions.

[4]Fonseca (2016).

[5]See Bourgoignie (2006).

[6]See Fonseca (2014).

4.3 The Development of the Consumer Society

The development of the consumer society[7] at the end of the nineteenth century favoured access to a multitude of goods and services, a reduction in their price and an increase in their quality. It provoked an immediate benefit: tremendous gains in life expectancy. From the mid-1960s to 2005, average global life expectancy rose from 56 to 65 years; in developed countries, from 71 to 76 years.[8] Furthermore, there has been 35% in the world population since 1992, which corresponds to an additional 2 billion people. In a manifesto signed by 15,364 scientists from 184 countries published in November 2017 in the journal *BioScience*, the authors contend that:

> We are jeopardizing our future by refusing to moderate our intense material consumption, geographically and demographically unequal, and to realize that rapid and continued population growth is one of the main factors of environmental and even societal threats.

The document signed by the scientists claims for a revision on the economy to reduce inequality and to ensure that prices, taxes and incentive systems consider the real costs that consumption patterns impose on the environment.[9] The imbalance between consumers and professionals has generally increased, as will be discussed further below.

4.4 Current Modes of Consumption

Consumption levels in developed countries—which most developing societies aspire to achieve—have come at the cost of plundering the planet's resources. They rest on a model that aggravates social disparities not only between the so-called Global North and Global South, but even within countries. The gaps between developed and developing countries are growing and those same gaps occur also within developed countries. Much of the population is expected to be content with the leftovers: the waste of that development.

Already in 1998, the numbers were impressive. From a United Nations report in the same year, inequalities in consumption were extreme: 20% of the world's population in high-income countries captured 86% of total private consumption expenditure; the poorest 20%, just 1.3%. The richest fifth consumed 45% of all

---

[7]Lipovetsky (2006) analyses the history of consumer society through three distinct phases. The first phase was from 1880 to the end of World War II: the birth of mass markets. The central idea here was consumption–seduction. The second phase was from 1950 to 1980: mass consumer society. The central idea in this period was desire. The third phase is from 1980 to the present: hyper-consumption society. The central idea in this period is emotional consumption.

[8]United Nations (2002a). There was extraordinary growth in world population in the twentieth century. In 1800, the world population was 1 billion people; 200 years later, in 2011, that figure had risen to 7 billion. See: United Nations (2011).

[9]Shields (2017).

meat and fish, while the poorest fifth consumed 5%. The richest 20% of the world's population consumed 58% of total energy; the poorest, less than 4%. The top 20% had 74% of all telephone lines; the bottom 20% had 1.5%. The richest fifth consumed 84% of all paper; the poorest fifth, just 1.1%. Around one-fifth of the world's richest people owned 87% of the vehicles; the poorest fifth had less than 1%.[10] Current consumption levels undermine the base of natural resources and exacerbate inequalities. The dynamics of the connection between consumption, poverty, inequality and the environment accelerates.[11]

With the definitive emergence and, by its very nature, constantly evolving virtual technology, social disparity between the North and South assumes yet another facet because of the advent of the virtual age. From the point of view of income distribution and access to consumer goods, technology in general and specifically virtual technology is an income concentrator, and not an instrument for achieving social justice.

To get out of the economic crisis of the 1930s, manufacturers put in place strategies to encourage consumers to buy low-cost products. Therefore, industrial production was guided by the desire to provide the population with products of limited life-span or, put simply, low-end products. If companies use weaker materials, the argument goes that this is because consumers always want more without being willing to pay more. Economic factors are the strongest influence on consumer decisions. With the price of appliances falling everywhere in the industrialised countries, it costs less to buy new one than to repair an existing one, the costs of parts and labour being too high.

Currently, there is a need to develop programmes that make it easier to refurbish products and expand the second-hand market, as well as to encourage consumers to rent, share or donate and not only focus on recycling. According to the Observatoire de la consommation responsable de l'Université du Québec à Montréal (UQÀM),[12] it is crucial to help rehabilitate products and expand the used market, which implies that there remains a lot of educational work to do. Reuse is a step before recycling. Promoting eco-design, using sustainability as a selling point and designing more easily repairable products—these are the principles that, in an ideal world, should guide companies.[13]

In the 1990s, computers had a useful life of about 10 years. Today, they are considered obsolete after 3 years. Although they still work, one cannot install the latest operating systems without slowing down the machine or updating its applications. The same issue arises with smart phones and tablets, whose new models—always more efficient—follow one another every 6 months. This is technological

[10]United Nations Development Programme (1998).

[11]Ibid, p. 1.

[12]The *Observatoire de la consommation responsable* is a research unit recognised by the *École des sciences de la gestion* (ESG UQÀM), attached to the Responsible Consumer Research Team. For more information, see https://ocresponsable.com/.

[13]Decrop (2017).

obsolescence. It is thus important to distinguish between service life and usability.[14] Technological obsolescence is an environmental disaster because electronic devices contain several highly polluting materials. According to a survey developed by the United Nations University, nearly 42 million tonnes of e-waste were discarded in 2014.[15] The top three African countries with the highest e-waste generation in absolute quantities in 2014 were Egypt (0.37 million tonnes), South Africa (0.35) and Nigeria (0.22). The only African countries that have enforced national e-waste related legislation are Cameroon and Nigeria.[16]

4.5 The Limits of Development

The developed countries are worried about the seemingly unlimited population growth of the countries of the Global South and simultaneously starting to realise that the resources of the planet are limited. People seem to becoming slowly aware of the destructive effects of capitalism both on the environment and on social relations.[17] Those effects are reinforced by the globalisation of the economy and by the fragility of ecosystems. Serious environmental disasters—such as the 1984 Indian Bhopal gas tragedy,[18] the 1986 Chernobyl nuclear accident[19] and the recent Samarco crime that occurred in Brazil in 2015[20]—contribute to the recognition of the close link between human actions and the environment.

Western countries are beginning to question the limits of development paradigm in effect since the industrial revolution. To curb the consumerist frenzy, some call for transforming the current economy, based on growth, into a different kind of economy—this does not seem realistic, however. According to this theory, environmental initiatives that do not dispense with development, but simply make it greener are no longer enough. For instance, Mead sustains that it is already too late for sustainable development. According to him, those steps only endorse the approach of the economic orthodoxy. The failure to convince decision-makers of the criticisms of economic development projects leads to an actual need to define the foundations of new socio-economic systems that abandon the paradigm of growth.[21]

[14]Caza (2015).

[15]Baldé et al. (2015).

[16]Ibid, p. 22.

[17]Bonnin (2009).

[18]See: Eckerman (2011).

[19]See: Dolk and Nichols (1999).

[20]See: Fonseca and Fonseca (2016).

[21]Mead (2017).

4.6 Collaborative Consumption

Yet, others believe people must focus on collaborative consumption, a concept that dates back to 1978, and originally referred to 'events in which one or more persons consume economic goods or services in the process of engaging in joint activities with one or more others'.[22] The collaborative economy (also known as the sharing economy) generally refers to an economic model built around the exchange, sharing or sale of goods between individuals, where the use of a good or access to a service prevails over property rights. It involves revisiting the production, organisational and remuneration methods of companies and initiating new usage behaviours at both the corporate and consumer levels. Collaborative consumption is generally considered a form of collaborative economy, with three other related dimensions: collaborative production,[23] collaborative financing[24] and collaborative knowledge.[25] Specialists maintain that the financial crisis of 2008, the secular decline of household purchasing power, increased environmental and social sensitivity of consumers and especially the rapid development of digital technology could explain the rapid growth of collaborative consumption.[26]

Collaborative consumption is considered a new consumer pattern boosted by the development of the internet and social networks. It is a trend that renews consumption logics via pooling, exchange, barter or sharing. Three main categories of collaborative consumption form are generally identified: (a) product–service systems (this category takes normally the form of a conventional lease or leasing organised by an intermediary, such as a company, like Uber and car2go); (b) redistribution markets, like the exchange of second-hand goods between individuals and; (c) collaborative lifestyles (this category gathers sharing formulas of immaterial resources such as space, skills, time and money, such as couchsurfing).[27] Among the different expressions of collaborative consumption,[28] the most oft-cited are eBay, Airbnb, Couchsurfing, Bixi, car2go and Uber, although some of these are not collaborative at all (e.g., car2go).

---

[22]Felson and Spaeth (1978).

[23]As an example, manufacturing workshops open to the public and equipped with various computer-controlled machines and tools, like the so-called Fab Labs.

[24]Any form of financing for project promoters from other citizens without the participation of a bank, such as fundraising (crowd funding donations), raising commercial or solidarity loan between individuals or even alternative currencies like Bitcoin.

[25]The most eloquent example is Wikipedia, where knowledge is shared interactively.

[26]Ertz and Duriff (2017).

[27]Ibid.

[28]Co-working is also usually mentioned, although it is clearly more relevant to the world of work and to collaborative production. It questions the classical distinction between production and consumption and establishes a new organisation of labour. The workspace is virtual. For instance, in the work platform Askwonder.com anyone can become an analyst. The first analyst (or member of the research team) shall find five virtual trustable sources of information, who then sends it to another analyst who answers the request.

Although collaborative consumption is often presented as a new form of alternative consumption, even falling within the logic of decay and optimisation of non-consumption, actually it reinforces the ideology of hyper-consumption,[29] thanks to a continuous renewal of products. For example, the exchange of second-hand products on internet platforms accelerates the desire for consumption. Considered sustainable consumption, platforms of exchange of second-life products do not cease to accelerate the vicious circle of desire that restarts from the disillusion caused by acquisition and subsequent use. So desire is unceasingly born anew.

For-profit commercial platforms encourage the dominance of platform owners, who capture a large part of the wealth through tax avoidance and rely on a mass of self-employed micro entrepreneurs who do not participate in any kind of 'collaboration' at all. The large-scale development of web platforms and marketplaces seems to encourage more individualistic behaviour that goes against the logic of sustainability. Based on the functionality of products, collaborative consumption falls within this framework of flexibility, lightness and practicality reinforced, far removed from environmental objectives. Thus, individual utilitarian or economic motivations predominate over altruistic, social and environmental values. Even considering that there has been a re-appropriation of the microeconomic sphere by citizens, of which collaborative consumption is but one example,[30] the acceleration of desire generated by the pleasure of novelty that comes with easy, continuous renewal of products sees collaborative consumption perpetuate the repetitive consumption of products that are regularly because of their technological or psychological obsolescence. Besides, collaborative consumption is a difficult pattern to adopt in some countries in the Global South because of various kinds of deficiencies faced there.

4.7 The Evolution of Consumer Protection Rules

Consumer law has been transformed and updated by a number of factors, including those resulting from concerns related to sustainable development. Numerous international, regional and national texts have followed. The recent evolution of the consumer protection legislation indicates that Consumer Law is also intended to rationalise the behaviour of consumers. Through Consumer Law, consumers have been increasingly encouraged to adopt sustainable consumption patterns.

In the international legal system, it was Chapter 8 of Agenda 21[31] that first called on countries to adopt national strategies for sustainable development (NSDSs). The United Nations Millennium Declaration[32] adopted by the United Nations General

[29]Lipovetsky (2006).

[30]Cannon and Chun (2014).

[31]United Nations Sustainable Development (1992).

[32]United Nations Millennium Declaration (2000). A/RES/55/2.

Assembly in the year 2000 pressured countries to meet their treaty obligations on sustainable development, with a series of Millennium Development Goals (MDGs). MDG 7A called on countries to integrate sustainable development principles into programmes and country policies. The 2002 Johannesburg Plan of Implementation (JPOI)[33] through paragraph 162 (b) pressured United Nations Member States to take immediate steps towards making progress in formulating and elaborating NSDSs. The 2030 Agenda for Sustainable Development[34] endorses the commitment of Member States to accomplish sustainable development for all.

From the regional perspective, the European Union adopted some very interesting initiatives in the area of sustainable consumption: Eco-label; the European Eco-Management and Audit Scheme; green public procurement; eco-design, energy labelling; the Multi-stakeholder Dialogue on Environmental Claims; the Product Environmental Footprint and Organization Environmental Footprint methodologies; the Product Safety and Market Surveillance Package; the Rapid Alert System for dangerous non-food products; the Rapid Information System ('RAPEX'); the car labelling directive; the Retail Forum for Sustainability; the Food Sustainable Consumption and Production Round Table; and organic farming and activities against food waste. Integrating environmental concerns into national consumer laws is far from complete. According to the 2013 report *Good Practices in the National Sustainable Development Strategies of OECD Countries* produced by the OECD,[35] not many countries had legal frameworks for their national sustainable development strategies until then, such as Korea, which apply to sustainable consumption. Nevertheless, in the Canadian province of Quebec, the sustainable development rule dates from 2006,[36] whereas the federal law dates from 2008.[37] Some countries like France, Norway and Switzerland have legal provisions relating to the right to environmental information. In fact, as stated by United Nations Conference on Trade and Development (UNTRAD), most national consumer policies promote sustainable consumption patterns in an indirect manner, through information and educational campaigns.[38] However, considering the endless asymmetry of information access, and its direct effect on consumers' freedom of choice, those campaigns are important but insufficient. Japan enacted the Fundamental Law for Establishing a Sound Material-Cycle Society, which reinforces its sustainable consumption and recycling initiatives. The United States has Guides for the Use of Environmental Marketing Claims.[39] According to an UNTRAD report, sustainable consumption

---

[33]United Nations (2002b).

[34]United Nations (2015). A/RES/70/1.

[35]OECD (2006).

[36]Loi sur le développement durable, LR Q 2006, c D.8.1.1 (Québec, Canada).

[37]Loi fédérale sur le développement durable, LC 2008, c 3 (Canada).

[38]United Nations Conference on Trade and Development (1985–2013), p. 4.

[39]Federal Trade Commission, Title 16, Commercial Practices Guide and other United States statutes, covering, inter alia, automotive fuel rating, labelling for biodiesel fuel and energy efficiency of various products (United States of America).

falls outside the mandate of consumer protection agencies in most cases.[40] According to the same document, except for El Salvador,[41] Mexico[42] and the United States,[43] most national legislation reserves powers concerning sustainable consumption for environment ministries or a group of institutions.

4.8 Conclusions

Environmental Law, as much as Consumer Law, is a discipline with a very precise objective. These two bodies of law are aimed at the defence, respectively, of the consumer and the environment. The formulation of sustainable consumption theory presupposes the integration of these two spheres of protection, establishing related rights and obligations. It has not been the aim of this chapter to discuss the responsibilities of the other social actors but clearly the promotion of sustainable consumption patterns is the duty of all. Indeed, the responsibility of consumers,[44] suppliers and the State must be met in a differentiated way. Conversely, differentiated and staggered responsibility must be shared by all for a sustainable consumption.

The search for new consumption patterns must be accompanied by a definition of sustainable production methods, as the consumer remains in a situation of vulnerability that means he or she depends on the functions of production and distribution in the markets. To change consumer behavior, it is therefore necessary to modify the production standards themselves. The target of most of the initiatives pointed out through the chapter is almost always the consumer, the argument often being made that it is hard for the State to intervene in production and distribution chains, especially considering that these chains are becoming more and more transnational. Hence, intervention on the demand side may seem more accessible than on the supply side, precisely because of global dynamics of the organisation of production and marketing circuits. Nevertheless, it is unreasonable to expect the consumer to bear responsibility for actions that lie in the domain of the production system.

Although much remains to be done concerning the social and ethical aspects, sustainable consumption has undeniably incorporated environmental protection as one of its core values. This progress needs to be further strengthened so that sustainable consumption can be seriously considered as a legal concept that has

---

[40]United Nations Conference on Trade and Development (1985–2013), p. 13.

[41]Decreto No. 776 de la Asamblea Legislativa de la Republica de El Savador, Ley de Protección al Consumidor, artículo 4 (k) (El Salvador).

[42]Ley Federal de Protección al Consumidor, publicada en el Diario Oficial de la Federación el 24 de diciembre de 1992, artículos 8 bis, 24 (Mexique).

[43]Federal Trade Commission, Title 16, Commercial Practices Guide and other United States statutes (USA).

[44]Binninger and Robert (2008).

become effective. What seems indispensable for the effective integration of the concept of sustainable consumption into Consumer Law is the recognition and consequent strengthening of the collective dimension.[45] This is a common and essential feature of both Consumer Law and Environmental Law. In countless situations, illicit conduct towards the consumer affects not only a single individual or group of consumers, but the whole collectivity.

Environmental law, in turn, is a collective phenomenon by nature, as is sustainable consumption, as a derivation of it. An example of this collective dimension is the global consequences of unbridled consumption. Therefore, the only possible way for the consumer to create a demand for the environmental quality of products or services and for waste prevention is beyond the knowledge of the overall environmental quality of the product.[46] Organic farming is an example of a successful form of sustainable consumption that has a clear collective dimension. Consumers are becoming increasingly aware of the link between the products and services they buy and the fundamental issues of location and conditions which includes social and production processes.

Beyond the concern to inform the consumer—a mandate of Consumer Law that is more important than ever—it seems that strengthening the legal instruments that protect the collective rights of consumers is the most effective way to ensure the adoption and integration of new consumption patterns in a dynamic market environment.[47] A more accurate perception of the reality of consumer relations requires the multiplication of modes of representation, including jurisdictional, of the collective interests of consumers. Recognition of the collective dimension of consumer interests undoes the epistemological link between the individual and the act of consumption, better reflects the socio-economic environment in which today's consumers operate and better prepares Consumer Law to deal with novel consumption issues as they arise.[48] To focus only on the individualistic aspects of Consumer Law makes the necessary opening to the change of modes of consumption impossible.

---

[45]Fonseca (2013).

[46]von Weizacker et al. (1997).

[47]According to Bourgoignie (cited in Brovelli et Sancy, 2016, p. 128): 'Les objectifs, les fondements et les valeurs de la discipline sont interpellés et les vrais enjeux du droit la consommation réaffirmés avec force. La vision s'écarte de celle d'un droit vecteur d'un mouvement <consumériste> prônant l'individualisme, le matérialisme et l'hédonisme des individus. Les aspirations sociales et collectives du mouvement de défense des consommateurs sont rappelées, auxquelles s'ajoutent dorénavant les attentes et les impératifs liés au besoin d'assurer le développement durable et la protection de la nature'.

[48]Fonseca (2017).

References

Baldé CP, Wang F, Kuehr FR, Huisman J (2015) The global e-waste monitor—2014. United Nations University, IAS—SCYCLE, Bonn, Germany

Binninger A-S, Robert I (2008) Consommation et développement durable. Vers une segmentation des sensibilités et des comportements. La Revue des Sciences de Gestion 229:51–59

Bonnin CA (2009) L'intégration de la consommation durable en droit français et canadien. Ph.D. Thesis presented at Faculté des études supérieures de l'Univerité Laval, Québec

Bourgoignie T (2006) Un droit de la consommation est-il encore nécessaire en 2006? In: Bourgoigne T (dir.) Regards croisés sur les enjeux contemporains du droit de la consommation. Yvon Blais, Cowansville

Bourgoignie T (2016) Protection du consommateur et développement durable Consommateur souverain, pollueur, responsable ou victime? In: Brovelli G, Sancy M (dir.) Environnement et développement durable dans les politiques de l'Union européenne, actualités et défis. Presses Universitaires de Rennes

Cannon B, Chun H (2014) A framework for designing co-regulation models well-adapted to technology-facilitated sharing economies. Santa Clara Comput High Technol Law J 23 (31):1–75

Caza P-E (2015) «Obsolescence programmée: vraiment?» Actualités UQÀM, 12 April 2015, on line, <https://www.actualites.uqam.ca/2015/obsolescence-programmee>. Accessed 25 Nov 2017

Decreto No. 776 de la Asamblea Legislativa de la Republica de El Salvador, Ley de Protección al Consumidor, artículo 4 (k) (El Salvador)

Decrop A (2017) La consommation collaborative Enjeux et défis de la nouvelle société du partage. Deboeck Supérieur, Louvain-la-Neuve

Dolk H, Nichols R (1999) Evaluation of the impact of Chernobyl on the prevalence of congenital anomalies in 16 regions of Europe. EUROCAT Working Group. Int J Epidemiol 28(5):941–948

Eckerman I (2011) Bhopal Gas Catastrophy 1984: causes and consequences. In: Nriagu JO (ed) Encyclopedia of environmental health, vol 1. Elsevier, Burlington, pp 302–316

Ertz M, Duriff F (2017) Définition de la consommation collaborative et des concepts associés. In: Alain Decrop (dir.) La consommation collaborative Enjeux et défis de la nouvelle société du partage. Deboeck Supérieur, Louvain-la-Neuve

Federal Trade Commission, Title 16, Commercial Practices Guide and other United States statutes, covering, inter alia, automotive fuel rating, labelling for biodiesel fuel and energy efficiency of various products (USA)

Felson M, Spaeth JL (1978) Community structure and collaborative consumption a routine activity approach. Am Behav Sci 21(4):614–624

Fonseca P (2013) Origens e fundamentos do Direito Brasileiro do Consumidor. In: da Fonseca PG, Ramos FD, Bourgoignie T (dir.) A proteção do Consumidor no Brasil e no Quebec: Diálogos de direito comparado. Eduff, Niterói

Fonseca P (2014) Ruptura de confiança e novas iniciativas. Revista de Direito do Consumidor 94:113–156

Fonseca P (2016) Droit de la protection du consommateur au Québec et au Brésil Une analyse comparée. IEIM, Montréal

Fonseca P (2017) Direito do Consumidor. Eduff, Niterói (Brazil)

Fonseca P, Fonseca I (2016) Brazil's greatest environmental catastrophe–Samarco's Fundão Tailings Dam. Environ Policy Law 46(5):334–337

Friant-Perrot M (2008) La consommation durable et la protection des consommateurs : réflexions sur les nouveaux rapports entre le droit de la consommation et le concept de développement durable. In: Parent G (dir.) Production et consommation durables : de la gouvernance au consommateur-citoyen

Ley Federal de Protección al Consumidor, publicada en el Diario Oficial de la Federación el 24 de diciembre de 1992, artículos 8 bis, 24 (Mexique)

Lipovetsky G (2006) Essai sur la société d'hyperconsommation. Gallimard, Paris
Loi fédérale sur le développement durable, LC 2008, c 3 (Canada)
Loi sur le développement durable, LR Q 2006, c D.8.1.1 (Québec, Canada)
Mead HL (2017) Trop tard La fin d'un monde et le début d'un nouveau. Ecosociété, Montréal
«Obsolescence programmée: vraiment?» in <https://www.actualites.uqam.ca/2015/obsolescence-programmee>. Accessed 20 Nov 2017
Organisation for Economic Co-Operation and Development (2006) Good Practices in the National Sustainable Development Strategies of OECD Countries
Pieratti G, Prat J-L (2000) Droit, économie, écologie et développement durable: des relations nécessairement complémentaires mais inévitablement ambiguës. RJE (3):421–444
Shields A (2017) «L'humanité court à sa perte, préviennent 15 000 scientifiques à travers le monde» by Alexandre Shields, *Le Devoir*, 14th of November, 2017, on line, <http://www.ledevoir.com/environnement/actualites-sur-l-environnement/512875/cop23-hausse-des-emission-de-ges>. Accessed 15 Nov 2017
United Nations (1987) Development and International Economic Co-Operation: Environment. Report of the World Commission on Environment and Development. A/43/427 (also known as the Brundtland Report)
United Nations (2002a) World Populations Prospect
United Nations (2002b) Plan of Implementation of the World Summit on Sustainable Development
United Nations (2011) The Determinants and Consequences of Population Trend, New York
United Nations (2015) Transforming our world: the 2030 Agenda for Sustainable Development. A/RES/70/1
United Nations Conference on Trade and Development (1985–2013) Implementation report on the United Nations Guidelines on Consumer Protection
United Nations Development Programme (1998) Human Development Report 1998: consumption for human development. UNDP, Oxford University Press, New York
United Nations Millennium Declaration (2000) A/RES/55/2
United Nations Sustainable Development (1992) United Nations Conference on Environment & Development Rio de Janerio, Brazil, 3 to 14 June 1992. Agenda 21
von Weizacker EU, Lovins L, Lovins AB (1997) Factor 4: duplicar el bienestar con la mitad de los recursos naturales (trans: Kovacsics A). Barcelona, Galáxia Gutenberg

Chapter 5
Sustainable Consumption and Brazilian Consumer Behaviour

Diógenes Faria de Carvalho and Vitor Hugo Do Amaral Ferreira

Contents

Abstract This chapter offers some reflections on the idea of sustainable consumption, considering the emotional, cultural and moral influences that affect consumer decision-making. We have chosen a model of sustainable consumption that avoids the usual strategy of transferring the regulatory responsibility of the State and other enterprises to individual consumers. As a result, we highlight aspects of the 'culture of immediacy' that guides modern society, especially in Brazil, notably the tendency to waste and the indifference to future needs. We focus on the individual's strong need to justify the acquisition of goods. The biology of the human brain also affects consumption—namely, the tendency to overvalue immediate gains. Given this scenario, sustainable consumption requires more than individual behavioural change—information and education, and changes in product design are called for. Sustainable consumption thus combines consumer responsibility with an approach that calls on the State to regulate production and pursue legal action against infringements by producers that cause environmental destruction and offend the public morality.

D. F. de Carvalho (✉)
BRASILCON (Brazilian Institute for Policy and Consumer Law), Goiânia, Brazil

Pontifical Catholic University of Goiás (PUCGO), Goiânia, Brazil

V. H. Do Amaral Ferreira
BRASILCON (Brazilian Institute for Policy and Consumer Law), Goiânia, Brazil

Franciscan University (UFN), Santa Maria, Brazil

A. d. Amaral Junior et al. (eds.), *Sustainable Consumption*,
https://doi.org/10.1007/978-3-030-16985-5_5

5.1 Introduction

Contemporary society is experiencing a paradoxical situation. There is, on the one hand, a desire to consume, encouraged by the capitalist system, because it is the basis for it and is intensified by individual and immediate interests, and on the other hand, the need to preserve natural resources and maintain an environmental balance, to guarantee the survival of the human species.

If we are not always consuming, the capitalist economic model runs the risk of being extinguished. If we maintain current levels of consumption without a real concern for the environment, then we, as a species, are at risk of extinction. For this reason, it can be affirmed that:

> for millennia, man's task has been to defend himself from nature, but at the end of the twentieth century the focus has changed, now we need to defend nature from man himself, for there is a progressive awareness of the destructive effect of our controlling capacity.[1]

Often the solution to the problem is the idea of sustainable consumption, based simply on consumer awareness, equating the situation of people as both consumers and citizens, as if these were synonyms. However, this does not seem to be the most adequate solution for Brazilian society, for cultural, moral and emotional reasons.

In the world we live in, marked by the phenomenon of globalisation, we observe the introduction of structural models where human and consumer relations with their inherent economic–financial logic put personal satisfaction first, to the detriment of the collective interest. This therefore challenges basic ethical norms and encourages indifference to the scarcity of natural resources indispensable to life and our responsibility to limit our depletion of them. It is a consumer society based on short-term thinking and the desire for immediate gain.[2]

According to Campbell, consumption is related to the purpose of existence. In this way, the act of consuming is, ontologically speaking, the self-assertion of the consumer as a human being. Hence, 'wanting' and 'desiring' a particular good becomes necessary for the integration of the individual into his group, into a society characterised as a 'consumer society'. Consumption acquires in our society, the contours of a social mechanism that produces meaning and identities. Individuals identify with brands, technologies and multimedia, affirming themselves through them, thus differentiating themselves from others. Non-access to consumption can mean social exclusion, marginalisation and frustration. It is a lifestyle that cultivates the present, the ephemeral and the satisfaction of all needs.

In this context of habits harmful to the environment, the need for environmental preservation is inserted as a way to also preserve the human species. The notion of 'sustainable consumption' emerges. It can be defined as:

> [T]he use of products and services that meet the basic needs of individuals and bring about an improved quality of life, equating the use of natural resources, reducing the use of toxic

---

[1]Tadeu (2005).

[2]Azevedo (2008).

> materials, as well as pollutant emissions and the generation of waste in order to ensure the survival of future generations.[3]

The proposal of sustainable consumption is marked by analyses that consider that if consumers are well informed and have sufficient knowledge, they will have the necessary 'environmental awareness', which will lead them to assume environmentally correct attitudes and behaviours.[4] However, this does not seem to be a correct conclusion, because the articulation of environmental responsibility by individuals—who often try to justify their consumption behaviours and minimise their harmful effects through discourse rhetoric—is weak. The idea of sustainable consumption cannot therefore be reduced to changes in the behaviour of individuals, or to changes in product design. State action needs to focus on public policies for sustainable consumption; after all, it is the State's duty to seek efficient means in the implementation of policies of regulation and environmental protection.

The Brazilian Constitution of 1988 foresees the need for economic development to be pursued in conjunction with the implementation of environmental policies. It imposes on the State the role of regulator with respect to the environment, as provided for in item VI of Article 170:

> Article 170. The economic order, based on the valorization of human labour and free initiative, is intended to guarantee to everyone a dignified existence, according to the dictates of social justice, observing the following principles:
>
> [. . .] VI–defence of the environment, including differentiated treatment according to the environmental impact of the products and services and their processes of elaboration and provision.

In this sense, the objective of this chapter is to analyse sustainable consumption from the perspective of the responsibility of the State, not merely from that of consumer education or information, addressing the behaviour of the Brazilian consumer and the way the purchasing decision is made within the culture of immediacy. According to Helleringer and Sibony, consumers do not always use deliberation to make decisions; instead, they prefer mental shortcuts, relying on intuition.[5] For this reason, we seek to show how the common strategy currently used—based on the transfer of regulatory activity from the State and the market to the consumer—is ineffective and must be rejected.

---

[3] de Andrade (1998), p. 63.

[4] Portilho (2005).

[5] Helleringer and Sibony (2016–2017), pp. 611–612.

5.2 Consumer Society and the Consumer Decision-Making Process

In consumer society, the concept of well-being is linked to the possession and acquisition of goods and services as a form of identity affirmation and the preservation or acquisition of social status.[6] For Barbosa, contemporary consumer society can no longer be explained by the old theories, proposed by economic science, in which people consume only to satisfy their physical and biological needs according to the laws of the market. With the advent of globalisation and the shift from what Zygmunt Bauman calls producer society to consumer society, forms of acquisition and the use and disposal of goods are no longer explained by the mere physical and biological need to consume.[7]

The level and style of consumption have become the main source of cultural identity and participation in collective society. Thus, society has come to classify goods in a hierarchical manner and consider some more lawful than others, and thus, the goods and social practices acquire certain cultural meaning. People are classified by what they consume. To illustrate, Barbosa and Campbell argue that 'it is more lawful to consume books and CDs generically than clothes, shoes and purses... what matters is that in the first case we are intellectuals and in the second, futile and empty'.[8]

We also classify goods as 'basic' and 'superfluous', and consider the purchase of those morally legitimate, while the purchase of superfluous items requires of us rhetoric and justifications that lessen our guilt. Thus, we develop criteria of legitimacy and rhetoric to justify what, when and why we consume. As Barbosa and Campbell elaborate:

> [W]ords like 'if I buy now I'll be saving later due to the low price at this moment', or 'it was a great opportunity because I really needed it'... When this kind of rhetoric runs out, the usual discourse of 'I deserve it' is used. I deserve it because 'I work a lot', 'because I have not bought anything for myself for a long time, or even for others', 'because life cannot be just about work, it has to have pleasure' etc.[9]

Another root of this culture of consumption, according to Bauman, is the stimulation of emotion and the lack of cultivation of human reason.[10] Referring to the work *Roots of Brazil*, Sérgio Buarque de Holanda outlines his theory of the 'cordial man'—a unique characteristic of Brazilian society, where we notice an extremely rich and overflowing emotive background, intrinsic to social relations, which he calls the Brazilian contribution to civilisation. The focus on the irrationality and emotion

---

[6]Alcoforado and Silva (2009).

[7]Bauman (2008).

[8]Barbosa and Campbell (2006), p. 40.

[9]Ibid, p. 37.

[10]Bauman (2008).

of 'cordial' consumers may be related to the low saving rates and the high level of indebtedness among Brazilian consumers.

Our brain contains a defect, related to emotions, such that we tend to overvalue immediate gains and ignore future costs. Emotions thrill and provide immediate rewards, but they misunderstand the consequences of their long-term decisions. On the subject, Jonah Lehrer observes that: 'Of course, the emotional parts of the brain will give less value to the future—'life is short', and we want pleasure now so we all end up spending too much today and putting off the costs for tomorrow (and the day after tomorrow and so on).'[11]

Thus, the immediacy and consequent overvaluation of gains or immediate rewards, which are characteristic of the Brazilian culture, cause people to spend a lot of time living with the consequences of their decisions, and little time actually taking them. Considering Giuzi's affirmation that 'deciding is a process of choosing between valid and competing alternatives', it is worth highlighting the spontaneous and improvised nature of the Brazilian decision-making process. The 'now' or 'hasty' culture, as expressed by sociologist Stephen Bertman, is apt to define a consumerist society marked by the renegotiation of the meaning and value of time.[12] We live the present culture: consumer life is a life of fast learning and fast forgetting. On this subject, Rejane Guedes Pedrosa states that:

> Forgetfulness has a redeeming function, discarding old concepts to give rise to the new, and this must occur permanently. In this culture of immediacy, patience and perseverance are not welcome, unlike the speed and capacity to promote new beginnings, each with new avatars full of resources that favour consumption.

Economies of scale, which seek maximum utilisation of productive factors to lower production costs and increase the volume of goods and services produced, require the constant invention of new needs, an increase in purchases and the production of disposable and non-durable products, together with the creation of waste. Excessive consumption is also stimulated by the promise of achievement through consumption.

Therefore, when we think about what drives excessive consumption, the notorious influence of emotions on the consumer's decision to buy comes in to play. Moreover, bearing in mind Brazilian cultural issues related to immediacy and a certain disregard for the long-term costs involved, it becomes logical to hold that a sustainable consumption strategy based on simple consumer awareness may prove to be ineffective. Discussing sustainable consumption is then 'mere talk', a way to satisfy our daily consumption needs without undermining the ability of future generations to meet their own needs, because acting in a sustainable way is, above all, thinking about future generations. This means the consequences that would control sustainable behaviour are in the distant future, and, in general, today's consumer will not have to encounter the harmful future consequences arising from the problems generated by current unsustainable behavioural patterns. For Pedroso:

---

[11] Lehrer (2010), p. 111.

[12] Kargel (2009).

> It is very clear from this conception in the current scenario, that where consuming more is reinforcing for the individual who behaves more immediately, in the long run the consequences will be disastrous. However, if a possible cancer in the future is not a strong enough potential consequence to make smokers stop, it is possible to imagine that concern for future generations will have no or very little effect on the increase in the number of individuals who will exhibit sustainable consumption behaviour, since consuming is reinforcing for the individual who consumes, and generally has immediate consequences.[13]

He continues:

> To say that we must change the way we consume because of the possibility of future generations not having the natural resources to satisfy their own needs, can be a slogan that will have no effect now on those who consume, since a long-term consequence would have little effect on the current consumer, and, in general, those who behave in such a manner today will probably not come into contact with the scarcity of natural resources in the future.[14]

The choices we make regarding the time variable are called inter-temporal choices. Thus, when we make choices the future can be underestimated, just as the present can be overestimated. In many cases, the preference for smaller rewards, which come faster, to the possibility of larger rewards in the future, is noted. In behavioural language, rewards closer in time, although smaller in value, may be preferred to deferred rewards, which are greater in value. This is because the delay leads to the loss of the reinforcing value of a given reward. As Oliveira and Jesus have noted:

> Future utility is thus discounted at a rate that remains constant over a period of time. If the increase in utility of the time-deferred object does not exceed the rate at which it is discounted, the time-deferred object continues to depreciate against the object allotted to the present moment, this preference remaining stable as time progresses.[15]

Empirical checks developed by interdisciplinary researchers in psychology and economics, concerning distortions as an inter-temporal choice have the function of facilitating immediate satisfactions, even at higher costs. The normative model of inter-temporal choices presupposes that the loss of value of a delayed amount follows an exponential model, in the form:

$$v = V - kD \tag{5.1}$$

where v is the present value equivalent to an amount V to be received after a delay D and k refers to the discount rate of V per unit time. In turn, studies with non-humans[16] and humans[17] have shown that the exponential model fails to predict

[13]Pedroso (2016), p. 65.

[14]Ibid, p. 65.

[15]Oliveira and Jesus (2013), p. 43.

[16]Mazur (1987).

[17]Kirby and Markovic (1995).

how delayed quantities lose value and have proposed a general hyperbolic discount model:

$$V = V/(1 + kD) \tag{5.2}$$

where the discount rate (or the loss of the value of the amount V) is an inverse function of the delay. This model and other variants with exponents in the denominator have been tested in recent years and its descriptive power of the discounted value in inter-temporal choices has been attested to, mainly with respect to the inversion in preferences with the increase in the delay, which is a given not predicted in the exponential model. In contrast, a power function of the inverse of the delay, which also predicts the inversion in the preference, has been tested and its predictive power has been superior to that of the hyperbole with Brazilian samples.[18] This can be systematised as:

$$V = V/aDb \tag{5.3}$$

This model can be described as:

$$V = V \cdot aD - b \tag{5.4}$$

with b being a negative value, corresponding to the measure of loss of value per unit of delay and a measure of the value of the amount when the delay approaches 0, tending towards V. The so-called hyperbolic discount is nothing more than an evaluation at the present moment of a time further away, in which a devaluation as a function of the degree of withdrawal imposes itself, characterised by intervals and a marked sensitivity to time intervals between the present and future moments. The sensitivity decreases as the interval represents a greater distance. This model has described the loss of the value of delayed rewards and presents advantages over the exponential model as it predicts a reversion in preference with increasing delay.

It can thus be seen that sustainability, as something effective, is now difficult, because the consequences in question are delayed, and the harmful effects are in effect delayed consequences. The fact is that when people are invited to engage in current behavioural patterns where detrimental consequences are likely to occur at some point in the distant and uncertain future, change in consumer behaviour is, unfortunately, very unlikely.

[18]Coelho et al. (2003).

5.3 From Green Consumption to Sustainable Consumption

The beliefs concerning what is responsible for the environmental crisis have undergone a process of change. Until the 1970s, the crisis was attributed to population growth in underdeveloped countries. With the Stockholm Conference, it has come to be understood as the process of industrialisation and production within developed countries. From the beginning of the 1990s, with Rio92, the discussion took a new direction, with the lifestyles and consumption patterns of capitalist societies being identified as responsible for the ecological emergency. It is in this context that the idea of 'green consumption' arose, reflecting a sense of co-responsibility of individuals as consumers who are concerned about the environmental impact of their demands. Green consumption later morphed into the more descriptive notion of 'sustainable consumption'.

The 'green consumer' is therefore concerned not only with price and quality variables, but also with the environmental variable, preferring products that are less or not at all damaging to the environment. The green consumer strategy has become common within business and marketing as well. The consumer–citizen discourse, responsible for the healthy environment, is repeated by several authors, who insist on the transfer of regulatory activity to individuals. However, such a discourse has several flaws. First, there is confusion between the terms 'citizen' and 'consumer' within the conscious consumer discourse. These identify equivalence between the terms; however, the citizen–consumer relationship must be rethought, before the concept of citizenship is outdated. When citizens and consumers are considered synonyms, then education, housing, leisure and health cease to be social rights, becoming rather, personal achievements. Thus, if the citizen is reduced to the notion of consumer, his actions are focused on the private sphere of consumption. The liberal discourse uses the notion of citizenship to say that only those who participate in the consumer market can be considered 'a citizen'.

According to Portilho, the concept of 'green consumption' fails to address the processes of production and distribution, as well as transferring the responsibility of the regulatory State and, of the producing and distributing companies, to the individual consumer.[19] He holds that the green consumption strategy can be analysed as a kind of transfer of regulatory activity, in two aspects: (a) from the State to the market, through self-regulation mechanisms; and (b) the State and the market for the citizen, through consumption choices.

Green consumption does not touch on important themes either, such as the consumer culture and the need to change the paradigms of consumer society. It is for these reasons that we contend that the strategy of sustainable consumption based on the transfer of regulatory activity should be debunked. Public consumption policy should be focused on the State and on free initiative, and it is also up to the judiciary to strongly repress the causes of environmental damage.

[19]Portilho (2005).

Thus, recognising the problematic issues of green consumption, a new proposal emerges—a model of sustainable consumption. With this model, the environmental issue is confronted through collective public actions and political interventions. Actions such as reducing the consumption of plastic bags, recycling and reusing raw materials, approving the Solid Waste Act, etc. become part of a complex system that shares responsibility across sectors.[20]

The 'sustainable consumption' model is still under construction and is the result of a continuous and multilateral process that offers itself as an innovative paradigm for individual and collective practices in consumer relations. Portilho emphasises the need for civil society, acting in an organised way or not, to find solutions to the problems arising from the globalisation process, and considers the possibility of politicising the private sphere, believing that the role of consumers will change to become political subjects seeking a sustainable development project for society.[21]

Thus, to achieve sustainable consumption, there must be a change in consumer behaviour patterns currently driven by the culture of consumerism, but such change requires regulatory intervention by the State. According to Santos, consumption asserts itself as an individualising practice, driven by programmed obsolescence, whose environmental effects and impacts are unevenly distributed.[22] Bauman, in turn, explains the term consumerism as the attribute of a society where its members are themselves saleable commodities.[23] As we said earlier, consumer life is ruled by speed and excess, and contempt for yesterday's needs and for what has become planned obsolescence shows the irresponsible culture of immediacy and waste.

Therefore, there is no way to talk about change in the pattern of behaviour and, consequently, in the pattern of consumption without passing through a cultural change first.[24] The current situation demands a revaluation of society in its multiple aspects, emphasising the need for ethical conduct that is in accordance with the current problems that we face. We must, therefore, shatter the idea of development observed only from the economic point of view, and understand it in a more comprehensive way, adding ethical values, and socio-cultural aspects as a whole, to this concept. The crux of the matter is that cultural changes do not happen from one day to the next. They take time, and therefore, there is a need for state action. Thus, eco-labelling strategies and other stimuli that aim to transfer the burden of choice to the consumer tend to lose space in the face of collective actions that do not concentrate all responsibility on the individual.

Increasingly, public policies are needed that intercede in the private sphere and lead society to new socio-behavioural paradigms based on responsibility and sustainability. Additionally, Helleringer and Sibony claim that 'the issue is to take into account that consumers will not read much'.[25] Thus, we believe that state actions and

[20]Santos (2005) and Portilho (2005).

[21]Portilho (2005).

[22]Santos (2005).

[23]Bauman (2008).

[24]Santos (2005).

[25]Helleringer and Sibony (2016–2017), p. 629.

interventions, such as fiscal policies that reduce taxes on environmentally friendly products, would be more effective than simply focusing on consumer awareness. We do not, however, detract from education and consumer information. In the end, they have an effect and are extremely beneficial. However, considering the present culture, the immediacy, the waste, it is difficult to defend that the notion that the '*consumer–citizen*' could, by her own initiative, soften the environmental crisis. The crux of the matter is that cultural changes do not happen from one day to the next. They take time, and therefore, there is a need for state action. Thus, eco-labelling strategies and other stimuli that aim to transfer the burden of choice to the consumer tend to lose space in the face of collective actions that do not concentrate all responsibility on the individual.

5.4 A Regulatory Model for Environmental Protection

Article 174 of the Brazilian Constitution of 1988 establishes that the State, as a normative agent and regulator of economic activity, 'shall exercise, in the form of the law, the functions of inspection, incentive and planning, being a determinant for the public sector and indicative for the private sector.' Therefore, it posits three ways of acting as a regulating State: inspection, incentive and planning.

The supervisory or inspection function implies verification of the economic sectors to avoid abusive forms of behaviour. The incentive function, in turn, represents the stimulus that the government must offer for the economic and social development of the country, through the establishment of measures such as tax exemptions, the increase of quotas for imports, etc.[26] Finally, planning can be understood as 'a technical process charged with transforming existing reality towards previously established objectives'.[27]

In this context, the 1988 Constitution incorporated the need for economic development to oversee the implementation of environmental policies, as set forth in item VI of Article 170, so that the State should exercise the role of regulatory agent in this regard as well. State regulation of environmental preservation must be seen as a prerequisite for the exercise of economic activities. The State must take the necessary measures to avoid the indiscriminate degradation of natural resources, implementing policies of regulation and environmental protection, to encourage the production and consumption of ecologically sound products. Discussing the state regulatory activity regarding environmental protection, Gabriel Placha states that:

> One instrument of awareness and implementation of environmental policies, which is incipient and little-used in Brazil, since it is restricted to a few cases, is the Contribution of Intervention on the Economic Domain established by § 4, Article 177 of the Constitution

[26]Carvalho Filho (2010).

[27]Silva (1989).

> of the Republic, according to Constitutional Amendment No. 33/2001, is more focused on the environment... Environmental taxation appears as a viable instrument for increasing economic efficiency in a sustained manner and should not be used for purely tax purposes.[28]

Taxation can therefore be considered as a state regulatory instrument for the protection of the environment, in that it transfers the amount required to repair environmental damage to the production cost side. However, care must be taken, as laws on taxation will still have to observe the principles of legality, contributory capacity, etc.

The issue of environmental protection is not, however, limited to the institution of environmental taxes. Comparative law demonstrates that it is efficient when tax benefits are granted to taxpayers who adopt effective environmental management policies and create environmental structures.[29] Under the terms of the 1988 Constitution, the State, in the exercise of its powers, may issue rules for the protection of the environment. The regulatory activity in this regard is essential in view of the need to preserve natural resources, which are limited. This is verified by a controlled balanced between the logic of competition and social logic.[30]

Thus, the State cannot adopt a position of neutrality before the constitutional dictates of the economic and social order, transferring the responsibility for environmental preservation to the consumers and their limited capacity of awareness. Constitutional norms demonstrate the need for specific action by the State, which must act to promote social justice and protect the environment and not simply correct market failures or suppress economic abuses.

5.5 Final Considerations

The idea of sustainable consumption is neither limited to changes in the behaviour of the individual, nor to changes in product design. There must be state action and interventions to promote the protection of natural resources and to encourage the purchase and production of ecologically sound products. It is also necessary to have effective judicial action to punish strongly the simple causes of environmental damage. The transfer of regulatory activity to the consumer is unfair and ineffective because it requires a mingling of the concepts of citizen and consumer, whose relationship must now be rethought. The concept of 'citizen–consumer' ignores the fact that these are two distinct categories and that the decision-making process of the consumer starts from an emotional deficit by which purchases are driven by emotional impulses and unconcerned with the consequences of the act. It also ignores cultural issues, linked to immediacy, waste and programmed obsolescence.

---

[28]Placha (2007).

[29]Placha (2007).

[30]Gabardo (2003).

Thus, if the problem of the environmental crisis effects all of us, it is important to emphasise the urgency of adopting new measures. The discourse of the need for a well-informed consumer has not proved effective, as today's culture of the individual and egoistic thoughts has prevailed. If someone throws a bottle on the ground, then it is forgiven because it is just a bottle, very different from the big companies who dump their waste in rivers. The absence of the notion of collectivity, characteristic of today's society, requires state interventions to enable ecologically appropriate consumer decisions. Moreover, the Brazilian Constitution itself demonstrates the need for specific State action in environmental protection.

Finally, consumption is linked to moral and cultural issues. On moral grounds, our society feels the need to justify the acquisition of goods, starting with phrases such as 'I deserve' or 'I work so hard', and so on. In the same way, we categorise goods according to established criteria, such as 'basic' and 'superfluous'. In the first case, the purchase is justified morally, while in the second, we reduce the guilt of those who buy with trite rhetoric and easy justification. From a cultural perspective, we seek immediate pleasure; we get used to constant product innovation, and we cannot resist it. We make decisions influenced more by emotion than by reason, and we ignore the results of this behaviour. Furthermore, empirical research proves that when people are invited to engage in current behavioural patterns where the damaging consequences are likely to occur at some point in the distant and uncertain future, change is very unlikely.

How can we disregard moral and cultural questions to simply state that consumer information is enough to alleviate the problem of the environmental crisis? How can we insist that the consumer, even if informed, consider the environmental variable, which almost always raises the price of the product, if he himself does not see the impact of his actions (individually) on the environment? Or, if he does not receive enough to meet his basic needs and those of his family?

It is quite clear that, in Brazilian society, the concern of companies is to facilitate and incentivate purchasing, not with the impact of their products on the environment. They act through advertising which encourages the purchase in instalments or any other method that minimises the effect of the purchase on the pocket of consumers. These, in turn, are guided by the moral and cultural issues that we have already detailed at length. Therefore, the argument that the State should be the focus for public consumption policy is reinforced.

References

Alcoforado MG, Silva PR (2009) Reflexão sobre o estilo de vida e o padrão de consumo numa sociedade sustentável. In: II SBDS – SEMINÁRIO BrasilEIRO DE DESIGN SUSTENTÁVEL, São Paulo. Anais do 2° Seminário Brasileiro de Design Sustentável–RBDS. São Paulo: s. ed., 2009

Azevedo (2008) Plauto Faraco de. Ecocivilização: Ambiente e direito no limiar da vida. São Paulo: RT

Barbosa L, Campbell C (2006) Cultura, consumo e identidade. Editora FGV, Rio de Janeiro

Bauman Z (2008) Vida para consumo: a transformação das pessoas em mercadorias. Tradução de Carlos Alberto Medeiros. Rio de Janeiro: Zahar

Carvalho Filho JdS (2010) Manual de Direito Administrativo. 23. ed. rev. e atual. Lumen Juris, Rio de Janeiro

Coelho C, Hanna ES, Todorov JC (2003) Magnitude, atraso e probabilidade de reforço em situações hipotéticas de risco. Psicologia: Teoria e Pesquisa 19(3):269–278

de Andrade NLS (1998) Consumo sustentável. Justitia, São Paulo, n. 59 (181/184), jan./dez. Disponível na Internet em: <http://bdjur.stj.gov.br/xmlui/bitstream/handle/2011/23666/consumo_sustentavel.pdf?sequence=1>. Acesso em 16 abr. 2011

Gabardo E (2003) Eficiência e Legitimidade do Estado. Manole, Barueri

Helleringer G, Sibony LS (2016–2017) European consumer protection through the behavioral lens. Columbia J Eur Law 23:607–646

Kargel AM (2009) Sucesos Puntuales. Disponível na Internet em <http://sucesospuntuales.files.wordpress.com/2009/01/sucesospuntuales3.pdf>. Acesso 20 abr. 2011

Kirby KN, Markovic NN (1995) Modeling myopic decisions: evidence for hyperbolic delay-discounting within subjects and amounts. Organ Behav Hum Decis Process 64(1):22–30

Lehrer J (2010) O momento decisivo. Tradução de Marcelo Schild. Best Business, Rio de Janeiro

Mazur JE (1987) An adjusting procedure for studying delayed reinforcement. Commons, ML.; Mazur, JE.; Nevin, JA, pp 55–73

Oliveira M, Jesus F (2013) Arquitetura situacional do crédito: tempo, cognição, afeto e decisão. Revista Crítica de Ciências Sociais–101/2013, colocada online no dia 17 de fevereiro 2014. Available at: http://rccs.revues.org/5356

Pedroso R (2016) Valor subjetivo do consumo sustentável. Goiânia: PUCGO. 2016. 207f. Tese (Doutorado em Psicologia)—Programa de Pós-Graduação em Psicologia, Pontifícia Universidade Católica de Goiás, Goiânia

Placha G (2007) A atividade regulatória do Estado. Dissertação (Mestrado em Direito) – Faculdade de Direito, Pontifícia Universidade Católica do Paraná, Curitiba, 2007. Disponível na Internet em <www.cipedya.com/web/FileDownload.aspx?IDFile=171930>. Acesso em 18 abr. 2011

Portilho F (2005) Consumo Sustentável: Limites e possibilidades de ambientalização e politização das práticas de consumo. Cadernos EBAPE, Rio de Janeiro, FGV, Edição Temática 2005

Santos LL (2005) Comunicação e consumo sustentável: das entrelinhas do capitalismo leve ao enquadramento da sustentabilidade na mídia. Revista fronteiras VII(3):223–233

Silva JA (1989) Direito Constitucional Positivo. Malheiros, São Paulo

Tadeu SA (2005) As dimensões do consumo: reflexões para uma teoria compreensiva. Revista de Direito do Consumidor. São Paulo (56):202–219

Part II
Traditional Legal Disciplines and Sustainable Consumption

Chapter 6
The Role of Sustainable Consumption and Disaster Law in Climate Risk Management

Délton W. de Carvalho and Fernanda D. L. Damacena

Contents

Abstract Because of several factors (including climate change, technological development, decisions regarding land use, modern economic conditions, population growth, and demographic trends), climate-related disasters have significantly increased in recent decades. Climate change can be considered a common global problem. This is because, among other reasons, the benefits obtained by an individual from an action that generates greenhouse gas emissions impose a cost on others—the environment, future generations, and individuals from other countries (particularly the least developed countries). Advances in climate science have provided a clearer understanding of the phenomenon and its response to human and natural influences. However, it is known that the effects of climate change on the environment and society will depend on not only the response of the Earth system but also humanity's behaviour in the face of change. In this context, this article asserts that there is a relationship between unsustainable consumption, climate change, and disasters, and then analyses this relationship. From this perspective, it proposes a reflection on the role of unsustainable consumption, in what will be called

D. W. de Carvalho (✉) · F. D. L. Damacena
Universidade do Vale do Rio dos Sinos (UNISINOS), São Leopoldo, Brazil
e-mail: delton@deltoncarvalho.com.br; fdamacena@unisinos.br

A. d. Amaral Junior et al. (eds.), *Sustainable Consumption*,
https://doi.org/10.1007/978-3-030-16985-5_6

a negative feedback process, with the potential for vulnerability to disasters linked to extreme climatic events.

6.1 Introduction

Because of many factors (including climate change, technological development, decisions regarding land use, modern economic conditions, population growth and demographic trends, etc.), weather-related catastrophes have drastically increased in recent decades. In line with worldwide patterns, Brazil has confronted an increase in such events, particularly flooding, landslides, and droughts. In the past, Brazil was mostly confronted with periodic severe droughts in the northeast, but this is no longer the only type of natural disaster it must deal with. There has been an intensification of disasters overall, particularly flooding and landslides.[1]

In meteorological or climatological terms, major shifts in climate (hereinafter 'extreme events') have occurred on a scale that may range from days to millennia. Short-term extreme events (related to meteorology) and medium-term events (related to climate) are of the utmost relevance for human relationships, because of their potentially significant impacts. Studies of climate impacts are based on the findings of atmospheric circulation models and may project evolution of climate variables such as temperature and precipitation, grounded in the assumption of increased concentration of greenhouse gases (GHGs) in the atmosphere. The GHG effect is a natural phenomenon that makes human life on Earth possible. When there is an equilibrium between incident solar energy and the energy reflected by the terrestrial surface in the form of heat, the climate stays practically unchanged. However, the energy balance can be altered in different ways.[2] One is a change in the concentration of GHGs in the atmosphere (which has also occurred because of the unsustainable increase in anthropic emissions of these gases).[3] In this context, anthropic participation occurs in terms of production and consumption.

---

[1]In 2015, the UN singled out Brazil as one of the ten countries with the greatest number of people affected by disasters in the past 20 years. See: United Nations (2015), Pivetta (2016), Castro (2017) and Evaristo (2017).

[2]According to Brasil (2017): '(1) By changes in the amount of energy reaching the surface of the Earth; (2) by changes in the orbit of the Earth or of the sun itself; (3) by changes in the amount of energy reaching the surface of the Earth that is reflected back to space, due to the presence of clouds or particles in the atmosphere (also called aerosols, resulting from burnings, for example); and (4) thanks to alterations in the amount of longer wave energy reflected back into space, due to changes in the concentration of greenhouse gases in the atmosphere.'

[3]According to Brasil (2017): 'Emissions occur in practically all human activities and sectors of the economy: in agriculture, through the preparation of the soil for planting and fertilization; in animal husbandry, through the treatment of manure and the enteric fermentation of cattle; in transportation, through the use of fossil fuels such as gasoline and natural gas; in the treatment of solid residues, via

This chapter starts from the assumption that there is a relationship between unsustainable consumption, climate change, and natural disasters. From this perspective, we propose a reflection on the role of 'unsustainable consumption,' in what we shall call a negative feedback process, conducive to vulnerabilities and disasters. The text is organised into three sections. In the first part, we introduce the Brazilian legal framework concerning disaster risk management, highlighting the contribution of GHG emissions to global warming and the increased occurrence and magnitude of extreme events. In the second part, the connection between unsustainable consumption and climate is examined, along with the anthropogenic contribution therein. Emphasis is placed on the following points: (a) sustainable consumption as one of the pillars of effective disaster risk management; (b) the normative semantics of sustainable development in Brazil; (c) the role of sustainable consumption in the process of climate risk management; and (d) the prospects for Brazil in the area of energy efficiency and consumption, in the face of a scenario of climate adaptation. In the final section, we concentrate on the normative contribution to the process of induction of sustainable behaviour, an important factor in reducing vulnerability to disaster.

6.2 The Foundations of Disaster Law in Brazil

Since 2010, Brazil has had to face atypical, severe, and recurrent disaster scenarios. The first executive measure in this context was Decree No. 7513/2011, ordering the development of a National System for Monitoring and Alerts of Natural Disaster. Among the elements necessary for the system to function was the creation of the National Centre for Monitoring and Alerts of Natural Disasters (CEMADEN). Tied to the Secretariat for Policy and Programs of Research and Development (SEPED) within the Ministry of Science, Technology and Innovation, CEMADEN has the following responsibilities: (a) to develop, test, and implement a system of forecasting of the occurrence of natural disasters in areas susceptible to risk throughout Brazil; (b) to assist in prevention and make possible the identification of vulnerabilities in land use and occupation, with an emphasis on urban planning and infrastructural installation; (c) to promote increasing awareness and consequent preparedness of the population at risk, inducing effective actions of prevention and damage reduction before the disaster; and (d) to manage information emitted by radar, pluviometers, and climate forecasts.

This information, for its part, has the following functions:

1. to prepare alerts of relevant natural disasters, for protection and civil defence within the Brazilian national territory;

the way trash is treated and disposed of; in the forests, through deforestation and forest degradation; and in industry, through production processes, for example of cement, aluminium, iron and steel.'

2. to prepare and disseminate studies aimed at producing the required information for the planning and execution of actions against natural disasters;
3. to develop innovative scientific and technological capabilities to continually upgrade natural disaster warnings;
4. to develop and implement observation systems for the monitoring of natural disasters;
5. to develop and implement computer models of natural disasters;
6. to operate the computer systems required for preparing natural disaster alerts;
7. to promote capacity building, training, and support for post-graduate activities in the respective areas of activity; and
8. to issue natural disaster warnings for the National Centre for the Management of Risks and Disasters (CENAD), in the Ministry of National Integration, assisting the National Civil Defence System.[4]

The aims and responsibilities of CEMADEN are grounded in the assumption that in systemic terms, disasters result from natural, technological, or socio-political circumstances. This combination of factors external (exogenous) and internal (endogenous) to the social system, when cumulatively and synergistically executed, is capable of producing a loss of systemic stability. Thus, compromising systemic stability has repercussions for the breakdown of the routines that are collectivity inherent to communities and to society and the need for urgent (and generally unplanned) measures to administer (re-establish) the situation.[5] Disasters are extreme phenomena capable of affecting social stability, in a process of irradiation and feedback of its causes and effects in different contexts—economic, political, juridical, and scientific.

After a couple of national legislative initiatives (Decree No. 7257/10 and Law No. 12340/10) solely centred on disaster response, Law No. 12608/12, which instituted the National Civil Protection and Defence Policy (PNPDC), emphasises prevention. The priority for preventive actions aimed at minimising disasters is established in its guidelines,[6] while reduction of disaster risks is one of the goals[7] of national policy.

From the general provisions, it may be perceived that the law innovates in comparison with traditional legal dogma concerning the environment, because it brings together in one chapter the concepts of risk and uncertainty. Even if that combination were not enough, from the viewpoint of common law, Article

[4]Brasil (2011). To meet such targets, the first step is to identify at-risk areas, vegetal coverage, and soil composition on relief maps. In the federal project, the mapping of about 500 at-risk areas on slopes and 300 inundated areas is anticipated. The project also anticipated investment in structure and monitoring equipment, such as radar and pluviometers.

[5]Porfiriev (1998), p. 62.

[6]As textually expounded in Art. 4, III, of Law 12608/12: 'Art. 4. The PNPDEC guidelines are as follows: III. The priority given to preventive actions related to reducing disasters to a minimum.'

[7]Under Article 5, I, of Law No. 12608/12: 'Among the goals of the PNPDEC are:. . . I–to reduce the risk of disasters.'

2, paragraph 2, gives its seal of approval to one of the most fundamental bases for an anti-disaster system: precaution. This concept is doubly qualified by the legislation. First, it represents the duty of the states in relation to the adoption of measures required for reducing the risk of accidents. Second, it holds that preventive and mitigatory efforts must be undertaken regardless of the uncertainty regarding the magnitude and probability of occurrence of a potentially catastrophic event. Conversely, in terms of Brazilian law, uncertainty regarding the risk of disaster is no reason not to act.[8] In the final analysis, there is an unequivocal duty of care expressly stated in Brazilian law.

Thus, departing from the tradition centred on merely corrective activities, Brazilian legislation began to emphasise the centrality of preventive and mitigatory measures, representing risk management in all phases of the circle of disasters. Risk management has acquired relevance, whether in the development of prevention strategies or in response to disasters, mitigating the disaster in question or even preventing new occurrences. It may be deduced from a reading of the present law (Law No. 12608/12) that it is grounded in a 'systemic approach to actions of prevention, mitigation, preparation, response, and recovery.' The success of such a systemic approach within the scope of risk management is linked to the fundamental characteristics of disaster law in general, which do not diverge from Brazilian disaster law: interdisciplinarity and autonomy.

6.2.1 Interdisciplinarity and Autonomy

There is interconnectedness between disaster law and the other diverse branches of law.[9] This integration and articulation is necessary for attaining the goals of disaster law. For that to happen, the role of law in settlements and treatment of disasters must be of an interdisciplinary nature, with strong sensitivity to multidisciplinary information (from other areas of knowledge); it is within its scope to inform and improve decision-making processes related to disasters.

This multiplicity of branches encompassed by the law (interdisciplinarity) is not, however, capable of affecting the autonomy of disaster law vis-à-vis its constitutive specificity. This will be addressed below. A non-terminative analysis of the interdisciplinary intersection of disaster law and the specific absorption of other legal areas involved in performing the specific functions of the former is of fundamental importance in consolidating and conceptualising the function of law in disaster prevention or mitigation, and in providing emergency response, compensation, and reconstruction.

---

[8]Art. 20, § 20 of Law No. 12608/12.

[9]For deepening on the doctrinal discussion about the law of disasters in the Brazilian legal system, see: Carvalho and Damacena (2013), Carvalho (2015, 2017), Carvalho and Farber (2017).

Interdisciplinarity The first and perhaps most intense interaction between disaster law and other legal rationalities occurs with Environmental Law, because in a large majority of cases, environmental disasters are inextricably linked to failures of or compliance with the regulatory process, and to environmental enforcement (illegal occupation of specially protected areas, non-compliance with licensing standards, etc.). Many disasters are caused by environmental contamination, whether immediate or gradual, and by non-compliance with existing environmental protection regulations. Likewise, the relevance attributed by Environmental Law to risk management is increased exponentially by disaster law, having, as it does in this respect, the function of constituting its very identity. However, the processes of environmental risk assessment, already operationally treated by Environmental Law through legal instruments and mechanisms (such as environmental impact assessment, environmental risk assessment, and environmental control plan), acquire special importance when the risks managed are characterised as 'low probability, high consequence.'

Likewise, the intersections with urban law are obvious, especially because decisions regarding land use constitute one of the principal factors that expand the risk and cost of disasters.[10] Although occupation of at-risk areas seems to represent a merely private and individual decision, in a large majority of cases it is an 'option' designed by inadequate public planning and land occupation policy. Such occupation is also influenced by incentives (or lack thereof) provided by public authorities. This may easily be confirmed by observing that the low price or abandonment of areas situated in areas of risk (of flooding, landslides, or contamination, for example) tacitly facilitates and stimulates their occupation by those who already find themselves in a situation of great vulnerability, which is thereby made even greater.

Municipal jurisdiction over urban land use is derived from the Brazilian Constitution itself, considering that Article 30, VIII foresees municipal jurisdiction to 'promote, to whatever degree possible, appropriate territorial ordering, through planning and control of the use, subdivision and occupation of urban land.' Along these lines, the new normative system of prevention and response to disasters articulates various normative statutes, such as the Statute of the Cities (Law No. 10257/01) and the Law on Subdivision of Urban Land (Law No. 6766/70), under the orientation of the new structure and new wording of these texts, through Law No. 12608/12 (Articles 2, VI (h), 41, VI, 42-A and 42-B, of Law No. 10257/01; Article 12, § 1 of Law No. 6766/79).

Just as happens with urban law, disaster law is closely related to administrative law. Along these lines are outlined, as an example, the legal aspects inherent to the master plan, and the limits on land use derived from the technical and social findings of the mapping of areas at risk (at the municipal, state, or national level). There is also enormous relevance to the law on material disasters, such as the exercise of police authority and concession of licences and building permits. In this sense, the

[10]Concerning the factors of amplification of risk and costs of disasters, see: Farber et al. (2015) and Carvalho and Damacena (2013).

new normative paradigm forbids the competent public authorities to 'grant licences or building permits in areas at risk specified as unsuitable for construction by the master plan or the legislation derived from the same.'[11] Also inherent to this legal sphere are the administrative emergency and contingency plans for when a state of public calamity or emergency situation is decreed. Further, absorbing content inherent to administrative law is also a procedure for decreeing such exceptional situations, as well as obtaining financial transfers for emergency response and reconstruction in areas affected by disasters (by the municipality, the federal state, or the Federal District, vis-à-vis the federal government).

Autonomy All along the cycle of disasters, various branches of law are called upon to act, to prevent disasters and minimise, respond to, compensate, or reconstruct sites that have suffered catastrophic damage. Despite the fragmentation of legal doctrines and institutes that often affects the regulation of disasters, it is quite evident that this branch is autonomous. In the first place, this autonomy is reflected in the existence of a specific normative system formed by different legal statutes, such as those referred to in Decree No. 7257/10 and Law No. 12340/10 and Law No. 12608/12.

Another aspect of the autonomy of disaster law is that the legal rules interact in a unique manner, always in integration with the phases of the disaster. Thus, the availability of insurance and forms of government assistance may affect pre-disaster mitigation measures. It is not possible, in terms of the legal treatment of disasters, to treat planning of land use, disaster response, mitigation, and compensation in isolation from one another. Therefore, the other branches are unable to adequately integrate these strategies, making the functional specificity of disaster law necessary. Furthermore, the singular unit of this branch is in risk management, which, continually and uniformly, permeates all the phases of a disaster (the disaster cycle) and its treatment (circularity of risk management). Conversely, the law has the function of reducing vulnerabilities,[12] whether physical or social, and stimulating the resilience of communities hit by disasters. It is for no other reason that the formation of the meaning of disasters oscillates between highly specific and complex (a) causes and (b) consequences, converging to a description of socio-environmental phenomena of great media appeal[13] and multi-contextual spread (economic, political, legal, and environmental), capable of compromising the (c) stability of the social system.

A traditional conception of disasters points to human and social impacts because of 'nature'.[14] This naturalistic conception of catastrophes tends to link disasters to natural events, which provoke large-scale human and property damage. Underlying this is an even more traditional perspective on disasters, the Cartesian distinction

---

[11]Under Article 23 of Law No. 12608/12.

[12]For a careful analysis of the role of law in the reduction of vulnerabilities, see: Handmer et al. (2007). And for reflection on vulnerability in the Brazilian legal system, see: Damacena (2017).

[13]Sugerman (2007), p. 3.

[14]Ségur (1997), p. 1693.

between man and nature. Disasters are conceived of as those non-habitual natural events of irresistible intensity.[15]

However, the technological and scientific evolution of modern society, mainly because industrialisation, has unleashed an expanded human capacity to intervene in nature.[16] One of the greatest human interferences with the atmosphere is the scientifically proven release of GHGs, which are the primary cause of climate change. This phenomenon, which has consequences on a global scale, is the point of greatest intersection between Environmental Law and disaster law. This connection can be seen more clearly when one observes that one of the factors of greatest impact on GHG emissions is unsustainable consumption. From this relationship may be detected the strong nexus between disasters, especially those that originate in climate, and production and consumption relationships.

6.3 Unsustainable Consumption and Climate: The Human Dimensions of Change

Academic research[17] and conferences[18] have concentrated on the relationship between consumption and the environment. However, little attention has been paid to the connection between climate change and unsustainable consumption. The increasing strength of this relationship exponentially increases the negative loop of risk management, which contributes to increasing vulnerability, imbalances in earthly life, and the occurrence of disasters, such as extreme climate events.

It is important to note that impacts capable of contributing to global warming originate in some degree of excessive consumption.[19] Regarding climate change, excess consumption may be expressed by the level of atmospheric GHG concentration it produces in levels that substantially increase the risk of catastrophic climate change. In view of the goal of efficiently managing this risk, there is an emerging consensus that atmospheric concentrations must not exceed ca. 450 to 550 parts per million (ppm), and that by the year 2050, a reduction in recent levels on the order of 60–80% will be required to attain atmospheric concentrations of 450–550 ppm.[20]

---

[15]Ibid.

[16]According to Depuy (2007), pp. 1192–1193: 'The totally unprecedented fact that characterizes our science- and technique-based societies is that we are now capable of unleashing such processes in and upon nature. Tomorrow's droughts, cyclones and tsunamis, or what time will tell (this time, which has always been a metaphor of nature) will be the product of our actions. . . . They will be the unexpected products of the irreversible processes that we will have unleashed, most of the time without having wanted to do so or even known.'

[17]Axelrod et al. (2014); O'Rourke and Lollo (2015), pp. 233–259; Spaargaren (2003), pp. 687–701; Salzman (1997), p. 1243; Weiss (1990), pp. 6–31.

[18]United Nations (2017a, b, c).

[19]Kysar and Vandenbergh (2008).

[20]Vandenbergh and Steinemann (2007), p. 1673.

The relationship between unsustainable consumption and climate change is obvious; however, the approach taken commonly involves rapid growth in population and the system of production, highlighting the latter. Recently, the MERCOSUR-European Union Dialogue Conference on Climate Change and Energy and Food Security, held in Brazil, discussed this question.[21] On that occasion, the Representative of the United Nations Fund for Population Activities (UNFPA) in Brazil, Jaime Nadal, stressed: 'It is not possible to disassociate climate change from demographic growth, the population's patterns of consumption and the processes of industrialization of production.' Nadal asserted the following:

> [T]he impacts of population's patterns of consumption and demographic growth on resource availability, models of production and climate change are visible. . . . When one compares the increase in GHG emissions with demographic growth, one perceives a certain parallel. They have grown concomitantly over the years.[22]

The Intergovernmental Panel on Climate Change had already, in its most recent report, detected this relationship in a systemic and transdisciplinary manner. In the words of the document, 'the efforts to mitigate and adapt to climate change imply a growing complexity of interaction, particularly in the intersections between water, energy, land use and biodiversity.'[23] The report highlights the urgent need for alternative approaches as well: (a) greater energy efficiency and sources of clean energy that contribute to reduced emission of atmospheric pollutants, which alter the climate and are prejudicial to health; (b) reduction in energy and water consumption in urban areas, through building ecological cities and recycling water; (c) sustainable agriculture and silviculture; and (d) protection of ecosystems to store carbon and provide other ecosystem services.[24]

Above and beyond the eco-benefits of better management and use of natural resources, the Intergovernmental Panel on Climate Change expounds upon the scientific consensus that among the dominant forces for anthropogenic emissions are the following:

1. population;
2. the structure of the economy;
3. income and the distribution of income;
4. policy and the state of the art of technology;
5. the availability of energy resources and the changing use of land; and
6. individual and social patterns of consumption, or investment decisions.[25]

With a great deal of evidence and concordance, the report concludes that deep cuts in emissions will require a diversified portfolio of policies, institutions, and

[21] Fundação Konrad Adenauer (2017).
[22] United Nations. ONUBrb (2017).
[23] Fleurbaey et al. (2014).
[24] United Nations. ONUBrb (2017).
[25] United Nations. ONUBrb (2017), p. 1.

technologies, along with changes in human behaviour and patterns of consumption.[26]

6.3.1 Sustainable Consumption As a Pillar of Effective Disaster Risk Management

The seed of needed changes in patterns of life and consumption was planted internationally at Eco 92, which, grounded in Agenda 21, represents the first great alert regarding the need to rethink sustainably-based consumption, as a kind of a map to attaining sustainability. Consequently, the United Nations Environmental Program (UNEP) and the United Nations Department for Economic and Social Affairs (UNDESA) were appointed to lead the transition process. In 2003, Marrakech hosted the first meeting and launched an initiative known as the Marrakech Process, which aims to implement Sustainable Production and Consumption (SCP) globally, in synthesis:

> The Process responds to the call of the Johannesburg Plan of Implementation (World Summit on Sustainable Development 2002) to support the regional and national initiatives to accelerate the shift towards SCP patterns, thus de-linking economic growth from environmental degradation.[27]

The basic guidelines of sustainable consumption, internationally and nationally, are connected to the expectations for a disaster risk management cycle; that is, one where prevention is effectively manifest in all phases. Compliance with Environmental Law (whether concerning the norms for regulation of emissions or the decoupling of negative natural resource utilisation), with the aim of not only promoting better patterns of consumption and sustainable production but also leading the transition to a more ecological and socially inclusive global economy, must be the intention of a society that seeks to avoid disasters.

As if to reinforce this line of reasoning, international negotiations in 2015 culminated in the adoption of the Sustainable Development Goals (SDG) at the United Nations Summit for Sustainable Development. Originating in a mandate of the Rio+20 Conference, the SDGs are intended to orient national policies and international cooperation activities for the next 15 years. They replace and update the Millennium Development Goals (MDG). In addition to hosting the conference, Brazil participated in all the sessions of intergovernmental negotiation. The agreement includes 17 goals and 169 targets across a variety of topics.[28] In the area of consumption, goal 8 and its correlate 8.4 stand out:

---

[26] IPCC (2014), p. 114.

[27] United Nations (2019).

[28] 'How: eradication of poverty, food security and agriculture, health, education, gender equality, reduction of inequalities, energy, water and sanitation, sustainable patterns of production and consumption, climate change, sustainable cities, protection and sustainable use of the oceans and

8. To promote sustained, inclusive and sustainable economic growth, full and productive employment and decent work for all . . . .

8.4 To progressively improve, through 2030, global resource efficiency in consumption and production and endeavour to decouple economic growth from environmental degradation, in accordance with the 10-Year Framework of Programmes on Sustainable Consumption and Production, with developed countries taking the lead.[29]

Thus, it may be seen that the notion of sustainable consumption is one of the pillars of sustainable development—which in turn represents one pathway to the desired sustainability. Ferrer et al. defines it as a process aimed at building a global society capable of perpetuating itself in time, under conditions that ensure human dignity.[30] The idea that sustainability is a must is linked with the fundamental human right to an ecologically balanced environment. The failure of the attempts at sustainability engenders environmental unsustainability, which is the link between Environmental Law and disaster law.

6.3.2 The Normative Semantics of Sustainable Development in Brazil

In Brazil, the standardisation of the concept of sustainable development is oriented by the constitutional semantics of guardianship of quality of life for the present and future generations. To wit, under the terms of Article 3, XIII, a sustainable pattern of production and consumption is tied to 'the production and consumption of goods and services in such a manner as to meet the needs of the present generation and improve living conditions, without compromising environmental quality and the needs of future generations.'

Although unsustainable consumption represents one of the variables most worthy of consideration in matters of disaster, especially when derived from extreme climate events, the law governing the National Civil Protection and Defence Policy in Brazil does not expressly mention the issue. Law No. 12608, of 2012, was responsible for numerous alterations of legislation in Brazil (for example, the City Statute and environmental education) but did not directly alter relations of consumption. At any rate, this connection may be observed in the definition of 'shared responsibility for product life cycle,' anticipated in the National Policy on Solid Waste, understood as follows:

The set of individual and interlinked responsibilities of manufacturers, importers, distributers and merchants, *of consumers* and of the heads of public urban cleaning and solid waste management services, to reduce to a minimum the volume of solid waste and

---

terrestrial ecosystems, inclusive economic growth, infrastructure and industrialization, governance, and means of implementation' (see United Nations 2016).

[29]United Nations (2016).

[30]Ferrer et al. (2014), pp. 1433–1464.

> refuse generated, and to reduce the impacts on human health and environmental quality derived from the product life cycle, in the terms of this law.[31]

The Brazilian legislators' option for the division of labour is oriented by the understanding that *production only grows because consumption levels accompany it*. Thus, any decrease in the pattern of unsustainable production is also related to consumers' choices and behaviour. Conversely, the unsustainable process of market supply and demand plays an important role in the cycle of disaster risk management. After all, the emission of gases that contribute to the greenhouse effect and the whole process that encircles it is one form of unsustainability, the consequence of which is global warming, as reflected in increasing numbers of extreme climate events.

Today, this relationship is sustained by attribution science, which examines the degree to which climate change increases the probability of the occurrence of a specific extreme event. Current advances of this science are because of two main factors: (a) improved understanding of climate and the climate mechanisms that produce extreme events; and (b) advancement in the methods employed to detect the attribution of events.[32] According to the American National Academy of Science,[33] the rapid development of scientific capabilities makes it possible to 'make and defend quantitative declarations regarding the degree to which man-made climate change has influenced the magnitude or probability of specific types of events or classes of events.' Several studies corroborate this evolution. Since 2011, the American Meteorological Society[34] has published several reports that attempt to explain the role of climate change in extreme events, the most recent it was launched in 2018. Furthermore, international research programs have joined together to undertake the project *World Weather Attribution*,[35] which has the same goal.

This relationship, not catastrophic but realistically demonstrated by science, may be observed in both the developed and the developing countries. In the case of Brazil, São Paulo has been singled out as one of the most chaotic cities in the country in various aspects, but the levels of consumption and pollution, encompassing emission of pollutant gases, stand out. The forecast was that by 2017, the city would reach a landmark of six million registered automobiles. In this, the biggest municipality in Brazil, the number of cars is increasing faster than the number of inhabitants, at an annual rate of two new vehicles for each new resident. This index is symptomatic of a city (and a country) which, whether directly or indirectly, has stimulated the use of individual transport, particularly through the lack of priority given to collective means of locomotion such as subways, trains, and buses.[36] In

[31]Article 3, XVII. Brasil (2010) (emphasis added).

[32]Stocker et al. (2014); Shindell et al. (2009), pp. 716–718.

[33]National Academies of Sciences, Engineering, and Medicine (2016).

[34]American Meteorological Society (2018).

[35]Information about the project. Climate Central (2017).

[36]Statistics from Detran-SP (Departamento Estadual de Trânsito de São Paulo), the body responsible for distributing licence plates of vehicles that circulate in the state, show that from January to December 2016, the city of São Paulo acquired 114,800 cars. See: Valle (2017).

2017, the Inventory of Atmospheric Emissions of Road Passenger Transport in the Municipality of São Paulo, conducted by the mayor's office, concluded that the preference of the population of São Paulo for private cars has an impact on not only the city's traffic but also its air quality and global warming. An analysis of the contribution of each mode of transportation to the emission of pollutants revealed that automobiles are responsible for 72.6% of the GHG emissions in the transportation sector, although they carry only 30% of passengers.[37]

Following the orientations of the Brazilian National Climate Change Policy, since 2009, the municipality of São Paulo has enacted legislation requiring fuel-burning buses to be removed from circulation. The plan is to replace them with clean technologies by 2018. It is quite probable that the target will not be met by the deadline; but in June 2017, the first battery-powered electric bus in the city was launched, signalling an inclination to 'adapt' in the times of climate change.[38] Adaptation is the word that best summarises the principles, objectives, guidelines, and instruments of Brazilian climate legislation.[39] Studies and diagnoses play a fundamental part in the successful leadership of this process. Making available data and scenarios need not outline or justify the relationship between consumption and energy; this is a fundamental point for designing climate risk management to have greater potential efficacy in the future.

6.3.3 Energy Efficiency and Consumption: Brazilian Perspectives in a Climate Adaptation Scenario

One of the Brazilian documents that best demonstrate the country's change of stance in relation to energy management is the National Plan for Energy Efficiency (PNEf), which outlines plans for the development of the energy sector. The plan has its origins in the definition of a scenario of economic growth in Brazil from 2010 to 2030.

> Through this scenario, the production capacity and demand for each energy matrix was projected. With these data, a simplified energy balance sheet was formulated, in order to identify in which sources of energy Brazil will have a supply deficit; in the future, energy is

[37]Girardi (2017). Detailed information about the study available at: Inventário de Emissões Atmosféricas do Transporte Rodoviário de Passageiros no Município de São Paulo. Instituto de Energia e Meio Ambiente (2017).

[38]'CPFL Energia, in partnership with the Rede Graal, installed 25 charging stations for electric vehicles at certain points in the city of Campinas, on the Bandeirantes highway and in Anhanguera. Public recharge is still free of charge, because the charges for this type of refill have not been set by the National Electrical Energy Agency (ANEEL). The idea is to get a head start and understand how the market may develop, starting with the creation of this infrastructure. The organization's projection is that by 2030, between 4 and 10 million electric vehicles will be circulating in Brazil.' (AmCham Brasil 2017).

[39]Brasil (2009).

to be supplied with the participation of alternative sources. The energy derived from petroleum and hydroelectric plants presents short-term prospects for development and a need for investment. Second, natural gas will tend to increase its participation in the supply of energy. Third, wind and solar energy are to increase their participation in the Brazilian power grid.[40]

The steps to increase energy conversion encompass the industrial, transportation, building, sanitation, and education sectors, among others. For the transportation sector, the PNEf proposes a real change in Brazil's current road-dominated transportation system, to a mixed road, rail, and waterway model. Complementing diversification of the transportation grid is the need to promote technological development to improve engines, including options for hybrid and electric motors. Another point of emphasis is expansion of the Brazilian Vehicle Labelling Program (PBE Vehicular) to include a greater number of types and models of vehicles. In relation to energy use in edifications, it is necessary, according to the PNEf, to invest in the technical training of construction professionals in the area of energy efficiency. Furthermore, the plan stresses the need to incorporate concepts of energy efficiency in urban planning studies and the work codes of Brazilian municipalities, and to make obligatory the certification of the energy efficiency of buildings.[41]

According to the data of the plan, there is a mean potential for application of energy efficiency measures of around 2.9–7.3% for the year 2020 and 4.4–10.9% for 2030, compared to 2010, considering the different Brazilian sectors and macroeconomic scenarios. However, according to one report:

> [T]o attain the autonomous potential energy efficiency outlined in the National Plan for 2030, in the range of 4.4% to 8.7% in relation to 2010, effective planning to promote the rational use of energy inputs by the different sectors of the economy and the general population will be required.[42]

In this context, it is important to stress the need to upgrade the legal incentives for the rational use of energy in the country, together with promoting medium and long-term planning. History has shown that it is necessary to diversify the Brazilian energy grid to reduce the risk of a power shortage in the country, such as has occurred in the past. In 2015, in the midst of a drought, Brazil was also confronted with the worst energy crisis in history.[43] In recent years, the city of São Paulo has faced a variety of extreme events—first the drought, then several floods followed by mudslides, which caused massive economic damage and human mortality. Several examples of the relationship between unsustainable consumption and climate change

[40]Bronzatti and Iarozinski Neto (2008).

[41]Brasil (2011).

[42]Altoé (2017), pp. 285–297.

[43]According to Garcia (2017): 'Besides the drought, two other facts explain the crisis. First, the government reduced the tariff at the very moment when costs were rising, encouraging wasteful use and creating an enormous deficit among the distributors, which had to be bailed out with federal money. Second, the government delayed the generation and transmission works and decided to auction off power, privileging low price and not providing the rates of return that the market was asking for.'

can already be observed in Brazil, but at least for now, it does not seem that people have learned their lessons about this. There is still a deficit in terms of risk management, and the priority course of action continues to be after the fact.

The Brazilian legal framework concerning energy efficiency is quite extensive and provides guidance to the principal government agencies, with voluntary and obligatory measures. Nevertheless, certain aspects of the framework need to be improved, particularly increasing the efficacy of the National Energy Plan 2030. The main aspect seems to be the need to establish a regulatory framework for ensuring the sustainability of energy efficiency. The climate change scenarios in Brazil point to the need for prevention measures, whether because of a shortage of water[44] or floods. Furthermore, the limitations of the funding mechanisms for energy efficiency actions should be stressed, along with the urgent need to develop stimuli for the power concessionaries and distributors to invest in energy efficiency.

In this context, the role of post-consumption must not be neglected. In Brazil, based on the responsibility shared throughout the product life cycle,[45] the National Policy on Solid Waste establishes the duty of the final recipient (the consumer) to return a product's leftover waste after it has been used. Regarding companies, the process must be conducted through reverse logistics, defined as:

> an instrument of economic and social development characterized by a set of actions, procedures and means designed to make possible the collection of solid waste and their restitution to the business sector for reutilization, in its own cycle or in other productive cycles, or for another environmentally appropriate final destination.[46]

These duties have as yet received little examination, despite the fact that they are a key factor in ensuring that the recycling chain leads to the desired sustainability. The sustainability of consumption has a direct bearing on a country's climate vulnerability, and consequently, its increased or decreased susceptibility to disasters.

6.3.4 The Role of Sustainable Behaviour in the Process of Climate Risk Management

The relationship between consumption and global environmental impact is complex. As a rule, this complexity manifests within one of the essential dimensions of climate change, namely the connection between production and consumption. To be sure, reducing the consumption of goods the production of which emits a large amount of

---

[44]It is important to mention that since 2009, the National Water Agency (ANA) has periodically organized the Conjuntura Water Resources Report in Brazil, in which data on water and its management in the national territory are systematized. The most recent version was published in 2017. Agência Nacional de Águas (2017).

[45]Article 3, XVII. Brasil (2010).

[46]Article 3, XII. Brasil (2010).

GHGs into the atmosphere also involves active behaviour.[47] Consumption of cheap clothing from China, production of SUVs, and the expansion of packaged processed foods are examples of GHG-producing actions.

In recent years, the responsibility of industrialised nations has been at the centre of discussions on climate change. Nevertheless, the role of the consumer, whether in developed or developing countries, cannot be ignored in this context. A study published in the *Journal of Industrial Ecology*[48] highlights the contribution of households, whose consumption contributes more than 60% of global GHG emissions, and between 50% and 80% of the total land, material, and water utilisation. Mobility, shelter, and food are the most important categories of consumption in all the environmental footprints. Globally, foodstuffs represent 48% and 70% of household impacts on land and water resources, respectively; consumption of meat, dairy products, and processed foods increases rapidly with income. Shelter and mobility stand out, with a high intensity of carbon and materials, while the importance of services' footprints is related to the large amount of domestic expenditures associated with them. The study suggests a solid and significant relationship between household expenditures, driven by the growing demand for consumption of non-primary goods, and their environmental impact. In fact:

> Almost all environmental threats have some link with consumer behavior: water is used in the cultivation of food for consumers or to water the lawn; biodiversity is threatened by the destruction of habitats on behalf of housing or agriculture; air and water pollution comes from automobiles, power generators or factories that provide consumers with goods, power or services.[49]

Climate change is just the most recent manifestation (although by far the most complex) in humanity's collision course with planetary limits. It is obvious that ecological stress takes many forms, from the destruction of species, atmospheric pollution, and deforestation to the disappearance of coral reefs, the exhaustion of fishing areas, and the destruction of swamplands. Earth's capacity to absorb waste and pollutants is becoming increasingly more overburdened. Faced with this reality, the questions raised by Daniel Farber are of great relevance: can we 'shift to a more sustainable path? Would that require an unacceptable sacrifice of social welfare?' For the author the answer is yes, but 'in a free society, it is preferable to change individual lifestyles by creating sustainable infrastructure, informing individuals and providing incentives, not by coercing individuals into choices that society prefers them to make.'[50] The human and environmental benefits are consequences of this process. Further, according to Farber, 'Changes in legal rules can reduce barriers to sustainable consumption and give more people the opportunity for sustainable, satisfying lives.'[51] Conversely, sustainable consumption and green communities

[47] Anderson (2012), p. 1.

[48] Ivanova et al. (2016), pp. 526–536.

[49] Farber (2012), p. 1479.

[50] Ibid.

[51] Ibid.

are large-scale goals that will not be easy to achieve. However, they are not utopian, and lawmakers can take significant steps in the short term.[52]

Sustainable consumption should be understood as a strategy that requires public, private, and institutional engagement. The decisions taken by institutions have public and private repercussions and consequences. In fact, this line of thinking is one of the points of adhesion between consumer relationships and disaster law. Furthermore, attaining a sustainable level of consumption requires the stabilisation of environmental impacts; this coincides with the ambitions of the risk management circle, which has climate change as its most significant point of connection with Environmental Law.

For this reason, some studies[53] propose that greater attention be paid to reducing individuals' contributions. Policymakers have a wide range of regulations and economic and social measures with this aim at their disposal, although there remains a great deal of work to be done to evaluate their applicability.[54] The notion of *individual behaviour* has been defined in several different ways. Vanderbergh and Steinemann offer the following concept:

> Individual behavior includes only those behaviors that are under the direct, substantial control of the individual and that are not undertaken in the scope of the individual's employment. As a result, we include emissions from personal motor vehicle use, personal air travel, and mass transport. We exclude emissions from motor vehicle use and air travel undertaken in the course of employment (e.g., driving for a delivery service or flying on a business trip). Similarly, we include emissions attributable to household electricity use, but we exclude emissions attributable to the industrial production of household goods (e.g., the emissions resulting from the production, shipping, and retailing of appliances and food).[55]

Vanderbergh and Steinemann[56] suggest that psychological questions are highly relevant to the recognition and observance of norms when non-compliance does not imply loss or sanctions. Along these lines, certain individuals are strongly influenced by what they perceive to be the behaviour of others. Other people, in turn, are influenced by the perception that others are also involved in pro-social behaviour. Therefore, if people perceive behaviour change to reduce carbon as 'just for eccentrics' or environmentalists, they will have a lower propensity to engage in those behaviours. However, if behaviours aimed at reducing emissions are perceived and disseminated as something that can affect everyone, it is more probable that other people will adopt them. This tendency warrants thorough examination in public policy and the legislation that regulates it.

Controlling individual sources of pollution is an extremely difficult task, because regulation of individual behaviour may be seen as intrusive, even if that is not its intention, and this generates a fear of the political reaction it may provoke. In the

---

[52]Ibid, pp. 1479, 1482.

[53]Vandenbergh and Steinemann (2007), pp. 1673–1745.

[54]Ibid, p. 1704.

[55]Vandenbergh and Steinemann (2007), p. 1690.

[56]Ibid, p. 1705.

current circumstances of climate crisis, change in customs of a social and personal nature should be considered as customary transformations that occur during such a crisis. What Daniel Farber[57] refers to as the 'Republican moment' was a public response to environmental disasters in the sixties and seventies that resulted in broad public support for a range of legislative and regulatory measures and the emergence of what is called the 'environmental citizen.'

For Babcock,[58] this moment is occurring again today, and the global crisis we are facing is climate change. The challenge for this generation is to broaden the spectrum of custom to specifically include individual responsibility for climate change, with a view to transform personal environmental behaviour. It is known that most personal behaviour reflects each person's lifestyle and is not incorporated into the content of environmental norms. Even if there did exist rules to regulate such behaviour, their efficacy would be a serious problem, particularly because of their cost, and because they would also result in enormous political resistance, representing, in the opinion of some, an invasion of privacy and government interference with individual freedom. For these reasons, it is improbable, albeit desirable, that the legislative branch will approve environmental norms dealing with personal behaviour—doing so would amount to political suicide. Be that as it may, change in individual and industrial environmental behaviour does need to occur, even if it is derived from other means.[59]

If a sense of the urgency in calls for environmental protection is necessary to influence behaviour, climate change and disasters do well represent this urgency. The seriousness of the situation, which sooner or later, with greater or lesser emphasis, will affect everyone, should revolutionise democratic participation. In a way, the problem of global warming has lent impulse to the migration from a period of apathy in relation to environmental issues to one in which there is greater interest in the subject. Problems such as hurricanes, melting of the polar icecaps, increasing temperatures and the threat of extinction of polar bears are causes of public concern. In countries like the United States and Australia, this concern has manifested itself in lawsuits aimed at calling to account the major emitters of GHGs, and even the public authorities. In Brazil, litigation from climate events has been of a more individualistic nature, aimed at calling to account public authorities. It should be pointed out, however, that the duty of the central, state, and municipal governments vis-à-vis climate risks in Brazil is a part of a process that must be complemented by the participation of private initiatives and the citizenry.

Concern about the impact of consumption on the planet is not a recent phenomenon. The novelty is the evolution of the scientific knowledge that has been able to connect these two points. Long before the first research studies of anthropogenic contributions to global warming, Agenda 21 (in its Title IV, *Changing Consumption Patterns*) first expressed the recognition by the developed countries of their primary

[57]Farber (1992), p. 66.

[58]Babcock (2009), pp. 1–18.

[59]Babcock (2009).

responsibility to reduce the environmental impact of consumption. This stance did not produce very great effects, particularly regarding mitigation of GHG emissions, because its efficacy depends on the resolution of a broader socioeconomic impasse between the northern (hyper-consumerist) and southern (overpopulated) hemispheres.

In this context, Salzman[60] makes two relevant observations. The first concerns this posture of calling one another to account, which may be considered useless. After all, a society's total ecosystem load does not depend merely on the size of its population, but also on its level of consumption and its technological base.[61] In fact, the north does impose a heavier burden, not only by the magnitude of its consumption and its impact, but also because the lifestyle of the developed countries and their patterns of consumption are taken as models for the developing countries, even the poorest ones.

The second is tied to the semantics of the term sustainable consumption. For the author,[62] to have the capacity to influence public policy and legislation, the term 'sustainable consumption' must be more substantial than simply providing for the needs of future generations. Along those same lines, Ferrer mentions that an expanded notion of sustainability, understood as going beyond the concept of sustainable development (the content of which is basically economic), is required. Thus:

> [B]uilding a global community of citizens is indispensable to the progress of sustainability. Among other measures, this requirement demands that we surpass the partial 'Western' vision of the world in order to make possible a dignified life for all the inhabitants of the planet, doing away with unjustifiable inequalities. Hunger and poverty are not sustainable.[63]

The branches of a democratic state under the rule of law play an important role in this context, starting with the legislative branch, as we shall see below.

6.4 The Regulatory Contribution to the Process of Inducing Sustainable Behaviour

In recent decades, environmental protection policy has paid great attention to two particular questions: (i) what is an acceptable level of pollution? and (ii) what type of legislation would be most adequate to reduce pollution to that level? Sustainable consumption has received much less attention.[64] One possible reason for that is that facing this question implies questioning what we consume, and in what quantity. It is

---

[60] Salzman (1997), p. 1252.

[61] Ibid.

[62] Ibid.

[63] Cruz and Ferrer (2015), pp. 239–278. See, also: Ferrer et al. (2014); pp. 1433–1464.

[64] Lazarus (1993), p. 787.

true that legislation is more effective in regulating patterns of consumption than levels of consumption.

Along these lines, it is important to differentiate the concepts of sustainable production and sustainable consumption. For Salzman,[65] this difference is extremely relevant because of the way each of these concepts defines the problem, establishes a solution and applies political, economic and legal tools in the search for solutions. Thus, in relation to sustainable production, we know what the problem is (pollution), and that it can be administered by industrial policies to reduce to a minimum the discharge of pollutants. This is generally oriented by rules of command and control present in legislation and resolutions. Regarding this point, it seems that the problem has been identified. Perhaps the greatest difficulty is in the sphere of political will and the disposition to pay the price. In relation to sustainable consumption, there is no common understanding regarding the problem and its solution. Unlike sustainable production, sustainable consumption does not have defined objectives, and there is disagreement concerning the appropriate measures that should be taken to achieve it. Few legislatures in the world have turned their attention to the impacts of consumption. This difficulty may be explained by the fact that unsustainable consumption affects the security of future generations, and this has implications for people's lifestyles.

This imbalance seems senseless, because *a priori*, public policies focusing on consumption should be just as robust as those focusing on production. After all, products have an impact throughout their life cycle (extraction, transportation, production, distribution, use and disposal). Production more specifically impacts two of these stages: production and disposal. Therefore, regulation of products via provisions for sustainable consumption does not reduce impacts throughout the life cycle of the product.[66] One of the justifications for the weakness of legislation of this type is political. After all, if they were different, they would directly affect social norms involving lifestyles, equality, and questions of cultural identity—matters that for many people cannot easily be resolved by laws and courts.[67]

Certain examples demonstrate that it is possible to count on public policy and legislation to drastically reduce unsustainable consumption, and that this diminishes per capita carbon footprint without reducing standards of living. Most European cities are urbanised and have a different model of infrastructure and organisation. The dense population clusters around the city core have good systems of mass transportation. Turin, Italy and Bordeaux, France invest billions of euros in streetcar networks linked to the high-speed European intercity railway network. In Bordeaux, traffic has been reduced by 30%. With fewer cars on the road, bicycling is much more agreeable, and the number of bicyclers has tripled. Meanwhile, Turin has restored the city area's architectural prominence.[68]

---

[65]Salzman (1997), pp. 1243–1292.

[66]Salzman (1997), p. 1256.

[67]Ibid.

[68]The data were obtained from the article: Ajl (2017).

On a smaller scale, the city of Güssing in Austria[69] has reduced its carbon emissions by more than 90% simply by prohibiting the use of fossil fuel in public buildings. It uses an innovative technology to convert residual wood into natural gas:

> The transformation of Güssing began when the city was struggling to pay its electricity bill, and through legislation, ordered all public buildings to stop using fossil fuel. Since then Güssing has promoted a whole industry of renewable energy, with 50 companies creating more than 1,000 jobs and producing heat, power and fuel from the sun, sawmills, corn and cooking oil.[70]

There is another lesson here: command and control measures can be effective. In Germany, the city of Freiburg uses a command and control directive[71] to make energy efficiency obligatory in houses. The directive provides for a maximum use of 75 kW-h per cubic metre. The inhabitants of the city have streetcars and bicycles to move around the city, using cars only when necessary. Other German cities have simply banned old cars and trucks. Amsterdam has also taken this approach. The absence of older vehicles means less particulate matter in the air, which makes the cities much more habitable and healthy. Further, in 2017, Milan[72] approved municipal anti-congestion legislation, authorising a fee to enter the downtown area. The mayor of Milan has publicly anticipated a 30% reduction in pollution and a 10% reduction in traffic. Electric and hybrid cars can enter the restricted zone without paying any fee. The taxes from the fees are channelled to fund buses, bicycle pathways and green vehicles.

If, on the one hand, each country has different environmental legislation, varying from state to state and from municipality to municipality, to respect local diversity and circumstances, then on the other hand, global environmental legislation as a whole tends to be directed towards the new reality imposed by climate change. This new reality needs to be able to be engaged by public policy, but not one single alternative. This option seems more appropriate for countries with a strong and well-structured state, especially economically, where policy does not run the risk of remaining a dead letter. More and more, in the developing countries, both legislators and public administrators must pay attention to private action and public-private partnerships, as an alternative both in the area of sustainable consumption and in climate risk management.

6.5 Final Considerations

The current Brazilian legal system adopts a systemic approach to deal with disasters in their different phases, structured around a cycle of risk management. Legal strategies and best practices in terms of disaster prevention and mitigation have

---

[69]Tirone (2017).

[70]Ibid.

[71]European Commission (2016).

[72]BBC (2008).

received greater attention, becoming the main goal of the National Civil Protection and Defence Policy, since the publication of Law No. 12608/2912. Nevertheless, that legislation lacks important regulations and certain upgrading options, especially concerning emergency and compensatory questions. Generally speaking, consumption relationships are also regulated by federal legislation. Law No. 8078 of 1990 provides for consumer protection. There is also the law instituting the National Policy on Solid Waste.[73]

This chapter has sought to demonstrate that unsustainable consumption on the part of both companies and consumers generates a process of negative feedback, exponentially increasing global warming, which contributes to increasing vulnerability and the occurrence of extreme weather events. Alteration of the greenhouse effect is from series of factors, including human contribution, which is responsible for atmospheric pollution, one of the worst kinds of pollution, and which has the capacity to seriously affect ecosystems and the human beings in them.

Taking as its starting point the relationship between unsustainable consumption and climate vulnerability, this chapter has sought to demonstrate the relevance not only of action by public authorities but also the behaviour of citizens, in the process of adaptation to climate change and reduction of the risk of disaster occurrence. The association between consumption and climate change has received little attention in existing academic literature, despite the fact that one of its consequences is its contribution to increasing extreme events, and that this trend is continuing. Even if adaptation to climate change is a major subject of writings and international meetings at present, mitigation of weather damages is fundamental. Mitigation of GHGs emissions related to consumption is an arduous task as well. Among other measures, it depends on budgetary planning and demands an adaptation phase for consumers and businesses; accessible public transport; changes in the energy grid; and legislation containing incentives, complementary to command and control, and facilitating the construction of a culture of more sustainable consumption.

References

Agência Nacional de Águas (ANA) (2017) Conjuntura dos recursos hídricos no Brasil. Available via: http://www.snirh.gov.br/portal/snirh/centrais-de-conteudos/conjuntura-dos-recursos-hidricos/conj2017_rel-1.pdf. Accessed 02 May 2018

Ajl M (2017) Europe's cities show how to lower emissions while raising living standards. Inside Climate News, Available via https://insideclimatenews.org/news/20090721/europes-cities-show-how-lower-emissions-while-raising-living-standards. Accessed 12 Nov 2017

Altoé L (2017) Políticas públicas de incentivo à eficiência energética. Estudos Avançados 31:285–297

AmCham Brasil (2017) Mobilidade urbana limpa: é possível imaginar um futuro de carros e ônibus elétricos no Brasil? Estadão, São Paulo. 16 Oct. 2017. Available via http://economia.estadao.

[73]It deals with consumer protection and makes other provisions. See: Brasil (1990).

com.br/blogs/ecoando/mobilidade-urbana-limpa-e-possivel-imaginar-um-futuro-de-carros-e-onibus-eletricos-no-brasil/. Accessed Nov 2017
American Meteorological Society (2018) Explaining extreme events in 2017 from a climate perspective. Available via https://www.ametsoc.org/ams/index.cfm/publications/bulletin-of-the-american-meteorologicalsociety-bams/explaining-extreme-events-from-a-climate-perspective/. Accessed 13 Jun 2019
Anderson B (2012) Everyday environmentalism: law, nature and individual behavior. Electronic Green J 33:1
Axelrod RS, Vandeveer SD, Downie DL (eds) (2014) The global environment: institutions, law, and policy. CQ Press, Thousand Oaks
Babcock HM (2009) Global climate change: a civic republican moment for achieving broader changes in environmental behavior. Pace Environ Law Rev 26:1–18
BBC (2008) Milan introduces traffic charge, BBC News online, January 2. Available at: http://news.bbc.co.uk/2/hi/europe/7167992.stm. Accessed 15 Nov 2017
Brasil (1990) Lei n° 8.078, de 11 de setembro de. Dispõe sobre a proteção do consumidor e dá outras providências. Available at: http://www.planalto.gov.br/ccivil_03/Leis/L8078.htm. Accessed 12 Nov 2017
Brasil (2009) Lei n° 12.187, de 29 dezembro de 2009. Institui a Política Nacional sobre Mudança do Clima – PNMC e dá outras providências. Available at: http://www.planalto.gov.br/ccivil_03/_ato2007-2010/2009/lei/l12187.htm. Accessed 12 Nov Jan 2019
Brasil (2010) Lei n° 12.305, de 2 de agosto de 2010. Institui a Política Nacional de Resíduos Sólidos; altera a Lei n° 9.605, de 12 de fevereiro de 1998; e dá outras providências. Available at: http://www.planalto.gov.br/ccivil_03/_ato2007-2010/2010/lei/l12305.htm. Accessed Nov 2017
Brasil (2011) Ministério de Minas et Energia. Plano Nacional de Eficiência energética: premissas e diretrizes básicas. Available at: http://www.mme.gov.br/web/guest/publicacoes-e-indicadores/plano-nacional-de-eficiencia-energetica. Accessed Nov 2017
Brasil (2017) Ministério do Meio Ambiente. Efeito Estufa e Aquecimento Global (Greenhouse Effect and Global Warming). Available at: http://www.mma.gov.br/informma/item/195-efeito-estufa-e-aquecimento-global. Accessed Dec 2017
Bronzatti FL, Iarozinski Neto A (2008) Matrizes energéticas no Brasil: cenário 2010–2030. Encontro Nacional de Engenharia de Produção v. 28:1–15
Carvalho DW (2015) Desastres ambientais e sua regulação jurídica. RT, São Paulo
Carvalho DW (2017) Gestão jurídica ambiental. RT, São Paulo
Carvalho DW, Damacena FDL (2013) Direito dos desastres. Livraria do Advogado, Porto Alegre
Carvalho DW, Farber DA (2017) Estudos aprofundados em direito dos desastres. Prismas, Curitiba
Castro F (2017) Mapa expõe risco de inundação costeira. Estadão, São Paulo. 19 May Available via https://noticias.bol.uol.com.br/ultimas-noticias/brasil/2017/05/19/mapa-expoe-risco-de-inundacao-costeira.htm. Accessed 10 Dec 2017
Climate Central (2017) The world weather attribution project. Available at: https://www.climatecentral.org/. Accessed Dec 2017
Cruz PM, Ferrer GR (2015) Law, sustainability and the technological premise as a way to expand its foundations. Sequência 71:239–278
Damacena FDL (2017) Limites e possibilidades da prestação jurisdictional na redução da vulnerabilidade. In: Carvalho DW, Farber D (eds) Estudos aprofundados em direito dos desastres. Prismas, Curitiba, pp 447–478
Depuy J-P (2007) Are there still natural catastrophes? Análise Social 41:1192–1193
European Commission (2016) Directive of the European parliament and of the council amending directive 2012/27/EU on energy efficiency. Available at: https://ec.europa.eu/energy/sites/ener/files/documents/1_en_act_part1_v16.pdf. Accessed 10 Dec 2017
Evaristo SA (2017) Em quatro anos, secas e inundações afetaram 55,7 milhões de brasileiros Zero Hora. Dec 1, 2017 Available at: https://gauchazh.clicrbs.com.br/ambiente/noticia/2017/12/em-quatro-anos-secas-e-inundacoes-afetaram-557-milhoes-de-brasileiros-cjanuf86900ju01mkkli12zgq.html. Accessed Dec 2017

Farber DA (1992) Politics and procedure in environmental law. J Law Econ Org 8:59–58
Farber DA (2012) Sustainable consumption, energy policy, and individual well-being. Vanderbilt Law Rev 65:1479
Farber DA et al (2015) Disaster law and policy, 3rd edn. Aspen, New York
Ferrer GR, Glasenapp MC, Cruz PM (2014) Sustentabilidade: um novo paradigma para o direito. Revista Novos Estudos Jurídicos 19:1433–1464
Fleurbaey MS et al (2014) Sustainable development and equity. In: Edenhofer O et al (eds) Climate Change 2014: mitigation of climate change. Contribution of Working Group III to the Fifth Assessment Report of the Intergovernmental Panel on Climate Change. Cambridge University Press, Cambridge, pp 283–350
Garcia G (2017) Brasil enfrenta a pior crise energética da história. O Globo Available at: http://noblat.oglobo.globo.com/geral/noticia/2015/01/brasil-enfrenta-pior-crise-energetica-da-historia.html. Accessed 10 Nov 2017
Girardi G (2017) Carros levam 30% dos passageiros, mas respondem por 73% das emissões em SP. São Paulo, O Estadão. 23 May Available at: http://sustentabilidade.estadao.com.br/noticias/geral,carros-transportam-30-dos-passageiros-mas-respondem-por-73-das-emissoes-em-sp,70001806416. Accessed 09 Dec 2017
Handmer J, Loh E, Peng Choong W (2007) Using law to reduce vulnerability to natural disasters. Georgetown J Poverty Law Policy 14:13–28
Instituto de Energia e Meio Ambiente (2017) Emissões Atmosféricas: Inventário de emissões atmosféricas do transporte rodoviário de passageiros no município de São Paulo. Available at: http://emissoes.energiaeambiente.org.br/. Accessed 12 Dec 2017
IPCC (2014) Chapter 1: introductory chapter. In: Climate Change 2014: mitigation of climate change. Contribution of working group III to the Fifth Assessment Report of the Intergovernmental Panel on Climate Change. Cambridge University Press, Cambridge, pp 111–150
Ivanova D et al (2016) Environmental impact assessment of household consumption. J Ind Ecol 20:526–536
Konrad-Adenauer-Stiftung (2017) Diálogo Mercosul-união europeia sobre mudanças climáticas, segurança energética e alimentar. Political Foundation Archive. Available at: http://www.kas.de/energie-klima-lateinamerika/pt/publications/49876/. Accessed 20 Dec 2017
Kysar DA, Vandenbergh MP (2008) Climate change and consumption. Environment Law Reporter 38. Available at: https://papers.ssrn.com/sol3/papers.cfm?abstract_id=1286886##. Accessed 20 Dec 2017
Lazarus RJ (1993) Pursuing 'environmental justice': the distributional effects of environmental protection. Northwest Univ Law Rev 3:787–785
National Academies of Sciences, Engineering, and Medicine (2016) Attribution of extreme weather events in the context of climate change. National Academies Press, Washington, D.C.
O'Rourke D, Lollo N (2015) Transforming consumption: from decoupling, to behavior change, to system changes for sustainable consumption. Ann Rev Environ Resour 40:233–259
Pivetta M (2016) Um Brasil mais vulnerável no século XXI. Pesquisa Fapesp. n. 249, Nov. Available at: http://revistapesquisa.fapesp.br/2016/11/17/um-brasil-mais-vulneravel-no-seculo-xxi/. Accessed Dec 2017
Porfiriev BN (1998) Issues in the definition and delineation of disasters and disaster areas. In: Quarantelli EL (ed) What is a disaster? Definition and delineation of disasters. Routledge, New York, pp 56–72
Salzman J (1997) Sustainable consumption and the law. Environ Law 27:1243–1292
Ségur P (1997) La catastrophe et le risqué naturels: Essai de definition juridique. Revue du droit public et de la science politique en France et a l'etranger 6:1693–1716
Shindell DT et al (2009) Improved attribution of climate forcing to emissions. Science 326:716–718
Spaargaren G (2003) Sustainable consumption: a theoretical and environmental policy perspective. Soc Nat Resour 16:687–701

Stocker T et al (eds) (2014) Climate Change 2013: the physical science basis: working group I contribution to the Fifth Assessment Report of the Intergovernmental Panel on Climate Change. IPCC, Bern

Sugerman SD (2007) Roles of government in compensating disaster victims. Issues Leg Scholarsh 6(3). Article 1

Tirone J (2017) 'Dead-end' Austrian town blossoms with green energy. New York Times, 28 Aug. Available at: http://www.nytimes.com/2007/08/28/business/worldbusiness/28iht-carbon.4.7290268.html?_r=1. Accessed 20 Nov 2017

United Nations (2016) Objetivos de desenvolvimento sustentável. Available at: http://www.itamaraty.gov.br/images/ed_desenvsust/ODSportugues12fev2016.pdf. Accessed 10 Nov 2015

United Nations (2017a) COP23: UN climate change conference. Available at: http://www.un.org/sustainabledevelopment/cop23/. Accessed 10 Nov 2015

United Nations (2017b) Goal 12: Ensure sustainable consumption and production patterns. Available at: http://www.un.org/sustainabledevelopment/sustainable-consumption-production/. Accessed 10 Nov 2015

United Nations (2017c) Sustainable consumption and production policies. Available at: https://www.unenvironment.org/explore-topics/resource-efficiency/what-we-do/sustainable-consumption-and-production-policies. Accessed 02 Nov 2017

United Nations. ONUBrb (2015) ONU: Brasil está entre os 10 países com maior número de afetados por desastres nos últimos 20 anos. Available via: https://nacoesunidas.org/onu-brasil-esta-entre-os-10-paises-com-maior-numero-de-afetados-por-desastres-nos-ultimos-20-anos/. Accessed Dec 2017

United Nations. ONUBrb (2017) Agência da ONU ressalta conexão entre mudanças climáticas e padrões de consumo. Available via https://nacoesunidas.org/agencia-da-onu-ressalta-conexao-entre-mudancas-climaticas-e-padroes-de-consumo/. Accessed 10 Nov 2017

United Nations (2019) About the marrakech process. Available at: https://sustainabledevelopment.un.org/topics/sustainableconsumptionandproduction/process. Accessed 13 Jun 2019

Valle C (2017) São Paulo está perto de ter 6 milhões de carros. Por que isso é um problema. 23 Feb. Available at: https://www.nexojornal.com.br/expresso/2017/02/23/S%C3%A3o-Paulo-est%C3%A1-perto-de-ter-6-milh%C3%B5es-de-carros.-Por-que-isso-%C3%A9-um-problema. Accessed 12 Nov 2017

Vandenbergh MP, Steinemann AC (2007) The carbon-neutral individual. N Y Univ Law Rev 82:1673–1745

Weiss EB (1990) In fairness to future generations. Environ Sci Policy Sustain Dev 32:6–31

Chapter 7
Sustainable Public Procurement in Brazil

Teresa Villac and Maria Cecília Loschiavo dos Santos

Contents

Abstract This chapter examines the inclusion of sustainability in the system of public procurement in Brazil. We present the law as a tool for environmental and social change. In this approach, we go beyond the reductionist view of administrative law as a mere operational tool for public procurement and the provision of government services and constructions. Further, we present development as incorporating fundamental ethical values, in contrast to the traditional paradigm that casts it as synonymous with economic growth. We detail the changes in Brazilian administrative law since 2010 that have introduced sustainability criteria into public procurement and the provision of public services and construction. The challenge of implementing these criteria in public procurement must be met by effective and collaborative public governance that acts in a network to support sustainability.

T. Villac (✉)
Federal Office of the Attorney General (AGU), São Paulo, Brazil

M. C. L. dos Santos
School of Architecture and Urbanism at the University of São Paulo, São Paulo, Brazil
e-mail: closchia@usp.br

A. d. Amaral Junior et al. (eds.), *Sustainable Consumption*,
https://doi.org/10.1007/978-3-030-16985-5_7

7.1 Introduction

This chapter analyses the inclusion of sustainability in Brazilian public procurement and the challenges of ensuring effective public governance in its implementation. The chapter proceeds in four parts. After this introduction, we contextualise the subject and offer a survey of the existing legal institutional framework around public procurement in Brazil. The chapter then advances a critical analysis of public-sector management that emphasises the salience of 'network public governance' as a new paradigm of sustainability in public procurement. This analysis draws on three public-sector initiatives that have advanced the principle of sustainable public procurement within such a paradigm: (a) the 'Environmental Agenda in Public Administration' programme (AP3) of the Ministry of Environment (MMA); (b) the introduction of shared sustainable purchases; and (c) a series of actions undertaken by the Federal Office of the Attorney General (AGU).

The chapter concludes by proposing a Law that could act as a mechanism of social and environmental change. This proposal includes a framework for sustainable procurement so that development is no longer understood as simply synonymous with economic growth. Coming from a fundamentally ethical perspective, this legal framework is based on human rights, reduction of social and regional inequalities, preservation of the environment and intergenerational well-being.

7.2 Sustainable Public Procurement

Sustainable public procurement involves the use of state purchasing power to foster sustainable development, thereby exceeding the reductionist view of administrative law as a mere operational tool for procuring supplies for the State and the provision of government services and public works. Sustainable public procurement seeks to reach a higher level of principles and ethics, which turns government procurement into a positive mechanism in achieving sustainable development.[1] In so doing, it recognises that modes of production, consumption and disposal are triggers of environmental and social imbalances. Moreover, it adopts a particular legal ontology, one that views Law as a vector of paradigmatic change within a society that currently encourages hyper-consumption, disregards the social values of labour and sees human beings assume a predatory stance towards the natural environment.

Sustainable public procurement has moved onto the international public agenda, first in the 1992 Rio Declaration (Principle 8) and then under Agenda 21. This was followed by the Marrakech Task Force, an initiative launched by Switzerland in 2005 and hosted within the United Nations Environment Programme (UNEP). The focus of the taskforce was the implementation of sustainable public tenders across a

[1] Villac (2015) and Villac et al. (2014).

number of countries.[2] In 2015, there was a new emphasis on sustainable tenders, which was detailed in the United Nations Sustainable Development Goals (SDGs).[3]

In Brazilian law, although the Federal Constitution (1988) establishes that the State must preserve the environment, sustainable tendering was for many years restricted to initiatives of the federal states, particularly São Paulo and Minas Gerais, and the São Paulo city government. Only in 2010 did an initiative from the Federal Ministry of Planning (MPOG) introduce sustainable public procurement into the federal public administration. This was accomplished via an infra-legal rule from the Federal Secretariat of Logistics and Information Technology.[4] The use of an infra-legal rule to implement this policy was because of the difficulties and resistance encountered up to 2010 in realising the policy legislatively, either through amendment to the Brazilian Tender Law (Law No. 8666/1993), the main law regulating government procurement in Brazil, or through decree. Thus, through NI No. 1/2010—which all bodies within the Brazilian federal public administration must implement—the inclusion of sustainability in public contracts became mandatory.

In December of that same year, a legislative amendment to Article 3 of Law No. 8666/1993 endorsed the principle that public tenders should promote sustainable national development:

> Bidding processes will ensure compliance with the constitutional principle of equal protection, selection of the most advantageous proposal for the administration and promotion of sustainable national development and will be processed and judged in strict accordance with the basic principles of legality, impersonality, morality, equality, publicity, administrative probity and it will be conform to the invitation for bids, processed with objective judgement and other principles correlated.

On the eve of Rio+20 in 2012, Decree No. 7746/2012 was sanctioned to regulate the sustainability criteria for public acquisitions, services and construction. It was recently amended in specific aspects by Decree No. 9178/2017. Having presented the institutional framework, we examine the challenges that exist in implementing sustainable public tendering in Brazil.

---

[2]Sustainable procurement is a process whereby organizations meet their needs for goods, services, works and utilities in a way that achieves value for money on a whole life basis in terms of generating benefits not only to the organization but also to society and the economy, while minimizing damage to the environment. See: Marrakech Task Force on Sustainable Public Procurement (2011).

[3]SDG 12.7: To promote sustainable public procurement practices, according to national policies and priorities.

[4]Normative Instruction No. 1 of January 2010. In Brazil, the so-called *instruções normativas* (normative instructions) are infra-legal norms or administrative acts that regulate some aspect of a law, typically by specifying in detail how the rights, duties or mandates within that statute are to be applied in practice.

7.3 The Challenges of Public Governance in Support of Sustainable Public Procurement

Implementing sustainable public procurement in Brazil in the long term in such a way that does not emphasise economic development 'at all costs' and, in the case of public tenders, ensuring the lowest possible bid price, is a significant public governance challenge. According to Brazil's General Audit Court:

> Governance in the public sector refers, therefore, to the mechanisms of evaluation, direction and monitoring and to the interactions between structures, processes, and traditions, which determine how citizens and other stakeholders are heard, how decisions are made, and how power and responsibilities are exercised. It is concerned, therefore, with the ability of political and administrative systems to act effectively and decisively to solve public problems.[5]

It will therefore be through effective public governance that the introduction of a new paradigm in tenders (through norms) will prove more or less successful.

With this discussion in mind, in the present chapter we examine three recent Brazilian sustainable tender initiatives. In so doing, we adopt the perspectives of Castells[6] and Latour[7] concerning social actor–networks, as well as Dowbor's collaboration paradigm.[8] The three initiatives under study, which we will call 'actions', adopted the concept of Ricoeur.[9]

The chapter proceeds as follows: we first analyse the A3P—launched by the MMA in 2001—which operates through a national collaborative network. We then discuss the programme of sustainable shared procurement, which is a public-sector management initiative (now enshrined in law) that started with the aggregation of procurement management by eight federal public bodies under the leadership of the Botanical Garden Research Institute of Rio in 2010. In 2012, the MPOG standardised sustainable shared purchases as a general national regulation.

We also present the case of the Federal Office of the Attorney General (AGU), which combines voluntary efforts with institutionalisation actions. Common to all three actions is the fact that they began with—and are supported by—network formations consistent with Castells's approach, based on the effectiveness of the collaboration paradigm.[10]

---

[5]General Audit Court [TCU] (2014). See, also: Peters (2012).

[6]Castells (2007).

[7]Latour (2012).

[8]Dowbor (2016).

[9]'The action is mine; it depends on me; it emanates through the power of agency'. Ricoeur (2012), p. 85.

[10]Dowbor (2016).

7.3.1 *The 'Environmental Agenda in Public Administration' (AP3)*

The A3P is an official programme of the Brazilian government that aims to implement sustainability practices in public administration. According to the MMA, A3P is explicitly grounded on the need to

> rethink the patterns of production and consumption of the public sector, and in return, search for innovative strategies, associated with the adoption of social and environmental criteria, principles and guidelines.[11]

The MMA established the programme in 2001, and it operates according to six themes: rational use of natural resources and public goods; appropriate management of the resulting waste; quality of life in the work environment; awareness and training of employees; sustainable procurement and public works. In 2002, the United Nations Educational, Scientific and Cultural Organization (UNESCO) heralded the initiative as the 'best of its kind' in the environmental area.[12]

Another distinctive feature of the programme is that it is not compulsory and accepts voluntary accession of public bodies. Although there is a formal legal instrument, and Law is present in the relationship between the MMA and any public agency that joins the programme, it is not a traditional obligatory relationship, but rather implies compliance with a work philosophy that is merely disseminated by MMA.

By analysing the A3P Program, we identify some aggregating factors that relate to a less hierarchical and more integrative and cross-sectional public management that seeks to break up segmented visions in the governmental work environment. Bliacheris relates this to the effective achievement of socio-environmental responsibility within public administration.[13]

Batista and Costa have foregrounded a model of public administration that is in line with 'liquid modernity'[14]; namely, one that is more consensual and demystified, in which inefficiency does not prevail:

> On the other hand, the bureaucratic State reveals itself as a model full of deficiencies, presenting a series of dysfunctions, such as: excessive formalism; 'sacralization' of procedures and rules previously established, even losing, on the part of the public agents, the capacity to understand their meaning and purpose; exhibition of signs of authority; difficulties to meet the needs of the community; depersonalization of relationships, etc.
>
> The complex reality of the modern pluralistic and liquefied society has dismantled the possibility of success of a purely bureaucratic model that has proved to be impotent. If the bureaucratic model favours previously determined and inflexible procedures, the current reality requires a more result-oriented Public Administration, especially to maximize the

---

[11]MMA (2017).

[12]MMA (2017).

[13]Bliacheris (2013).

[14]Bauman (2005).

assistance of social needs. In this sense, it is necessary to simplify the entangled rules of the administrative order, which has been compared to a rigid guiding instrument to pursuit the common good.[15]

The programme merges social and environmental management practices with the dissemination of knowledge. Considering its innovative nature in the context of Brazilian public administration, several manuals were prepared to explain the thematic axes and guide implementation. The digital version of the materials and their availability on the A3P website are an example of information flow[16] as a feature of the new 'Network Society'[17] in public administration.

In addition to these published materials, there is an annual meeting between public managers, local training programmes and actions to stimulate and disseminate best practices (the so-called A3P Award). The network of actions is open to any public bodies and citizen and A3P operates on the principle of cooperation—without formal hierarchy—to empower local public agents.

7.3.2 Sustainable Shared Purchases

The second case that we examine is sustainable shared purchases. An initiative of the Botanical Garden Institute of Rio de Janeiro, the first instance of sustainable shared public purchases in Brazil brought, in 2010, the Institute and a range of other federal agencies together in the same tender process, to jointly acquire a large volume of sustainable office materials.

Silva notes that sustainable shared purchases refer to the

joint acquisition of goods and services that generate less environmental impact, more social justice and economic efficiency, with scale gains, done by public organizations from different sectors or between units of the same public organization, aiming to foster the sustainable production and consumption in the country.[18]

In 2010, a single tender process was organised to meet the demand of ten federal public agencies in 48 categories of office products (e.g., stationary, accessories and the like). The volume of products demanded resulted in economies of scale, putting the lie to the assumption that sustainable products are more expensive. Actions were development with collaborative integrations of public agents from different bodies. A new conception of interinstitutional teamwork was thereby established.[19]

---

[15]Batista and Campos (2014), pp. 33–34.

[16]Castells (2007).

[17]Castells (2007).

[18]Silva (2014), p. 76.

[19]Silva (2011).

The 2010 action led by the Botanical Garden Institute was a success. The network between federal managers based in Rio de Janeiro—the Forum of Executive Leaderships of Federal Agencies in Rio de Janeiro (GesRio)—developed in its implementation goes a long way to explaining that success. The management director of the Institute presented GesRio with the proposal to jointly bid for sustainable office products, and the collective action proceeded.[20] The motivation of the members of the collaboration was another success factor, given that managing collaboration across so many distinct agencies was both unprecedented and administratively challenging.[21] This is because managing collaboration across so many distinct agencies was both unprecedented and administratively challenging.

The Institute took the 100,000 items that it would ordinarily put to tender for its own use and added 700,000 more in the joint acquisition, resulting in scale economy reductions in per-unit prices of around 50%.[22] Considering the positive impact, MPOG, in 2012, acknowledged and institutionalised, sustainable, shared procurement through the aforementioned infra-legal norm,[23] which by 2010 had become applicable to the entire federal executive branch in Brazil.

Other sustainable, shared purchase initiatives followed. Silva has presented a case study of such action in 32 administrative units within one public agency, highlighting the importance of knowledge management, exchange of experiences and communication, which represent, in the words of the author, 'an intangible value added to the entire institution'.[24] She goes on to say that 'the proposed governance model, participative and collectively designed, helped to create an atmosphere of collaboration between the several actors involved' and had 'sustainability as the background'.[25]

According to Silva, this reflects no less than the building of a new paradigm in public administration through interdisciplinarity and the author explain that:

> There is still much to do. The interdisciplinarity and transversality of the actions aimed at consolidating the Public Power as a conscious consumer make up a rich field of research for those who venture along such a path and, which is, at the same time, a path of many achievements for managers who want to make a difference and contribute to sustainable development in Brazil.[26]

[20]Ibid.

[21]Dowbor (2010).

[22]Silva (2011).

[23]Normative Instruction No. 1/2010.

[24]Silva (2016), p. 25.

[25]Ibid, p. 26.

[26]Silva (2014), p. 83.

7.3.3 *The Work of the Office of the Attorney General to Support Sustainable Public Procurement*

The initiatives at the Office of the Attorney General (AGU)[27] to support sustainable tendering combine accession to the A3P Program in 2008 with voluntary actions. Accession to the AP3 Program took place through the initiative of the AGU's Ombudsman General, which shows the importance of the institutionalisation of ombudsmen in Brazilian public agencies, as a channel for the full exercise of citizenship.

According to Maria Augusta Ferreira, Denis Moreira and Érika Pires Ramos—participant actors in the institutionalisation and implementation of A3P at AGU—the core motivation was implement actions to be effectiveness the constitutional duty of preserving the environment[28] and a National Management Commission was set up, and its activities were formally recognised. Because of the many difficulties in changing the organisational culture of a sustainable public administration, as well as little knowledge and the internalisation of socio-environmental problems beyond the units that deal directly with Environmental Law, dissemination proceeds gradually.

Institutionalisation has increased, and the Office of the Adviser General acknowledged the relevance of the theme; in 2013, it set up the Specialized Centre on Sustainability, Bids and Contracts (NESLIC), with national coverage and competence on the topic. Public training continued, also through partnerships with other bodies, forming a collaborative network of actors.[29]

At the end of 2016, AGU launched the National Project on Sustainable Procurement to carry out public training in the Brazilian federal states on the following topics: sustainability and sustainable development; environmental ethics; the legal basis of sustainable procurement; judgements of the General Audit Court; the social inclusion of waste pickers; solid waste management and; sustainable public-sector management. The courses have minimum costs and the speakers, from universities and public administration, have agreed to participate voluntarily in 2-day events of public and free training. The collaborative network has guided this action, with partnerships with other public agencies and specialists on the addressed topics.

[27]The AGU's constitutional mission is to represent the Union (i.e., Brazil's federal government) in legal and extra-legal issues, as well as to consult and advise the executive branch through the Office of the Adviser General. Legal consulting and advice includes the provision of opinions and information, as well as official legal interpretations of the Constitution, laws, treaties and other normative acts. In summary, AGU interprets the Law for the federal public agencies of the executive branch, assisting authorities to check the legality of administrative acts, by examining the texts of auction notices and announcements regarding public procurement before publication. See: Complementary Act No. 73/1993, and Article 38 of the Public Tenders Act.

[28]Ferreira et al. (2009).

[29]Latour (2012).

7.3.4 *Governance Through Sustainability and Network Articulation*

The network articulation that exists in these three actions is a sign that the 'rupture of the traditional public space, expressed by the multiplication of new and non-institutionalized spaces'[30] has reached Brazilian public administration, albeit in an incipient way. It suggests that we are now observing the increased institutionalisation of sustainability within public agencies, which now accord it a high priority.

Based on Castells and Latour, we conclude that collaboration networks make up an internal social movement in public administration, which tries to minimise the deficiencies in the implementation of sustainable public procurement in Brazil. Although expressly foreseen in the normative framework of tenders since 2010, there has so far been elaboration of a national inter-ministerial programme for its execution.

Baran[31] has developed a theory of networks system for information flows. The author observes that networks can operate in three distinct ways: (a) from a central point, which distributes all information flows; (b) in a decentralised way, with the multiplication of several points of distribution; or (c) in a distributed fashion, characterised by the absence of a central point of distribution and connection between points, which communicate throughout the system. Baran's diagram was developed to enable the flow of information in communication networks and has been used to analyse networks of social actors. By combining the possibilities of the diagram with Castells's network thinking[32] and Latour's actor-network,[33] network articulation in public administration operates according to the decentralised model, which is already part of the formal public structure.

Collaboration networks for sustainability are part of the structure of public administration in Brazil, through both institutional and voluntary or ad hoc connection points. The voluntary connections of civil servants have been a collective response to the absence of a consistent and articulated national policy on sustainable procurement. They are about exchanging experiences and best practices with the aim to strength the network system and keep actors' motivation.

Voluntary connections of actors in this network occur internally in the public agencies or in relationships with other bodies. The institutional connection points stem both from voluntary articulations and from the difficulties of meeting the social–environmental requirements of institutional frameworks for sustainability. Institutionalisation of sustainability occurs through public training in partnership and terms of cooperation, to strengthen actions for sustainable public management.

---

[30]Faria (2012), p. xi.

[31]Baran (1964).

[32]Castells (2007, 2015).

[33]Latour (2012).

Collectors of recyclable materials—sometimes called 'wastepickers'—are significant social actors in sustainable public contracting in Brazil because they are responsible for the transport of waste and recyclable packing of federal public entities to final destinations. At present, recognition and dialogue with this group has been limited and their position is quite precarious. While there is some evidence that collectors have been brought into spaces of dialogue that give them legitimacy and voice,[34] notably in one collaborative academic-vocational project in a cooperative located in São Paulo,[35] we saw no evidence of this happening in large scale in our study. Moreover, the challenge is to adopt a humane and ethical approach in a public administration that prioritises reduction of social inequalities, according to the National Policy for Solid Waste.

The perspective of network performance in favour of sustainable tendering has the challenges of keeping the motivational factor because of the resistance of the Brazilian public bureaucracy, which is still unaware of sustainable public management as a structuring axis of public policy, as presented by Burgos.[36] Another challenge is empowering local actors and the lack of institutional dissemination beyond those involved with the subject in public agencies. In this sense, Bosselmann suggests ways to build a governance for sustainability that involves Law, ecological citizenship, and civil society.[37] The challenge of the dissemination of sustainable tendering in Brazil is the setting, networking and conservation of a governance for sustainability in public administration, by considering good governance as a fundamental right, which is legitimised by transparency, sustainability, dialogue, impartiality, probity, social participation and full accountability.[38]

7.4 Sustainable Public Procurement in Brazil: Ethical Emphasis on the Future

The insertion of sustainability in Brazilian national public procurement introduced the challenge of change the traditional mindset in public sector, and advance for news paradigms, not compatible with the maintenance of one indolent reason.[39] New scenarios are still developing[40] and the objective is, with sustainable public procurement, to increase a model of development that considers sustainability as a constitutional value and principle.

[34]Santos (2014).
[35]Santos (2005).
[36]Burgos (2013).
[37]Bosselmann (2015).
[38]Freitas (2014).
[39]Santos (2000).
[40]Villac (2017).

Considering that law can be a mechanism of contention to retain the hegemony of regulation,[41] the introduction of sustainability in public procurement was a rupture and the law assumed emancipatory content. In this sense, starting from the perspective that there is always an articulation between the fields of power and knowledge,[42] the practice of managers and public agencies, connected through networks, has been responsible for the dissemination and implementation of sustainable public procurement in Brazil. Thus, it is precisely the knowledge that has spread beyond the academy that has provided the scarce national literature on sustainable tenders and theoretical information and operational tools for public managers, enabling access to the new sustainable perspective for government procurement, in an action approach that Castells calls 'Network Society'.[43]

Based on a non-physical *locus* that allows information to flow, 'Network Society', of which electronic media is an example, can also be created in territories other than virtual networks, allowing for new interaction between public agents to achieve objectives, including sustainable tendering. The issue of sustainable public procurement, the content of which is not exclusively legal, takes, therefore, an integrated and 'strongly empirical' approach.[44] Thus, as the meaning of sustainable tendering in Brazil is not fixed and stable and we can expect is that it will continue to evolve and develop.

Further, we should try to expand this still inchoate idea beyond the strict walls of administrative law, giving it a configuration that relates to value of sustainability within the framework of public-sector management. Furthermore, the Law, whose foundation has been historically the conservation of a Brazilian development focused on GDP growth, has innovated and exercised its role of social change. In fact, the adjective 'sustainable' applied to development is not just a precise concept introduced and applied within the Public Tenders Law. It is also a speech act in Dijk's perspective,[45] allowing for both critical analysis and rupture of the power structures that reproduce themselves. It is precisely at this point that the possibility of social change through discourse emerges[46] in the legal case through norms and interpretation of sustainability as a constitutional value and principle.

If we consider that Law is, at the same time, a complex and partial instrument of power,[47] and that sustainable tenders are mainly the result of public management practices, their conception will have implications. These include effects on the processes that emphasise a new model of development, not merely economic but based on environmental ethics, are inclusive in social and human terms, and are attentive to environmental preservation.

[41]Chauí (2013).

[42]Foucault (1975, 1996).

[43]Castells (2007).

[44]Juarez (2014), p. 33.

[45]van Dijk (2008).

[46]Fairclough (2001).

[47]Foucault (1977).

To consider sustainable procurement as a socio-environmental public management mechanism, in a perspective that articulates state consumption not just with other fields of knowledge but with values and attitudes,[48] will result in changes in the supplier market. They will also determine whether environmental ethics and citizenship in the public sector are strengthened. The relationships between knowledge and power also take place here, as Foucault notes:

> I have the impression, and I tried to make it visible, that there is an endless articulation of power with knowledge and of knowledge with power. We cannot be satisfied by saying that power needs such or such a discovery, of such or such way of knowing, but rather that exercising power creates objects of knowledge, makes them emerge, accumulates information, uses them. We cannot understand anything of economic knowledge if we do not know how power, and economic power, are daily exercised. The exercise of power perpetually creates knowledge and, conversely, knowledge causes effects of power.[49]

Sustainable public procurement is a new theme in Brazil, and connected with a concept of sustainable development that has well-being, environmental preservation and human freedoms as the base of public policies in the country. There are relationships between the knowledge to be developed both in academic circles and in the praxis of the public sector, and the resulting power can break with the dominant paradigm of the lowest price and its exclusive focus on the economic features of public procurement.

Law moves beyond the sphere of the strictly legal–formal and regards pressing themes, in search of the real sense of justice, the cultivation of virtue and the concern for the common good.[50] The perspective indicated by this research refers to social–environmental public policies that have in Public Law a solid ethical base by considering the principles and multidimensionality of Sustainability.

Sustainable public procurement will only cement as a pillar of sustainable Brazilian public management—one that provides a broader perspective of daily work in administration, in a systemic, articulated and integrated view—if organisational culture in the governmental sector, where collaboration paradigm prevails, takes socio-environmental responsibility seriously as well. Such a move would allow what Abramovay considers essential to a new economic perspective:

> As important as the application of these forms of cooperation and reciprocity is the fact that they are not restricted to parallel, alternative and minor universes, but they enter strongly into the very sphere that obeys to the logic of prices. And it is precisely there, in a mixture of domains that until recently were waterproof and hostile (the market and direct social cooperation) that we find one of the most promising routes, even in a decentralized economy, for social actors to guide part of their behaviors towards goals where ethics and respect for ecosystems have a prominent place. This is where the key to transition into a new economy lies.[51]

---

[48]Crocker (2016).

[49]Foucault (1975).

[50]Sandel (2013).

[51]Abramovay (2012), p. 131.

This is not a theoretical perspective limited to academic circles and publications. The practice of Brazilian public management in sustainable purchases already shows that collaboration in the public sector, based on the collective experience of environmental ethics, is not only possible, but also successful. The challenge, then, is to build in Brazil a concept of sustainable public procurement that, academically and in public practice, can grow and strengthen with an interdisciplinary vision, with environmental ethics as its core.

References

Abramovay R (2012) Muito além da economia verde. Ed. Abril, São Paulo

Act No. 8666 of 21 June 1993 (1993) Regulates Art. 37, clause XXI, of the Federal Constitution, with norms for biddings and contracts in the Public Administration. Retrieved from http://www.planalto.gov.br/ccivil_03/Leis/L8666cons.htm

Baran P (1964) On distributed communications networks. Part I. Introduction. Research memorandum RM-3420-PR. The Rand Corporation, Santa Monica, CA. Retrieved from http://www.rand.org/content/dam/rand/pubs/research_memoranda/2006/RM3420.pdf

Batista OA Jr, Campos S (2014) A administração pública consensual na modernidade líquida. Fórum Administrativo 14(155):31–43

Bauman Z (2005) Identidade. Zahar, Rio de Janeiro

Bliacheris MW (2013) A responsabilidade socioambiental na administração pública. In: Rossi AM, Meurs JA, Perrewé PM (eds) Stress e qualidade de vida no trabalho: melhorando a saúde e o bem-estar dos funcionários. Atlas, São Paulo, pp 125–139

Bosselmann K (2015) O princípio da sustentabilidade: transformando direito e governança (P. G. França, Trad.). Revista dos Tribunais, São Paulo

Burgos F (2013) A atuação dos governos locais na redução das desigualdades socioeconômicas. (Tese de Doutorado). Escola de Administração de Empresas de São Paulo, Fundação Getúlio Vargas

Castells M (2007) A Sociedade em Rede. Fundação Calouste Gulbenkian, Lisboa

Castells M (2015) O Poder da Comunicação. Paz & Terra, Rio de Janeiro

Chauí M (2013) Saudação a Boaventura de Sousa Santos. In: Santos BS, Chauí M (eds) Direitos humanos, democracia e desenvolvimento. Cortez, São Paulo

Constituição da República Federativa do Brasil de 1988 (1988) Brasília. Retrieved from http://www.planalto.gov.br/ccivil_03/constituicao/constituicao.htm

Crocker R (2016) Somebody else's problem – consumerism, sustainability & design. Greenleaf Publishing Limited, Yorkshire

Dowbor L (2010) A formação do capitalismo no Brasil (ensaio teórico). Brasiliense, São Paulo

Dowbor L (2016) Democracia econômica: alternativas de gestão social, 2nd edn. Vozes, São Paulo

Fairclough N (2001) Discurso e mudança social (I. Magalhães, Trad.). Ed. UnB, Brasília

Faria JE (2012) Prefácio. In: Campilongo CF (ed) Interpretação do Direito e Movimentos Sociais. Elsevier, Rio de Janeiro

Ferreira MASO, Moreira D, Ramos EP (2009) Advocacia cidadã e a agenda ambiental da administração pública. Revista Virtual da AGU, ano 9(92). Retrieved from http://www.agu.gov.br/noticia/n-92-set-2009

Foucault M (1975) Entrevista sobre a Prisão: O Livro e seu Método (com J.-J. Brocheir). Coleção Ditos e Escritos, v. IV: Estratégia, Poder-Saber. Editora Forense Universitária, Rio de Janeiro. 2003

Foucault M (1977) Poderes e Estratégias (entrevista com J. Racière). Coleção Ditos e Escritos, v. IV: Estratégia, Poder-Saber. Editora Forense Universitária, Rio de Janeiro. 2003

Foucault M (1996) A verdade e as formas jurídicas. Nau, Rio de Janeiro
Freitas J (2014) Direito fundamental à boa administração pública. Malheiros Editores, São Paulo
Juarez RP (2014) Transparencia en la contratacion pública: análisis de los ayuntamientos de la Comunitat Valenciana y de Galicia. Revista de la Escuela Jacobea de Posgrado (7):143–168. Retrieved from https://www.jacobea.edu.mx/revista/numero7.php
Latour B (2012) Reagregando o social – Uma introdução à teoria do Ator-Rede. Edufa, Salvador
Marrakech Task Force on Sustainable Public Procurement (2011) Activity Report, May 2011
Ministry of Environment (2017) A história da A3P. Retrieved from http://www.mma.gov.br/destaques/item/8852
Ministry of Planning, Budget and Management (2010) Instrução Normativa n° 01, de 19 de janeiro de 2010. Retrieved from https://www.governoeletronico.gov.br/documentos-e-arquivos/INSTRUCAO%20NORMATIVA%20N.%2001%20de%202010%20-%20Compras%20Sustentav.pdf/view
Peters BG (2012) Governance as political theory. In: Yu J, Guo S (eds) Civil society and governance in China. Palgrave Macmillan, New York, pp 17–37
Ricoeur P (2012) O Discurso da Ação. Edições 70, Lisboa
Sandel MJ (2013) Justiça. O que é fazer a coisa certa. Civilização Brasileira, Rio de Janeiro
Santos BS (2000) A crítica da razão indolente – contra o desperdício da experiência. Cortez, São Paulo
Santos MCL (2005) Re-shaping design. A teaching experience at COOPAMARE. Listen to the recyclable collector's voice. Cumulus Working papers, Helsinki. Retrieved from http://www.cumulusassociation.org/wp-content/uploads/2015/09/WP_Utrecht-13_04.pdf
Santos MCL (2014) Design, waste & dignity. Research supported by CNPq–National Council for Scientific and Technological Development. Editora Olhares, São Paulo. Retrieved from http://biton.uspnet.usp.br/residuos/wp-content/uploads/2014/11/DesignWasteDigntiy_ENG.pdf
Silva RC (2011) Compras públicas sustentáveis: uma experiência de compra sustentável compartilhada no Instituto de Pesquisas Jardim Botânico do Rio de Janeiro (Monografia). Escola Nacional de Administração Pública [ENAP]. Retrieved from http://repositorio.enap.gov.br/bitstream/handle/1/378/Compras%20P%C3%BAblicas%20Sustent%C3%A1veis.pdf?sequence=1&isAllowed=y
Silva RC (2014) Compras compartilhadas sustentáveis: construindo um novo paradigma. Revista do 5o Congresso Brasileiro de Gestão do Ministério Público, 75–84. Retrieved from http://www.cnmp.mp.br/portal/images/Revista_5_congresso_CNMP_2.PDF
Silva RC (2016) Compras públicas sustentáveis: uma experiência compartilhada (Monografia). Retrieved from http://www.ecg.tce.rj.gov.br/premio-ministro-gama-filho-20161
Tribunal de Contas da União [General Audit Court] (2014) Referencial básico de governança aplicável a órgãos e entidades da administração pública (Versão 2). Secretaria de Planejamento, Governança e Gestão, Brasília
van Dijk TA (2008) Discurso e poder. Contexto, São Paulo
Villac T (2015) Direito Internacional ambiental como fundamento principiológico e de juridicidade das licitações sustentáveis no Brasil. In: Santos MG, Villac T (eds) Licitações e contratações públicas sustentáveis, 2nd edn. Forum, Belo Horizonte
Villac T (2017) Sustentabilidade e contratações públicas no Brasil: Direito, Ética Ambiental e Desenvolvimentto. PhD thesis. Instituto de Energia e Ambiente, Universidade de São Paulo, São Paulo. Retried from http://www.teses.usp.br/teses/disponiveis/106/106132/tde-08112017-141101/
Villac T, Bliacheris MW, Souza LC (2014) Panorama de licitações sustentáveis-direito e gestão pública. Fórum, Belo Horizonte

Chapter 8
The Preventive Function and Sustainable Consumption: A Creative Challenge for Attorneys

Claudia Lujan Oviedo

Contents

Abstract It is a challenge for an attorney to establish the causal link between the causal fact and the threat of health risks associated with non-sustainable consumption. We must use various sources of scientific information to do this and achieve the highest certainty possible. Under the doctrine and constitutional jurisprudence of Argentina since the reform of the National Constitution in 1994, human rights have been at the forefront in repairing damages caused to people. The possibility of protection 'beyond repair' has also arisen in this context. However, its corresponding *ex post* restitution is in many aspects insufficient. It is therefore necessary to take preventive action *ex ante* and protect future victims of possible damages. Through the Argentinian Civil and Commercial Code (2015), the State safeguards human rights, whether subject to breach or the threat of being breached. The code refers to the so-called 'principle of prevention', which should be classified

C. L. Oviedo (✉)
Professor of Facultad de Derecho de la Universidad de Mar del Plata, Buenos Aires, Argentina

A. d. Amaral Junior et al. (eds.), *Sustainable Consumption*,
https://doi.org/10.1007/978-3-030-16985-5_8

as a general principle of law. Effective legal protection establishes the precaution of the mere possibility of a future illicit and the threat of violation of fundamental rights, such as, for example, a risk of harm to the consumer's health.

8.1 Introduction

A basic element of civil liability is the relationship between causation and effect that must exist between an action or lack of action and the resulting harm. Causation is foreseeable and avoidable, and determines the so-called adequate causality or reasonably sufficient to generate the corresponding civil liability. In Guido Calabresi's work *The Costs of Accidents*,[1] the term 'nexus' is translated into Spanish as the perfect concatenation between the damaging fact and the damage itself. According to Tony Honoré,[2] when scrutinizing the notion of causality, it is possible to assimilate it to a multi-purpose tool having the key function to predict what is about to happen (the notion of cause is considered as a process).

Like any other modern legal system, the State of Argentina provides rights for compensation to individuals that are subjected to property damage, personal injury or psychiatric harm. However, the new Argentinian Civil and Commercial Code of 2015 (hereinafter the 2015 Code) goes further; the legislator introduced the duty to prevent damages applied to general civil law matters. Book III, title V, second section of the new Code introduces four sections into the Argentinian legal system that extend the 'preventive function' to any person that could prevent or diminish damages to other individuals. This so-called 'principle of prevention' ought to be classified as a general principle of law.

The principle of prevention had already been identified as a general principle of Argentinian civil law under the General Environmental Law No. 25675. To accomplish this, it is necessary to focus on the system regulated by Sections 1710–1714, 1727, 1736 and 1744 of the 2015 Code.

The prevention principle is essential within the framework of sustainable consumption, as it makes an appeal to prevent the production of environmental damages by suspending practices that are known and have foreseeable negative environmental impacts. General Environmental Law No. 25675 defines this principle by stating that the causes and sources of environmental issues shall receive priority and comprehensive treatment to prevent the negative effects they might have on the environment. Prevention, in this sense, trumps cure, as restitution for environmental damage is usually late and sometimes impossible.

The 2015 Code constitutes a tool that allows regulators to address the inequalities between consumption, consumer and supplier. It allows greater interaction between

[1]Calabresi (1970).

[2]Honore (2002).

public and private law and it integrates the consumption and environmental fields. Through adequate information, the consumer can manage her consumption and contribute to sustainable production.

The purpose of this chapter is to analyze the preventive function within the 2015 Code. The chapter begins with a brief discussion of the legal precedent and case law associated with the preventive function followed by a discussion of how the Code applies to budgets and its legitimacy will be stated. Then, potential harm will be explained together with prevention and sustainable consumption. Likewise, a case study will be discussed herein, which is related to the inadequate production of vegetables and harm prevention towards health, because of its consumption. Moreover, it also represents a challenge for those who are authorized to exercise this profession, as well as teaching the law.

8.2 The Principle of Prevention Before and After the New Argentinian Civil and Commercial Code

A duty of prevention applies to all individuals generally. Section 1710 of the 2015 Code establishes that all individuals, within their capabilities, shall:

(a) Avoid the production of unjustified damages.
(b) Adopt reasonable measures, in good faith and according to their circumstances, to avoid damages or to diminish their scale [...];
(c) Not aggravate existing damages.

The duty to avoid and not aggravate damages already existed in the previous Argentinian Civil and Commercial Code of 1871 as part of the general duty to not cause damage (*alterum non laedere*). However, the prevention duty—as it was stated before 2015—was a distinctive obligation of due diligence imposed on individuals carrying out certain functions or holding a distinct position or professional authority, such as a doctor or lawyer. In contrast, Section 1710 goes further and enlarges the obligations of preventing damages towards every individual subject to the corresponding situations.

New social developments and economic models prompt behavioral changes, often rapid ones, that directly affect production and consumption. Therefore, the legal system must have answers for certain issues that neither were contemplated nor can be resolved by existing law. The tremendous expansion of production and consumption that results in the multiplication of situations that could become hazardous for human and environmental safety is a case in point. Argentinian civil and commercial law also faced this challenge, and so a new Code was introduced in 2015 to update the law in light of new circumstances.

Under the traditional tort law, civil liability starts with an actual damage and ends with compensation for the harm produced. This concept creates an inequality between the damage and its compensation and it is therefore necessary to create new legal remedies to act *before* the damage actually takes place. Thus, preventive action emerges as the optimal tool to avoid infringement of fundamental rights. Political and social circumstances during the crafting of the legal changes in Argentina paved the way for the prevention principle to be incorporated into the 2015 Code.

International environmental law is usually mentioned as an antecedent of this preventive principle, as shown by Section 3 of General Environmental Law No. 25675. Here, the prevention principle is considered as a right/duty to prevent damages protected by the preventive anticipatory action included in the new Code in Section 1710 as the 'preventive function'. In common law, the duty to mitigate damages had a significant breakthrough with the case of *Ethyl Corp. v. Environmental Protection Agency* (541 F 2d 1, 1976) where it was argued that limiting the amount of lead additives in gasoline because of the potential health hazards could be applied. This antecedent clearly connects potential hazards with non-sustainable consumption.

Having a legal system that permits *ex ante* action constitutes a real protection of the fundamental human right to health. The 2015 Code includes the preventive principle among its assessment aspects and its application becomes effective through the preventive action established by Sections 1710 through 1715.

By constitutionalizing private law, the 2015 Code introduced a significant change in the existing paradigm that was focused on the protection of a 'person's patrimony' to the protection of the 'person' herself. Consequently, public and private law became connected. The prior code—known as the Velez Sarsfield Code after its principal author—did not include the idea of prevention. This principle became evident with specific laws sanctioned to protect fundamental private rights during the process of de-codification that has taken pace in Argentina during the last two decades. Among them, the Consumer Protection Law No. 24240, sanctioned in 1993, and the aforementioned General Environmental Law, sanctioned in 2002.

Each resulting microsystem has its own principles, which, in turn, possesses superior hierarchy within its specific scope. The Consumer Protection Law No. 24240 is limited to consumers, users and suppliers of goods and services and does not have any connection with Law No. 25675. The same occurs with the General Environmental Law, which does not consider the natural capital required for production. Both microsystems—environmental and consumption—thus do not convergence in any way.

Moreover, consumers are not provided sufficient tools to access the necessary information to detect infringements of the existing environmental protection law. Serious damage to health could be prevented with such knowledge. Law No. 24240 establishes that whomsoever manufactures, imports, distributes or markets products or renders services must provide consumers or users with accurate and objective information that is truthful, detailed, efficient and sufficient regarding intrinsic characteristics, and should do so in such a way that those goods or services—under foreseeable and/or normal usage conditions—do not represent any hazard for the health or physical integrity of consumers or users. On the other hand, Law No. 25675 includes among its priority objectives certain changes in social values and behaviors to enable sustainable development and improvement of the quality of life for present and future generations.

However, a better quality of life cannot be expected without considering current requirements of consumption and production. Therefore, the objectives established by each specific law cannot be attained in isolation. The 2015 Code succeeds in connecting the above-mentioned laws with rights guaranteed by the Constitution, thus protecting fundamental human rights that might become subject to harm or threat.

Regarding sources and interpretation of the law, the Supreme Court of Argentina has spoken authoritatively when it determined that 'the rule of interpretation provided by Section 16 (of the prior civil code) exceeds the limits of private law' and established that sources should be interpreted under the general principles enforced throughout the domestic legal system.

According to the court's ruling:

Duty of Harm Prevention

Under Section 1710: Every person has the duty, in the event that such duty depends on him or herself:

(a) To avoid unjustified harm causation;
(b) To adopt, in good faith and under the circumstances, reasonable measures to prevent the occurrence of damage, or diminish its magnitude; if such measures prevent or diminish the magnitude of an injury for which a third party would be liable, it has the right to reimburse the value of the expenses incurred, under the rules of unjust enrichment;
(c) Not to aggravate the harm, if it has already occurred.

I believe that this description is generic and responds to the general duty not to cause unjustified damage within the limits of reasonableness. Therefore, the description included in this section is unnecessary, as it repeats the basic principles of acting in good faith of any contractual or extra-contractual relationship. That is, without the existence of this section the same result would be reached, because whoever acts in breach of the legal system by the general rules is considered obliged to pay for taxes.

The preventive function with respect to the legitimation in Section 1712 acquires relevance when it establishes the following: 'they are legitimized to claim those who prove a reasonable interest in preventing the damage.' It is a jurisprudential task to interpret or verify that the real importance of this lies in granting the corresponding tool of the preventive function to anyone who proves him or herself having a reasonable interest, whether individual or collective. In my opinion, the enumeration made in Section 1710 in subsections (b) and (c) should be included in Section 1712 as guidance or as a pillar for the application of the latter. The reading of subsection (b) of Section 1710 reinforces my interpretation that the preventive function is activated when it is exercised by someone other than the person who is obliged not to harm. This paragraph enables anyone who takes measures to avoid or reduce the magnitude of a damage to obtain reimbursement of the value of the expenses incurred.

Jimena Marquez in her chapter on 'The Preventive Guardianship of Damage' points out four elements that constitute the reasonable interest in preventing damage:

1. Threat of damage. This refers to the serious and reasonable possibility of the occurrence of an illegal act, even if a harmful result does not occur. All that is required is that there is a 'risk of illegality' regardless of what actually happens because what is intended is to avoid its consequence.[3]
2. Illegality. This refers to the existence of an act or omission contrary to the duties generically imposed by law or specifically by a contract.
3. Causal relationship between the threat of damage and the action or lack of action from the passive subject refers to the possibility that a behavior contrary to law could result in a damage, its repetition or continuation (Section 1736).
4. The possibility that through a restraining order (*tutela inhibitoria/accion*), a court order to proceed/execute or not to proceed or execute the consequences of the illegal act could be reduced, avoided or stopped altogether, thus preventing irreparable damage.

Under the doctrine and constitutional case-law of Argentina since the reform of the National Constitution in 1994, human rights principles have emerged as central to the compensation of damages suffered by people. The Argentinian legal system has reinforced protection that goes beyond mere compensation of damages. This is because such an approach is in many ways insufficient to protect individual, collective, and non-transferable rights, such as a person's honor, labor rights and the right to a healthy environment. This is the reason for acting *ex ante* to protect future victims from potential damages.[4]

To understand the legal instrument that *ex ante* protects individuals' rights from potential damages, it is necessary to interpret the provisions on preventive functions in Sections 1710–1713 in harmony with Section 1736 of the 2015 Code. Section 1736 establishes that the burden of proof regarding the relationship of causality falls on whomsoever alleges it, whereas Section 1713 establishes that the burden needs to be balanced in the least restrictive and most appropriate manner to ensure the effective attainment of its goal.

The question becomes twofold: firstly, what are the conditions to be attributed to the person for his action or intervention to be considered causal? Secondly, must the legitimized party as per Section 1712 prove two extremes, such as the foreseeability of a damage, as well as the possibility of preventing, diminishing or not aggravating the damage through inhibitory action?

[3]Regardless of the concept that Jimena Márquez gives us, in my opinion it is not necessarily the existence of a threat to start an action from an illegal fact, but it could be a lawful act that becomes a damage. In our case, it constitutes a threat to health.

[4]Cossari (2017).

8.3 The Features of the Breach of Prevention Duties

The element that activates this action is the simple possibility of preventing likely harm, threat of a violation of fundamental rights, with a threat being sufficient. It is also included herein, the possibility that illegal activity will continue causing damage.

8.3.1 Party Legitimation: The Plaintiff and the Defendant

On the one hand, concerning Section 1710, a person can conclude that the duties to prevent damages apply to whomever is capable of averting or diminishing the harm. This clearly enlarges the prevention duty far beyond persons carrying out certain functions in which 'due diligence' is determined by law. Everyone can thus act as a plaintiff and must act under the principles of prevention.

On the other hand, the legal framework becomes relevant whenever a breach of prevention duty is brought to the Supreme Court of Justice through the defendant. That is, whenever a subject, individual or collective rights, requires judicial measures to avoid the commission of a damage and/or the continuation of a damage. The active legitimation to invoke the rights for prevention against another person is broad. Section 1710 includes not only those affected directly, but also those who might eventually become affected by the harm as end users, consumers, or citizens. When a damage to a right of collective impact is alleged, the legitimation should be extended under the 'Halabi' ruling from the Argentinian Supreme Court.[5] Unquestionably, this is an extremely broad legitimation.

The Halabi Legal Case

Ernesto Halabi, an Argentine lawyer, brought an injunction against the Argentinian State, asking for Sections 1 and 2 of Law No. 25873 and Regulatory Decree 1563/04 to be declared unconstitutional. Such regulation of the National Telecommunications Law (Law No. 19798) included the obligation to the providers of telecommunications services to capture and derive the communications they transmit for remote observation, under what was established by the applicable law, at the request of the Judiciary or the Public Prosecutor's Office. Conversely, we saw that a wide range of intervention was established as providers of telecommunications services had the obligation to register and systematize personal and location data of their users and customers together with records of communication. They are obliged to preserve the information for a period of 10 years.

The injunction was based on the fact that the provisions of Law No. 25873 and its regulatory decree violated the guarantees established in Sections 18 and 19 of the

[5]H. 270. XLII. Halabi, Ernesto vs. P.E.N.—Act No. 25.873, dto. 1563/04 of legal injuction 16986, Buenos Aires, 24 February 2009 (Judgements 332:111).

National Constitution, as they authorize intervention in telephone and internet communications without a specific law that would determine in which cases and on which grounds the aforementioned action could be undertaken. Likewise, Mr. Halabi stated that this intervention was a clear intrusion in his personal privacy, which constituted a violation of his rights to personal privacy and privacy as a user and also a violation of his privilege to confidentiality that, as a lawyer, obtains in communication with clients. In both courts, the Lower Court and the Court of Appeals, Sections 1 and 2 of Law No. 25873 were declared unconstitutional and it was also determined that the legitimacy of the plaintiff did not exclude the collective impact of the effect under the second paragraph of Section 43 of the National Constitution. Consequently, the decision issued under such conditions, should take advantage of all users who have not participated in the trial.

On 24 February 2009, the Supreme Court of Justice reaffirmed the unconstitutionality of the aforementioned sections of Law 25873. The Court ruled on a question of procedural legitimacy and, in so doing, recognized three categories of rights: (1) individual rights; (2) group rights pertaining to common or 'collective' goods and; (3) group rights, referring to homogenous individual interests. Group rights included in the second category—common or 'collective' goods—are, under Section 43 of the National Constitution, exercised by the Ombudsman of the Nation, the associations that make up the group or collective interest and the affected party.

Although there is no law in the Argentinian legal system that regulates the effective exercise of so-called group legal actions, Section 43 of the National Constitution is operative and establishes that the judges is obliged to give efficacy when clear evidence is provided on the pretension of a fundamental right and access to justice for its holder.

Furthermore, the Supreme Court of Justice introduced the requirements that a group legal action or 'group rights referring to homogeneous individual interests' should meet. We can identify three conditions from the court's corresponding analysis:

1. The first condition is harm. However, for group legal actions, the fact of a single event that causes an injury to a relevant plurality of individual rights has been added.
2. The second condition is that the harm caused to a group of individual rights should not make differences in the nature of damage caused in each subject. Nevertheless, it must have homogeneous elements within that plurality of subjects to be affected by the same fact.
3. The third condition is enforceability. Meaning, for this type of legal action, the Supreme Court holds that it must be enforceable that the individual interest, considered in isolation, does not justify the promotion of a claim as a small amount of such claim related to the cost of the litigation. Consequently, the access to justice could be affected. However, this clarifies that the action will result anyway in those cases where other aspects related to matters such as the environment, consumption or health, or when they affect groups that have

traditionally been weakly protected. In such circumstances, the nature of those rights exceeds the interest of each party.

Regarding the active legitimacy of group legal actions, Section 43 of the National Constitution, in terms of group or collective rights concerning personal or property rights or those arising from effects on the environment and competition, establishes the rights of users and consumers, as well as the rights of subjects who have suffered discrimination. However, in Argentina's positive law there is no law that regulates the effective exercise of group legal actions. Faced with this scenario, the Supreme Court establishes that what is regulated in Section 43 of the National Constitution is fully operative and that judges must provide the appropriate effect according to such words. For this reason, we can establish that the group legal action (as long as it meets the corresponding requirements) has a broad active legitimacy, because it can be filed by the affected party, by the Ombudsman and by the associations registered under the valid regulations for those purposes.

8.4 Effects of the Judgement

Verily, the judgements issued only produce effects concerning those who have the corresponding parties' status for the trial (and possibly third parties), and these cannot benefit or harm third parties outside the legal action. However, this ruling cannot be applied to judgements where group rights of collective incidence are at stake, because in such cases, the same body of group legal actions would be considered 'meaningless'. Therefore, in this case of group legal actions, a change is verified in the traditional concept of the scope of the effects of a judgement and the scope of the *res judicata*, expanding its effects to all members of the affected class, having the sentence, in relation to such effects, i.e., '*erga omnes*'.

8.4.1 Prevention and Horticultural Production: Rules Applicable to Environmentally Sustainable Production

Sustainable production is a model of production of goods and services that minimizes the use of natural resources, the generation of toxic materials and waste and polluting emissions without putting human health at risk.

8.4.2 Potential Damage

The principle of prevention should be approached from different angles, as attorneys face a truly creative challenge at the time of completing the elements of the preventive function. The individual or collective claimants must prove the existence of a reasonable correlation between the activity (or lack of activity) of the respondent and a certain damage that causally derives from it as an immediate consequence. The possibility of avoiding an activity may affect its consequences (whether the corresponding actions have been performed or are in progress.) The complexity lies in the fact that this is only possible when the damage did not occur completely; if the damage had occurred prior, the appropriate legal remedy is compensation.

In this scenario, it is necessary to identify the key concepts of the normative structure to lay out the claim that will be used as a source of jurisdictional judgement at the time of deciding whether to apply the inhibitory action. This approach is based on the following extremes of various guidelines that need to be combined:

1. Distinguish between the possibility of future damage and the possibility of not aggravating an already existing damage. In the event of the damage already being in progress, there will be greater scientific possibility to prove the extremes required by the rule. Hence, the exactitude required to determine the causality depends on the situation being faced: either a future damage or a damage in progress. To understand the notion of 'aggravation', it is important to keep in mind that risks are always contained by limits that identify and assign them a certain danger or tendency to become damage.
2. In turn, it should be proven that the inhibitory action would make it possible to avoid, diminish or not aggravate the damage. This represents a different analysis that should be under Section 1727, which proposes a framework of reference by describing the type of consequences resulting from a fact that usually occurs in the natural and ordinary course of events.

Prevention can be classified as either general or specific, or provisional or definitive. When the preliminary injunction is exercised by a consumer, it will always be specific and provisional.

In summary, a general prevention refers to actions performed by the potential perpetrator of an activity that may cause a financial penalty. It is a mere general potentiality of coercion or threat penalty that would lead to the non-compliance of the corresponding norms. This coercion has the ultimate purpose of deterring an individual from undertaking legally prohibited behavior, so this person, knowing the negative consequences that a certain behavior would imply, refrains from breaking the provisions of the legal system.

On the other hand, specific prevention has its legal precedent in the criminal sphere where the main goal of this kind of prevention is to prevent those who have already committed an illegal act from repeating this action in the future. Thus, special prevention is not addressed to society as a whole, but to those who have already violated the legal system. This concept can be applied to the law of torts and it is initiated through preliminary guardianship. The consumer, facing a threat of damage, can request that the judge apply the action to avoid it. The judge is the one who will adopt the safety measures or will apply restraining orders. Therefore, when the consumer exercises a preliminary injunction it will be against the specific prevention.[6]

Moreover, preliminary guardianship may consist of a provisional or definitive prohibition order. This distinction is related to the magnitude of the costs that the defendant will have to face because of the application of the penalty adopted. The provisional penalty has a specific temporary characteristic. Its temporality is explicitly, purposefully limited. The definitive penalty comes from the Latin word *definitīvus*, which means to qualify everything that concludes, ends, resolves or consumes something. Anything definitive cannot be reversed or left without effect. The consumer will obtain the temporary penalty of a temporary nature considering that he/she will request guardianship based on a 'potential harm'.

The sense of the preventive function is to avoid greater costs to the supplier through reasonable measures that avoids or diminishes the magnitude of the damage. The plaintiff must merit the efforts and monetary costs that the provider must support and explain them clearly in the bases of his/her claim. A temporary penalty will always be more reasonable. The temporary closure of an industrial establishment is less burdensome than its final closure.

I intend to analyze the key words that attorneys must consider: Prevention. Avoid. Risk. Causality. All these terms have more than one meaning in the Spanish language.

8.4.3 Prevention

From the Latin *praeventio*, prevention is the action and effect of preventing. It implies preparing in advance all that is necessary to achieve an end, to anticipate a difficulty, to foresee a damage and to notify as a warning. Therefore, prevention is arrangements made in advance to minimize risk of negative consequences. The objective of prevention is ensuring as much as possible the non-occurrence of an eventual damage. It is possible to link the notion of prevention to care or precaution. Preventing could also be the action of warning about a specific danger.

8.4.4 Avoidance

The term avoid derives from the Latin *evito, evitas, eitare, evitavi, evitatum*, all made up by the prefix *ex*—separation from what is inside—and by the verb *vito, vitare, vitavi, vitatum,* which translates as to evade or to withdraw from. The etymology means to preclude the occurrence of something unpleasant. It can also mean attempting to evade a certain situation or person. Although an intention to avoid is

[6]Acciarri (2015).

possible, it is not always viable. When it is not possible to avoid what is about to happen, it must be faced.

It is then that the aforementioned action is undertaken with the purpose of overcoming, eliminating or sidestepping something. It may also imply an attempt to circumvent the certain responsibility expected of a person by denying or refusing that obligation.

8.4.5 Risk

Its oldest etymological origin lies in the Arabic word *rizq*, which means 'what Providence offers'. The term is related to the proximity of a possible damage and to the vulnerability it causes to those who suffer it. Risk prevention refers to the action and effect of preventing something. Therefore, among other meanings, the concept includes to prepare in advance for a certain purpose, to foresee a damage or to anticipate a difficulty.

8.4.6 Causality

Causality is the relationship between cause and effect. One of the fundamental axioms of thought is that: *each phenomenom has a cause*. The notion of causality assumes two distinct logical structures:

1. The causal relationship is rational. This is the idea of causality accepted by classic physics. All effects are consequence of a preceding cause; there is always an occurrence that causes another.
2. The causal relationship is probabilistic. The concluding theme of the probabilistic approach is that—rather than directly producing an effect—causes change the probabilities that an effect will occur.

'Causation', then, is the necessary connection of two phenomena, one of which (called the cause) conditions another (called the effect).

A distinction can be made between an absolute cause and a specific cause. An 'absolute cause' is the combination of all circumstances whose presence necessarily determines the effect. It is only possible to establish an absolute cause in relatively simple cases. A 'specific cause' is the set of circumstances already existing at the time of the occurrence of a given situation (i.e., before the effect takes place) that shapes the conditions for the cause to act and to produce the consequence. Moreover, the specific cause requires scientific research focused on proving the occurrence of a given phenomenon. The specific cause combines the most essential components of the absolute cause and the remaining components are considered as *reasons* of the action giving place to the specific cause.

The causality problem is a field of bitter dispute between materialism and idealism. Materialism supports the thesis that causality has an objective and a universal nature that exist outside and independently from conscience. Accordingly:

> [S]ubjective idealism negates causality in general, either reducing it to a mere succession of sensations that are habitual in men (Hume), or else, albeit acknowledging that causality constitutes a necessary nexus, it considers that it is provided to the world of phenomena by the knowing subject (aprioristic nature of causality, Kant).[7]

8.5 Prevention Duties and Sustainable Consumption: Prevention and Horticultural Production—Rules Applicable to Environmentally Sustainable Production

Sustainable production is a model of production of goods and services that minimizes the use of natural resources, the generation of toxic materials, waste and polluting emissions without putting human health at risk.

Argentina is an important agricultural producer globally. In terms of extensive production, the largest export products are soybeans, cotton and corn and, in terms of intensive production, fruits and vegetables. Intensive agriculture develops mostly around urban areas. Chemical products are used to achieve pest control and in this way obtain a greater yield from the harvest. The use and application of these agrochemical substances are strictly regulated in unique and regional laws. These rules establish the application method and the corresponding areas where its use is enabled. Non-compliance with the application methods can cause alarming levels of agrochemical pollution in the soil, air and water that, absorbed by vegetables, can be a risk for human consumption.

From the perspective of the new paradigms of the 2015 Code and the development goals regarding sustainable consumption, Argentina is going through different legal conflicts caused by the application of agrochemicals and the threat that consumption of products treated in this way can cause to health. This is how a recent ruling in the Buenos Aires Province becomes interesting, because it makes it possible to visualize the conflict that arose from the claim of potential harm to health because of non-compliance with the regulations for the application of agrochemicals.

The relationship between the prevention principle and sustainable consumption is clearly activated. The ruling *Picorelli Jorge Omar y Otros c/ Municipalidad de General Pueyrredon S/insconst. Ord. No. 21296* [Picorrelli and Others vs. the Municipality of General Pueyrredon] of 25 November 2014, in summary, expresses the following:

> The case occurred within an area of intensive cultivation of vegetables, mainly tomatoes, peppers and lettuce, located in the territory of the Buenos Aires Province. The individual

[7]www.filosofia.org/filomat/df144.htm.

intent was to bring a legal action complaining about fumigations related to the illegal application of agrochemicals constituting an illegal action since the applicable rules established in the provincial Act No. 10699 were not respected. This law contemplates the budgets of climatic conditions and wind factors that guarantee the absence of risks to health and the environment.

The application in the protected ecological zone constitutes an imminent risk for the population, directly before the possible or indirect contact with the contaminated water or soil through the consumption of vegetables that have been exposed to this fumigation and taken to the market without respecting the grace period necessary for the agrochemical products to be safe. In this case, the consumer faced the imminent threat of consuming tomatoes, peppers or lettuce risking his or her health. This situation could be avoided by the application of Section 1712 of the Civil and Commercial Code, which requests the suspension of sale of these products until they have been proven by means of expertise the existing degree of toxicity.

Section 1710, subsection (b), enables the third party to perform all preventive acts, considered as appropriate and, afterwards, requesting the expenses reimbursement. The eligibility requirements are those persons or groups or persons legitimated for the action that can demonstrate the following:

Unlawfulness (Illegality)

The unlawful action or omission of it makes foreseeable the production of damage, its continuation or aggravation. The actor must demonstrate that in the production of vegetables fumigated without respecting the application protocol contained in Law No. 10699, which expresses the following:

1. 'The objectives of this law are the protection of human health, natural resources and agricultural production through the correct and rational use of the products mentioned in the following section, as well as to avoid contamination of food and the environment.' (Section 1)
2. 'Every natural or legal person, with the exception of carriers, application and warehouses or storage companies lessors, whose activity is included in Section 2 of this law, will have the obligation to count, in accordance with pertinent regulations, with an advisor or professional technical director, agronomist engineer or other qualifying degree enrolled in the Professional Council of provincial jurisdiction, as determined in the respective Regulations.' (Section 5)

The application of pesticides on crops, especially horticultural crops, must be suspended with a corresponding notification made in advanced that for each case specifies the regulation of this law. Any food product contaminated with pesticides in quantities greater than the tolerance levels specified in the regulations of this law will be confiscated and destroyed, without prejudice to fines or other penalties or actions that may apply (Sections 11 and 12 of the aforementioned law).

Threat of Harm

The right to information established in the National Constitution and concordant laws is of essential importance. The consumer must have access to public information linked to the production processes. Knowledge of the protocols that must be respected in the management of crops must be communicated clearly. Human rights are basic values that include every person because of their human condition, and are necessary to live with dignity: food, health, education, employment, a healthy environment and freedom of expression. These rights are recognized as historical, inalienable, indefeasible, universal, indivisible, progressive, interdependent, which evolves in time. Additionally, being interdependent, the violation of only one of them has multiple violations. The National Constitution enacted in 1994 incorporates in its Section 42 the right to consumer information.

On 16 September 2016, National Law No. 27275 on the Right of Public Access to Information was published. Its main objectives are to guarantee the effective exercise of the right of public access to information, to promote citizen participation and transparency of public management, based on the principles of transparency and maximum disclosure. Access to public information may only be limited in the event of any of the exceptions provided within this law, according to the needs of democratic and republican societies, proportional to the interest that justifies them.

From the aforementioned guarantees, consumers should have easy access to both production protocols and possible damage to health. The *Picorelli vs. Municipality* ruling represents an important judicial precedent by virtue of its demonstration of how a citizen can activate defense mechanisms when she warns, through observation, of non-compliance with safety standards in a production process. In this case, the actor could verify the unlawful act through his senses. Through smell, smelling pesticide and, by sight, watching fumigation in prohibited areas.

Nevertheless, to bring a preventive function, the plaintiff must demonstrate the connection of causality between this unlawful act (i.e., fumigation in a prohibited area and the smell of a pesticide, with possible harm to health). In an attempt to achieve this connection, the legal operator must focus its work on scientific research to obtain the necessary evidence demonstrating the threat to human health that would constitute the consumption of these products sprayed with agrochemicals. Conversely, there is a need to demonstrate that, with the application of the inhibitory guardianship, a damage towards health by the consumption of these vegetables can be avoided, reduced or at least not aggravated.

Herein, the lawyer faces a new challenge when designing and substantiating the claim. The action must contain an objective analysis. Section 1726 of the 2015 Code establishes that the relationship between the harmful event and the harm must be an 'adequate' causal connection. Under the theory of adequacy, not every condition of the result is a cause in its legal sense, but only that cause which is normally adequate to produce the result. The adequacy judgement is made up of the objective probability or predictability of production of the result.

How can the plaintiff demonstrate the adequacy between a 'possible' damage and a current event? The lawyer will necessarily have to build a demonstration hypothesis with scientific argumentation. He/she will have to venture into different disciplines of science, which cannot always provide a unique answer. In the case of analysis, the plaintiff will have to gather scientific evidence demonstrating the harmful consequences that consumers may suffer when ingesting vegetables sprayed with chemical products. Statistics that relate agrochemical use to health damage are an important tool in this context. Establishing the criteria of least possible restriction, for example, in this case, the suspension of the sale of vegetables could overlap with the time needed to perform the chemical and biological analyzes required to demonstrate the presence of toxins in the plants.

The objective of an environmental policy for a company/companies should be to guide the planning, organization and control of all its activities incorporating the environmental dimension within them. Thus, in this way, the basis of the environmental policy will be to promote the implementation of sustainable production with actions that anticipate any type of environmental impact resulting from the actions arising from the activity. The main environmental risks include insufficient water supply, inadequate sanitation, air pollution, and chemical waste.

The corporate/company actors related to consumers and the environment have the responsibility to promote public participation and debate on sustainable consumption, and to provide broad information to consumers and businesses towards the implementation of activities that achieve patterns of sustainability in consumption. The preventive function thus has a double purpose, to avoid possible damage and to contribute to sustainable production.

8.6 Conclusion: Impact on the Practice and Teaching of Law

The methodology for teaching law is a subject that will have to be reconsidered drawing on new research tools and advice from interdisciplinary scientific sources. Such tools will also be of great professional benefit for legal practitioners. It is necessary to analyze and assess different educational strategies that have been designed and applied to generate knowledge within the social sciences:

> Teaching to research is a complex process and a widely diversified activity; therefore, research education using as a reference specific practices and effective processes for the generation of knowledge is more fruitful and produces better results. When the goal is to teach research as a trade, it is not sufficient to base the proposed academic schedule on a mere description, analysis and review of the job, but also the learner needs to become involved in all the tasks involved in this practice supported by a more experienced person and immersed in an institutional exercise promoting scientific research.[8]

It is not about submitting a new theory of science but a different method for teaching research within the science of law. Creativity is needed to develop an explanatory scheme that will enable the judging body to reach the same conclusion that is being proposed. Furthermore, originality and a description of novel stances are fundamental when arguing the preventive action. Considering the dynamics of the 2015 Code based on specific principles, the preventive function in this particular

[8]www.unne.edu.ar/unnevieja/Web/cyt/cyt/2002/01-Sociales/S-053.pdf.

case requires us to rethink the techniques for arguing and applying the aforementioned set of rules that contain a procedural microsystem as part of a substantive law.

The main environmental risks are insufficient water supply, inadequate sanitation, air pollution and chemical waste. The parties within our society, which are related to the environment and consumers, have the responsibility to promote social advertising and debate on sustainable consumption. Moreover, it is essential to obtain comprehensive information for consumers and all type of businesses towards the implementation of activities that achieve patterns of sustainability in consumption.

The preventive function has a double purpose. The first consists in avoiding possible damage and the other in contributing to sustainable production. For the consumer to be effective, numerous mechanisms must be activated, including the following: adequate information and scientific research. These are essential tools for the lawyer when supporting his/her case.

References

Acciarri Hugo A (2015) Elementos del Análisis Económico del Derecho. La Ley ISBN 9789870329473. Buenos Aires

Calabresi G (1970) The costs of accidents. A legal and economic analysis. Yale University Press

Cossari M (2017) Prevención y Punición en la Responsabilidad Civil ('Prevention and penalty of civil liability'). Book Collection of the Civil and Commercial Code of Argentina, February 2017, Buenos Aires, 'El Derecho' Publisher, pp 33–55

Honore Anthony M (2002) Responsibility and fault 1st.publishing ISBN 9781841133997. EE. UU

Chapter 9
Taking Care of Business: Engaging Dialogue on Solutions to Unsustainable Commercial Practices

Sergio Sebastián Barocelli

Contents

Abstract The purpose of this chapter is to analyse one of most important relationships between consumption and the environment: unsustainable business practices. We define these as an example of unfair practices. Unsustainable business practices infringe the sustainability principle and consumers' right to access consumption on a sustainable basis. Unsustainable business practices may appear at different stages along the value chain: production (or provision, in the case of services), distribution, marketing, after-sales service and waste disposal. Assessment of the sustainability of business practices demands analysis of the full range of ecological, economic and social effects. The United Nations has underscored the importance of combating unsustainable business practices in a number of documents, including the UN Guidelines for Consumer Protection. The guidelines promote sustainable business practices and discourage those that are environmentally disruptive (points 51, 59, 62 and 71). Furthermore, the UN 2030 Agenda for Sustainable Development's twelfth goal is to ensure sustainable consumption and production patterns. To evaluate the degree of sustainability of commercial practices, it is important to establish an adequate 'dialogue of sources' between Environmental Law and Consumer Law. This 'dialogue' should include considerations of the environmental impact of products and services, consumer access to adequate information, freedom

---

S. S. Barocelli (✉)
Universidad de Buenos Aires, Buenos Aires, Argentina

Universidad del Salvador, Buenos Aires, Argentina
e-mail: sbarocelli@derecho.uba.ar

A. d. Amaral Junior et al. (eds.), *Sustainable Consumption*,
https://doi.org/10.1007/978-3-030-16985-5_9

of choice, proper disclosure, protection of the economic and social interests of consumers and compulsory standards for the safety and quality of consumer goods and services.

9.1 Introduction

This chapter analyses one of most important relationships between consumption and environment: unsustainable business practices. The present analysis will focus on Argentinian legislation, drawing broader links with Latin America, without prejudice to the fact that many of its postulates have global application. The chapter proceeds as follows. Firstly, we will detail the key relationships between the environment and consumption. Secondly, we will analyse the implications of the principle of sustainability. We will then reflect on the theme of commercial practices and consumption relations and subsequently focus on the analysis of unsustainable practices as a variety of abusive practices. Finally, we will summarise the core points and draw some conclusions.

9.2 Relationships Between Consumption and the Environment

Relationships between consumption, development and the environment are intimate and complex and have their correlation in the legal field in the way in which it has delimited and outlined its approach. In this understanding, we can establish various bridges and common bonds between Environmental Law and Consumer Law, namely:

1. Their contemporariness. Although each discipline has what we could call its 'prehistory', that is to say, there are precedents in the matter from ancestral times and their emergence in both cases towards 1960s reflects the reality of a post-war, industrial, 'welfare' and consumer society;
2. The protective perspective; of consumers (in the case of Consumer Law) and of the environment (Environmental Law), notwithstanding the anthropocentric or geocentric conceptions that exist on this point;
3. The fact they embody 'third generation' human rights;
4. Their interdisciplinary character within the legal field, breaking the classic dyad of public-private right, and transdisciplinary with respect to other sciences and fields of enquiry;
5. The recognition within the scope of each of both individual and collective rights;

6. The important role they have played in enhancing the visibility of environmental and consumerist movements and the struggles of these to bring questions of protection to public agenda to advance the cause of rights;
7. The common purpose they share, namely quality of life;
8. Their international dimension, through recognition in different international documents and forums, at global, regional or intergovernmental levels;
9. Their constitutionalisation, in the Argentine case, under the section on 'new rights and guarantees' through the 1994 amendment of the National Constitution (CN), mainly in Sections 41, 42 and 43, also finding in the procedural protection common tools;
10. The way they encourage and incentivise public authorities in the active defence of these rights;
11. The role they have within the distribution of competences, concurrently, between federal and local jurisdiction;
12. The number of legal principles they share: prevention and precaution, responsibility, solidarity and sustainability. We will focus on this work on the latter.

9.3 The Principle of Sustainability

The principle of 'sustainability', as we pointed out, is one of the key bridges between Consumer Law and Environmental Law. That principle, further, subsumes two themes: the right of access to consumption and the right to sustainable consumption. This is what the Argentina National Civil Law Conference has recently held, in its stated understanding of the principles of access to consumption and sustainable consumption as two autonomous principles, although adequately articulated by Section 1094 of the Argentinian National Civil and Commercial Code (CCC).[1]

The right of access to consumption is the first right of 'entry' into the system and constitutes a minimum right of inclusion of every human being. For this reason, access to the consumption of certain basic goods and services guaranteed by the State (food, housing, health, education, and the like) that allow a decent standard of living and to cover essential needs is crucial. Although not expressly enumerated in the constitutional text, the right of access to consumption is implicit and arises from a systemic interpretation of various constitutional norms (Sections 41, 42, 33 and 75, paragraphs 19 and 125 CN) and international treaties with constitutional hierarchy, under Section 75, paragraph 22 CN. These include Section 11 of the International Covenant on Economic, Social and Cultural Rights, Section 11 of the American Declaration of the Rights and Duties of Man, Sections 22 and 25.1 of the Universal Declaration of Human Rights (among others) and the United Nations Guidelines for Consumer Protection. The right of access to consumption is mentioned at the provincial level as well, in Section 4 of the Code of Implementation of

[1]XXVI National Conference of Civil Law.

the Rights of Consumers and Users of the Province of Buenos Aires (Act No. 13133). In this way, everyone is guaranteed a reasonable standard of quality of life and satisfaction of their basic needs and those of their family, as essential elements of individual autonomy that allows achieving human development, understood as the 'development of the full potentiality of being'.[2]

Access to sustainable consumption is also linked to the right to protect the economic interests of consumers, whereby adequate regulations ensure essential goods and services—such as food, clothing, home utilities, transportation, and the like—are available and prohibit arbitrary denials of contracting and regulate prices, standard terms and conditions, and so on.

Sustainable consumption[3] implies, as prescribed by the United Nations Guidelines for Consumer Protection, that the needs of goods and services of present and future generations are met in such a way that they can be sustained from an economic, social and environmental point of view. It is known that sustainable development requires the elimination of the two poles of the market society—consumerism and poverty—as they are contrary to the ecological, economic and social equilibrium that underpins sustainability.[4]

On this first point, the elimination of poverty finds its connection with the aforementioned right of access to consumption as a minimum level of social inclusion for all inhabitants. As for consumerism, the issue is more complex. Consumer satisfaction, Bauman notes, is a program of life in the consumer society.[5] It requires, therefore, social awareness concerning the consumption of goods and services and their rational use, as well as the search for effective tools for the equitable and sustainable use of natural resources without affecting the needs of future generations, throughout the chain of production, distribution, marketing and consumption. It also demands regulation of market behaviours that promote it, such as illicit advertising and inappropriate marketing techniques and other indirect media such as the cultural industry.[6]

We believe that sustainable consumption should not be focused exclusively on the ecological dimension, but should be viewed much more broadly, incorporating the multiple aspects of conceptions of environment and sustainable development, namely:

---

[2]Pérez Bustamante (2011), p. 159.

[3]This right is expressly recognised in the United Nations Guidelines on Consumer Protection and in some provincial constitutional texts, such as the Constitution of Entre Ríos, Sections 30, 78 and 83 or the Provincial Code of Implementation of Rights of Consumers and Users of the Province of Buenos Aires (Act No. 13133), in its Articles 3f, 8 and 9. Likewise, it is implicit in the Argentinian Constitution, through a systemic interpretation of the preamble and Sections 33, 41, 42, 43, 75 paras. 2, 18 and 19 and 125 CN and constitutes one of the attributions of the authority of federal application, under Section 43 para. a) of Argentinia's Consumer Protection Act.

[4]Pérez Bustamante (2011), p. 159.

[5]Bauman (2007).

[6]Barocelli (2015), p. 63.

1. An *ethical* dimension that promotes rational and responsible decision-making by consumers when consuming or opting for a product or service. We place special emphasis on austerity as a way to reduce the ecological impact of consumption, but also as a way to achieve well-being and happiness in the face of unrestrained economic growth and consumerism;
2. An *ecological* dimension that includes the famous 'Rs' of the ecological movement: reduce, reuse and recycle. In this dimension, several other elements—such as organic agriculture and livestock, artisanal production and marketing, waste treatment and disposal, and the effects of logistics and transport on the use of energy, among others—are also contemplated;
3. An *equity* or *social* dimension, whereby decent work standards are promoted and slave and child labour and abusive, aggressive and discriminatory business practices eliminated. The fight against poverty, hunger, disease and deprivation also goes in this direction;
4. An *economic* dimension, which brings attention to the consequences of the organisation of the factors of production, which aims to strengthen social alternatives and seeks a new international economic order. The promotion of policies that promote a more equitable distribution of wealth, fair trade, social solidarity, 'green' taxation, and 'green contracting' in the public sector, among other initiatives, are also within this dimension.

As we can see, the principle of sustainability plays a programmatic role in guiding the formulation, implementation and evaluation of public economic and social policies, especially regarding production and consumption, as established in Section 43 of Law No. 24240 on the Protection of Consumers, but also as a limitation on private autonomy.

9.4 The Problem of Commercial Practices in Consumer Relations

Commercial practices are at the heart of the sustainability principle. The doctrine has presented a range of definitions in this context. Business practices have been characterised variously as: the techniques and mechanisms ordered to the development of consumption, whether seeking their promotion, incentive or support[7]; the procedures, mechanisms, methods or techniques used by suppliers to promote, maintain, develop or guarantee the production of goods and services to the final recipient[8]; the set of acts that the suppliers of products and services direct to the market, in such a way as to position their product, or their service, or constitute themselves as a preferred option in front of potential consumers[9]; and the

[7]Frustagli and Hernández (2015a, b), p. 593.

[8]Lorenzetti (2009), p. 158; Japaze (2009), p. 164; Kemelmajer de Carlucci (2015), pp. 105–134.

[9]Santarelli (2015), p. 370.

mechanisms, techniques and methods that serve, directly or indirectly, to facilitate production. Thus, we have an extremely broad concept that includes marketing, guarantees, after-sales services, contract execution and the extinction of obligations arising from contracts.[10]

Argentine legislation contains no normative conceptualisation of commercial practices. There are, however, normative sources that have inspired the national legislator, such as the Community Directive No. 29/2005 of the European Union, which characterises business practices as 'any act, omission, conduct or demonstration, or commercial communication, including advertising and commercialization, coming from a merchant and directly related to the promotion, sale or supply of a product to consumers.' In this sense, we understand that commercial practices can be characterised as any behaviour deployed in the market by a supplier of goods and services as background, during execution, as a consequence of or in the context of a consumer relationship, including production, import, distribution, promotion and marketing and information about the products or services themselves.

The vertiginous dynamics of the market and developments in the field of science and technology produce constant changes in the commercial practices of suppliers, as well as the emergence of new modalities. They can be repeated over time and more or less continuously follow policies, strategies, protocols, codes, procedure manuals, pre-set standards, and the like; or, they may constitute conduct deployed occasionally or only once in the market. The habitual, systematic or occasional character will simply be a differentiating element when it comes to the merits of a particular behaviour or conduct.

It is impossible to list every example of business practices. Indeed, while the most common are techniques of extraction and production, trade and distribution, marketing and intermediation of goods and services, as well as advertising, marketing and brand building, these are but a sample of the full range of commercial practices. Furthermore, on the marketing side there are season offers, promotions and discounts, terms, customer service practices and loyalty programs to consider as well. At the back end of the product life cycle, commercial activity naturally includes the termination of contractual relationships, after-sales service and care, provision of technical services, spare parts and accessories and the guarantee or warranty regime and the method of waste disposal and treatment.

Commercial practices constitute legal acts when they are lawful and legal facts in cases of wrongfulness, which tie legal relationships between those who deploy in the market (suppliers) and those who are recipients or are affected by whatever is supplied (consumers). As the long list of activities given above suggests, commercial practices surround all the vicissitudes of consumer relations: before setting the contractual or other relationship, during the period of the contract, when it is closed or even after. Likewise, business practices can affect, as we said before, who is the principal and who is on the margins or outside of a consumer relationship, even as they are potentially strongly affected by it.

---

[10]Stiglitz (2014), p. 103.

As a general principle, we can argue that suppliers of goods and services are free to carry out the commercial practices that they deem necessary to run their enterprises. The grounds of this freedom can be found in the rights to free commerce, property, trade and expression recognised constitutionally (Sections 11, 12, 14, 17, 19, 20, 22, 33 and 75 CN). However, these rights, like all constitutional rights, are not absolute and find their limits in non-interference with other rights and duties. Among these are the duty of the State to protect work, social security, family and provide adequate housing (Section 14 CN); the rights of consumers (Section 42 CN); transparency in market and free competition, prohibitions against abuse of dominant positions and other anti-competitive practices and control of monopolies (Section 42 CN); environmentally sustainable development (Section 41 CN); prosperity, well-being and progress (Section 75, paragraph 18 CN); human development, economic progress with social justice, productivity of the national economy and the generation of employment (Section 75, paragraph 19 CN); community property and the human development of indigenous people (Sections 75 paragraph 17 CN); measures of positive action that guarantee real equality of opportunity and treatment, and the full enjoyment and exercise of the rights of children, women, the elderly and persons with disabilities (Section 75 paragraph 23 CN, the Convention on the Elimination of all forms of Discrimination against Women, and the Convention on the Rights of the Child and Convention on the Rights of Persons with Disabilities). Additionally, other rights are constitutionally protected through international treaty, including quality of life, universal access to adequate food, clothing and housing and a continuous improvement of the conditions of existence and protection against hunger (Section 11 of the International Covenant on Economic, Social and Cultural Rights) and defence of the social function of property and prohibition against exploitation of individuals (Section 21 of the American Convention on Human Rights), among others.

In Consumer Law, the foundation is mainly based on the constitutional clause protecting consumers,[11] which allows for prohibition against violation of good faith and consumers' fundamental rights,[12] as well as regulation of pre-contractual, contractual and post-contractual business–consumer relations[13] and any marketing techniques or policies that could be the result of an abuse of economic power.[14] Therefore, if the commercial practices developed in the market by suppliers of goods and services violate these constitutional provisions or any standards that are based on them, they may be subject to state regulation and, in some cases, be considered as illicit.

While the general principle is, as detailed above, the freedom of suppliers to deploy all kinds of commercial practices in the market, their limit will be given by compliance, as a legal act of the requirements of lawfulness, that is, not to contradict

[11]XXIV National Conference on Civil Law.

[12]Santarelli (2015), p. 370.

[13]Hernández and Frustagli (2012), p. 807; Frustagli and Hernández (2015a, b), p. 594.

[14]Frustagli and Hernández (2015a, b), p. 363.

the law, morality, good customs, public order, rights of others or human dignity, under Section 259 and 279 of the CCC. Otherwise, we will be facing an illicit commercial practice, as a legal fact.

The statement of motives of the draft CCC established that:

> The code foresees a series of behaviours, contractual or not, that abuse the good faith of the consumer, as well as his situation of economic or technical inferiority. It is understandable, therefore, that such practices are considered unlawful per se, regardless of whether they cause harm or not. For them, the absolute presumption of illegality rules. They are practices that appear both in the field of recruitment and outside of these.

In the present work, we use the expression ‘illicit commercial practices’ or ‘unfair practices’ as synonyms. Some authors use the term ‘unfair practices’ to refer generically to all cases of illicit commercial practices and others limit it exclusively to cases of illicit commercial practices that violate the right of consumers to fair and dignified treatment. In that sense, we privilege the use of the term ‘unfair practices’ in the first meaning that we detailed, because it is the most widespread in Argentine doctrine. In this understanding, we believe that illicit commercial practices can be characterised as all conduct, by action or omission, that affects the rights of consumers and users displayed in the market by a supplier antecedent to or as a consequence of, or in the context of the activity aimed at providing goods or services in the market. These, then, are de facto situations, because if they were incorporated in contractual clauses we would be in the presence of ‘unfair terms’ and not ‘unfair practices’, notwithstanding that in many cases both forms will apply.

We do not share the view that for a practice to carry the aforementioned designation that it needs to be ‘institutionalized’, namely, that it occurs repeatedly over time or as part of the modus operandi of the firm.[15] Undoubtedly, if its generalisation or standardisation has been proven in internal codes or protocols, the standard of conduct for administrative sanctions or the corresponding punitive damages will be aggravated. It suffices that for an illicit conduct to configure an abusive practice, as a legal fact, it must simply link a consumer relationship with the victim and legitimise the latter to claim in the framework of Consumer Law for the defence of his or her rights and interests.

Nor do we believe that abusive practices require, to be understood as such, the existence of an intention to commit fraud by the responsible supplier, or the actual materialisation of the contracting or configuration of a material damage to any consumers that are the recipients thereof. The mere verification of conduct contrary to the rights of consumers constitutes illegality, in our view.[16] Thus, as long as the practice violates the principle of good faith, the rights of consumers, whether illegal, contrary to morality, good customs, public order or human dignity, we will be in the presence of an abusive practice. This is also true, without prejudice to the existence of other illicit commercial practices outside of Consumer Law that violate the rights of third parties (competitors, workers, the State or society in general). At the time of

[15]Chamatropulos (2015).

[16]Rusconi (2013).

framing a commercial practice as abusive, we believe that damage having been caused does not constitute the limit of its illegality, but rather in having the potential to produce it. Thus, the preventive function that Consumer Law must play is based on this assessment.

Nonetheless, when evaluating wrongfulness from a subjective perspective we believe that hypervulnerability should be the basic standard to be considered to determine whether there is abuse. In the classic law of the nineteenth century and in some perspectives of present-day Consumer Law, totalising categories—'bonus pater familias', 'the good businessman', 'the reasonable man', the 'average man or consumer', the 'rational or smart consumer', etc.—prevailed to set standards for evaluating behaviour. We differ from the criterion set out in Directive 2005/29/CE of 5 November 2005 on unfair commercial practices. This Directive prohibits unfair commercial practices, understood as those 'contrary to the requirements of professional diligence'[17] or that

> [distort] or can substantially distort, with respect to the product in question, the economic behaviour of the average consumer which affects or to which the practice is directed, or of the average member of the group, if it is a commercial practice directed to a specific group of consumers.[18]

Thus, the 'average or rational consumer' seems to be a 2.0 version of the classic 'bonus pater familias'.

These categories make invisible the differences, particularities and situations of many groups—the most vulnerable, disadvantaged and disadvantaged of society, above all—for which we believe the administrative and judicial authorities should focus consumer protection efforts and the measures to prevent damage. Hypervulnerability, that is, the most aggravated protection, must be the measure to assess the appropriateness of the full validity of consumer rights when analysing the legality of commercial practices. This seems to be the criterion that the Argentinian Supreme Court of Justice has determined in its ruling in *Ledesma vs. Metrovias*.[19] On this point, the jurisprudence also argued that the test of the 'diligent man' should not be applied, but instead less rigid standards, according to the characteristics of the audience to which the commercial practice is directed. For the court, the key point was that consumers are not perfectly rational when making choices in the marketplace, that their bounded rationality makes them vulnerable given that commercial marketing practices tap the psychology of consumers by means of cleaver sales

[17]Section 2 (h) of the directive defines professional diligence as: 'the level of special competence and care that can reasonably be expected of the merchant in his dealings with consumers, in accordance with honest practices of the market or with the general principle of good faith in the field of activity of the merchant'.

[18]Directive 2005/29/EU, Section 5.2.

[19]Supreme Court of Justice of the Argentine Nation, 04/22/2008, *Ledesma, María Leonor vs. Metrovías S.A.* Fallos: 331:819.

techniques, affecting their free determination with respect to the choices they make about purchases of goods and services.[20]

As we pointed out, because the dizzying dynamics of the market and the developments in the field of science and technology produce constant transformation and the emergence of new forms of illicit practices, it is useful to frame the assumptions of illicit commercial practices according to the right that is violated under open rates, without prejudice to the fact that some of them, because of their generality or public interest, deserve express regulation.

Regarding assessments of illegality, we argue that when categorising practices, the provisions of the CN, human rights treaties with constitutional hierarchy (Section 75 paragraph 22 CN) and other international instruments, the legal rules governing MERCOSUR integration, the LDC, the CCC and other special and indirect regulations should be kept in mind. On this basis, we can classify abusive or illicit practices as:

1. Misleading commercial practices, when they contradict the right to information, the principle of good faith or go against the education of consumers (s 4 LDC and Section 1100 CCC);
2. Unsafe commercial practices, when they violate the right to health and safety protection (Sections 5 and 6 LDC);
3. Aggressive commercial practices, such as those that are contrary to the right to fair and equitable treatment (Section 8 LDC and Sections 1097 and 1098 CCC);
4. Commercial practices contrary to the economic interests of consumers (Section 42 CN);
5. Commercial practices contrary to the right to free choice of consumers and free competition in the market (Section 1099 CCC);
6. Unsustainable commercial practices, with respect to those that contravene the right of access to consumption and/or sustainable consumption (Section 1096 CCC);
7. Other illicit commercial practices, when they contradict other rights, such as the obstruction of access to justice, the organisation and participation of consumers or other rights or legal interests recognised by the legal system.

It should also be noted that the proposed classification is merely indicative as many abusive practices may include the simultaneous violation of two or more consumer rights.

---

[20]Civil and Commercial Appeals Chamber of La Plata (Argentina), room II, 09/27/2.005, *Tomas, Héctor A. vs. Hipermercado Carrefour*, 27/09/2005, LLBA 2006, 358.

9.5 Unsustainable Commercial Practices as Unfair Practices

The United Nations has highlighted the importance of combating unsustainable business practices through several instruments, chief among them the United Nations Guidelines for Consumer Protection. The guidelines promote the establishment of sustainable business practices and discourage those with environmentally disruptive effects (points 51, 59, 62 and 71).

Moreover, the UN 2030 Agenda for Sustainable Development includes ensuring sustainable consumption and production patterns as its twelfth goal (SDG12). Some of the targets of SDG12 are:

1. Achieving sustainable management and efficient use of natural resources;
2. Halving *per capita* global food waste at the retail and consumer levels and reducing food losses along production and supply chains, including post-harvest losses;
3. Achieving environmentally sound management of chemicals and all wastes throughout their life cycle, under the agreed international frameworks, and significantly reduce their release to air, water and soil to minimise their adverse impacts on human health and the environment;
4. Substantially reducing waste generation through prevention, reduction, recycling and reuse;
5. Encouraging companies, especially large and transnational companies, to adopt sustainable practices and to integrate sustainability information into their reporting cycle;
6. Promoting public procurement practices that are sustainable, under the national policies and priorities;
7. Ensuring that people everywhere have the relevant information and awareness for sustainable development and lifestyles in harmony with nature;
8. Supporting developing countries to strengthen their scientific and technological capacity to move towards more sustainable patterns of consumption and production;
9. Developing and implement tools to monitor sustainable development impacts for sustainable tourism that creates jobs and promotes local culture and products;
10. Rationalising inefficient fossil-fuel subsidies that encourage wasteful consumption by removing market distortions, under the national circumstances. This includes by restructuring taxation and phasing out those harmful subsidies, where they exist, to reflect their environmental impacts, considering the specific needs and conditions of developing countries and minimising the possible adverse impacts on their development in a manner that protects the poor and the affected communities.

The UN has made explicit reference to certain unsustainable practices: wasting food and energy or releasing chemicals and waste into the atmosphere, water or soil.

We believe that unsustainable commercial practices constitute a supposition of illicit commercial practices that violate the principle of sustainability, recognised in Section 1094 CCC. That is, they violate the right of consumers to access sustainable consumption, fundamentally with respect to the environmental impacts of production, distribution, marketing, after-sales service and disposal of waste and on consumption models and its sustainability in economic, ecological and social terms.[21] Thus, it could be considered an assumption of commercial practice contrary to access to consumption that would suspend the supply of water and sanitation or the provision of electricity to people dependent on it.

In relation to sustainable consumption, we can mention as examples the possibility of action by consumers against commercial practices that violate environmental standards and principles throughout production, distribution, marketing, consumption and disposal of waste. Consumers are also able to act in response to advertising, marketing techniques or other commercial practices that promote unsustainable behaviour or violate rules of environmental protection, natural or cultural heritage or biological diversity. Here it is also interesting to connect this to the right to information, especially through labelling, for example, of genetically modified products, those tested on animals or those that may have some animal product contained as an ingredient. Labelling on the use of packaging, including elements that are difficult to recycle, etc. are also pertinent here.

It is also interesting to evaluate the illegality of the application of the principles of non-regression and progressiveness. Therefore, regressivity or lack of progressivity in practices linked to sustainability may result in a violation of the right to development.[22] In local law, we find regulations such as those that cover the supply of polyethylene bags[23] or the waste management of electrical and electronic equipment (WEEE).

Another pertinent issue is product obsolescence. Obsolescence implies the end of the utility of a product, which is then usually discarded. There are different types of obsolescence:

1. Programmed obsolescence, or the manufacture of products with defects calculated to cause the end of useful life to accelerate consumption of replacement products in an abusive manner;
2. Indirect obsolescence occurs when the products are impossible to repair because the design does not allow it, spare parts are not manufactured or the consumer is not informed about the possibility of repairing the product. Then, when a product fails, the consumer has no choice but to discard it;
3. Obsolescence as incompatibility usually occurs with new operating systems that are not compatible with the products that have reached the end of their useful life;

[21] For more on this, see Pérez Bustamante (2008); Sozzo (2012), pp. 139–166.

[22] Pérez Bustamante (2008), pp. 236–237; Garrido Cordobera (2013).

[23] See, for e.g., Act No. 3147 of the Autonomous City of Buenos Aires and Act No. 13868 of the Province of Buenos Aires.

4. Psychological or perceived obsolescence arises from marketing campaigns that suggest to consumers that certain products should be perceived as obsolete.[24]

Here the regulations on guarantees (Section 11 to 18 LDC) play a fundamental role, especially the existence of technical services and access to spare parts and accessories (Section 12 LDC). To combat obsolescence, in addition to promoting corporate social responsibility and reviewing the legal regimes of product guarantees, from the economic point of view it is recommended to promote the ecological design of products, a circular economy and an economy of functionality that replaces that of the possession.[25] Regarding perceived obsolescence, promoted by advertising and other commercial practices, a broader debate is necessary to focus on a case-by-case basis to determine illicit commercial practices.[26]

The issue also has an important relationship with advertising. Being a type of commercial practice, we argue that advertising that leads to behaviours that are harmful to the environment (Sections 70 and 81 of Law No, 26522 on Audio-visual Communication Services) are abusive (Section 1101 CCC). Moreover, under Section 1102 CCC, consumers may claim for the cessation of illicit advertising and compensation for damages suffered. We thus find an interpretation consistent with the other rules of Latin American legal systems that go in that direction (such as Brazil's Code of Consumer Defence and Paraguay's Consumer Protection Act, among others). We can also highlight here some initiatives of self-control and self-regulation in this line, such as the Code of Ethics and Advertising Self-regulation of the civil association Consejo de Autorregulación Publicitaria (CONARP), under the auspices of the Asociación Argentina de Publicidad (AAP) and the Cámara Argentina de Anunciantes (CAA). Section 14 of the code states:

Advertising Must Take Care of the Environment.

1. Consequently, it shall not, directly or indirectly, generate or stimulate:
   a) The material and visual contamination of the environment.
   b) The alteration of natural and urban landscapes.
   c) The obstruction of the view of public places, tourist attractions and monuments.
   d) Excess noise that threatens the health and well-being of people.

2. In advertising and other forms of commercial communication, such as labels and product packaging, containing environmental statements:
   a) Reference should be made to the current characteristics of the product and its impact on the environment, duly verified by technical and scientific methods.

---

[24]European Economic and Social Council (2013).

[25]The closed cycle or circular economy is a production system that maximises the useful life of the products and minimises the waste and loss of valuable resources. In the opposite direction to the current linear use-and-throw paradigm, closed cycle manufacturing achieves life cycle efficiency by facilitating: maintenance and repair; the reconditioning and re-manufacturing, and; dismantling and fine recycling. See: Salcedo Aznal (2014).

[26]For more on this, see: Estevarena (2016), p. 281.

b) Ambiguity will be avoided, explaining its real scope with precision.
c) The statements must be clear and legible.
d) They must clearly indicate the life cycle of the product, with the corresponding justification.
e) Those statements that refer to the absence or reduction of a particular ingredient can be used only if the level of said substance does not exceed recognised limits of contamination.

9.6 Conclusions

From the foregoing discussion, we propose several conclusions, which reflect the fact that the principle of sustainability is a key principle of Consumer Law.

1. The principle of sustainability includes the right of access to consumption and the right to sustainable consumption.
2. The principle of sustainability plays a programmatic role in guiding the formulation, implementation and evaluation of public economic and social policies, especially regarding production and consumption. The principle of sustainability also plays a role limiting private autonomy.
3. Unsustainable business practices constitute one form of illicit business practice.
4. Advertisements that induce behaviours that are harmful to the environment are a form of abusive advertising.

The problems of the environment and consumption are of topical relevance and deserve adequate solutions in different areas and perspectives. As Pope Francis has maintained, the rhythm of consumption, waste and alteration of the environment has surpassed the possibilities of the planet, in such a way that the current lifestyle, being unsustainable, can only end in catastrophe.[27] As he notes:

> Large masses of the population are excluded and marginalized: without work, without horizons, without exit. The human being is considered in himself as a consumer good, which can be used and then thrown away. We have started to promote a 'throw away' culture. It is no longer simply a matter of the phenomenon of exploitation and oppression, but of something new: exclusion. And with exclusion, the individual's belonging to the society in which he lives is affected to the core, since he is no longer merely marginalized or peripheral but left out altogether. The excluded are not merely 'exploited' but are seen as waste, 'leftovers' . . . unbridled consumerism together with inequality is doubly damaging to the social fabric.[28]

[27]Francis (2015).

[28]Francis (2013).

References

Barocelli SS (2015) Principios y ámbito de aplicación del derecho del consumidor en el nuevo Código Civil y Comercial. DCCyE 2015(febrero):63

Bauman Z (2007) Vida de consumo. Fondo de Cultura Económica, Buenos Aires

Chamatropulos DA (2015) Ley de Defensa del Consumidor comentada. La Ley, Buenos Aires

Estevarena E (2016) Un diálogo de consumo y desarrollo sustentable. In: Barocelli SS (ed) Impactos del nuevo Código Civil y Comercial en el Derecho del Consumidor. Diálogos y perspectivas a la luz de sus principios. Facultad de Derecho (UBA), Buenos Aires, p 281. Available at http://www.derecho.uba.ar/investigacion/investigadores/publicaciones/barocelli-impactos-del-nuevo-Codigo-civil-y-comercial-en-el-derecho-del-consumidor.pdf. Accessed 22 Feb 2018

European Economic and Social Council (2013) Towards more sustainable consumption: industrial product lifetimes and restoring trust through consumer information. CCMI/112-EESC-2013-1904. Available at: https://www.eesc.europa.eu/en/our-work/opinions-information-reports/opinions/towards-more-sustainable-consumption-industrial-product-lifetimes-and-restoring-trust-through-consumer-information. Accessed 22 Feb 2018

Francis (2013) Apostolic Exhortation 'Evangelii Gaudium'. Available at http://www.vatican.va/holy_father/francesco/apost_exhortations/documents/papa-francesco_esortazione-ap_20131124_evangelii-gaudium_sp.html. Accessed 22 Feb 2018

Francis (2015) Encyclical letter 'Laudato si'. Available at http://w2.vatican.va/content/francesco/es/encyclicals/documents/papa-francesco_20150524_enciclica-laudato-si.html. Accessed 22 Feb 2018

Frustagli SA, Hernández CA (2015a) Prácticas comerciales abusivas. In: Stiglitz GA, Hernández CA (eds) Tratado de Derecho del Consumidor, vol I. La Ley, Buenos Aires, p 593

Frustagli SA, Hernández CA (2015b) Comentario al Artículo 1096. In: Garrido Cordobera L, Borda A, Alferillo PF (eds) Código Civil y Comercial. Comentado, anotado y concordado, vol II. Astrea, Buenos Aires, p 363

Garrido Cordobera LMR (2013) El desafío ambiental del SXXI: la aplicación de los principios de no regresión, de solidaridad y pro homine. Available at http://www.acadervs.org.ar/doctrina/el-desafio-ambiental-del-siglo-xxi.-la-aplicacion-de-los-principios-de-no-regresion-de-solidaridad-y-pro-homine/at_download/file. Accessed 22 Feb 2018

Hernández CA, Frustagli SA (2012) Aspectos relevantes de la relación de consumo en el Proyecto de Código Civil y Comercial de 2012. Proyecciones de sistema sobre el régimen estatutario de daños al consumidor, JA 2012-IV–807

Japaze MB (2009) La publicidad comercial y las prácticas comerciales. In: Rusconi D (ed) Manual de Derecho del Consumidor. Abeledo–Perrot, Buenos Aires, p 164

Kemelmajer De Carlucci A (2015) Prácticas abusivas en los contratos de consumo. Análisis desde la nueva legislación argentina. Revista de Derecho. Segunda época. Año 10. N.° 11 (julio 2015), 105–134. Available in http://revistas.ucu.edu.uy/index.php/revistadederecho/Section/viewFile/726/716. Accessed 22 Feb 2018

Lorenzetti R (2009) Consumidores, 2nd edn. Rubinzal-Culzoni, Santa Fe, p 158

Pérez Bustamante L (2008) Los derechos de la sustentabilidad. Desarrollo, consumo y ambiente. Colihue Universidad, Buenos Aires

Pérez Bustamante L (2011) El consumidor frente a la protección de la salud. In: Regazzoni C (ed) Salud y Conciencia. Pública Fundación Sanatorio Güemes, Buenos Aires, p 159

Rusconi D (2013) Prácticas comerciales ilícitas en Brasil. La Ley 2013-E, 312; AR/DOC/2329/2013

Salcedo Aznal A (2014) Las nuevas actitudes hacia el consumo y la producción: las mejores prácticas en el ámbito del consumo colaborativo y la obsolescencia planificada (Un mundo en tránsito). Paper presented at IDEARIA–12° Encuentro de Economía Alternativa y Solidaria, Córdoba (Spain), 30 April–3 May 2015. Available at http://base.socioeco.org/docs/estudio_20v.f.pdf. Accessed 22 Feb 2018

Santarelli FG (2015) El contrato de consumo en el Código Civil y Comercial. In: Stiglitz RS (ed) Contratos en el nuevo Código Civil y Comercial. Parte general, vol I. La Ley, Buenos Aires, p 370

Sozzo G (2012) Consumo digno y verde: humanización y medioambientalización del Derecho del Consumidor (sobre los principios de dignidad del consumidor y de consumo sustentable). Revista de Derecho Privado y Comunitario N° 2012-3 Rubinzal-Culzoni. Santa Fe, pp 139–166

Stiglitz RS (2014) Lealtad comercial, prácticas comerciales abusivas y publicidad en el Código Civil y Comercial de la Nación. Sup. Esp. Nuevo Código Civil y Comercial 2014 (Noviembre), 103

XXIV National Conference on Civil Law, Dispatch B, point 3, Conclusions of commission n ° 8 'Consumer Law', held on September 26, 27 and 28th, 2013 in the Autonomous City of Buenos Aires

XXVI National Conference of Civil Law, Conclusions of the Commission N ° 6, 'Consumer Law', held at the National University of La Plata in 2.017. http://jornadasderechocivil.jursovs.unlp.edu.ar/wp-content/uploads/sites/10/2017/10/COMISION-N%C2%B0-6.pdf. Accessed 22 Feb 2018

Chapter 10
Consumer Law and Sustainability: The Work of the United Nations

Ana Cândida Muniz Cipriano

Contents

Abstract The topic of sustainable consumption was discussed during the second reform of the United Nations Guidelines for Consumer Protection in the 1980s. This reform and the inclusion of the theme of ecology and sustainable development in the UN Guidelines were probably a consequence of the discussions in the 1970s about the importance of protecting the environment and the future of the planet (Conference of Stockholm). Today, the technological developments have so expanded the scope of consumer choice that even the means to contract services and buy products has changed dramatically. Beyond contractual issues, consumers nowadays face other complex challenges: consuming sustainably, addressing the different types of waste, waste production and environmental impact. In view of the need to consider and practice sustainable consumption in our lives, what future actions in the field of consumer protection and sustainability through the work of the United Nations can we expect?

A. C. Muniz Cipriano (✉)
UNCTAD, United Nations Conference for Trade and Development, Geneva, Switzerland

A. d. Amaral Junior et al. (eds.), *Sustainable Consumption*,
https://doi.org/10.1007/978-3-030-16985-5_10

10.1 Introduction

Considering the development of policies to protect and preserve the environment, it is also relevant to ponder the worldwide advance of consumer protection policy and legislation. While some may contend that consumer protection constrains or inhibits sustainability efforts, this chapter defends the complementarity of the two areas and the importance of both in preserving the environment and protecting consumers. As consumers, we are all entitled to a safe and sustainable environment. This is perhaps the core of the conjunction of these issue areas. The advent of a variety of new technologies and tools that are expanding the range of products and services available to consumers (and the means to access these)—as well as the advance of consumer protection in developing countries—poses several challenges to both environmental preservation and consumer protection. Beyond contractual challenges, consumers face a range of complex facts in making informed decisions when purchasing a product or service, including the importance of considering sustainable consumption.[1] Moreover, consumer protection is vital for environmental protection as it relates to consumer health and safety. The disposal of unsafe and/or defective products has a serious impact on the environment, obliging both businesses and consumers to rethink the manufacture and marketing of these products.

In view of the pertinent links between sustainability and consumer protection (as well as the urgency in discussing them), this chapter explores this complementarity in detail and asks how they have played out within the ambit of the United Nations and what the future holds for progressing both. The chapter begins by detailing sustainability and consumer protection both conceptually and historically. It then discusses how these themes co-relate and why this is important. The final section surveys what future actions we might expect from the United Nations and its Member States in relation to both sustainability and consumer protection.

10.2 Historical Background

The topic of sustainable consumption was discussed during the second reform of the United Nations Guidelines for Consumer Protection (UNGCP) in the 1980s. The reform and the inclusion of the theme of ecology and sustainable development[2] were probably a consequence of the discussions in the 1970s on the importance of protecting the environment and the future of the planet.

---

[1] The right to be informed is one of the very basic rights consumers are entitled to, being enshrined in all consumer protection legislation around the world. Pre-purchase information about products and services allows consumers to make proper decisions, which nowadays includes awareness of a product's sustainability status.

[2] The UNGCP were expanded to include a new section on sustainable consumption, Section H of the current version (2015) in Resolution E/1999/INF/2/Add.2 of 26 July 1999.

The United Nations Conference on the Human Environment (the Stockholm Conference of 1972) was the first UN summit to focus exclusively on environmental issues. It was a result of United Nations General Assembly Resolutions 2398 (XXIII) and 2581 (XXIV), by which the member countries decided to convene a global conference 'to serve as a practical means to encourage, and to provide guidelines to protect and improve the human environment and to remedy and prevent its impairment.'[3]

The Stockholm Conference of 1972 launched the international agenda on ecology and, from its work, an official statement, the Universal Declaration on the Protection and Preservation of the Human Environment was published, initially based on ideas proposed by the United Nations Educational, Scientific and Cultural Organization (UNESCO). Years later, an independent commission, the UN World Commission on Environment and Development, was created to analyse and provide ideas on the theme of ecology and on problems related to the environment, waste and health. The commission was headed by Gro Harlem Brundtland, former prime minister of Norway, whose work in the field of public health and sciences was acknowledged internationally. The Brundtland Commission was dissolved in 1987 once the official report was presented. That report became a reference for future efforts, defining sustainability as 'development which meets the needs of current generations without compromising the ability of future generations to meet their own needs.'[4]

The Brundtland Report's salience and impact was underscored by the fact that it considered different aspects of society, a variety of needs and concerns, combining economic and social wellbeing with the preservation of the environment for the good of present and future generations. The international and national debates and activities in the field of ecology were very much influenced by the report's publication and mostly the launch of the new (at that time) concept of sustainability. Although based on scientific evidence and supported by academics, as well as civil society, the Brundtland Report's practical effect would become evident years later with the holding of the Rio 92 Conference or Eco 92.

The UN Conference on Environment and Development (Rio 92 Conference), held in the city of Rio de Janeiro, Brazil, is arguably one of the most important, relevant and successful conferences ever held under the auspices of the United Nations. It was perhaps the only conference ever to bring authorities, representatives, and heads of state from basically all Member States (172, 108 at level of heads of state or government and about 2400 representatives of NGOs; 17,000 people attended the parallel NGO Forum).[5] Apart from its impressive numbers, the Conference also had significant lasting impact in: (a) gathering most UN Member States to engage in concrete and effective actions to preserve the planet and its resources; (b) drafting and implementing a specific agenda, a so-called wide-ranging blueprint for action to achieve sustainable development worldwide; and (c) becoming a

[3]Report of the United Nations Conference on Human Environment (1972), p. 37.

[4]United Nations (1987), p. 16.

[5]United Nations Conference on Environment and Development (1992a).

reference for all subsequent conferences related to environment, human rights, commerce, development and the like.

The Rio 92 Conference saw three official documents adopted: Agenda 21, the Rio Declaration on Environment and Development, and the Statement of Principles for the Sustainable Management of Forests. Agenda 21 is a plan of action to be taken by all Member States, related stakeholders and organisations of the United Nations System. Ten years later, both implementation of Agenda 21 and the commitments to the Rio principles, were strongly reaffirmed at the World Summit on Sustainable Development (WSSD) which was held in Johannesburg, South Africa, in 2002.

Despite initial controversies surrounding the term 'sustainability', it is currently and commonly used as an important reference to environmental care and preservation. Following the trend established at Rio 92, the Johannesburg Conference reaffirmed the commitments made previously. The WSSD was followed by the launch of the United Nations Decade for Education for Sustainable Development starting in 2005, led by UNESCO. At that point, the context of a constantly changing world with economic, educational and environmental disparities, where there are billions of consumers eager to access products and services and where most of these consumers are to access such products and services for the first time (developing economies) came into sharp relief. It was thus deemed vital that consumers be informed of the full spectrum of their rights, including 'access to non-hazardous products, as well as the right to promote just, equitable and sustainable economic and social development and environmental protection.'[6]

10.3 Consumer Rights and Sustainability

Recent advances in information and industrial technology have produced novel products and services available to consumers, so that even the way services are contracted and products bought has changed considerably. Today, everything is easier and faster. Products and services are almost immediately available, accessible in so far as a purchase is simply one 'click' away. The challenge here is that consumer information and awareness has not kept pace with the development of new services, new products and new methods of purchasing them.

Beyond contractual challenges, these new tools and ways to buy products and services have an important impact in consumer's lives and lifestyles which, in their turn, have also an impact in the environment. The topic of sustainable consumption, different types of waste, waste production and environmental impact must also be considered a result of new ways to purchase products and services, as well as of consumer's choice.

Another topic of concern which has its risks increased by the mentioned new commercial methods is consumer product safety. There are still considerable

[6]United Nations Guidelines for Consumer Protection (2015).

differences among countries regarding the regulation of product safety. It is important to note that not all countries have specific legislation on hazardous products. Although this topic may be disregarded by some, it is relevant to recognise that consumer product safety has, as well, a significant impact in the environment. Depending on the product added value, consumers often simply dispose of defective and hazardous products regardless of the consequences of such waste to the environment. The double standard practised by businesses in developing countries (e.g., the Fiat Stilo case in Brazil, where the local model was fitted with fewer safety features than the same car sold in Europe) also poses important issues as products deemed to be unsafe and hazardous in developed countries—and are often consequently prohibited from sale—are commercialised in developing ones.

These challenges have been addressed by the United Nations, which has developed several actions and policies in favour of environmental protection combined with consumer protection. In the context of consumer protection, the UNGCP—targeted at both businesses and consumers—consider that attention must be given to certain needs, to consumer education, access to information and to the promotion of sustainable consumption patterns. The text of the UNGCP states that unsustainable patterns of production and consumption are the main cause of the continuing deterioration of the environment, which is why all Member States should focus on promoting sustainable consumption patterns. The text also highlights that the policies for sustainable consumption patterns should consider the needs of developing countries.[7]

Furthermore, on 25 September 2015, the United Nations launched 17 Sustainable Development Goals (SDGs) under the banner Transforming our World: The 2030 Agenda for Sustainable Development (the Global Goals for Sustainable Development is another term used). Meant to succeed the Millennium Development Goals (MDGs) agreed at the Millennium Summit of the United Nations in 2000, the SDGs are a global call to action to eradicate poverty, protect the planet and ensure that we humans live in peace.

Among the 17 objectives, there are seven related to the topic of consumer rights and protection:

- SDG 4–Quality education;
- SDG 10–Reduced inequality;
- SDG 11–Sustainable cities and communities;
- SDG 12–Responsible consumption and production;
- SDG 13–Measures to combat climate change;
- SDG 16–Peace, justice and effective institutions; and
- SDG 17–Partnerships for achieving the goals.

---

[7]Here it is important to consider the status of developing countries and the impact of having new consumers, meaning citizens that are purchasing certain products and services for the first time. Such consumers deserve not only to access goods and services but also access to information, to the protection of their rights, such as the right to redress, to justice and to education.

Both the UNGCP and the SDGs envisage the promotion of consumer protection policies, considering environmental risks and environmental protection by Member States. Additionally, the themes of environmental protection and sustainable development are included in almost all the conferences and actions carried out by the United Nations.

The department of the United Nations Conference on Trade and Development (UNCTAD) responsible for Competition Law, Consumer Law and consumer policies, has developed several actions and projects including the theme of responsible sustainable consumption. Projects to develop consumer protection in Latin America, Africa, the Middle East, for instance,[8] include actions to promote education and access to consumer information raising awareness of the importance of responsible consumption to ensure the protection of the planet. Actions to promote the preservation of the environment are considered complex because of the different levels of development of Member States. This is undoubtedly a challenge to be overcome. But it is also undeniable that educational actions are the basic and core strategy to inform consumers about the products and services' risks and the different choices available in the market place.

As mentioned, the 2030 Agenda has 17 main goals that seek to achieve wellness for people and the planet, prosperity and universal peace. The 17 SDGs refer to strategic topics of an action plan to enable the achievement of such objectives. Indeed, by eradicating poverty, hunger, providing people access to education, health and decent work, we can achieve peace and prosperity for all.[9] Among the seven SDGs that bear on sustainability, there are four topics directly connected to consumer rights and sustainability: SDG 4 (education), SDG 11 (sustainable cities and communities), SDG 12 (responsible consumption and production) and SDG 13 (climate). Although there are other topics also related to consumer rights, these four directly seek to achieve sustainable goals considering people's actions and responsibilities as citizens and as consumers. This means that these four goals have one common actor: *the consumer*.

10.4 Education, Sustainable Cities and Communities, Responsible Consumption and Production, and Measures to Combat Climate Change

Sustainable consumption comprises both consumers' rights to safe and effective products and services and the preservation of our own planet. Joining these is the consumer right to participate in sustainability efforts through access to information and education. It is relevant to consider that Consumer Law is regarded as one of the 'new law topics' as it is a result of the development of globalisation and capitalism

[8]United Nations Conference on Trade and Development (2017).

[9]Ibid.

itself. In this sense, Consumer Law has been under development at least since the 1930s[10] and many developing countries still lack specific legislation on consumer rights. Consumer rights are considered, by some experts, part of human rights because very basic rights—such as access to water, consumer product safety (consumer health and safety), housing, pharmaceuticals, access to basic products and services (the so-called 'essential' goods and services such as water, for example)—are implied.

It is also important to consider that although access to consumption has been broadened through globalisation, access to information has not followed the same tendency. Worse still, the use of consumer data has been increasing without consumer knowledge or awareness. Yet, even in the age of the internet and the digital economy not all consumers have access to information or education, making consumer education and awareness even more relevant. For this reason, consumer protection authorities, businesses and other related stakeholders must face up to a clear set of important challenges that can no longer be ignored or postponed.

Increasing access to products and services provided by the development of new technologic tools are, because of consumption, producing even more waste than before.[11] When we refer to consumer rights we refer to the rights of individuals, as well as a collective right of a modern society—a society moved by consumption and driven by globalisation.[12] Therefore, consumers play a major role modern society as one of the most important drivers of economic activity. In an era when more and more individuals can access products and services for the first time (here we refer to developing economies), it also becomes a challenge to educate and inform these new consumers.[13] who, most of the times, are being acknowledged as citizens for the fact of being consumers, for the fact of being able to purchase products and services.

That is the reason why it is a relevant challenge to address consumer protection and consumer rights as the climate changes (SDG 13), because it also affects consumer rights, consumer welfare.[14] According to the United Nations, climate change affects every country on every continent, having serious impacts in national economies, in people's lives, in people's health and wellbeing. Such changes and impacts are mostly irreversible, provoking extreme weather conditions and events, such as rising sea level, changing weather patterns, drying areas.[15]

Considering that such impacts and disasters are the result of individual activities, how are we to reverse this situation? How are we to guarantee people's consumer rights (including both the right to access products and services and basic welfare)? As global warming and climate change affects us all, regardless of nationality,

---

[10]It is important to note that the North American Federal Trade Commission was founded in 1914.

[11]European Environment Agency (2016).

[12]Bauman (2007) and Santos (2001).

[13]Muniz Cipriano (2012).

[14]UNCTAD (2017).

[15]UNCTAD (2016), p. 149.

economic status or borders, methods and actions must be coordinated in international cooperation.

The UNGCP are a clear reflection of the concerns related to the impact of climate change and the urgency of developing and implementing sustainable actions. The UNGCP, annexed to UNGA Resolution 70/186, make specific reference to the needs of developing countries, including the setting of the SDGs and the MDGs:

> Guideline 1 of the United Nations Guidelines for Consumer Protection (UNGCP), which states that: 'Consumers should have the right to promote just, equitable and sustainable economic and social development'.

Setting the scope of application and general principals of the UNGCP, its first section clearly recognises that consumer rights are collective rights, therefore rendering importance to ethics, transparency, sustainability, as well as international cooperation.

We may ask, then, how it is possible to fulfil consumers 'legitimate needs' (Section III on General principles, Guideline 5) such as access to essential goods and services, the protection of vulnerable and disadvantaged consumers and the protection from hazards to consumer's health and safety? The answers are provided, in fact, within the guidelines text, as follows:

> Guideline 5: e) Access by consumers to adequate information to enable them to make informed choices according to individual wishes and needs; f) Consumer education on environmental, social and economic consequences of consumer choice; i) The promotion of sustainable consumption patterns

10.5 The Answers: What to Expect in the Future?

The United Nations agenda for sustainability and the recent Conference on Trade and Development mandate on consumer protection and policy foresees a strategic plan that Member States—as well as stakeholders such as businesses—are expected to develop in Section H of the document as follows:

> **H. Promotion of sustainable consumption**
>
> 49. Sustainable consumption includes meeting the needs of present and future generations for goods and services in ways that are economically, socially and environmentally sustainable.
>
> 50. Responsibility for sustainable consumption is shared by all members and organizations of society, with informed consumers, Member States, businesses, labour organizations and consumer and environmental organizations playing particularly important roles. Informed consumers have an essential role in promoting consumption that is environmentally, economically and socially sustainable, including through the effects of their choices on producers. Member States should promote the development and implementation of policies for sustainable consumption and the integration of those policies with other public policies. Policy making by Member States should be conducted in consultation with businesses, consumer and environmental organizations and other concerned groups. Business has a responsibility for promoting sustainable consumption through the design, production and distribution of goods and services. Consumer and environmental organizations have the responsibility for promoting public participation and debate on sustainable consumption, for

informing consumers and for working with Member States and businesses towards sustainable consumption.

Guideline number 50 is a comprehensive path that all stakeholders must follow if consumer rights and environmental protection are to be guaranteed. Education and information are the key measures.

The implementation of awareness campaigns and education tools depend heavily on the cooperation of stakeholders—Member States, representatives of consumer protection and environment civil society groups, businesses and of course consumers. The work that has been done by the United Nations engages all actors as no matter which field one operates in or whether one is a producer, a retailer, a civil servant or a consumer, all are subject to the impacts of climate change. This is the reason why most of the foreseen programs, tools and actions include cooperation among all possible stakeholders, as laid out in the following guidelines:

51. Member States, in partnership with businesses and relevant organizations of civil society, should develop and implement strategies that promote sustainable consumption through a mix of policies that could include regulations; economic and social instruments; sectoral policies in such areas as land use, transport, energy and housing; information programmes to raise awareness of the impact of consumption patterns; removal of subsidies that promote unsustainable patterns of consumption and production; and promotion of sector-specific best practices in environmental management.

52. Member States should encourage the design, development and use of products and services that are safe and energy and resource efficient, considering their full life cycle impacts. Member States should encourage recycling programmes that encourage consumers to both recycle waste and purchase recycled products.

53. Member States should promote the development and the use of national and international environmental health and safety standards for products and services; such standards should not result in disguised barriers to trade. . . .

56. Member States should promote awareness to the health-related benefits of sustainable consumption and production patterns, bearing in mind both direct effects on individual health and collective effects through environmental protection.

While Agenda 21 was an action plan that reinforced the sustainability concept in the public mind, the UNGCP provide a detailed path for all related stakeholders to follow to ensure proper regulatory mechanisms and the necessary technologies, policies and incentives (for market access, for instance) to reduce pollution and of the depletion of resources are developed.

Access to a proper, and safe environment—and one that continues for future generations—may be a challenge when we consider: (a) the development of new technologies (always encouraging consumers to buy new versions and updates of products); (b) that rapidly falling prices for certain products and services facilitates access for new consumers; and (c) that developing countries are the largest source of new markets and new consumers. Such facts highlight the importance of promoting researches on consumer behaviour and consumer patterns. By knowing consumer behaviour, proper policies and education mechanisms can be developed to inform consumers of their rights and actions (impact in the environment) so that they can make informed decisions when purchasing products and/or services.

Another very sensitive topic is related to the fact that the 'new consumers' are mostly vulnerable and disadvantaged. They are therefore entitled to more protection given the specific circumstances they may face, such as illiteracy and poverty, Rural consumers, the elderly and children, among others, may be particularly vulnerable. It is these consumers in particular that are subject to information asymmetries (as against large business and institutions) and also may lack general capacity to understand their rights or the impact of their consumption decisions. They are, in addition, particularly vulnerable in so far as they may not be fully aware of contract conditions or the implications of particular policies or regulations.

For instance, the reality of life for a consumer from a rural area in a developing country means complete information regarding her rights as a consumer or citizen is likely to be sorely lacking. It is also important to consider the relevance of environmental preservation for a rural consumer whose subsistence depends on the surrounding environment, so that she can pursue her household's economic objectives to the fullest. These distinct circumstances justify the development of policies which consider all consumers' rights. It is unfair to expect that consumers from developing countries, for instance, will simply follow developed country policies on sustainable consumption that might limit access to certain goods and services. This is why education is important, if not crucial, in the process of implementing sustainable consumption patterns.

Consumer access to products and services in developing countries will—in the first instance—focus on necessities like food or pharmaceutical products, rather than, say, electronics. Nevertheless, as economies develop new products and services become popular as they become more affordable to the average household, meaning more and more consumers have access to products and services previously considered to be 'luxuries'. This can account for the disparities among consumers, and mainly explains why consumers' vulnerabilities and disadvantages cannot be ignored or overlooked.

At this juncture, it is also important to note that the UNGCP urge Member States, businesses and stakeholders to engage in international cooperation actions to develop and implement policies to

> promote sustainable consumption, taking into account social needs, the need for disincentives for unsustainable practices and incentives for more sustainable practices, while avoiding potential negative effects for market access, in particular for developing countries (Guideline 59).

Moreover, the UNGCP acknowledges not only consumer vulnerabilities, but also the disparities among Member States by referring to *common but differentiated responsibilities* (CBDR),[16] a principle of International Environmental Law first implemented at the Rio 92 Conference. It establishes that all Member States are responsible for addressing environmental destruction regardless of its cause. The CBDR principle establishes a shared obligation or duty, while acknowledging differences in economic and social development between Member States. It

[16]United Nations (1992b), p. 4.

therefore seeks to balance countries' responsibilities for environmental impacts, such as climate change.

Once disparities are recognised, a path is opened to foster education initiatives. In view of the mentioned disparities between countries, international cooperation promoting education and awareness is the key to inform all consumers, citizens and businesses of the importance of preserving our environment for the sake of present and future generations—and for the sake of our own health and existence. Sustainable consumption is one of the key actions to preserve and protect our environment and as such, it can only be achieved by promoting education.

The UNGCP dedicates an entire chapter—Section VI (Guidelines 79–94)—to international cooperation. In fact, the differences between countries are an advantage when it comes to sharing experiences. The exchange of information and experience certainly enrich the development of actions and policies. Thus, the UNGCP encourage cooperation including examples, such as the 'setting up or joint use of testing facilities, common testing procedures, exchange of consumer information and education programmes, joint training programmes and joint elaboration of regulation (Guideline 79)'.

Among the various topics that should be addressed when discussing sustainable consumption, the themes of essential goods, health products and safe products are of most importance. Consumers have the right to be informed about the risks of products and services, about the implications of discarding products as well. The exact definition of 'essential goods' may vary but the right to access them is undeniable. Likewise, it is undeniable that such products should be safe and free from attributes that might negatively impact the environment. Therefore, the more information on hazardous products (and products that have been prohibited in developed countries) is disclosed and shared, the better. After all we cannot underestimate the fact that products cross borders, not least by tourism.

Cooperation among specialised agencies (consumer, health, metrology, food) is also relevant to combat fraudulent and deceptive cross-border commercial practices, to promote investigations and awareness campaigns. Sometimes just the fact of existing cooperation between authorities prevents businesses from committing fraud, causing harm to the environment or infringing consumer rights. To be successful, any cooperation should engage all related stakeholders—consumers above all. This can be achieved by the participation of consumer protection NGOs, strengthening awareness and capacity.

10.6 Conclusion

The key: education

The means: international cooperation

For some, consumer protection and consumer rights may be considered an incompatible with sustainability or sustainable consumption. However, consumer rights also imply the right to equitable and sustainable economic and social development, and environmental protection contributes directly to this. After all, *we are all consumers*[17] and are all subject to the impacts of climate change, no matter the field we operated in and no matter whether we are a provider, a retailer, a civil servant or a consumer. These effects also transcend borders and ignore economic and other forms of social status. For this reason, consumer protection is entirely compatible with environment preservation.

As consumers, we also have the right to be informed about the risks and consequences of our choices when buying a product or a service. We all, as consumers, have the right to be instructed about the products and services distributed in the market place; about product life cycles and how products are discarded as waste that is potentially very harmful. Therefore, methods and actions must be taken, urgently, and the key here is education, information and transparency.

There are certainly many challenges to be overcome when developing measures and actions related to consumer protection and sustainable consumption. The development of new technologies, rapidly falling prices for certain products and services (facilitating access by new consumers, especially in developing countries) and the promotion social inclusion in developing countries are challenges which must be considered in the process of fostering sustainable consumption. This is the reason why, only by instructing and informing consumers about their rights to access information, to their rights to a safe and sustainable environment, it is possible to revert to environmental harm. Consumers must be informed so that they can be aware of the environmental impact of their choices, so that they can make informed decisions and then new sustainable consumption patterns can be forged. It is a process to change consumption patterns and consumer education is the answer to create new consumption habits.

The key is education and the means are initiatives of international (bilateral, regional, and global) cooperation. International cooperation can overcome discrepancies between countries, fostering the exchange of experience as we are all inserted in different realities. These differences can be reverted into positive results by the exchange of consumer information and education programmes, the setting up or the joint use of testing facilities, common testing procedures, joint training programmes and joint elaboration of regulations and initiatives. At this point of urgency to protect our environment and rethink waste disposal, there cannot be any excuse regarding available means. That is when the exchange of experiences becomes valuable among nations.

There is no doubt that unsustainable consumption patterns are one of the major causes of environmental harm. There is no doubt that the consumption of unnecessary goods and the distribution of hazardous products in the marketplace have a significant negative impact on the environment. In this case, industrialised countries

[17] Kennedy (1962).

must take the lead in developing sustainable consumption patterns. International cooperation is the means to overcome discrepancies and foster the exchange of experience. International initiatives focused on education programmes is the basic tool to achieve the SDGs, engaging all actors to guarantee basic human needs, reduce inequality and thus preserve our environment.

References

Bauman Z (2007) Consuming life. Polity Press, London

European Environment Agency (2016) The Dobris assessment. https://www.eea.europa.eu/publications/92-826-5409-5/page036new.html. Accessed 12 Jan 2018

Kennedy JF (1962) Special message to the congress on protecting the consumer interest by. http://www.presidency.ucsb.edu/ws/?pid=9108. Accessed 10 Jan 2018

Muniz Cipriano AC (2012) La sociedad brasileña y los cambios económicos a partir de la perspectiva de la Sociología del Derecho. In: Reflexiones sobre Derecho Latinoamericano.: Estudios en homenaje a la Profesora Lídia M. Rosa Garrido Cordobera, vol 6. Editorial Quorum, Buenos Aires

Santos M (2001) Por uma outra globalização: do pensamento único à consciência universal, Editora Record 13ª Edição, São Paulo

UNCTAD (2016) Manual on consumer protection. UNCTAD, Geneva, p 149

UNCTAD (2017) Achieving the sustainable development goals through consumer protection. http://unctad.org/en/PublicationsLibrary/ditccplp2017d2_en.pdf. Accessed 16 Mar 2018

United Nations (1972) Report of the conference on human environment, 1972. http://www.un-documents.net/aconf48-14r1.pdf. Accessed 10 Dec 2017

United Nations (1987) Report of the World Commission on environment and development: our common future, 1987. http://www.un-documents.net/our-common-future.pdf. Accessed 10 Dec 2017

United Nations (1992a) Conference on environment and development, 1992. http://www.un.org/geninfo/bp/enviro.html. Accessed 10 Dec 2017

United Nations (1992b) Agenda 21, the Rio Declaration on environment and development, 1992. https://sustainabledevelopment.un.org/outcomedocuments/agenda21. Accessed 10 Dec 2017

United Nations Guidelines for Consumer Protection (2015) Item 5, letters a, b and c

Further Reading

Bauman Z (2005) La Globalización, Consecuencias Humanas. Fondo de Cultura Económica de España, Buenos Aires

Castells M (2005) A Sociedade em Rede, Era da Informação: Economia, Sociedade e Cultura, Volume I, 8ª Edição. Editora Paz e Terra, São Paulo

Lehtonen M (2004) The environmental social interface of sustainable development: capabilities, social capital, institutions. Ecol Econ 49:199–2014

Chapter 11
International Trade in Environmental Goods and Services and Sustainable Production and Consumption

Guilherme José Pastana de Figueiredo

Contents

Abstract Recent decades have been marked by a growing awareness of the importance of environmental issues. The interrelation between trade liberalization and protection of the environment is becoming increasingly clear, although a balance between them has proved difficult to strike. The desire for sustainable consumption can play an important role in the search for this balance, helping in the adoption of eco-friendly production and consumption patterns. In this scenario, environmental goods and services (EGS) have great potential because they are directly related to the generation of positive externalities (e.g., water, soil and air decontamination) and the reduction of negative externalities (e.g., reuse of materials and correct disposal of chemical products). The expression 'environmental goods and services', however, is not clearly defined; its meaning varies according to the context, as well as the emergence of new technologies. This chapter takes up the question of EGS in the

G. J. P. de Figueiredo (✉)
University of São Paulo Law School, São Paulo, Brazil

A. d. Amaral Junior et al. (eds.), *Sustainable Consumption*,
https://doi.org/10.1007/978-3-030-16985-5_11

context of trade. In so doing, it explores the following five themes: (1) the various definitions of EGS; (2) environmentally preferable goods and services and sustainable consumption; (3) the characteristics of the environmental goods and services sector; (4) the treatment of the subject matter in international trade agreements and; (5) future trends.

11.1 Introduction

Recent decades have been marked by growing consciousness about environmental issues at the international level. Statements about the importance of environmental protection have been replicated in several international declarations of United Nations, but not only. International trade agreements have also enacted legal provisions granting exclusive rights to goods and services produced or performed through sustainable practices. Having in mind the enforceability of trade agreements contrary to international non-binding instruments, International Economic Law may be an important driver in enhancing sustainable production and consumption. The chapter details the concept of environmental goods and services (EGS) in the field of International Economic Law and sheds light on the growing number of free trade agreements that grant benefits to EGS.

The importance of enhancing sustainable practices gained prominence in 1972 because of the United Nations Conference on the Human Environment held in Stockholm, Sweden. Twenty years later, the discussion deepened at the United Nations Conference on Environment and Development held in Rio de Janeiro. The outcome of this Conference, the Rio Declaration on Environment and Development, highlighted the importance of promoting sustainable production and consumption patterns. Noteworthy initiatives that have been recently adopted towards the same end are the 10-Year Framework of Programs on Sustainable Consumption and Production (2012) and the 2030 Agenda for Sustainable Development (2015).

Incentives to sustainable consumption and production patterns may play an important role in the search for sustainable development. In this scenario, EGS have great potential to contribute to the achievement of sustainability because they are directly related to the generation of positive externalities (e.g., water, soil and air cleaning, remediation and decontamination) and the reduction of negative externalities (e.g., reuse of materials and correct disposal of chemical products). However, given that the EGS sector involves a number of essential public services, it is important to assure their universalization and their provision at accessible prices, goals that are not always easy to achieve considering the possible conflicts with some provisions of international trade agreements.

The chapter is divided in two. The first part will address different definitions of EGS, environmentally preferable goods and services and sustainable consumption (and their interrelationship) and, finally, the characteristics of the EGS sector. The second part will then focus on the legal provisions in several relevant trade agreements.

11.2 The Multiple Conceptualizations of Environmental Goods and Services

The term EGS may designate different goods and activities depending upon the context, which may cause some confusion for those unfamiliar with the subject. Thus, it is useful to first make a brief analysis of the various conceptualizations of EGS that have been applied in other contexts to then later distinguish those from the one adopted in the present work.

In the field of Environmental Law, environmental services are usually treated as synonymous with ecosystem services. Ecosystem services can be understood as the benefits obtained from ecosystems, such as the capture of carbon dioxide and its conversion into oxygen (photosynthesis), the maintenance of relative humidity, the prevention of erosion and silting of rivers by the roots of trees, to name but a few. Whereas some authors consider environmental and ecosystem services to be synonymous, others have defined environmental services as based on ecosystem services, but propose a differentiation. According to this second current of thought, environmental services consist only of human activities that favor the provision of ecosystem services, that is, of actions to improve or maintain an ecosystem.

The two concepts of environmental services are often associated with the idea of payment for environmental services (PES), which involves an economic quantification or valuation of the positive externalities generated by ecosystem services. The appropriateness of paying for environmental services or the very concept of ecosystem services *qua services* are controversial issues. Within the context of PES and ecosystem services, environmental goods are understood as natural resources or as goods that compose an ecosystem (e.g. trees, flowers, insects, wild animals etc.).

Despite the recognized conceptualizations of EGS mentioned above, the approach adopted in this chapter is neither connected to the idea of PES, nor does it have a direct relation with the notion of ecosystem services. In the present work, environmental services stand for activities that have a positive impact on the environment by preventing, evaluating or remedying some form of pollution or contamination. From this perspective, some examples of environmental services are sewage, elimination of residues, urban cleaning, control of air pollution and remediation and cleaning of soil and water. Environmental goods, in turn, are goods that have a positive impact on the environment or are essential for the provision of environmental services (e.g. solar water heaters, solar cells and devices for the removal of heavy metal ions for industry uses). Considering that there is no consensus on the meaning of EGS, both conceptualizations above should not be regarded as conclusive definitions of EGS, but merely as a parameter that provides an outline for further discussion in the subsequent sections.

11.2.1 The Close Interrelation of Environmental Goods and Environmental Services

The performance of most environmental services depends upon the use of goods. For the treatment of sewage, for example, a lot of equipment, filters, chemicals, etc. are required. Even simpler services such as air pollution control may depend upon some specific instruments for monitoring and measuring the emission of pollutants.[1]

As much as environmental services depend on goods, the same condition holds the other way around. Many environmental goods directly involve the provision of services, and it is quite common for this provision to entail the sale of goods, to the advantage of companies who offer them in the market. In this sense, a study from the Swedish National Board of Trade considered the environmental goods sector a good example of the intertwined relationship between goods and services, given that manufacturers of environmental goods need to be able to deliver a number of services to sell their stocks and maintain market competitiveness.[2] The Swedish National Board of Trade also clarifies that environmental services are not always specifically identified in statistical studies, as it is common for companies not to charge separately for the goods and the services related to them.[3]

Given the strong link between environmental goods and services and the difficulties in establishing a clear division between them, it is common for these themes to be dealt with together. Similarly, in many cases, statistical studies and data jointly analyze environmental goods and services. The trend can be seen in studies by the World Trade Organization (WTO), the successor to the General Agreement on Tariffs and Trade (GATT), the United Nations Conference on Trade and Development (UNCTAD), the Organization for Economic Co-operation and Development (OECD) and several other organizations, as well as in ongoing trade negotiations, such as the Environmental Goods Agreement (EGA) and the Trade in Services Agreement (TiSA), etc.

The OECD/Eurostat Informal Working Group defines the environmental goods and services industry as 'activities which produce goods and services to measure, prevent, limit, minimize or correct environmental damage to water, air and soil, as well as problems related to waste, noise and eco-systems.' Under this definition, goods and services that are less dangerous or less harmful to the environment are included in this industry.[4] This definition is close to the one adopted by Asia-Pacific Economic Cooperation (APEC), which resulted in the 2012 Vladivostok Declaration, and by the Parties that negotiate the EGA and the TiSA. Meanwhile, this

---

[1]According to Steenblik et al. (2005): '[M]any of the goods included on the APEC or the OECD list of environmental goods are used in the performance of environmental services. These include, in particular, items for holding, conveying, treating and filtering liquids, and instruments for monitoring and measuring' (p. 2).

[2]National Board of Trade (Kommerskollegium) (2014), p. 3.

[3]Ibid, p. 4.

[4]See: OECD (1999).

concept sets a broader view of this industry than the definition adopted in the classifications proposed within the United Nations Central Product Classification and the GATT/WTO.

11.2.2 Environmental Goods and Services, Environmentally Preferable Products and Sustainable Consumption

Most environmental issues—such as pollution in its most varied forms (air, water, sound, visual and soil pollution), the excess of litter produced, the scarcity of natural resources, the destruction of the ozone layer, and global warming, among others—have become much more serious in recent decades. With increasing environmental awareness and the adoption of stricter environmental standards, companies have found greater incentives to seek solutions to environmental problems.

Until the end of the twentieth century, solutions to combat environmental issues used to focus on the end of the production lifecycle ('end-of-pipe' solutions). They dealt mainly with the waste/pollutants generated, as with sewage treatment and the installation of filters in industrial chimneys. Throughout the time, new technologies were developed, as much as goods and services began to focus on seeking energy efficiency, reducing the use of raw materials, increasing productivity and using less environmentally harmful materials.

Conversely, there was a growing concern about sustainable consumption and production, defined in a study by the United Nations Environment Program (UNEP) as the use of services and products that respond to basic needs and bring a better quality of life in a manner that does not hamper the needs of future generations. This concept considers the whole lifecycle of the service or product. Resource use and the emissions of waste and pollutants are the main aspects considered.[5] Therefore, a change of perspective took place. Companies no longer concerned themselves only with dealing with existing environmental problems, but also began to search for the means to avoid or minimize the emergence of such problems. Preventive action became fundamental.[6]

These new solutions have, in many cases, a more beneficial effect on the environment than end-of-pipe solutions, besides the fact that they are more efficient (it is better to avoid pollution than to pollute and seek ways to clean up). In this

---

[5]See: UNEP (2010), p. 44.

[6]According to a study by the OECD, 'the ability of some service providers to offer an integrated package of environmental technologies to address complex environmental problems may be spurring the move away from end-of-pipe solutions to those based on prevention'. See Steenblik et al. (2005), p. 19. This trend had already been observed by the OECD in studies from the 1990s: 'Description and discussion of the [environmental] industry which focuses mainly on "end-of-pipe" technologies and clean-up services will not capture structural change in the industry as regulation and incentives move towards reducing resource use, minimizing waste, and adopting cleaner technologies and production processes'. OECD (1996), p. 5.

context, it emerges the possibility of differentiating between, on the one side, environmental goods and services and, on the other side, environmentally preferable products (EPP).

Although there is also no consensus on the definition of EPPs, it can be deduced from the very term that these products are those with better (or less damaging) consequences to the environment, when compared to similar ones. In this sense, the US General Services Administration (GSA) defines EPPs as products less harmful to human health and the environment when compared with other products with the same purposes. The criteria for comparison may involve 'raw materials, manufacturing, packaging, distribution, use, reuse, operation, maintenance, and disposal' of products. Under the GSA, EPPs need to have more than one environmentally friendly attribute.[7]

Simply put, EPPs are 'goods or services where the environmental benefits are derived in the course of their production, use and disposal'.[8] That said, products derived from sustainable agriculture, sustainable forestry products and ecotourism services would be examples of EPPs,[9] although they do not qualify as environmental goods and services under many classifications. While some studies propose a joint analysis of goods and services, which is justifiable from an environmental perspective, their distinction proves to be relevant when interpreting the commitments undertaken by countries in international agreements.

The existence of a clear definition of the concept of environmental goods and services is relevant in as much as international negotiations must seek adequate solutions for the emergence of new services. Considering that the search for environmentally preferable solutions can take place in practically any area—including tourism and leisure, transport, civil construction, etc.—another issue that deserves attention is the heterogeneity of the sector. Thus, in addition to the challenges posed by the emergence of future services in the category of environmental services, the existing services themselves are not easy to classify.

On the one hand, to consider all new goods and services with more favorable environmental consequences as environmental goods and services would be inaccurate. On the other, ignoring such advances and not promoting some form of differentiated treatment of these new solutions does not seem to be an adequate response either. Thus, part of the difficulty in international negotiations concerns the delimitation of the environmental services sector, as there is a growing movement for the inclusion of environmental solutions in other sectors of services. Some services offer so many benefits for the environment that it would not be unreasonable to classify them both under their original sector, as well as under new categories of environmental services. Sustainable energy supply, for instance, can be classified under different categories. It is worth noting that APEC already considers solar panels to be environmental goods.

---

[7]See: US General Services Administration (2017).

[8]Lendo (2005), p. 10.

[9]See: Ferrier (2004).

This was a trend in the negotiations between the United States and the European Union for the conclusion of the Transatlantic Trade and Investment Partnership Agreement (TTIP). In this sense, under the EU Position Paper, published on 7 January 2015, which involved the trade and sustainable development chapter of the TTIP:

> **III. Trade and Sustainable Development—Environmental aspects...**
> 2. Climate change and green goods and services
> Building on the article on Multilateral environmental governance, this article would cover the following key elements: ...
> 4) the shared objective of the Parties to facilitate and promote trade and investment in environmental goods and services, such as renewable energy goods and related services and energy efficient products and services, including through addressing nontariff barriers related to such goods and services, the adoption of policy frameworks conducive to the deployment of best available technologies, and through the promotion of standards that respond to environmental, climate, and economic needs.[10]

In some cases, the provision of an environmental service itself may result in other goods or services. For example, biomethane—a high-power fuel which can generate electricity and even replace natural gas—is produced from the collection and treatment of urban solid waste, sugar and alcohol waste, sanitary effluents, animal waste or various industrial effluents (typical examples of environmental services).[11]

Finally, it must also be considered that some of the main environmental services are essential services, which is why opening up these markets can have a direct (positive or negative) impact on the quality of life of the population and the environment. This issue also affects the very definition of environmental services, especially regarding the inclusion of water treatment and supply in this sector. On this matter, it is worth noting that the Services Sectoral Classification List[12] does not explicitly include water collection, purification and distribution services under any specific sector or subsector, and this raises the question of whether such services are comprehended by the WTO/GATS negotiations or not. Although it is reasonable to classify water treatment and supply as an environmental service when considering the water cycle, given the connection between this service and the collection and treatment of sewage,[13] many countries, especially developing ones, are fairly

---

[10]See: European Union (2015).

[11]See Associação Brasileira de Biogás e Biometano (2015).

[12]The Services Sectoral Classification List (MTN.GNS/W/120 or W/120) is a list of services divided in 12 sectors and several sub-sectors covered under the General Agreement on Trade in Services (GATS). It was compiled in 1991 during the eighth round of multilateral trade negotiations (Uruguay Round) conducted within the framework of the GATT. Its main purpose was to facilitate the negotiations.

[13]It is worth adding that this is the position advocated by the European Union in its trade negotiations.

According to Waglé et al. (2005), p. 17: 'the E.U. has been actively pursuing the liberalisation of environmental services, and also sectors that have bearing on water distribution and consumption, arguing that its proposal views the entire water cycle holistically with both water distribution and waste-water treatment falling under "resource use".'

apprehensive to regard them as such, as they involve a strategic natural resource directly associated with human life and health.[14] This is one of many examples that demonstrate that negotiations regarding the definition of environmental services involve not only technical criteria but also specific characteristics of the services rendered, which makes the analysis of the topic even more complex.

11.3 International Trade Agreements, Environmental Goods and Services and Sustainable Consumption

In 1944, at the United Nations Monetary and Financial Conference (Bretton Woods Conference), the regulation of the post-World War II international economic and financial order was put in place. The International Bank for Reconstruction and Development (IBRD),[15] the International Monetary Fund (IMF)[16] and the GATT[17] were the most significant outcomes of this Conference.

The eighth GATT trade round (the Uruguay Round) was launched in September 1986 at Punta del Este, Uruguay, and the agenda involved discussion of 15 subjects (tariffs, non-tariff barriers, natural resource products, textiles and clothing, agriculture, tropical products, GATT articles, Tokyo Round codes, anti-dumping, subsidies, intellectual property, investment measures, dispute settlement, the GATT system and services). In December 1993 most negotiations were resolved and on 15 April 1994 the final minutes were signed by ministers from 109 of the

[14]According to Celli Junior (2009), p. 199: 'Ao incluir em sua proposta a "água para uso humano", as CE suscitaram, porém, várias preocupações relativamente ao controle da água como um recurso, bem como questões relacionadas ao acesso equitativo de água limpa entre as camadas mais pobres das populações dos países em desenvolvimento. A reação predominante, como não poderia deixar de ser, foi de cautela, já que, ao aceitarem tal premissa os Membros em desenvolvimento poderiam acabar assumindo compromissos que incluiriam implicitamente qualquer serviço que tivesse ligação com um 'uso final' ambiental, tais como engenharia e construção, dentre outros'. On this, see also Figueiredo (2017).

[15]The IBRD Articles of Agreement became effective in December 1945 and the bank began its operations in 1946 with 38 members. The IBRD is one of the five institutions that compose the World Bank Group. The World Bank currently has 189 members. See: World Bank (2017).

[16]The IMF Articles of Agreement, in their turn, also became effective in December 1945 and the organization began its operations in March 1947 with the participation of 29 countries. The IMF currently has 189 members. See International Monetary Fund (2017).

[17]Signed by 29 countries in 1947, the General Agreement on Tariffs and Trade—GATT/47 was intended to be a temporary agreement until the approval of the Havana Charter, a much more comprehensive agreement which negotiations were subsequently abandoned, turning the GATT/47 into a definitive agreement. The GATT/47 focused mainly on the reduction and elimination of trade tariffs. Originally, it was a treaty without an administrative structure, because this structure was contemplated in other chapters of the Havana Charter that never came into force. Nevertheless, through practice and a series of later rounds of negotiations, the GATT evolved gradually into a de facto institution with a Council, a Secretariat and Committees. See: Hunter et al. (1998), pp. 1180–1181.

123 participating governments at a meeting in Marrakesh, Morocco. The most important result of the Marrakesh Agreement was the establishment of the WTO, but it also comprised several other agreements, such as the General Agreement on Trade in Services (GATS), the Agreement on Trade-Related Investment Measures (TRIMs) and the updated version of GATT (GATT/94).

Proponents of the Bretton Woods economic model argued that liberalization of international trade would promote a more efficient allocation of natural resources, means of production and labor, as each country would produce goods in which it had the greatest comparative advantage, which would lead to economic growth and job creation. This view is foreseen in the preamble of the GATT/47:

> Recognizing that their relations in the field of trade and economic endeavor should be conducted with a view to raising standards of living, ensuring full employment and a large and steadily growing volume of real income and effective demand, developing the full use of the resources of the world and expanding the production and exchange of goods
>
> Being desirous of contributing to these objectives by entering into reciprocal and mutually advantageous arrangements directed to the substantial reduction of tariffs and other barriers to trade and to elimination of discriminatory treatment in international commerce.

Environmental issues were addressed only incidentally and superficially by Article XX, items ‘b’ and ‘g’ from GATT/47, which dealt with exceptions to the application of the agreement. Accordingly:

> **Article XX—General Exceptions.**
>
> Subject to the requirement that such measures are not applied in a manner which would constitute a means of arbitrary or unjustifiable discrimination between countries where the same conditions prevail, or a disguised restriction on international trade, nothing in this Agreement shall be construed to prevent the adoption or enforcement by any contracting party of measures:...
>
> (b) necessary to protect human, animal or plant life or health;...
>
> (g) relating to the conservation of exhaustible natural resources if such measures are made effective in conjunction with restrictions on domestic production or consumption.

The provisions on the subject were limited, and there was no clear distinction between legitimate measures allowed by Article XX and arbitrary or unjustified measures. Consequently, it was up to the GATT panels (the bodies responsible for assessing possible conflicts between the Contracting Parties) to interpret and to define the scope of that article.[18]

The perception of the intertwined relationship and the apparent conflict between economic growth, social development and protection of the environment has

[18]Beurier and Kiss (2000) systematize the main questions that should be addressed by the panel: ‘En principe, trois questions devaient être posées lorsqu'il s'agissait de juger une mesure prise par un État invoquant l'article XX, al. b: - la politique ou la mesure en cause vise-t-elle essentiellement la protection de la santé et de la vie des personnes et des animaux ou la protection des végétaux? - la mesure en faveur de laquelle l'exception est invoquée, est-elle nécessaire pour la protection de la santé et de la vie des personnes ou la préservation des végétaux? - cette mesure cherche-t-elle à éviter d'une façon systématique toute discrimination arbitraire ou injustifieé ou bien une restriction déguisée au commerce international?’ (p. 357).

increased over time. This has been reflected in several of the principles set forth in the Declaration of the United Nations Conference on the Human Environment,[19] adopted in Stockholm, on 16 June 1972 (Stockholm Declaration), which was latter followed by the Rio Declaration on Environment and Development,[20] from the United Nations Conference on Environment and Development held in Rio de Janeiro in 1992. Among the recent international initiatives concerning sustainability, we can highlight the 10-Year Framework of Programs on Sustainable Consumption and Production, adopted at the United Nations Conference on Sustainable Development (Rio+20) in 2012 and the 2030 Agenda for Sustainable Development, adopted by the United Nations General Assembly in 2015.

In 1971, because of the growing concern about environmental issues, the GATT Council of Representatives agreed to set up a Working Group on Environmental Measures and International Trade (EMIT Group), which would be responsible for analysing the compatibility between international trade rules and environmental preservation policies. However, because of the strong opposition from developing countries, which feared the limitation of national sovereignty and the reduction of the scope of internal government actions,[21] and because of the lack of political will in general, it was not until 1991 that the members of the European Free Trade Association asked for the EMIT Group to be convened.[22]

It is worth mentioning that important advances in the environmental area marked the Uruguay Round. In this sense, the Agreement on Technical Barriers to Trade (TBT) and the Agreement on Sanitary and Phytosanitary Barriers (SPS) legitimized trade-restrictive measures to protect the environment. The Committee on Trade and Environment was also created, which shows the complementarity between these themes. Furthermore, the liberalization of trade in environmental goods and services also falls within the scope of GATT/WTO negotiations. Finally, the preamble of the agreement establishing the WTO itself incorporates the notion of sustainable development and recognizes the importance of preserving the environment:

> Recognizing that their relations in the field of trade and economic endeavour should be conducted with a view to raising standards of living, ensuring full employment and a large and steadily growing volume of real income and effective demand, and expanding the

---

[19]Some examples of this relationship are: Principle 2 (safeguard of natural resources for the benefit of present and future generations through careful planning or management), Principle 6 (struggle against pollution) and Principle 24 (international cooperation for the protection and improvement of the environment).

[20]Some examples are: Principle 1 (right to a healthy and productive life in harmony with nature), Principle 3 (present and future generations' right to development); Principle 4 (environmental protection as an integral part of the development process); Principle 7 (international cooperation to ensure the health and integrity of Earth's ecosystem with common but differentiated responsibilities); Principle 8 (reduction and elimination of unsustainable patterns of production and consumption); Principle 12 (promotion of an international economic system that leads to economic growth and sustainable development in all countries); and Principle 16 (internalization of environmental costs).

[21]See: Amaral Júnior (2011), p. 149.

[22]See: World Trade Organization (2017a).

production of and trade in goods and services, while allowing for the optimal use of the world's resources in accordance with the objective of sustainable development, seeking both to protect and preserve the environment and to enhance the means for doing so in a manner consistent with their respective needs and concerns at different levels of economic development.

11.3.1 The WTO and the Doha Round

The Doha Round is the latest round of trade negotiations within the WTO. It was officially launched at the WTO's Fourth Ministerial Conference in Doha, Qatar in November 2001.[23] The Doha Ministerial Declaration[24] provided an ambitious negotiating mandate dealing with relevant issues such as agriculture, services and intellectual property. Three paragraphs have been devoted specifically to the relationship between trade and the environment:

> 31. With a view to enhancing the mutual supportiveness of trade and environment, we agree to negotiations, without prejudging their outcome, on:
>
> (i) the relationship between existing WTO rules and specific trade obligations set out in multilateral environmental agreements (MEAs). The negotiations shall be limited in scope to the applicability of such existing WTO rules as among parties to the MEA in question. The negotiations shall not prejudice the WTO rights of any Member that is not a party to the MEA in question;
>
> (ii) procedures for regular information exchange between MEA Secretariats and the relevant WTO committees, and the criteria for the granting of observer status;
>
> (iii) the reduction or, as appropriate, elimination of tariff and non-tariff barriers to environmental goods and services.

However, more than 16 years after the launch of the negotiations, little progress has been achieved in a multilateral level, not only in relation to the specific objective of reducing or, where appropriate, eliminating tariff and non-tariff barriers to environmental goods and services,[25] but also in practically all other sectors. Notwithstanding the various proposals put forward by a number of countries and international organizations, consensus could not be reached on the mere concept of environmental services.

Although there are still discussions by WTO members on the subject,[26] major negotiations for the opening of environmental goods and services markets have left

[23] See World Trade Organization (2017b).

[24] WT/MIN(01)/DEC/1.

[25] Although there are some initiatives for tariff reduction of environmental goods, they are much more independent initiatives from the members themselves or from negotiations at the regional level than the result of multilateral negotiations. Moreover, non-tariff barriers and barriers to trade in services continue to form a major obstacle to trade in environmental goods and services.

[26] In this regard, at the meeting of the Trade and Environment Committee held in November 2016, a number of environmental issues were discussed, among them chemical and waste management initiatives (concerning mainly electronic waste) and the proposal presented by Canada, South

the multilateral framework, migrating to large preferential agreements (the Trans-Pacific Partnership—TPP, Regional Comprehensive Economic Partnership—RCEP, Comprehensive Economic and Trade Agreement—CETA, among others) and plurilateral agreements (EGA and TiSA) under negotiation or recently concluded.

11.3.2 The Environment Goods Agreement (EGA)

On 24 January 2014, 14 WTO members (Australia, Canada, China, Costa Rica, European Union, Hong Kong, Japan, New Zealand, Norway, Singapore, South Korea, Switzerland, Taiwan/Chinese Taipei and United States) made a joint statement announcing their commitment to achieve global free trade in environmental goods.[27] Besides highlighting the importance of tariff elimination of goods needed to protect the environment and combat climate change, the statement also served as a milestone to report on the intent to conclude an agreement on environmental goods. According to the statement, the agreement would strengthen the rules of the multilateral trading system and would benefit all members of the WTO by applying the most-favored-nation principle. Such an agreement would take effect from the participation of a 'critical mass' of WTO members. The negotiations were officially launched on July 8, 2014, according to a new joint statement presented in Geneva, Switzerland.[28] While dealing primarily with the reduction of tariff barriers, some negotiators argue that a second phase of the negotiations may involve non-tariff barriers and even environmental services.[29]

The conclusion of the EGA would play an important role in the negotiations on environmental services. Firstly, by establishing a possible concept of environmental goods (less likely) or by bringing in a list of goods that have this characteristic (most likely), it would help guide the definition of environmental services, thus facilitating the scope of multilateral negotiations. Secondly, the liberalization of environmental goods directly assists in the provision of environmental services because of the interdependence between sectors. While the EGA's product coverage is still under negotiation, some of the environmental technologies under discussion are those related to renewable and clean energy generation, air pollution control, energy efficiency, water and wastewater treatment, solid and hazardous waste treatment and environmental monitoring and analysis. Even if it is not concluded, the EGA

Korea, Costa Rica, Mexico and Taipei China to deepen discussions on climate change and trade. See: World Trade Organization (2016).

[27]See: Australia, et al. (2014a).

[28]See: Australia, et al. (2014b).

[29]At the time of submission of these declarations in 2014, it had not yet been determined whether the environmental services would be part of the Environmental Goods Agreement or the Trade in Services Agreement (TiSA). See: Vossenaar (2014), p. ix. However, it is more likely that environmental services will integrate TiSA.

discussions represent a clear trend of the Parties, which will also assist in multilateral, regional and bilateral negotiations on environmental goods and services.

11.3.3 The Trade in Services Agreement (TiSA)

The TiSA is a plurilateral agreement negotiated by 23 WTO members that account for 70% of world trade in services. Its negotiation was officially launched in March 2013 and its participants agreed with its base text in September 2013.

In a joint statement issued on 14 April 2016, the Global Services Coalition (GSC) described TiSA as 'the best opportunity to overcome that impasse [in multilateral negotiations] and establish a template for 21st century services trade, for the benefit of the world economy.' It went on to say:

> As the 17th round of the Trade in Services Agreement (TiSA) negotiations is currently underway in Geneva this week, the Global Services Coalition (GSC), representing services enterprises and sector-specific services associations in our respective countries and regions, call on the trade negotiators to intensify efforts towards a high ambition agreement.... For much too long, services in multilateral negotiations have been hostage to other negotiating topics. The TiSA offers the best opportunity to overcome that impasse and establish a template for 21st century services trade, for the benefit of the world economy.[30]

Other documents released by the European Commission reveal that environmental services are part of the negotiations and that the subject was discussed in several rounds of negotiations:

> This week, the EU will encourage a high level of ambition and particular progress on:... Environmental services—The EU will seek to end discrimination against foreign suppliers of environmental services. This means removing the existing barriers – not just abstaining from introducing new restrictions.[31]

> Environmental services. The discussion focused on market access in environmental services, based on an analysis of offers prepared by the proponent of the related Annex. Participants explained sensitivities, including those related to public services. There was no objection to the issue that participants should have the possibility to maintain or 'remunicipalise' services. The EU explained that our public utilities reservation on Market Access achieves precisely that: the ability to keep and introduce public monopolies at any level of government. Some participants highlighted their ambition in the sector and its importance in view of the global climate agenda.[32]

> Other topics.... Further market access discussions addressed Parties' commitments in energy and mining related services, environmental services and delivery services.[33]

[30]Global Services Coalition (2016).

[31]European Commission (2014), p. 1.

[32]European Commission (2015), p. 5.

[33]European Commission (2016a), p. 4.

In 2016, Brazil formally requested to participate in the negotiations. By November 2016, 21 negotiation rounds had taken place. As of July 2018, negotiations are 'on hold and are expected to resume when the political context allows'. There is no formally set deadline for ending the negotiations.[34]

11.3.4 The Trans-Pacific Partnership (TPP)

In 2005, Brunei Darussalam, Chile, Singapore and New Zealand signed the Trans-Pacific Strategic Economic Partnership—P4. In 2008, the United States and a number of other countries declared their interest in participating in a broader version of this agreement, which later resulted in the TPP. The TPP negotiations were initiated in 2010 and concluded in October 2015. In February 2016, the agreement was signed by 12 countries (Australia, Brunei Darussalam, Canada, Chile, the United States, Japan, Malaysia, Mexico, New Zealand, Peru, Singapore and Vietnam) corresponding to a GDP of US$27 trillion (37.4% of world GDP).[35]

In the version signed in February 2016, the TPP was composed of a preamble, 30 chapters/articles and four annexes. In its chapter 20, which brings forward predictions about the environment, there was an express reference to the importance of environmental goods and services:

Article 20.18: Environmental Goods and Services

1. The Parties recognize the importance of trade and investment in environmental goods and services as a means of improving environmental and economic performance and addressing global environmental challenges.

2. The Parties further recognize the importance of this Agreement to promoting trade and investment in environmental goods and services in the free trade area.

3. Accordingly, the Committee shall consider issues identified by a Party or Parties related to trade in environmental goods and services, including issues identified as potential non-tariff barriers to that trade. The Parties shall endeavour to address any potential barriers to trade in environmental goods and services that may be identified by a Party, including by working through the Committee and in conjunction with other relevant committees established under this Agreement, as appropriate.

4. The Parties may develop bilateral and plurilateral cooperative projects on environmental goods and services to address current and future global trade related environmental challenges.

Under Article 30.5 of this version, the agreement had to be ratified by at least six countries representing 85% of GDP in this group to enter into force. However, with the election of Donald Trump to the presidency of the United States, the negotiations suffered a setback. The United States sent a letter (dated January 30, 2017) to New Zealand, as Depositary of the TPP, and to TPP signatories, stating its intention

[34]See: European Commission (2017).

[35]See: Australian Department of Foreign Affairs and Trade (2018).

not to become a Party to the TPP and that they have no legal obligations arising from its signature on 4 February 2016.[36]

Notwithstanding the United States' withdrawal, the remaining countries continued negotiations. On 8 March 2018 the Comprehensive and Progressive Agreement for Trans-Pacific Partnership (CPATTP or TPP-11) was signed by ministers of Australia, Brunei Darussalam, Canada, Chile, Japan, Malaysia, Mexico, Peru, New Zealand, Singapore and Vietnam.[37] This agreement incorporated most of the provisions of the TPP Agreement., as follows:

> Pursuant to its Article 1, para. 1: 'The Parties hereby agree that, under the terms of this Agreement, the provisions of the Trans-Pacific Partnership Agreement, done at Auckland on 4 February 2016 ('the TPP') are incorporated, by reference, into and made part of this Agreement *mutatis mutandis*, except for Article 30.4 (Accession), Article 30.5 (Entry into Force), Article 30.6 (Withdrawal) and Article 30.8 (Authentic Texts).

11.3.5 The Transatlantic Trade and Investment Partnership (TTIP)

The Transatlantic Trade and Investment Partnership (TTIP) was a trade and investment agreement negotiated between the United States and the European Union. The final agreement would have 24 chapters, grouped into four parts: (1) market access; (2) regulatory cooperation; (3) rules and; (4) institutional.[38] One of the chapters would be devoted to sustainable development with a number of relevant environmental issues, such as multilateral environmental governance, climate change and green goods and services, international cooperation and waste and chemicals[39] Despite the progress in negotiations, the future of TTIP remains uncertain. Although official statements from the United States confirm their interest in continuing negotiations with the European Union, Donald Trump's position on the matter remains in doubt.

11.3.6 The Regional Comprehensive Economic Partnership (RCEP)

The Regional Comprehensive Economic Partnership (RCEP) is an agreement negotiated between the ten members of the Association of Southeast Asian Nations (ASEAN)[40] and six countries with which ASEAN has free trade agreements

---

[36]Ibid.

[37]See: Australian Department of Trade and Foreign Affairs (2018).

[38]See: European Commission (2016b).

[39]See: European Union (2015).

[40]Brunei, Cambodia, Philippines, Indonesia, Laos, Malaysia, Myanmar, Singapore, Thailand and Vietnam.

(Australia, China, South Korea, India, Japan and New Zealand). Together, they account for a GDP of US$23.8 trillion and a population of 3.5 billion people.[41] Negotiations began officially on 20 November 2012 at a meeting held in Phnom Penn, Cambodia. As of July 2018, 23 rounds of negotiations had been held.[42] Under its Guiding Principles and Objectives, the RCEP will involve the following topics: trade in goods, trade in services, investment, technical and economic cooperation, intellectual property, competition, dispute settlement and other issues:

> **II. TRADE IN SERVICES.**
> The RCEP will be comprehensive, of high quality and substantially eliminate restrictions and/or discriminatory measures with respect to trade in services between the RCEP participating countries. Rules and obligations on trade in services under the RCEP will be consistent with the General Agreement on Trade in Services (GATS) and will be directed towards achieving liberalization commitments building on the RCEP participating countries' commitments under the GATS and the ASEAN+1 FTAs. All sectors and modes of supply will be subject to negotiations.[43]

Regarding trade in services, all sectors and modes of provision will be negotiated. Environmental services are therefore included in the scope of the negotiations.[44]

11.3.7 The Comprehensive Economic and Trade Agreement (CETA)

On 30 October 2016, the European Union and Canada signed CETA. The agreement comprises a preamble, 30 chapters and several annexes. It addresses several relevant trade issues, such as national treatment and market access for goods, technical barriers to trade, sanitary and phytosanitary measures, investment, cross-border trade in services, government procurement, intellectual property, trade and sustainable development, trade and labor, trade and environment, transparency, dispute settlement, etc. The agreement has a series of provisions dealing with environmental protection and sustainable development (Article 22.1) and explicitly recognizes the precautionary principle (Article 24.8, para. 2).

Besides addressing general issues that influence the demand for environmental goods and services, CETA provides more directly about the sector as follows:

> **ARTICLE 22.3 Cooperation and Promotion of Trade Supporting Sustainable Development**
> 2. The Parties affirm that trade should promote sustainable development. Accordingly, each Party shall strive to promote trade and economic flows and practices that contribute to enhancing decent work and environmental protection, including by:
> (a) encouraging the development and use of voluntary schemes relating to the sustainable production of goods and services, such as eco-labelling and fair trade schemes; . . .

---

[41] See Australian Department of Trade and Foreign Affairs (2018).

[42] Ibid.

[43] Association of Southeast Asian Nations, et al. (2012).

[44] Ibid.

ARTICLE 24.9 Trade Favouring Environmental Protection

1. The Parties are resolved to make efforts to facilitate and promote trade and investment in environmental goods and services, including through addressing the reduction of non-tariff barriers related to these goods and services.

2. The Parties shall, consistent with their international obligations, pay special attention to facilitating the removal of obstacles to trade or investment in goods and services of particular relevance for climate change mitigation and in particular trade or investment in renewable energy goods and related services.

ARTICLE 24.12 Cooperation on Environment Issues

1. The Parties recognize that enhanced cooperation is an important element to advance the objectives of this Chapter, and commit to cooperate on trade-related environmental issues of common interest, in areas such as: ...

(e) trade-related aspects of the current and future international climate change regime, as well as domestic climate policies and programmes relating to mitigation and adaptation, including issues relating to carbon markets, ways to address adverse effects of trade on climate, as well as means to promote energy efficiency and the development and deployment of low-carbon and other climate-friendly technologies;

(f) trade and investment in environmental goods and services, including environmental and green technologies and practices; renewable energy; energy efficiency; and water use, conservation and treatment.

On 21 September 2017, CETA provisionally entered into force. As of July 2018, most of the agreement now applies.[45]

11.4 Conclusion

The promotion of sustainable production and consumption patterns and the liberalization of trade in EGS are themes addressed by several international agreements and declarations and they are closely related. There is a growing concern about increasing efficiency, reducing waste and minimizing negative effects on production chains and environmental goods, as well as about how services liberalization can serve as an important tool for promoting sustainable consumption and production patterns. Nevertheless, it is necessary to approach these initiatives with caution.

Precisely because of the importance of the environmental goods and services sector, the risks involved should not be overlooked, especially when it comes to the provision of essential public services, such as water supply and sewage collection and treatment. It is essential that liberalization does not hamper its universalization and its accessibility, in particular for the excluded segments of the population. Nevertheless, trade liberalization, conclusion of agreements and international trade commitments can serve as catalysts for the encouragement of research and development and for the use of the most appropriate and advanced technologies for each situation.

[45]'National parliaments in EU countries – and in some cases regional ones too – will then need to approve CETA before it can take full effect.' See: European Commission (2018).

Deadlocks in multilateral negotiations within the WTO do not mean an absence of international advances. The negotiation and conclusion of several agreements on environmental goods and services show the relevance of the subject and the growing association between trade and the environment.

References

Amaral Júnior A (2011) Comércio Internacional e Proteção do Meio Ambiente. Atlas, São Paulo

Associação Brasileira de Biogás e Biometano (2015) O que é biogás? Available at: http://www.abiogas.org.br/biogas-e-biometano. Accessed 20 Nov 2017

Association of Southeast Asian Nations, et al (2012) Guiding principles and objectives for negotiating the regional comprehensive economic partnership. Available at: http://www.mofa.go.jp/announce/announce/2012/11/pdfs/20121120_03_02.pdf. Accessed 20 Nov 2017

Australia, et al (2014a) Joint statement regarding trade in environmental goods. Available at: http://trade.ec.europa.eu/doclib/docs/2014/january/tradoc_152095.pdf and https://ustr.gov/sites/default/files/EGs-Announcement-joint-statement-012414-FINAL.pdf. Accessed 20 Nov 2017

Australia, et al (2014b) Joint statement regarding the launch of the environmental goods agreement negotiations. Available at: http://eeas.europa.eu/archives/delegations/wto/documents/press_corner/final_joint_statement_green_goods_8_july_2014.pdf. Accessed 20 Nov 2017

Australian Department of Foreign Affairs and Trade (2018) Trans-pacific partnership agreement. Available at: http://dfat.gov.au/trade/agreements/not-yet-in-force/tpp-11/Pages/trans-pacific-partnership-agreement-tpp.aspx. Accessed 30 July 2018

Australian Department of Trade and Foreign Affairs (2018) Regional comprehensive economic partnership. Available at: http://dfat.gov.au/trade/agreements/negotiations/rcep/Pages/regional-comprehensive-economic-partnership.aspx. Accessed 30 July 2018

Beurier JP, Kiss A (2000) Droit International de l'Environnement, 2nd edn. Pedone, Paris

Celli Junior U (2009) Comércio de Serviços na OMC: liberalização, condições e desafios. Juruá, Curitiba

European Commission (2014) Quick facts: 10th round of talks on trade in services (TiSA). Available at: http://trade.ec.europa.eu/doclib/docs/2014/december/tradoc_152928.pdf. Accessed 20 Nov 2017

European Commission (2015) Report of the 14th TiSA negotiation round 6–13 October 2015. Available at: http://trade.ec.europa.eu/doclib/docs/2015/october/tradoc_153917.15.pdf. Accessed 20 Nov 2017

European Commission (2016a) Report of the 18th TiSA negotiation round 26 May–3 June 2016. Available at: http://trade.ec.europa.eu/doclib/docs/2016/june/tradoc_154626.pdf. Accessed 20 Nov 2017

European Commission (2016b) EU negotiating texts in TTIP. Available at: See http://trade.ec.europa.eu/doclib/press/index.cfm?id=1230. Accessed 20 Nov 2017

European Commission (2017) Trade in Services Agreement (TiSA). Available at: http://ec.europa.eu/trade/policy/in-focus/tisa/. Accessed 30 July 2018

European Commission (2018) Comprehensive Economic and Trade Agreement (CETA). Available at: http://ec.europa.eu/trade/policy/in-focus/ceta/. Accessed 30 July 2018

European Union (2015) EU position paper trade and sustainable development chapter/labour and environment. Available at: http://trade.ec.europa.eu/doclib/docs/2015/january/tradoc_153024.pdf. Accessed 20 Nov 2017

Ferrier G (2004) An examination of trade in environmentally preferable goods and services in the NAFTA region. Environmental Business International Inc., San Diego

Figueiredo GJP (2017) Atuação do Setor Privado no Fornecimento de Água. In: Purvin G et al (eds) Direito Ambiental, Recursos Hídricos e Saneamento: estudos em comemoração aos 20 anos da Política Nacional de Recursos Hídricos e aos 10 anos da Política Nacional de Saneamento. Letras Jurídicas, São Paulo

Global Services Coalition (2016) Trade in Services Agreement (TiSA) negotiations. Available at: https://servicescoalition.org/images/GSC_Joint_Statement_on_TiSA_-_April_2016_-_Final.pdf. Accessed 20 Nov 2017

Hunter D, Salzman J, Zaelke D (1998) International environmental law and policy. Foundation Press, New York

International Monetary Fund (2017) What we do. Available at: http://www.imf.org/external/about/whatwedo.htm. Accessed 20 Nov 2017

Lendo E (2005) Defining environmental goods and services: a case study of Mexico, ICTSD Trade and Environment Series Issue Paper No. 1. CEC and ICTSD, Geneva

National Board of Trade (Kommerskollegium) (2014) Making green trade happen: environmental goods and indispensable services. Kommerskollegium, Stockholm

Organization for Economic Cooperation and Development (1996) The global environmental goods and services industry. OECD, Paris

Organization for Economic Cooperation and Development (1999) The environmental goods and services industry: manual for data collection and analysis. OECD, Paris

Steenblik R, Drouet D, Stubbs G (2005) Synergies between trade in environmental services and trade in environmental goods. OECD Trade and Environment Working Paper N° 2005–01

U.S. General Services Administration (2017) Environmentally preferred products. Available at: http://www.gsa.gov/portal/category/27119. Accessed 20 Nov 2017

United Nations Environment Programme (2010) ABC of SPC: clarifying concepts on sustainable consumption and production. UNEP, Paris

Vossenaar R (2014) Identifying products with climate and development benefits for an environmental goods agreement. ICTSD, Geneva

Waglé S, Ramani KV, Qutub SA (2005) International trade in environmental and energy services and human development: contributing to well-being, growth and access for all. Discussion Paper. UNDP, Colombo

World Bank (2017) Who we are. Available at: http://www.worldbank.org/en/who-we-are. Accessed 20 Nov 2017

World Trade Organization (2016) WTO members discuss climate change, chemicals and waste management. Available at: https://www.wto.org/english/news_e/news16_e/envir_14nov16_e.htm. Accessed 20 Nov 2017

World Trade Organization (2017a) Early years: emerging environment debate in GATT/WTO. Available at: https://www.wto.org/english/tratop_e/envir_e/hist1_e.htm. Accessed 20 Nov 2017

World Trade Organization (2017b) The Doha Round. Available at: https://www.wto.org/english/tratop_e/dda_e/dda_e.htm. Accessed 20 Nov 2017

Part III
Packaging and (Eco-)Labelling: Beyond the Information Paradigm

Chapter 12
Regulating Green Marketing Claims in the United States

James P. Nehf

Contents

Abstract This chapter examines the regulation of green marketing in the United States. Environment-friendly marketing (eco-labeling) has emerged as an important approach to selling products and services to consumers in the global marketplace. Various claims, marks, labels, symbols and logos are being used to appeal to consumers who are increasingly concerned about the environmental effects of their consumption. Touting a firm's "green credentials" can improve a seller's image in the public eye and potentially increase sales. Surveys show that a large proportion of consumers, at least those in more advanced economies, are willing to pay more for green or environment-friendly products and services. Studies have also shown that green customers tend to be more affluent, better educated and more loyal to brands that they purchase. On the business side, there is evidence that environment-friendly businesses enjoy positive public reputations, generally supportive government policies and potential for higher sales revenues. The problem is that green marketing can be overstated, confusing, misleading or outright false advertising. "Greenwashing" occurs when a business promotes its eco-friendly attributes in a false or misleading way. When that happens, consumers may end up paying more for products and services without supporting a positive environmental impact. The result is a loss of consumer welfare, little or no environmental benefit, and no social benefit except increased corporate profits.

J. P. Nehf (✉)
Indiana University McKinney School of Law, Indianapolis, IN, USA
e-mail: jnehf@iupui.edu

A. d. Amaral Junior et al. (eds.), *Sustainable Consumption*,
https://doi.org/10.1007/978-3-030-16985-5_12

12.1 Introduction

Environmental-friendly marketing (eco-labelling) has emerged as an important approach to selling products and services to consumers in the global marketplace. Various claims, marks, labels, symbols and logos are being used to appeal to consumers who are increasingly concerned about the environmental effects of their consumption. Touting a firm's 'green credentials' can improve a seller's image in the public eye and potentially increase sales. Surveys show that a large proportion of consumers, at least those in more advanced economies, are willing to pay more for green or environmentally-friendly products and services.[1] Studies have also shown that green customers tend to be more affluent, better educated, and more loyal to brands that they purchase.[2] On the business side, there is evidence that environmental-friendly businesses enjoy positive public reputations, generally supportive government policies, and potential for higher sales revenues.

All of this sounds like a win–win for consumers, businesses and the environment. If consumers value eco-friendly products and services—and businesses can make greater profits by providing them—one would expect the market for such products and services to grow rapidly. This has certainly been the case—at least with respect to marketing behaviour. The problem is that green marketing can be overstated, confusing, misleading or outright false advertising. 'Greenwashing' occurs when a business promotes its eco-friendly attributes in a false or misleading way. When that happens, consumers may end up paying more for products and services without supporting a positive environmental impact. The result is a loss of consumer welfare, little or no environmental benefit, and no social benefit except for increased corporate profits.

In 2010 Underwriters Laboratory identified seven different types of greenwashing practices—the 'Seven Sins', as they see it—that green marketers have been committing, focusing more on the marketing of consumer goods than services:

1. **Hidden trade-offs**: suggesting that a product is 'green' based on an unreasonably narrow set of characteristics while ignoring other important environmental issues. A product may, for example, be 'biodegradable' but manufacturing it creates a large carbon footprint; or a biofuel made of palm oil may sound eco-friendly, but only if one ignores the destruction of rainforests that may be necessary to harvest it.
2. **Unsubstantiated claims**: making an assertion that cannot be substantiated by readily accessible supporting information or by a reliable third-party certification. Tissue products, for example, may claim to be made from recycled material

[1] Gingerich and Karaatli (2015). Another study found that nearly two-thirds of consumers felt 'a sense of responsibility to purchase products that are good for the environment and society,' and 70% would buy an environmentally friendly product over a standard one (GlobeScan, SustainAbility, and BBMG 2012).

[2] Kewalramai and Sobelsohn (2012).

without any evidence in support. Even if technically true, the percentage of recycled material in the tissues could be small.

3. **Vagueness:** making a claim that is so broad that its real meaning is likely to be misunderstood by the consumer. 'All-natural' does not necessarily mean healthy or environmentally friendly (many toxic substances occur naturally in the environment). Other examples include 'environmentally safe,' 'eco-friendly,' 'green,' and the like.
4. **Irrelevance**: making an environmental claim that may be truthful, but it is not material or helpful for consumers seeking environmentally preferable products. A product might be marketed as 'CfC-free,' but if the chemical is banned by law anyway the information is not helpful in consumer decision-making.
5. **Lesser of two evils**: making a claim that is true within the product category, but distracts the consumer from the greater environmental impacts of the category as a whole. Organic cigarettes might be an example, or marketing a truck as 'fuel-efficient' for its class.
6. **Fibbing**: making environmental claims that are simply false or wildly misleading.
7. **Misleading symbols**: marketing a product with words or images (e.g., an official-looking green logo) that gives the impression of third-party endorsement where no such endorsement actually exists.[3]

12.2 Response to Greenwashing at the National Level: FTC Green Guides

Section 5 of the US Federal Trade Commission Act[4] prohibits all persons from making 'unfair or deceptive acts or practices in or affecting commerce,'[5] which would include deceptive green marketing practices. Under this very broad provision, the FTC in 1992 promulgated the 'Green Guides' to 'help marketers avoid making environmental claims that are unfair or deceptive under Section 5 of the FTC Act[.]'[6] The agency revised the Green Guides in 1996, 1998, and most recently in 2012 to 'take into account recent changes in the marketplace,' and to 'provide new guidance on environmental claims that were not common when the Guides were last reviewed.'[7]

The FTC tries to keep up with various forms of greenwashing that marketers implement over time. As is often the case, however, the agency lags behind the

[3]Underwriters Laboratories (2010).

[4]15 U.S.C. § 45(a)(1).

[5]*Id.* Under FTC decisions, a representation, omission, or practice is deceptive if it is likely to mislead consumers acting reasonably under the circumstances and is material to the consumers' decision whether or not to purchase the product or service. In re Cliffdale Assoc., 103 FTC 110 (1984).

[6]16 C.F.R.260.1(a).

[7]Federal Trade Commision (2012).

marketplace as new marketing practices evolve to supplant those that the FTC targets as potentially misleading. More importantly, however, the Green Guides are not legally binding on anyone. As is the case with other 'Guides' that the FTC has created over the years, they are just guidelines that marketers can (and should) use to protect themselves from claims of unfair or deceptive practices.[8] The Guides 'consist of general principles, specific guidance on the use of particular environmental claims, and examples.'[9] But there are no direct consequences flowing from a marketer not following the guidelines. Moreover, compliance with the Guides does not necessarily preclude the FTC from bringing a law enforcement action against the firm for deceptive conduct under the FTC Act.[10] As a practical matter, however, complying with the Guides should protect a business from an FTC enforcement action, and not complying with the Guides is a risky practice that may invite enforcement of some kind.

Although the Guides are not binding on marketers (or the FTC), they provide important guidance to business that would like to advertise the eco-friendly qualities of their products, packaging, or services, so they are worth reviewing briefly. Specifically, the Guides offer guidance on 14 types of environmental marketing claims: (1) general environmental benefits; (2) carbon offsets; (3) certifications and seals of approval; (4) compostable; (5) degradable; (6) 'free-of'; (7) non-toxic; (8) ozone-safe and ozone-friendly; (9) recyclable; (10) recycled content; (11) refillable; (12) made with renewable energy; (13) made with renewable materials; and (14) source reduction.

The Guides are reasonably detailed and include examples of practices that the FTC would likely deem deceptive. The agency has provided a summary of the Guides as well, divided into the abovementioned categories of environmental marketing claims[11]:

1. **General Environmental Benefit Claims**. Marketers should not make broad or general environmental claims such as 'green' or 'eco-friendly.' The agency's reasoning is that general claims are difficult to substantiate. The FTC encourages marketers to amplify general claims with specific environmental benefits. Such amplifications should be clear, prominent, and specific. The Guides also counsel against emphasizing or exaggerating small or unimportant benefits. Moreover, if a general claim indicates that a product has an overall environmental benefit because of a specific attribute, the marketer should consider the trade-offs resulting from the attribute and present the overall claim in a manner that is not misleading.
2. **Carbon Offsets**. A carbon offset is a credit for greenhouse gas reductions achieved by one party (e.g., a wind farm) that can be purchased and used to

[8]In contrast to FTC Guides, the agency also publishes FTC Rules, which do have the force of law and are binding on businesses that are subject to the Rules.

[9]16 C.F.R. 260.1(d).

[10]16 C.F.R. 260.1(b).

[11]Federal Trade Commission (n.d.).

compensate (offset) the emissions of another party (e.g., an industrial plant). Under the Guides, marketers should have 'competent and reliable scientific evidence' to support carbon offset claims. They should use 'appropriate accounting methods' to ensure they measure emission reductions properly, and they should disclose whether the offset pays for emission reductions far into the future, i.e., that will not occur for at least 2 years. Moreover, they should not advertise a carbon offset if the law already requires the activity that is the basis of the offset.

3. **Certifications and Seals of Approval**. Third-party certifications, labels and seals have become very popular for eco-marketing and are discussed in more detail later in this chapter. The FTC's view is that third-party eco-certifications may be considered 'endorsements' and are therefore subject to the FTC's general guides on endorsements.[12] Under the endorsement guidelines, marketers should disclose any material connections to the certifying organization. A material connection is one that could affect the credibility of the endorsement. Marketers should not use environmental certifications marks if they do not clearly convey the basis for the certification. This is because the certification, standing alone, is likely to even convey general, much less specific, environmental benefits. To prevent deception, if a marketer uses a third-party seal or certification mark that does not indicate the basis for the certification, it should clearly and prominently identify the specific environmental benefits that the product offers. The Guides further state that if a certification is based on attributes that are too numerous to disclose, the marketer can say something like: 'Virtually all products impact the environment. For details on which attributes we evaluated, go to [a website that discusses this product].' Finally, a marketer with a third-party certification still must substantiate all express and implied claims about the environmental impacts of its product or service.
4. **Compostable**. If a product is claimed to be compostable the marketer must have 'competent and reliable scientific evidence' that all materials in the product or package will break down into usable compost safely and in about the same time as the other materials with which it is composted. If the product cannot be composted at home safely or in a timely manner, the marketer should so indicate. If a claim is that a product can be composted in a municipal or commercial facility, such facilities must be available to a substantial majority of consumers targeted by the marketer.
5. **Degradable**. The Guides say that a marketer may make a degradable (biodegradable) claim only if has evidence that the 'entire product or package will completely break down and return to nature within a reasonably short period of time after customary disposal.' The 'reasonably short period of time' for com-

[12]The FTC Endorsement Guides are codified at 16 C.F.R. Part 255.

plete decomposition of solid waste products in the Guides is presumed to be 1 year, but the FTC has not been strict in applying this presumption.[13]

6. **'Free-of'**. Marketers can make a 'free-of' claim for a product although it contains some amount of the substance, but only if the product does not have more than trace amounts or background levels of the substance; the amount of substance present does not cause harm that consumers typically associate with the substance; and the substance was not added to the product intentionally. Further, the marketer cannot claim that a product is 'free-of' one particular substance if it contains another that poses a similar environmental risk. And a marketer should not claim that a product is 'free-of' a substance if that product or category of products has never has been associated with the substance or harm in question.
7. **Non-Toxic**. If a product claim asserts that it is non-toxic the marketer must have 'competent and reliable scientific evidence' that the product is safe for people and the environment.
8. **Ozone-Safe** and **Ozone-Friendly**. Without much specificity, the Guides provide that it is deceptive to misrepresent that a product is ozone-friendly or safe for the ozone layer or atmosphere if it contains any ozone depleting substances. Moreover, it is deceptive to assert that a product is ozone-friendly if its ingredients contribute to the creation of smog and ground-level ozone formation.
9. **Recyclable**. Marketers are cautioned about making recyclability claims when recycling facilities are not available to at least 60% of the consumers where a product is sold. If that threshold is not met, the marketer can state, 'This product may not be recyclable in your area.' If, however, recycling facilities for are available to only a few consumers, the marketer should use stronger qualifying language.
10. **Recycled Content**. Marketers can make recycled content claims only for 'materials that have been recovered or diverted from the waste stream during the manufacturing process or after consumer use.' Moreover, they should qualify claims for products or packages made partly from recycled material, e.g., 'Made from 30% recycled material.' If a product contains used, reconditioned, or re-manufactured components, the marketer should qualify the recycled content claim clearly to avoid deception about the components.
11. **Refillable**. Marketers should not make refillable claims unless there is a way to refill the package. For example, the business can provide a system to collect and refill the package or it can sell a product that consumers can use to refill the original package.

[13]On 20 October 2015, the FTC issued an opinion and order holding that ECM BioFilms, Inc., a manufacturer of plastic additives, had made false or misleading claims about the biodegradability of plastics containing the additive, but in its order the FTC forbade ECM BioFilms from making biodegradable claims if the product would not break down within five years of customary disposal. This followed a decision by an administrative law judge earlier in the year that rejected the one-year presumption. See: Federal Trade Commission (2015a) and *ECMBiofilms, Inc. v. FTC*, 851 F.3d 599 (6th Cir. 2017).

12. **Made with Renewable Energy**. Marketers can make renewable energy claims based on energy derived from fossil fuels if they purchase renewable energy certificates (RECs) to match the energy use. The Guides say that one way to minimize the risk of misunderstanding is to specify the source of renewable energy ('wind' or 'solar energy'). They can make unqualified 'made with renewable energy' claims if all or nearly all the significant manufacturing processes are powered with renewable energy (or non-renewable energy matched by RECs). Further, if a marketer generates renewable energy but sells RECs for all the renewable energy it generates, the marketer should not claim to 'use' renewable energy.
13. **Made with Renewable Materials**. Claims about using 'renewable materials' can be deceptive because they may imply that a product is recyclable, made with recycled content, or biodegradable. To avoid deception, the Guides suggest stating the material used clearly and prominently, and explain why it is renewable. If an item is not made entirely with renewable materials (except for minor and incidental components), the marketer should qualify the claim to avoid any misleading conclusions.
14. **Source Reduction**. Marketers sometimes claim that a product or package is lower in weight, volume, or toxicity, but such claims can be misleading if they are not placed in context. To avoid deception the Guides suggest prominent disclosure about the amount of reduction and the basis for any comparison. For example, rather than saying the product generates '10 percent less waste,' the marketer could say the product generates '10 percent less waste than our previous product.'

12.3 Enforcement of the FTC Guides

As the Green Guides do not have the force of law, they cannot be enforced directly by the FTC, consumer organizations, state officials, or anyone else. If the FTC wishes to bring an enforcement action against a marketer based on misleading environmental claims, it is not enough to show that a marketer has violated language in the Guides. To make its case that the FTC Act has been violated, the FTC must show that the practice in question is deceptive under its general standard for deceptive conduct in the FTC Act.[14] The Guides can assist the FTC in proving that claim, but if a court does not believe that a practice is deceptive there is no violation of law even if the marketer has not complied with specific language set forth in the Green Guides. A similar outcome would result if a state official brought an action against a marketer under the state's own deceptive practices law. The state

[14]The FTC defines deception as a practice involving a material representation that is likely to mislead consumer acting reasonably under the circumstances. The FTC articulated the standard in: In re Cliffdale Assoc., 103 FTC 110. (1984). The Guides are intended to provide guidance on what constitutes deception in making green marketing claims.

official could use the Guides as evidence that the marketer was engaging in deceptive conduct, but that alone would not win the case. A court would have to find that the marketing practice violated the standard for deceptive conduct set forth in the state statute or other applicable law.

The FTC has brought several enforcement actions under the Green Guides. Enforcement actions usually do not result in a court decision or a finding of guilt or innocence. Instead, the result is typically a consent order, a negotiated agreement whereby the alleged offender agrees not to make certain marketing claims in the future but does not agree that it did anything deceptive in the past. This 'remedy' is not always effective because it is often the case that the marketer has already moved on to a different marketing plan by the time the consent order is reached with the FTC. To illustrate how the Green Guides have worked in practice, some important FTC enforcement actions are reviewed below.

In 2013 the FTC brought enforcement actions against several marketers of so-called 'biodegradable' plastic products (bags, food containers, and bottles) where there was insufficient scientific evidence to support claims that the products were completely biodegradable when placed in a landfill. According to the FTC's complaints, approximately 92% of total municipal waste in the United States goes either into landfills, incinerators, or recycling facilities, which '[do not] allow [the plastics] to completely break down and decompose into elements found in nature within a reasonably short period of time.'[15] Further, the FTC alleged that the testing methods employed by the companies did not replicate the physical conditions of landfills, and thus could not substantiate the 'biodegradable' claims.[16]

One of the defendants was ECM Biofilms, Inc., which marketed plastic product additives under the trade name MasterBatch Pellets. According to the complaint, ECM issued its own 'Certificates of Biodegradability of Plastic Products,' which ECM used to convince its customers that its additives make plastic products biodegradable. ECM claimed, for example, that 'plastic products made with [its] additives will break down in approximately nine months to five years in nearly all landfills or wherever else they may end up.' The FTC's complaint charged ECM with violating the FTC Act by misrepresenting that: (1) ECM plastics made with the additives are biodegradable and will completely break down within a reasonably short period of time after customary disposal; (2) ECM plastics are biodegradable in a landfill; (3) ECM plastics are biodegradable in a stated qualified timeframe; and (4) that various scientific tests prove ECM's biodegradability claims.[17] Another defendant was Clear Choice Housewares, Inc. that sold what it claimed are biodegradable, reusable plastic food storage containers on its website, as well as in retail stores

[15]In re American Plastic Manufacturing Inc., Docket No. C-4453 (May 2, 2014) (Complaint ¶ 6).

[16]Id. at ¶ 8.

[17]In the Matter of ECM BioFilms, Inc., a corporation, also doing business as Enviroplastics International, FTC Matter/File Number: 122 3118, Docket Number: 9358, available at: https://www.ftc.gov/enforcement/cases-proceedings/122-3118/ecm-biofilms-inc-also-dba-enviroplastics-international-matter.

nationwide. Clear Choice's marketing materials claimed its products were biodegradable based on the application of a Bio-Tec product called Eco Pure. The FTC alleged that Clear Choice made false and unsubstantiated claims that Eco Pure made its products 'quickly biodegradable in landfills.' Another company, Carnie Cap, Inc., marketed plastic rebar cap covers. It claimed, with no qualification, that the Eco-One product makes it plastic rebar cap covers '100% biodegradable.'

The proposed consent orders were essentially the same. They prohibited the companies from making biodegradability claims unless the representations are true and supported by competent and reliable scientific evidence. To make an unqualified biodegradability claim, the companies must have evidence that the entire plastic product will completely decompose into elements found in nature within 1 year after customary disposal. This is consistent with language in the Green Guides.

In April 2018, the FTC approved final consent orders against four paint companies that misled consumers through claims that their products were free of emissions and volatile organic compounds (VOCs). The companies claimed their paints would not emit VOCs and other chemicals, including during and immediately after application. Some promotions also made explicit safety claims regarding babies, children, pregnant women, and other sensitive populations. The FTC maintained that the companies had no evidence to support these claims and that the companies marketed their paint using environmental seals without adequately disclosing that they had awarded the seal to their own products. The final consent decrees barred the companies from making unqualified emission-free and VOC-free claims unless, at all times during paint application and after, both content in and emissions from their paints are actually zero, or emissions are at 'trace' levels.[18]

In November 2015, the agency approved a consent order with a toilet tissue maker, requiring it to stop advertising moist toilet tissue and cloth as flushable or safe for sewer or septic systems unless it can substantiate those claims. The company misrepresented that its wipes: (1) were safe for sewer systems; (2) were safe for septic systems; (3) break apart shortly after being flushed; and (4) are safe to flush. The consent decree prohibits the company from misrepresenting that any wipe is safe to flush, 'unless it can substantiate that the wipe will disperse in a sufficiently short amount of time after flushing to prevent clogging and/or damage to household plumbing, sewage lines, septic systems, and other standard wastewater treatment equipment.' Moreover, the substantiation must be 'based on the expertise of professionals in the relevant area and have been conducted and evaluated in an objective manner by qualified persons, using procedures generally accepted in the profession to yield accurate and reliable results.'[19]

Another type of greenwashing that has been litigated by both the FTC and private class action lawyers is when general language such as 'eco-friendly,' 'natural,' or 'non-toxic' is used in advertising without qualification, giving rise to consumer confusion about exactly what is meant by such terms. The FTC cautions marketers

[18]Federal Trade Commission (2018).

[19]Federal Trade Commission (2015b).

to avoid making such generic claims. The term 'eco-friendly' conveys that the product 'has far-reaching environmental benefits and may convey that the product has no negative environmental impact.'[20]

A variation of this type of greenwashing is when a marketer says its product is 'made of' a natural ingredient that is associated with a health or environmental benefit. In this regard, the FTC has focused on the labelling and advertising of products in the textile industry that are made of rayon fabric, which can be derived from a variety of plant sources including bamboo. The FTC has cautioned marketers and retailers from labelling rayon-based textiles as bamboo, noting that most 'bamboo products, if not all, really are rayon, which typically is made using environmentally toxic chemicals in a process that emits hazardous pollutants into the air.'[21] Unless the textile is 'made of actual bamboo fiber,' a rayon fabric should be labelled and advertised as either 'rayon,' or 'rayon made from bamboo.'[22]

The 'hidden trade-off' has also received some attention from the FTC. This marketing practice typically involves 'promoting a small act of environmental stewardship while ignoring broader environmental effects.'[23] The FTC has explicitly acknowledged in its Green Guides summary that ignoring the bigger picture, while advertising a product as environmentally-friendly, may be deceptive. For example, claiming a product is "green' or 'made with recycled content,' may be deceptive if the environmental costs of using recycled content outweigh the environmental benefits of using it.'[24]

12.4 ISO Private Sector Standards

Certainly, the FTC Green Guides do not stand in isolation. There are international standards and business in the United States are free to adopt them if they so choose. As part of its ISO 14000 series of voluntary environmental standards, the International Organization for Standardization (ISO) created a group of standards specifically governing environmental labelling.[25] Particularly, the ISO 14020 standards cover three types of labelling schemes: Type I is a multi-attribute label developed by a third party; Type II is a single-attribute label developed by the producer; Type III is

[20]16 CFR 260.4, Ex. 1.

[21]FTC, How to Avoid Bamboozling Your Customers, August 2009.

[22]Id. See 'Nordstrom, Bed Bath & Beyond, Backcountry.com, and J.C. Penney to Pay Penalties Totaling $1.3 Million for Falsely Labeling Rayon Textiles as Made of 'Bamboo," FTC Press Release, December 9, 2015, available at: https://www.ftc.gov/news-events/press-releases/2015/12/nordstrom-bed-bath-beyond-backcountrycom-jc-penney-pay-penalties.

[23]TerraChoice, The Sins of Greenwashing Home and Family Edition, at 10.

[24]FTC, Environmental Claims, Summary of the Green Guides, at 1.

[25]See generally: 'Environmental Labels and Declarations,' International Organization for Standardization, brochure available at: https://www.iso.org/publication/PUB100323.html.

an eco-label whose awarding is based on a full life-cycle assessment. The ISO standards include:

1. *Standard ISO 14020* (Environmental Labelling: General Principles) sets out nine general principles that apply not only to labelling schemes but to all environmental claims, designed to promote accurate, verifiable and relevant information.
2. *Standard ISO 14021* (Environmental Labels and Declarations: Self-Declaration Environmental Claims, Terms and Definitions) sets out requirements for Type II labels, i.e. environmental claims made for goods and services by the producer.
3. *Standard ISO 14022* (Environmental Labels and Declarations: Self-Declaration Environmental Claims, Symbols) promotes the standardization of terms and symbols used in environmental claims, e.g. 'recycled content.'
4. *Standard ISO 14023* (Environmental Labels and Declarations: Self-Declaration Environmental Claims, Testing and Verification) addresses testing and verification policies.
5. *Standard ISO 14024* (Environmental Labels and Declarations: Environmental Labelling Type I, Guiding Principles and Procedures) provides guidance on developing programs that verify the environmental attributes of a product via a seal of approval.

The standard that most closely addresses the subjects of the FTC Green Guides is ISO 14021, which specifies requirements for self-declared environmental claims, including statements, symbols and graphics, regarding products (Type II marketing claims).[26] It further describes selected terms commonly used in environmental claims and gives qualifications for their use.[27] These claims are often based on a single attribute (e.g., a manufacturer's claim that a product is 'biodegradable') without considering the environmental impact of a product's entire life cycle, and without independent verification or certification by a third party. However, these claims must be verifiable, accurate, meaningful, and reliable if consumers are to understand the value of the environmental information they represent (e.g., their ability to protect the environment). The credibility of these claims can be increased if companies and organizations are able to support their claims by providing reliable information to purchasers or potential purchasers seeking to make more informed purchasing decisions on products or services. Consumer expectation can be raised by such claims, given the heightened public interest and awareness in the environment.

According to ISO 14021, an environmental claim is '[a] statement, symbol or graphic that indicates an environmental aspect of a product, a component or packaging. An environmental claim may be made on product or packaging labels, through product literature, technical bulletins, advertising, publicity, telemarketing,

[26]'Environmental Labels and Declarations—Self-declared Environmental Claims (Type II environmental labelling),' International Organization for Standardization, summary available at: https://www.iso.org/standard/66652.html.

[27]ISO 14021:2016(en), Environmental Labels and Declarations—Self-declared Environmental Claims (Type II environmental labelling), available at: https://www.iso.org/obp/ui/#iso:std:iso:14021:ed-2:v1:en:sec:6.

as well as through digital or electronic media such as the Internet.' Self-declared environmental claims are subject to these requirements, which are more detailed than those set forth in the FTC Green Guides. Under ISO 14021, such claims must be:

1. accurate and not misleading; substantiated and verified;
2. relevant to that particular product, and used only in an appropriate context or setting;
3. presented in a manner that clearly indicates whether the claim applies to the complete product, or only to a product component or packaging, or to an element of a service;
4. specific as to the environmental aspect or environmental improvement which is claimed;
5. not restated using different terminology to imply multiple benefits for a single environmental change; unlikely to result in misinterpretation;
6. true not only in relation to the final product but also shall consider all relevant aspects of the product life cycle to identify the potential for one impact to be increased in the process of decreasing another;
7. made if, despite the claim being literally true, it is likely to be misinterpreted by purchasers or is misleading through the omission of relevant facts;
8. presented in a manner that clearly indicates that the environmental claim and explanatory statement should be read together;
9. specific and make clear the basis for the comparison if a comparative assertion of environmental superiority or improvement is made;
10. presented in a manner that does not lead purchasers, potential purchasers and users of the product to believe that the claim is based on a recent product or process modification, if based on a pre-existing but previously undisclosed aspect;
11. not made where they are based on the absence of ingredients or features which have never been associated with the product category;
12. reassessed and updated as necessary to reflect changes in technology, competitive products or other circumstances that could alter the accuracy of the claim; and
13. relevant to the area where the corresponding environmental impact occurs.[28]

The ISO, based in Geneva is perhaps the most influential international voluntary standards development body. Formed in 1946, its members represent standard-making bodies in a large number of countries. There are now 111 countries representing over 95% of the world's industrial output. The standards are available to the member countries to adopt or adapt as they see fit. They have not, however, been adopted by any governmental authority in the United States, nor are they widely used by industry trade groups. Thus, the FTC Green Guides, which are also not legally binding on anyone but which the FTC can enforce via legal

[28]This summary is from implementation of the ISO standard in Canada. See ISO (2008).

proceedings as discussed above, are the standards that marketers in the United States usually find most authoritative.

12.5 Green Certification Marks Have Not Been Effective

A particular type of green advertising tactic is the use of 'certifications,' 'seals of approval,' or 'labels,' either created by the marketer itself or, more typically, as an endorsement by a third-party organization. Greenwashing allegations in this category can come in many variations, but generally involve a label, seal, or certification that either gives the consumer the impression that a product has been certified by an independent third-party organization—when it has not—or conveys insufficient information regarding what specific environmental benefits the 'label' is meant to convey.

Because of the increased use of green certification marks, the FTC issued guidance on this subject in its 2012 revisions to the Green Guides. Notably, the first example listed in the revised Guides indicates that when a marketer awards a 'logo' or 'seal' to its own product, it must indicate that it has done so 'with clear and prominent language' accompanying the seal that at least summarizes what the seal is supposed to mean. In September 2015, the FTC sent warning letters to five providers of environmental certification seals and businesses that use them. The Director of the FTC's Bureau of Consumer Protection stated in the Press Release that '[e] nvironmental seals and certifications matter to people who want to shop green... [b]ut if the seals' claims are broader than the products' benefits, they can deceive people.'[29] In its Press Release, the FTC also provided guidance on the appropriate use of such marks. Marketers should clearly and prominently state the basis of the seal to avoid any 'implied' general environmental benefits claims that cannot be substantiated.[30]

Such voluntary green certification marks are not a substitute for legal mandates. Seal programs can help consumers gain confidence in the environmental practices of the sellers and marketers of consumer products, but the most popular certification programs do not require that remedies be readily available to the victims of

[29]Federal Trade Commission (2015c).

[30]Id. Although the FTC has brought few actions to date specific to internal or third-party endorsements to date, statements by the agency indicate that it intends to ramp up its enforcement of green claims arising from improper use of certifications. The FTC also provided additional guidance on the appropriate use of 'performing seals.' Specifically, the FTC emphasized the need for marketers to clearly and prominently state the basis of the seal to avoid any 'implied' general environmental benefits claims that typically cannot be substantiated. Green 'certifications' or 'seals of approval' have become more prevalent in recent years as marketers attempt to distinguish their products from those of competitors. Such marks are usually created either by the marketer itself or by a third-party organization. Greenwashing problems with certification marks generally involve a certification or seal that gives misleading impressions or insufficient information regarding what specific environmental benefits the mark is supposed to convey.

information misuse, and the scope of the green certification can be narrower than individuals might expect. Moreover, the most popular certification programs do not perform regular and rigorous audits on their client's to ensure that the environmental standards (whatever they may be) are being satisfied.

In theory, green certification marks are a potential market-based solution to greenwashing. Intermediaries that create meaningful marks can raise the saliency of environmental claims in a way that increases the accuracy of consumer decision making without increasing cognitive effort on the part of the consumer. Market-generated signals can attract the attention of consumers if the seals are conspicuously displayed on web sites, advertisements or product packaging. Such signals can reduce cognitive effort and take advantage of framing effects—consumers will accept the information conveyed by the green mark as displayed, without delving further into the details of what the mark stands for. Standardizing and consolidating information disclosure can convey information efficiently. For example, before Congress enacted the Truth-in-Lending Act in the late 1960s, lenders disclosed interest rates in a variety of different ways, requiring consumers to exert substantial effort (and do complex computations) to make comparisons.[31] With mandatory, conspicuous disclosure of APR according to the actuarial method, consumers could compare rates by looking at simple numbers in bold letters in roughly the same location on all credit agreements. This reduced the cognitive effort necessary to acquire the information, as well as the effort to process it. Standard, conspicuous disclosure of environmental information might be conveyed in a similar way.

Thus far, however, green marks and seals have done little to further the accuracy goal, and they can be more misleading than helpful. Seals signify that a product conforms to the standards set by the certification body, but those standards are neither uniform nor particularly strong. The most popular certification programs do little more than signify that the product has environmental benefits consistent with the certification standards and (to the extent the sponsor can audit subscribers), that the seller generally follows those standards. The mark says little about the content of the standards themselves. A product with a green certification seal may have worse environmental impact than a product that does not display (and pay for use of) the mark. Consumers who believe that a green mark confirms a better environmental impact can easily be misled. Indeed, if a mark creates a misleading impression, it may induce more disclosure of information and patronage by consumers. Thus, green marks may reduce cognitive effort, but they can lead to less accurate decision making if they do not convey information consistent with consumer assumptions about the mark's meaning.

For green certification marks to serve as accurate and meaningful signals of environmental practices, there must be incentives for a mark's sponsor to set strong standards. To date, incentives have moved such programs in the other direction. To maximize the number of paying subscribers, it is better to have standards that look impressive to the casual observer but do not impede the ability of subscribers to

[31] OEO (1972).

engage in strong environmental practices. What is needed is a market signal for the marks themselves, or standards imposed by law that would give greater meaning to green certification marks. A regulation might provide, for example, that a ‘five star’ mark means the product complies with certain environmental standards, with fewer stars signalling less environmental protection. Absent such a change, green certification marks will continue to give mixed signals at best.

12.6 Regulating Green Claims at the State Level

Although the FTC is the primary enforcement agency with respect to misleading green claims in the marketplace, firms should nonetheless be aware that state law—particularly in markets known to have stricter laws such as California—may differ and provide easier avenues for potential litigants to bring claims. Each of the 50 states has enacted its own set of consumer protection statutes. US state consumer protection statutes vary to a large extent because each state has frequently enacted and amended statutes in non-uniform ways, either to resolve ambiguities in statutory language, to clarify or supplement consumer rights, or to address new consumer problems. Because of differing statutory language in enacted laws and evolving court interpretations of those laws, consumer protection norms under the array of US state statutes continue to vary widely. Moreover, while there are occasional calls for uniformity or a more national and harmonized approach, the calls generally have not gained widespread support. Indeed, in the last decade or so, calls for harmonization typically originate in the business community, seeking adoption of a national law that will insulate members from liability under state laws that are perceived to be less business-friendly than they would prefer.

The consensus among consumer representatives is that a state-by-state approach to regulating green marketing claims might be preferred unless there is a strong need for uniformity at the national level and a Congressional leadership that is likely to adopt strong standards. There is some support for this state-centred approach. First, it allows for a healthy degree of experimentation and an evolutionary approach to consumer protection nationwide. As states adopt different laws and models to address similar problems, there is opportunity to evaluate which laws and models work best. Over time, the good ideas take hold in other US states, and the weak approaches get strengthened or jettisoned. Second, a particular form of greenwashing may be more prevalent in one state or region before it takes hold in others. A state where the problem is most serious may enact protective legislation that can serve as a guide to others, where the problem is only in its early stages. Third, US state legislatures have been able to react more quickly to emerging consumer problems than the US Congress. Proposed national legislation, or a proposed amendment to existing national legislation, is perceived as a high stakes political event, and competing interest groups are inclined to expend tremendous resources to get their views heard. This often results in legislation either stalling in Congress or being enacted in a diluted form that is the least objectionable to major stakeholders but does little to solve the problem.

The State of California has been particularly active in setting standards to protect consumers against deceptive environmental claims. One example of greenwashing claims (albeit unsuccessful) brought under California law was *Hill v. Roll International Corp.*[32] The plaintiff brought claims under California's unfair competition law, false advertising law, and Consumers Legal Remedies Act, as well as common law claims for fraud and unjust enrichment, alleging that Fiji Water's use of the 'green drop' logo and similar representations on its product 'led [consumers] to believe that those products are 'environmentally superior' when compared to products that lack such designations.' The plaintiff argued that the Fiji 'green drop' logo (and other statements) were deceptive because the plastic packaging itself caused 'as much, if not more, of an adverse environmental impact when compared to similar bottled waters,' and required the consumption of '46 million gallons of fossil fuel, producing approximately 216,000,000 billion pounds of greenhouse gases per year,' for its creation and distribution. The complaint further claimed that the 'green drop' logo was misleading because it 'look[ed] similar' to seals of approval conveyed by independent, third-party organizations.

The California Court of Appeals ultimately found that the claims lacked any merit and dismissed the case. The Court concluded that 'Hill's beliefs do not satisfy the reasonable consumer standard, as expressed in the FTC guides,' in part because the 'green drop on Fiji Water bottles [does not] convey to a reasonable consumer in the circumstances that the product is endorsed for environmental superiority by a third party organization.'[33] The court further concluded that the '[e]very drop is green' slogan, displayed above, which was allegedly used at some point in store displays and advertising, did not 'alter the overall impression conveyed by the green drop and Web site address.'

On January 31 2017, Wal-Mart Stores, Inc. (Walmart) settled a consumer class action lawsuit brought by a group of California state officials, alleging that Walmart sold plastic products that were misleadingly labelled as 'biodegradable' and 'compostable.' As part of the settlement, Walmart agreed to pay $875,000 in civil penalties and to pay $50,000 to CalRecycle, the state recycling agency, to support testing of plastic products that are marketed as compostable or degradable.[34]

This lawsuit was one of several in California since the state passed laws governing 'green marketing' claims, addressing the marketing of plastic products in particular. California's green marketing laws are more demanding than the FTC Green Guides and, unlike the Green Guides they impose mandatory standards. The FTC Green Guides address 'compostable' and 'degradable' claims but they are more general and less prescriptive than California laws. For example, the Green Guides

[32] *Hill v. Roll Internat. Corp.*, 195 Cal. App. 4th 1295, 128 Cal. Rptr. 3d 109 (2011).

[33] 195 Cal. App. 4th at 1305, 128 Cal. Rptr. 3d at 116.

[34] *People of the State of California v. Wal-Mart Stores, Inc., et al.* (Sup. Ct. Napa County, 2017, No. 89). See also: Alameda County (2017).

require only that claims regarding the compostable or biodegradable characteristics of products be based on 'competent and reliable scientific evidence to support those claims.'[35] And they require that for a marketing claim regarding compostability and degradability not to be misleading, the products must be able to be broken down in a 'safe and timely manner,'[36] or 'within a reasonably short period of time after customary disposal.[37] California's plastics green marketing law obligates the regulated community to meet what the state has deemed to be reliable technical standards before certain green marketing terms may be used. The legislature has identified standards for use of the term 'compostable.'[38] And marketing claims that plastic products are 'biodegradable,' 'degradable,' or 'decomposable,' or that 'in any way imply that the plastic product will break down, fragment, or decompose in a landfill or other environment' are banned unless they comply with strict standards.[39]

12.7 Conclusion

Green marketing claims in the United States are regulated in much the same way that any other misleading or deceptive claims are regulated: primarily under the FTC Act and state laws that prohibit misleading and deceptive practices. While the FTC has issued Green Guides as voluntary guidance to industry and consumers that specifically address green marketing claims, those guides are not directly enforceable by the agency or by private parties. They stand as an indication of what the FTC might think are misleading practices but violating the Guides is not necessarily a violation of the FTC Act or any state law. Thus far, international standards on environmental marketing, such as the ISO standards, have not been widely adopted in the United States, either by government authorities or industry groups. Moreover, private sector efforts such as the development of certification marks or eco-friendly seals have not proved effective in signalling to consumers that a particular product or service has significant environmental benefits. Consequently, it is fair to say that there is vast room for improvement in creating a framework whereby consumers who wish to purchase eco-friendly goods and services will have reasonably accurate and complete information that they need to make informed decisions.

---

[35]16 C.F.R. §§ 260.7(b), 260.8(b).

[36]*Id.* § 260.7(b).

[37]*Id.* § 260.8(b).

[38]*See* Cal. Pub. Res. Code §§ 42355, 42357(a), 42357(b).

[39]In California, it is unlawful to sell plastic products that are labelled 'compostable' or 'marine degradable' unless at the time of sale, the food or beverage container, utensil or bag meets the applicable standard. See, specifically: ASTM D6400 for Compostable Plastics; ASTM D7081 for Non-Floating Biodegradable Plastics in the Marine Environment; ASTM D6868 for Biodegradable Plastics Used as Coatings on Paper and Other Compostable Substrates. Pub. Res. Code, §§ 42356, 42357 (2004); Pub. Res. Code, §§ 42359.5, 42359.6 (2006); Pub. Res. Code, §§ 42355 and 42358.5 (2011).

References

Alameda County (2017) 'DA Announces Settlement with Walmart over 'Greenwashing' Claims,' Alameda County District Attorney's Office Press Release February 1, 2017 (Press Release). Available at http://www.alcoda.org/newsroom/categories/press_releases

Federal Trade Commission (n.d.) Environmental Claims: Summary of the Green Guides. Available at https://www.ftc.gov/tips-advice/business-center/guidance/environmental-claims-summary-green-guides

Federal Trade Commission (2012) FTC Issues Revised 'Green Guides. FTC Press Release, 1 October 2012. Available at: https://www.ftc.gov/news-events/press-releases/2012/10/ftc-issues-revised-green-guides

Federal Trade Commission (2015a) Administrative Law Judge Issues Initial Decision in the ECM Biofilms, Inc. Case. FTC Press Release, 5 February 2015. Available at https://www.ftc.gov/news-events/press-releases/2015/02/administrative-law-judge-issues-initial-decision-ecm-biofilms-inc

Federal Trade Commission (2015b) FTC Approves Final Order Requiring Wet Wipe Manufacturer to Substantiate 'Flushability' Advertising Claims. FTC Press Release, 2 November 2015. Available at: https://www.ftc.gov/news-events/press-releases/2015/11/ftc-approves-final-order-requiring-wet-wipe-manufacturer

Federal Trade Commission (2015c) FTC Sends Warning Letters about Green Certification Seals. FTC Press Release, 14 September 2015. Available at: https://www.ftc.gov/news-events/press-releases/2015/09/ftc-sends-warning-letters-about-green-certification-seals

Federal Trade Commission (2018) FTC Approves Final Consent Orders Settling Charges that Four Paint Companies Misled Consumers through Claims Their Products are Emission- and VOC-Free. FTC Press Release, 27 April 2018. Available at: https://www.ftc.gov/news-events/press-releases/2018/04/ftc-approves-final-consent-orders-settling-charges-four-paint

Gingerich E, Karaatli G (2015) 'Eco-Friendly' marketing: beyond the label. J Appl Bus Econ 17 (3):49

GlobeScan, SustainAbility, and BBMG (2012) Re: Thinking Consumption, Consumers and the Future of Sustainability, November 27

International Standard Organization (2008) Environmental Claims: A Guide for Industry and Advertisers, June 2008. Available at http://www.competitionbureau.gc.ca/eic/site/cb-bc.nsf/eng/02701.html#s4_3

Kewalramai D, Sobelsohn RJ (2012) 'Greenwashing': deceptive business claims of 'eco-friendliness'. Forbes, 20 March 2012

OEO (1972) Legal Services Training Program, Memorandum (April 1972), reprinted in John A. Spanogle et al., Consumer Law 108 (2d ed. 1991)

Underwriters Laboratories (2010). Greenwashing Report. Available at: http://sinsofgreenwashing.com/findings/greenwashing-report-2010/index.html

Chapter 13
Collective Valuation of the Common Good Through Consumption: What Is (Un)Lawful in Mandatory Country-of-Origin Labelling of Non-Food Products?

Suvi Sankari

Contents

Abstract This chapter aims to assess whether the concept of sustainable consumption would support a reinterpretation of relevant trade law, namely EU and WTO rules, allowing for robust and harmonious country-of-origin (COO) labelling. Some—but not all—consumers have a bias towards goods and services produced locally, which has been confirmed by relatively recent polling results. Nevertheless, 40% of the EU population, when polled, signals a willingness to pay more for goods 'produced under certain social and environmental standards', and roughly a fifth claim that the origin of products affects their everyday transactional decisions. A product's COO arguably works as a proxy for the social and environmental standards of its production. COO is also material product information in itself, especially in light of product safety statistics. EU case law on (discriminatory) requirements of COO indication has traditionally been interpreted as denoting mandatory COO requirements as 'obviously illegal'. To uphold national COO labelling measures, defences based on consumer protection and the fairness of commercial transactions have been rejected as 'equally applicable in form only', although a duty to disclose COO arguably already exist in EU law and the Commission continues to pursue

S. Sankari (✉)
Faculty of Law, University of Helsinki, Helsinki, Finland
e-mail: suvi.sankari@helsinki.fi

A. d. Amaral Junior et al. (eds.), *Sustainable Consumption*,
https://doi.org/10.1007/978-3-030-16985-5_13

harmonised mandatory COO labelling rules for non-food products. Under WTO law, mandatory COO labelling, as information on processes and production methods (PPMs), is a suspect category of trade barriers, assessing the lawfulness of which may fall under Article XX GATT and Article 2 TBT Agreement. Hence, both EU and WTO legality of mandatory COO labelling remains unclear.

13.1 Introduction

What does trade law offer to consumers? The traditional free trade ideology, or economic rationality, that informs much of WTO law (or more precisely informed its predecessor GATT), as well as EU internal market law as disciplines builds on efficiency or productivity gains arising from comparative advantage or scale economics.[1] The very specific brand of traditional economic theory that conventionally underlies these disciplines perceives the benefit of free trade to consumers in narrow terms of price and quantity, and to society in terms of (global or at least national) economic surplus without any specific emphasis on consumer welfare. The entrenched economic (libertarian) free trade insight actually does not support the idea of having trade agreements at all, only free trade.[2] Instead of unilaterally opting for free trade, nations have opted for multilateral trade agreements. Trade law traditionally justified itself by internalising the externalities of national decisions ('terms-of-trade theory') or anti-protectionism ('commitment theory'). It first focused on banning quantitative restrictions and stabilising, as well as reducing, tariffs and continues to guard against trade barriers. Such barriers are, to WTO law, discriminatory national measures, whereas for EU law all national measures hindering free movement (e.g., protectionist domestic regulation) are barriers. Both the traditional economic as well as legal perspective build on an understanding dating from a remarkably different time than the present in terms of how products and services are produced. Supply or value chains that produce products have become global ('GSCs'/'GVCs') only after the foundations of both disciplines were already firmly carved in stone. Ideologically, the cornerstone of trade law is non-discrimination—of products but perhaps even more the producers, as well as countries they come from—the sole focus on which tends to, albeit inadvertently, in turn, limit the WTO law compliant regulatory choices available for importing

[1]Modern free trade economists (such as Graham, Balassa, Krugman, and Melitz) claim they build on Smith (1776) and David Ricardo (1821). For a concise WTO focused introduction to the strand of free trade economics underlying trade law for lawyers, see Jackson et al. (2013), p. 15ff. For another in a textbook of EU law see: Barnard (2016), pp. 4–8.

[2]For a recent literature review also discussing economic theories supporting the existence of trade agreements, see Rigod (2015), p. 80ff. For one critical of both 'commitment' and 'terms-of-trade' theories see Mavroidis (2012), p. 16ff.

Member States and, in turn, reduce ('legitimate') consumer interests to price rationality only.

As discussed in more detail below, when buying products, a large share of consumers arguably also considers variables other than price. If trade law offers choices to consumers, the choice is by design generally limited to product preference, that is, preference between (like) products as such and not for example between their place of origin, ecological footprint, mode of transport or process of manufacturing. In WTO law, the design may not specifically aim at excluding the possibility of consumer choice based on preferences as to the processes with which products are produced, however, this is the unintended side-effect or practical outcome of laying much emphasis on non-discrimination, lacking WTO AB case-law, and of much unresolved scholarly debate on how the law should be interpreted. Whereas scholarship maintains that EU law has moved beyond non-discrimination regarding the four market freedoms, especially goods, WTO law as a discipline struggles to accept the same.[3] It may be telling that, for example, market access means very different things to these two disciplines.

In WTO law, these other than physical qualities of products are referred to as either product-related or non-product-related processes and production methods (PPMs). Regardless of the GATT Agreement containing Article IX specifically on 'Marks of Origin', arguably under the TBT Agreement one such PPM—although debatable whether a product-related ('pr') or non-product-related ('npr') one—is indication of country of origin (COO). In WTO law, PPMs are not prohibited as such but national measures containing them are challengeable for their WTO compliance as (technical) barriers to trade under both GATT and TBT Agreements.[4] In EU law, national measures requiring mandatory marking of COO on product labels are, at least according to some accounts,[5] considered forbidden product labelling requirements.[6] Whereas in EU law, national (unharmonised) measures requiring mandatory COO labelling carry a stigma of illegality, in WTO law uncertainty concerning its 'correct' interpretation as to measures considered npr-PPMs being automatically

[3]The problems of WTO law seem to essentially relate to two specific issues illustrated by the WTO Appellate Body (AB) 'COOL' ruling from 2012, dealing with US COO labelling for beef (note 7 below): in the GATT context, the AB seems to identify discrimination from data on trade effects of measures (effect on quantity of imports)—such as market preference for US beef—and in the TBT context, the AB fails to distinguish between market-likeness and policy-likeness of products, see Mavroidis (2013). EU law suffers partly from similar issues. For a critical view on market access, see Snell (2010), on how the logic of the EU Court of Justice fails even applying non-discrimination, see Davies (2003), pp. 9–14, 42, 90, and 201 and beyond non-discrimination see Saydé (2017) and Davies (2017).

[4]As no consensus or definitive AB rulings exist on the relationship between the instruments, safest PPM measures comply with both GATT and TBT, see Durán (2015).

[5]Gormley (2009), p. 1212 cites EU Court of Justice Case 207/83 *Commission v United Kingdom*, EU:C:1985:161.

[6]In practice indications of origin take place on labels, and the assumption is that different (national) rules on disclosing COO require different, Member State specific labelling—although indistinctly applicable, still hindrance of trade and one considered to miss legitimate justification.

GATT/TBT illegal has led to WTO members being overcautious in adopting such national measures—a phenomenon witnessed also in the EU legislative context as to COO labelling.[7]

One of the SGDs is called 'responsible consumption and production'.[8] The definition for sustainable consumption and production is defined both by the UN and OECD by citing the Oslo Symposium definition from 1994: 'the use of services and related products, which respond to basic needs and bring a better quality of life while minimising the use of natural resources and toxic materials, as well as the emissions of waste and pollutants over the life cycle of the service or product so as not to jeopardise the needs of further generations'.[9] The goals of high-level trade law, EU and WTO, have also come to include sustainable development simultaneously with each other and the Oslo Symposium.[10] Both EU and WTO legal orders assign considerable weight to the aims or goals of the respective treaties, which translates into teleological, dynamic or purposive interpretation of the law.[11] Regardless of the 1993/4 additions, the value of freedom of trade perceived as freedom to make use of comparative advantage and protection against discrimination afforded to producers and traders in these disciplines remains strong, whereas any consideration for the consumer's role in making decisions on the market based on values other than price or brand remains weak.

More specifically, what is considered problematic about PPMs, in WTO law, is that such measures aim at unilaterally putting pressure on governments of other sovereign countries to amend their product, or production, regulation.[12] This goes against the idea of equality between sovereigns in International Law and could affect comparative advantages. In EU law, what is problematic about Member States'

---

[7]See proposal COM(2005) 661 final, which the Commission withdrew after the WTO AB report in the 'COOL case' (United States – Certain Country of Origin Labelling (COOL) Requirements, WT/DS384/AB/R / WT/DS386/AB/R, adopted 23 July 2012, DSR 2012:V, p. 2449), see COM/2012/0629 final.

[8]Responsible consumption and production is number 12 of the 17 goals as to sustainable consumption and production, possibly to avoid repeating the word sustainability. The UN website suggests consumers can in practice 'help' by '1. Reducing your waste and 2. *Being thoughtful about what you buy and choosing* a sustainable option whenever possible.' http://www.un.org/sustainabledevelopment/wp-content/uploads/2016/08/16-00055L_Why-it-Matters_Goal-12_Consumption_2p.pdf, accessed 15 May 2018 (italics added).

[9]OECD (2002), p. 9.

[10]Under the Treaty on European Union (TEU) Preamble (from 1993), the EU Member States will together 'promote economic and social progress for their peoples, taking into account the principle of sustainable development'. The Preamble of its founding agreement of 1994 states that parties to the WTO aim at 'expanding the production of and trade in goods and services, while allowing for the optimal use of the world's resources in accordance with the objective of sustainable development'—this is added to the goals inherited from the GATT Agreement (raise standards of living, full employment, growth, expanding production and trade).

[11]On international treaty law see Vienna Convention of the Law of Treaties (VCLT) Art. 31(1) and on interpreting EU law see Sankari (2013).

[12]Ankersmit (2017).

measures posing mandatory labelling requirements is that they are considered measures having equivalent effect to quantitative restrictions, prohibited by Article 34 TFEU.[13] Thus, such national measures are caught by Article 34 TFEU and their compliance with EU law is subject to the limited regime allowing for certain derogations from the general prohibition. Moreover, in EU law, facilitating consumer choice between different Member State COOs is considered to enable protectionism or, simply put, discrimination. What this means is that the traditional trade law view—hegemony of free trade and non-discrimination, as well as the WTO dichotomy between process-based and product-based measures—is difficult to reconcile with the idea that consumers should be provided a real chance at consuming more sustainably or responsibly, by basing transactional choices on a broad set of potentially relevant product information. In effect, the lodestars of free trade (law) work in the exact opposite direction than a regime would, had it been designed to protect the more holistic approach of the citizen–consumer. As Kysar (2004) reminds us, the multidimensional approach of the responsible consumer differs from the more compartmentalised approach of market regulators. The consumer can simultaneously consider multiple values related to transactional decisions, weigh the ecological, economic, political and social consequences of choices, and even be altruistic—rationally or not. Free trade (law) furthering responsible consumption should not prohibit but assist in making value-based transactional decisions on the market, the global ballot-box, by shifting focus to informing the consumer from fretting how the consumer might (not) use that information.

This chapter focuses on one PPM or labelling requirement only (COO) and examines the argument that mandatory disclosure of COO would provide the consumer valuable information on the products and services between which the consumer is choosing. Therefore information on COO should legitimately be perceived as material information on the product or service, and the consumer should have a right to this information. Such a right is best guaranteed by mandatory COO labelling requirements. Had the consumers access to this information, they could take it into account which, in turn, could further sustainable or responsible consumption—a concept aligned with the general goal of sustainable development, shared by the EU and WTO.[14] For example, COO is a signal—robust,

[13]For recent French and Italian examples on national labelling requirements (as measures having equivalent effect to import bans and quotas prohibited by Art. 34 TFEU) loosely termed 'gastronationalism', on food labelling measures not notified to the Commission, see https://www.politico.eu/article/belgium-calls-for-eu-help-against-french-gastronationalism/, accessed 15 May 2018 and https://www.politico.eu/article/italian-pasta-labels-test-limits-of-eu-law/, accessed 15 May 2018.

[14]The underlying assumption here is not that all consumers would integrate COO into their transactional decision-making matrix, even if information was available. However, were COO information available, they *could* and a segment of consumers *would* consider COO (ethnocentric ones as COO only but others as a proxy for something else).

although somewhat primitive and approximate, as well as open to manipulation—of the social, labour and environmental conditions under which goods have been manufactured.[15] As discussed in more detail below, COO also correlates with non-compliance with product requirements and product safety. Moreover, COO information is readily available, meaning that detailed information exists throughout the GVCs but is not necessarily given to consumers. The question is, if a valid justification for mandatory COO disclosure exists which in turn could further responsible consumption, could such measures either fit the established doctrines of WTO and EU law, or is there room (post-1993) to reinterpret the doctrines in a way that COO measures would comply with them?[16]

In general, from the point of view of trade law such as WTO and EU law, non-discrimination essentially boils down to treating like products like, and to products having traditionally been considered like regardless of differences in their manufacturing processes or production methods. However, from the point of view of responsible consumption, is it not part of consumers' global political responsibility—each within their economic means—to consider not just product-related but also non-product-related processes and production methods in their private transactional decisions which in turn affect the market? The emphasis in this chapter is on EU law instead of WTO law and on products instead of services. However, as EU law does not operate in a vacuum, WTO law is considered as part of the trade law framework which EU law must comply with. To denote a difference between individual and aggregate choices, this chapter refers with the concept of responsible consumption to the transactional decisions (to consume, or not, and what) made at the individual level (microbehaviour), whereas larger consumption patterns such as aggregate choices of consumers on the market are referred to as sustainable consumption (macrobehaviour).

[15] Only a few of the EU's Southeast Asian trading partners (The Philippines, Sri Lanka, Vietnam) are beneficiary countries of the EU's preferential tariff system ('Generalised Scheme of Preferences', GSP or GSP+), for having put into practice certain central international conventions. See http://ec.europa.eu/trade/policy/countries-and-regions/development/generalised-scheme-of-preferences/, accessed 15 May 2018.

[16] The legality of npr-PPMs has been widely debated in literature on WTO law, but the precise contours of the result remain vague. See Durán (2015), Mavroidis (2013), Charnovitz (2002) and Kysar (2004). Howse and Regan (2000), p. 251 note that: 'It is widely thought that all process-based measures not directly related to physical characteristics of the product itself are prima facie violations of GATT.'

13.2 Concerns Justifying Disclosure of COO to Consumers

Universally, a significant part of consumers has a bias for domestic products, especially in food.[17] Ethnocentric consumers, as they are sometimes referred to, prefer domestic goods also more generally as confirmed by relatively recent EU polling results.[18] Preference for domestic goods is but one of several preferences consumers hold, as both individual consumers, as well as consumers at large have an abundance of (mixed) preferences. For example, polling revealed that 40% of the European population signal willingness to pay more for goods 'produced under certain social and environmental standards'.[19] Additionally, roughly a fifth claim that the origin of products affects their everyday transactional decisions—although the poll does not explain what the impact of COO is in more detail.[20] However, for example in marketing research the correlation between consumers' willingness to pay and products' COO is no novel observation as literature review in the field draws on sources from 1981.[21] Suffice it to say that in their transactional decisions, consumers hold also other interests than solely price.[22] Moreover, such preferences can be endogenously held by consumers, or can be results of social conditioning or other mechanisms,[23] but are not necessarily results of (and neither need to result in) import bans or media campaigns against certain products or producers, governmental or not.

On a general level, EU legislation already divides products into two categories: food and non-food items. For the former category, several food and feed specific pieces of EU legislation already require mandatory disclosure of COO information.[24] For the latter category, no harmonised requirements as to COO disclosure exist.[25] An array of general ('European standards'), as well as sector specific mandatory EU rules and technical specifications (as well as voluntary standards) concern both categories of products. What the EU rules have in common is the aim, in general, of free movement, of high-level consumer protection and, more specifically, the aims of protecting the health, well-being, social and economic interests, as

[17]Lewis and Grebitus (2016). See also: https://ecommercenews.eu/uk-consumers-want-country-origin-labels-online-products/, accessed 15 May 2018.

[18]Eurobarometer (2010), Cappelli et al. (2017).

[19]More precisely, acccording to Eurobarometer (2010), p. 44: 'A significant proportion of Europeans are willing to pay more for products which help the environment, respect social standards, help developing countries or which are made in their countries'.

[20]Ibid.

[21]For example, including references to literature since 1981, see Aichner et al. (2016).

[22]The hegemony of price is taken at face value, whereas speculating on other values involved in consumer decision-making requires referencing. For one illustrative example of this, see Van den Bossche et al. (2007), p. 64.

[23]Schaefer and Crane (2005).

[24]For references to some, see the Preamble of Regulation (EU) No 1169/2011.

[25]Commission proposals—in more detail below.

well as safety of consumers, and protecting consumers against unfair commercial practices. More precisely, in existing EU law, mandatory disclosure of the COO of food-items is justified, for example, with references to omitting it misleading consumers, to securing a level playing field for businesses, to food-scares or crises, as well as to consumer interest and undisclosed 'particular interests'.[26] With the exception of drawing attention to food-scares and interests of the general public (including Commission impact assessments), the justifications for legislation do not differ between food and industrial products. Moreover, the means for reaching the desired ends do not much differ as legislation on both categories heavily relies on informing the consumer, although the approach can also be criticised building on behavioural insights.[27] In food, the COO rules follow the Union's non-preferential rules of origin established in its Customs Code which in turn are aligned with WTO law. Further, voluntary disclosure of the COO of food should comply with this harmonised COO criteria. Hence, one should ask why mandatory COO labelling of industrial products (in the EU) would be any different from that in food?

As to the enforcement of the (high) regulatory standards of the EU, legal academics have questioned whether the style in which the mutual recognition principle is applied on its internal market has led to a situation where all goods in circulation, especially those imported from outside the EU, must meet zero technical standards. This, instead of meeting the intended one standard—harmonised or not—or the banned dual burden of two sets of different standards, which is what mutual recognition aims to avoid.[28] Conversely, regardless of existing regulation setting detailed high-level technical requirements for products and the materials they are made of, a significant part of non-food products (i.e., industrial products) in circulation on the internal market do not comply with EU requirements for them. This is not only problematic from the point of view of immediate issues concerning consumer health and safety and level playing field for business, but also from the more environmental sustainability point of view: non-compliance with core sets of European standards and product requirements also has long-term effects on, for example, energy use and (toxic) waste.[29] The European Commission (Commission) has also recognised non-compliance as an issue. Hence, citing benefitting both

[26]See Recitals 29–33 of the Preamble to Regulation (EU) No 1169/2011.

[27]Helleringer and Sibony (2016–2017).

[28]More specifically Gormley (2008), pp. 1649–1650, argues: 'However, *Cassis de Dijon* contained two central errors: first, the ECJ spoke about goods 'lawfully produced and marketed' in another Member State, whereas it should have referred to goods 'lawfully produced or marketed in another Member State,' secondly, in the (illustrative) list of justifications for measures, the ECJ referred to the protection of public health. This latter error was subsequently corrected by the ECJ, although, surprisingly, the former never has been'. (References in the original omitted here.)

[29]An NGO report from 2017 ('Keeping it Clean: How to Protect the Circular Economy from Hazardous Substances'), commissioned by the European Environmental Bureau from Client Earth draws attention to the fact that the current EU legal framework does not adequately protect against hazardous chemicals entering the circular economy as materials recovered from waste for making new products, see http://eeb.org/publications/83/waste-and-recycling/1651/report-on-how-to-protect-the-circular-economy-from-hazardous-substances.pdf, accessed 15 May 2018.

businesses and citizens as the object, the Commission has recently proposed to strengthen market surveillance, because '[t]he increasing number of illegal and non-compliant products on the market distorts competition and puts consumers at risk.'[30] The Commission recognises that the main reasons for this development are economic operators' lack of knowledge of the rules, as well as their choice to *intentionally disregard the rules* to gain a competitive advantage.[31] Additionally, the choice of disregarding rules as a strategy is met—if not rewarded—on the EU internal market with lacking enforcement mechanisms: ineffective market surveillance and deterrence failure, especially as to the non-EU operators forming part of global value chains generating industrial products.[32]

13.2.1 The Case for COO

The Commission's view is at least in part empirically based, relying on empirical data on specific industrial sectors and the market generally.[33] Additionally, according to EU customs data,[34] in 2015 customs enforced 'sanitary, phyto-sanitary and veterinary technical standards' (i.e., a loose category of basic product requirements protecting human, animal and plant health) in 14,000 cases out of 347,000,000 imported articles (i.e., 166,000,000 customs declarations, known as SADs[35]). Plainly, this means that customs refused entry in less than 0.01% of imports on this ground.[36] Moreover, altogether 'more than 37 million items were identified as unsafe or uncompliant in terms of product safety' by the customs—meaning 10% of all imported items.[37] These aggregate figures say fairly little about the level of customs scrutiny or the COO of non-compliant products.

However, as to the first point, the customs part of market surveillance, the fact that 63% of all customs declarations (both import and export) are 'cleared' in under 5 min—91% within 1 h of reception and only 2% take more than 48 h—suggests

[30]COM (2017) 795 final, p. 1.

[31]COM (2017) 795 final.

[32](SWD(2015) 202 final, pp. 95–97, and references there).

[33]Commission cites some non-compliance numbers in its Inception Impact Assessment document on the initiative 'Internal Market for Goods – Enforcement and Compliance', of 13 May 2016, based on several inspections and studies cited in the document, http://ec.europa.eu/smart-regulation/roadmaps/docs/2017_grow_007_enforcement_compliance_en.pdf, accessed 15 May 2018.

[34]EU Customs Union – Facts and Figures https://ec.europa.eu/taxation_customs/sites/taxation/files/facts_figures_en.pdf, accessed 15 May 2018.

[35]Single Administrative Document (SAD).

[36]Assuming the cases were equally distributed between the some 2000 customs offices, or entry points, into the customs union, there would be 9.5 cases (out of 83,000 declarations) of such enforcement per entry point each year.

[37]EU Customs Union – Facts and Figures https://ec.europa.eu/taxation_customs/sites/taxation/files/facts_figures_en.pdf, accessed 15 May 2018.

relatively few or some relatively superficial actual (physical) customs checks are in fact being carried out. As to the second point—COO—in terms of what is in free circulation on the internal market, either as having originated in the EU or having passed customs onto the market, estimates of non-compliance with (a broad category of) standards vary generally from 5% to 53% for products, depending on the sector of the EU internal market.[38] Hence, it seems neither customs nor market surveillance authorities do not adequately detect or address non-compliant products, which all feeds into the deterrence failure. Therefore, presently, consumers are offered plenty of choice in non-compliant products on the market.

Despite lax customs and market surveillance, in detected cases of non-compliance the relationship between non-compliant products and certain countries of origin is straight forward. According to the Commission, 59% of all RAPEX[39] notifications between 2010 and 2016 concerned products of Chinese origin.[40] Comparing the origin of other goods circulating on the internal market, in 2014, yearly imports from China made up for 13% of all imports to the EU, whereas total intra-EU imports were roughly at the same level (15%). According to market surveillance statistics in 2015,[41] there is a difference between these two categories in the relatively inordinate amount of RAPEX notifications concerning products of Chinese origin (1262 notifications) when compared to goods produced in all EU countries combined (290 notifications). This is a trend which remained rather stable between 2004 and 2016, notifications on goods of Chinese origin increasing and EU origin decreasing.[42] However, in 2017, the respective numbers were 1167 (53% overall) and 574 (26% overall) for alerts, and the Commission explains rising number of alerts in goods of EU origin with alerts increasingly concerning motor vehicles (of EU origin).[43] Relying on the longer trend (2004–2016) and assuming the choice to inspect and notify is unbiased and builds on accurate information, then COO seems to matter regarding product safety. Moreover, IP protection most certainly relates to misleading consumers and to level playing field for businesses but may or may not relate to consumer safety as well. In 2012, 64% of all counterfeit or falsified

[38]Numbers cited by the Commission (note 29 above).

[39]The EU's Rapid Alert System for Dangerous Non-Food Products (RAPEX).

[40]According to Sajn (2018), p. 2: '[t]hese notifications refer only to products suspected of presenting a 'serious risk', and typically not to products whose non-compliance refers to administrative requirements, such as labelling or placing of warnings.'

[41]Rapid Alert System for dangerous products–2015 Results, p. 31, https://ec.europa.eu/consumers/consumers_safety/safety_products/rapex/alerts/repository/content/pages/rapex/reports/index_en.htm, accessed 15 May 2018.

[42]Rapid Alert System for dangerous products–2016 Results, p. 19, https://ec.europa.eu/consumers/consumers_safety/safety_products/rapex/alerts/repository/content/pages/rapex/reports/index_en.htm, accessed 15 May 2018.

[43]Commission press release, http://europa.eu/rapid/press-release_MEMO-18-1722_en.htm, accessed 15 May 2018.

goods seized upon EU borders were from China,[44] in 2014 80% of detained counterfeit goods originated from China, and the figure overall remains the same in 2016.[45]

As to other possible reasons consumers might have for COO preference, according to one recent survey of the general public 93% of French, 92% of German, and 88% of Spanish public consider China does not respect personal freedoms (human rights) of its people.[46] This is the consumer understanding, without going deeper into whether it is accurate or not, and presumably labour rights are not assumed to be better off. Thus far, we have established that consumers have broader than price preferences, fact-based or not, including reasons for rational and empirically-based COO related health and safety concerns, environmental worries, intellectual property right worries, and general unease with the societal circumstances under which products are made. Should the consumer have information on COO, these preferences might all play into consumers' transactional decisions: the question remains, should the consumer have that information?

13.3 Why Would Mandatory COO Information Be Allowed or Prohibited in WTO and EU Law?

13.3.1 EU Law Approach to COO

By acknowledging the general lack of compliance—and that market surveillance does not work—together with the more specific numbers above, one could argue that a COO indication on product labels would concern not just consumer protection in general but provides the consumer enhanced information directly related to also product safety. Viewed in this light, the latest move in consecutive Commission proposals for mandatory COO indication in non-food products is easier to understand. The Commission has in the past, in 2005, unsuccessfully pursued a specific instrument harmonising mandatory COO labelling, applicable *only* to products from third countries (i.e., non-EU countries). The reasons stated for the proposal included 'growing concern over the mounting incidence of misleading and/or fraudulent origin marks'—in a vein similar to Article IX GATT on Marks of Origin which refers to protecting consumers from fraudulent or misleading use of COO—as well as disadvantage concerning trade partners who require COO labelling and the

[44]European Commission http://trade.ec.europa.eu/doclib/docs/2009/september/tradoc_144591.pdf, accessed 15 May 2018.

[45]EU Customs Union—Facts and Figures https://ec.europa.eu/taxation_customs/sites/taxation/files/facts_figures_en.pdf, accessed 15 May 2018. and Report on EU customs enforcement of intellectual property rights: Results at the EU boarder in 2016, pp. 21–22. https://ec.europa.eu/taxation_customs/sites/taxation/files/report_on_eu_customs_enforcement_of_ipr_at_the_border_2017.pdf, accessed 15 May 2018.

[46]Pew Research Centre (2015).

differing national (EU Member State) rules on voluntary COO marking.[47] The Commission nevertheless withdrew the 2005 proposal in 2012 right after the WTO AB handed out its COOL decision, citing the following reasons: 'In addition to lack of agreement in the Council, recent developments in the legal interpretation of WTO rules by the organisation's Appellate Body have rendered this proposal outdated.'[48] More recently, in 2013, the Commission has pursued the same goal—although this time with *universal applicability* instead of limiting the measure (discriminatorily) to third country products—as part of consumer product safety, emphasising only product traceability for market surveillance purposes.[49] In addition to the scope of mandatory COO labelling becoming universal, the European Parliament (EP) in its amendments to the proposal specifically reduced the labelling requirements to stating COO only in English.[50] However, the proposal has faced political deadlock in the Council for unspecified reasons.[51] Regardless, the Commission has not withdrawn the proposal, as it considers the measure would be compatible with the 'goods package' it currently furthers (COM(2017) 787 final), should the necessary political agreement for accepting it be found. This proposal for a Regulation replacing the General Product Safety Directive (GPSD) which includes mandatory COO labelling of industrial products hence remains on the table, which may also indicate that with these modifications it is considered WTO law compliant.

Although not problematised by the Commission in either of the proposals just discussed, the question remains whether the proposed COO rule of the currently deadlocked proposal for a 'GPSD' Regulation is compatible with EU law itself. A following type of concern over protectionist aim of the EU measure—which needs to comply with WTO law—has been reported: 'There have also been criticisms that the proposed indication of origin obligation [Art. 7] is more a desire to protect European

[47]The 3rd recital of the preamble of the Proposal COM(2005) 661 final even states that: 'The economic significance of origin marking to consumer decision and trade is recognized by the practice of other major trading partners which have enacted mandatory origin marking requirements.'

[48]Communication (COM/2012/0629 final) including withdrawal of Proposal for a Council Regulation on the indication of the country of origin of certain products imported from third countries (COM(2005)661, 2005/0254/COD).

[49]Proposal for a Regulation (COM(2013) 78 final). Especially EP amendments 31–32 to the proposal continue to cite as reasons for the legislation: consumer access to supply chain information and increased level of awareness; reducing the risk of misleading the consumers as to the country of production; and similar practice being applied by many EU trade partners.

[50]Concerning the point of labelling requirements as trade hindrance, the EP amendments to COM (2013) 78 final added that COO can throughout the EU be indicated in English only (see Art. 3(a) of the proposed Regulation).

[51]Note from General Secretariat of the Council to Council (ST 8985 2016 INIT–2013/048 (OLP)). The political deadlock concerns Art. 7 of the proposal, which includes paragraphs on mandatory COO labelling for all industrial products, harmonising legislation to determine COO, the option to use either a Member State or the EU as COO, and authorisation to use English in doing this. Documents on the legislative process do not reveal which of these points gives rise to the deadlock, however, 11 Member States are willing to accept the proposal, whereas another 11 (BG, CY, EL, ES, FR, HR, IT, MT, PT, RO, SI) wish to delete Art. 7.

markets from competition from countries like China than having any necessary connection with safety.'[52] Whether the same suspicion of protectionism applies between different EU Member State COO is unclear. The traditional EU internal market approach, or interpretation of the law, dictates that *mandatory* national measures of Member States requiring COO labelling when applied (only) to *goods produced in other Member States* are distinctly applicable measures of equivalent effect to quantitative restrictions and serve no legitimate purpose, only give rise to discrimination and protectionism (between Member States).[53] Applying this interpretation based on national measure of an EU Member State that distinctly targets goods from other Member States to national measures requiring disclosure of COO universally (indistinctly) is highly questionable, however, uncertainty over whether indistinct national measures are both EU and WTO law compliant might serve as one reason to introduce harmonised EU legislation on COO labelling. The relatively few EU Court of Justice cases that exist to date do not address the distinction between different foreign countries of origin—if there is a difference between Member States and non-Member States as COO—and hence is mostly limited to assessing EU Member States (the internal market) as potential COO.[54]

As the EU Court of Justice (the Court) has not further clarified or revised its approach to justifying restrictions to free movement of goods within the internal market in case of indistinct and distinct national measures, the academic consensus remains that national mandatory COO labelling measures are caught by Article 34 TFEU.[55] As additional labelling requirements (producing costs), the measures are, first, restrictions to the free movement of goods between EU Member States and hence contrary to Article 34 TFEU.[56] Second, their purpose, the aim, is perceived inherently (not just arbitrarily or disguisedly) discriminatory in nature: 'the purpose of indications of origin or origin-marking... - ...is to enable consumers to distinguish between domestic and imported goods and that this enables them to assert any prejudices which they may have against foreign goods.'[57] Because the measures are

---

[52]Howells (2014), p. 531.

[53]For example Gormley (2009 at p. 430, citing Case 207/83) argues that '[w]hilst the defences of consumer protection and the fairness of commercial transactions in particular have been advanced to justify an equally applicable origin marking requirement, they have been rejected on the basis that such an origin marking requirement is equally applicable in form only.'

[54]Case 249/81 *Commission v Ireland*, Case 222/82 *Apple and Pear*, EU:C:1983:370, para. 18; Case 207/83 *Commission v United Kingdom*, EU:C:1985:161, para. 17, Case 325/00 *Commission v Germany*, para. 23, but cf. Case C-95/14 *Unione Nazionale Industria Conciaria (UNIC) and Unione Nazionale dei Consumatori di Prodotti in Pelle, Materie Concianti, Accessori e Componenti (Uni.co.pel) v FS Retail and Others*, EU:C:2015:492, paras 41–44.

[55]The prohibition of quantitative restrictions to the free movement of goods.

[56]Case C-95/14 (note 58 above), para 45: Furthermore, it is clear from the case-law in relation to Article 34 TFEU that language requirements such as those laid down by the national legislation at issue in the main proceedings constitute a barrier to intra-Community trade in so far as goods coming from other Member States have to be given different labelling involving additional packaging costs (judgment in Colim, C-33/97, EU:C:1999:274, paragraph 36).'

[57]Case C-95/14 *UNIC and Uni.co.pel*, EU:C:2015:492, para. 44.

discriminatory (distinctly applicable to imported products), the traditional interpretation of EU law is that the so-called 'mandatory requirements' cannot be successfully invoked to derogate from the right to free movement as in the case of indistinctly applicable measures, only the express derogations listed in Article 36 TFEU.[58] In spite of forceful criticism against this view,[59] the Court has thus far avoided developing its interpretation and is perceived at times to work around this omission by skipping the thorny issue of establishing discrimination (whether a measure is distinctly or indistinctly applicable—in law or in fact, or directly, indirectly or not at all discriminatory) and discussing justifying restrictions to trade only.[60] Measures adopted by EU institutions must abide by Article 34 TFEU, however, indistinctly applicable harmonised COO disclosure most likely could stand review against Article 34 TFEU if considered not to be a manifestly inappropriate measure. For the Court to find a measure manifestly inappropriate, the EU legislator must have made a manifest error as to the suitability or proportionality of the measure in terms of the objective it pursues.

The general exceptions included in Article 36 TFEU include national measures that aim to protect health and life of humans, however, the above problem remains whether a COO measure concerning industrial products can pursue some other objective than feeding prejudices (protectionism) as its *main* objective. Court case-law on national *universally applicable* (i.e., indistinct) mandatory COO indication measures does not exist. Whatever the standard against which national measures are reviewed, one would assume a harmonised EU measure on COO labelling, such as the universally (indistinctly) applicable measure proposed by the Commission in 2013, requiring no EU Member State specific relabelling of products could be compatible with EU law. The objective the measure pursues is, in the 2013 proposal, product safety. The proposed measure could have an effect in furthering responsible consumption and production, which in turn would contribute to reaching the EU goal of sustainable development.

As to the objective of COO measures, what seems to underlie the EU approach when consumers are concerned is hardly strict adherence to thinking consumers are only or even mainly guided by price-rationality. If consumers are the end-users of goods the 'protectionism view' has in mind, then the approach assumes consumers to be (illegitimately) ethnocentric which gives rise to their protectionist behaviour. The traditional EU approach—likely building on a certain branch of economics—seems to regard consumers *en bloc*, as prone to protectionism, and does not recognise there are different types of consumers who hold and balance plenty of values at the same time: there are also consumers who have legitimate reasons, for

[58]Gormley (2009) and Barnard (2016).

[59]Enchelmaier (2010), European Commission (2013) and Shuibhne (2017).

[60]In addition to indistinct and distinct measures, a third category exists, Case C-110/05 Commission v Italy ('Trailers'), EU:C:2009:66, para. 37: '[a]ny other measure which hinders access of products originating in other Member States to the market of a Member State'.

example, empirically based product safety concerns, for giving COO (positive or negative) weight among all the information they consider before purchase.

Further, one can argue that EU law already exists on which consumers could rely as to their right to COO disclosure. If considered material information on what is being purchased, a trader not disclosing this information would be breaching the rights granted to consumers by the Unfair Commercial Practices Directive.[61] Although the Directive textually allows for such an alternative interpretation, and this could be supported by referring to the overall goal of sustainability and the empirically-based information above, this is not how the text is currently interpreted. However, the legal challenges faced by a harmonised EU measure on mandatory COO disclosure do not stop here but include complying with WTO law.

13.3.2 The WTO Approach to Mandatory COO

The central arguments beyond non-discrimination that can be made from the WTO law perspective against mandatory COO disclosure relate to arguments against PPM labelling in general. As put concisely by Cheyne:

> [T]he desire to control consumer labelling is linked to two concerns: the idea that the use of non-product-related PPMs needs to be controlled because it is inappropriate and anti-competitive, and anxiety about the 'irrational consumer' who is misguided or misled about complex issues.[62]

The first, more producer-oriented concern revolves around costs and market effects of labelling requirements. The second, more consumer-oriented concern is sadly patronising and denies consumers access to a direct mechanism for collective valuation. Simply put, it makes responsible consumption difficult if not impossible.

Although the concerns giving rise to the law may be relatively straight forward, the law itself is not. All labelling requirements (both mandatory and not pr-PPMs and npr-PPMs) definitively fall under the Technical Barriers to Trade (TBT) Agreement. Scholarship maintains that the relationship between the TBT and GATT Agreements is unclear, and no definitive body of WTO AB case-law exists on the issue.[63] Mavroidis argues that 'the TBT should be regarded as substitute and not complement to the GATT, at least with respect to Articles III and XX', and hence the

[61]Cf. Art. 7 of Directive 2005/29/EC on the trader's duty to disclose material information on the product required for the 'average consumer' to take a 'transactional decision' on products and services. See also Howells et al. (2006).

[62]Cheyne (2014), p. 330.

[63]For example, the famous WTO AB US–Tuna and US–Shrimp cases deal with absolute import bans only, and the COOL case seems to be the first one dealing with lesser barriers—marks of origin—to trade under both TBT and GATT.

compliance with WTO law of technical regulations should be assessed against Article 2 TBT only.[64] Under Article 2 TBT, technical regulations can concern the product or PPMs, and to stand review need to have a legitimate aim, and be both necessary, as well as applied non-discriminatorily, to like products. Hence, for a measure requiring mandatory COO labelling to stand WTO law review, for sure, it would need to comply with both TBT and GATT Agreements.[65]

Although the analyses proceed somewhat differently, standing WTO law scrutiny requires—concerning both TBT and GATT—that the measure is not arbitrarily discriminatory or a disguised restriction of international trade and is necessary to protect a legitimate aim. Before crossing those hurdles, one central question in terms of PPMs as to whether discrimination exists revolves around what are determined 'like products'. The broader or narrower definition given to 'likeness' much affects the chances of a given measure to stand WTO review. Mavroidis (2013), for example, suggests that in the context of TBT, the AB should distinguish between 'market likeness' and 'policy-likeness' of products, which it at the moment does not do. As to 'policy-likeness', Andersen points out that the question is problematic in, for example, accepting that there is a market for products of child slave labour and for similar products that are not.[66] The PPM measure needs to have a legitimate aim to stand review. For the mandatory COO disclosure measure example, the GATT Art. XX exhaustive list of such aims includes, for example, protecting human, animal or plant life or health. More precisely, the objectives as stated in Article XX are those: '(b) necessary to protect human, animal or plant life or health' considered to cover also protecting the environment; and '(g) relating to the conservation of exhaustible natural resources if such measures are made effective in conjunction with restrictions on domestic production or consumption'; as well as to an extent sub-paragraph (e) 'relating to products of prison labour'. The legitimate objectives that justify technical barriers to trade under the TBT Agreement are non-exhaustively listed in Art. 2.2 TBT.[67] The list does not include consumer protection but does, *inter alia*, include protection of human health, as well as environment.[68]

However, it seems that although establishing legitimate objective under both GATT and TBT Agreement might succeed in our COO measure example, the hurdles of non-discrimination and the test of necessity (in EU parlance close to proportionality review) may be tricky.[69] Additionally, the COOL ruling of the WTO

[64]Mavroidis (2013), p. 524.

[65]Durán (2015), p. 112ff.

[66]Andersen (2015), p. 398.

[67]WTO AB has not provided further guidelines for extending the list, see Andersen (2015, p. 396).

[68]TBT 2.2 Article lists as legitimate objectives, *inter alia*: 'the prevention of deceptive practices; protection of human health or safety, animal or plant life or health, or the environment'. The non-exhaustive list of objectives is complemented by a non-exhaustive list of what is relevant to consider in assessing such risks. Both lists are open, the first for the general goal of sustainable consumption, the second for consumer preference (responsible consumption).

[69]Mavroidis (2012), pp. 355ff and 689ff.

AB from 2015—the Panel decisions in which case inspired the Commission to withdraw the EU COO measure proposal of 2005—seems to complicate things further. Problems essentially relate to two specific issues illustrated by the ruling that dealt with US mandatory COO labelling for domestic and foreign beef alike. Unlike with import bans such as were at question in US–Shrimp and US–Tuna, the labelling system merely informed consumers of the origin of meat. In the GATT context, the AB seems to identify discrimination (disguised restriction) from data on trade effects of measures (an effect on quantity of imports)—regardless of whether they emerge from presumed market preference for US beef or, as in this case, operative complications in the meat supply chain leading to, in effect, not processing as much meats of certain origin as before. In the TBT context, the AB fails to distinguish between market-likeness and policy-likeness of products, applying a broad understanding of likeness—opening a vast field for econometric research on potential market effects.

The aspect of necessity of the measure is evaluated similarly in both GATT and TBT reviews. The more important the protected aim and the less restrictive the measure effectively to achieve it, the stronger the chance for the national measure to stand review. Hence, based on the product safety example above, one could argue that disclosing COO of all products, domestic and foreign, to consumers is a necessary and non-discriminatory measure justified by the legitimate objectives of protecting the environment and human health. The aspects related to non-compliant toxic products finding their way into the circular economy could be one alternative way to justify the disclosure of COO—to curb the purchasing of products from countries of origin that struggle to meet EU product requirements. As argued above, this approach builds on two ideas. First is the idea that as COO in itself is not a wholly reliable indication of compliance or non-compliance but only an empirically-based proxy, hence it could justify employing a 'softer' market signal, a lesser barrier to trade than an import ban. Second is the idea that research has shown consumers are a diverse group with variable values affecting their transactional decisions and hence the market effects of their decisions based on more encompassing information will be diverse. It is difficult to foresee whether trade effects of such a measure would concern foreign products, but this would be rather incidental. Finding a way to interpret WTO law in a way that would allow mandatory COO labelling measures to stand would be supported by increasing the role of responsible consumption to further sustainable consumption and thereby contribute to reaching one of the WTO law goals: (global) sustainability. Further, several mandatory COO labelling regimes exist currently and their compliance with WTO law has not been challenged—however, this is no proof as to the WTO lawfulness of such measures.

13.4 Conclusions

The thorny bundle of issues that COO labelling is part of, the (non-)product-related processes and production methods (PPMs), relates to consumers holding preferences for more sustainable process-distinguished products. As described by Kysar:

> The heroic role of the consumer, then, is becoming even more heroic. Long understood to include a patriotic obligation continually to increase expenditure on material goods, the consumer's role also is being cast as an unwitting mechanism for collective valuation [by revealing through private market behavior their true level of support for human safety, the environment, and a host of other public goods].[70]

Politicians may be interested in tracking this market behaviour to gain knowledge of their voters' preferences, and economists may continue to rely on rational self-interested cost–benefit analyses to explain consumer behaviour. For the consumer, however, market choices may be more intrinsic and direct means to a preferred end, including the end of responsible or sustainable consumption, than is traditionally thought. This direct mechanism of responsible consumption market choices, as well as the more indirect signal that travels through politicians to shape decisions on societal collective good both require that the necessary information—such as PPMs—on which to base those choices is available.

From the consumer perspective, the 'more information' approach—that of mandatory COO disclosure—is warranted by increased autonomy and freedom of choice, as well as, in the light of the above, health, safety and environmental protection. From the market perspective, COO disclosure is supported by the goal to maintain fair competition between businesses (manufacturers), by rewarding costs of compliance.[71] This approach as to mandatory disclosure of COO is already widespread in EU food law.

This chapter argued that based on statistical information concerning correlation between product safety and COO, consumers have a rational reason to consider also COO when making transactional choices. With many other social, environmental and other variables, the COO of a product works as a good available proxy as to its PPMs. Moreover, consumers as a heterogeneous group balance many different values against each other and arrive at alternating results. In effect, it is through this mechanism that responsible consumption by individuals may turn into collective valuation of common good through consumption. The information legally required to be disclosed to the consumer plays a role in whether consumers can be responsible if they wish to and, in turn, whether their market choices lead to sustainable consumption.

If there is a problem with disclosure, it is not lack of information on COO of products generated by global value chains. The same goes for the parts from which products are assembled, if not courtesy of in-house electronic systems tracking the

[70]Kysar (2004), pp. 533, 635.

[71]Sajn (2018).

supply chain, then because of existing information at parcel-specific EAN code accuracy that is regularly required by customs whenever goods are transported. At least larger companies already hold COO information for a variety of reasons: customs rules on origin (RoO) are essential for accessing markets and for customs officials defining customs duties, moreover, providers of financial services, as well as clients may require this information. A PWC survey is said to have found that companies in fact are 100% aware of RoOs,[72] which in turn means the COO of items is relevant and known to economic operators.

Hence, COO disclosure seems to hinge upon its lawfulness. As argued above, although considered a thorny issue in trade law in the past, presently there are good reasons to suggest that an interpretation of relevant EU and WTO law that considers the goal of sustainable development should allow an EU measure of mandatory COO disclosure. Therefore, the remaining obstacle for mandatory COO disclosure in the EU context seems to presently be that either the Member States are reluctant to allow the EU to decide on rules concerning 'made in' rules instead of each Member State deciding for themselves, or there are interests involved that wish not to disclose COO to consumers, or both.

References

Aichner T, Forza C, Trentin A (2016) The country-of-origin lie: impact of foreign branding on customers' willingness to buy and willingness to pay when the product's actual origin is disclosed. Int Rev Retail Distrib Consum Res 27(1):43–60

Andersen H (2015) Protection of non-trade values in WTO appellate body jurisprudence: exceptions, economic arguments, and eluding questions. J Int Econ Law 18:383–405

Ankersmit L (2017) Green trade and fair trade in and with the EU: process-based measures within the EU legal order. Cambridge University Press, Cambridge

Anliker M (2016) Non-preferential rules of origin: high level assessment. Price Waterhouse Coopers, as cited by Hoekman B, Inama S in harmonization of rules of origin: an agenda for plurilateral cooperation? East Asian Econ Rev 22(1):3–28

Barnard C (2016) The substantive law of the EU: the four freedoms, Fifth edn. Oxford University Press, Oxford

Cappelli L, D'Ascenzo F, Natale L et al (2017) Are consumers willing to pay more for a 'Made in' Product? An empirical investigation on 'Made in Italy'. Sustainability 9:556–573

Charnovitz S (2002) The law of environmental PPMs in the WTO: debunking the myth of illegality. Yale J Int Law 27:59–110

Cheyne I (2014) Consumer labelling in EU and WTO law. In: Gaines S et al (eds) Liberalising trade in the EU and the WTO: a legal comparison. Cambridge University Press, Cambridge, pp 309–332

Commission Staff Working Document, A Single Market Strategy for Europe—Analysis and Evidence, Accompanying the document, Upgrading the Single Market: more opportunities for people and business (SWD(2015) 202 final)

[72] Anliker (2016).

Communication from the Commission to the European Parliament, the Council and the European Economic and Social Committee, The Goods Package: Reinforcing trust in the single market (COM(2017) 787 final)
Communication from the Commission to the European Parliament, the Council, the European Economic and Social Committee and the Committee of the Regions, 'Commission Work Programme 2013' (COM/2012/0629 final)
Consolidated version of the Treaty on European Union, OJ C 202 (2016)
Davies G (2003) Nationality discrimination in the European internal market. Kluwer Law International, The Hague
Davies G (2017) Between market access and discrimination: free movement as a right to fair conditions of competition. In: Koutrakos P, Snell J (eds) Research handbook on the law of the EU's internal market. Edward Elgar, Cheltenham, pp 13–28
Directive 2005/29/EC of the European Parliament and of the Council of 11 May 2005 concerning unfair business-to-consumer commercial practices in the internal market and amending Council Directive 84/450/EEC, Directives 97/7/EC, 98/27/EC and 2002/65/EC of the European Parliament and of the Council and Regulation (EC) No 2006/2004 of the European Parliament and of the Council ('Unfair Commercial Practices Directive') (OJ L 149, 11.6.2005, p. 22–39)
Durán GM (2015) NTBs and the WTO agreement on technical barriers to trade: the case of PPM-based measures following US—Tuna II and EC—seal products. In: Herrmann C, Krajewski M, Terhechte J (eds) European yearbook of international economic law. Springer, Berlin, pp 87–136
Enchelmaier S (2010) Article 36 TFEU: general. In: Oliver P (ed) Oliver on free movement of goods in the European Union. Hart Publishing, Oxford, pp 215–312
Eurobarometer (2010) SPECIAL EUROBAROMETER 357 'International Trade'
European Commission (2013) 'Free movement of goods Guide to the application of Treaty provisions governing the free movement of goods' (Ref. Ares(2013)3759436–18/12/2013)
Gormley LW (2008) Silver threads among the Gold. 50 years of the free movement of goods. Fordham Int Law J 31:1637
Gormley LW (2009) EU law of free movement of goods and customs Union. Oxford University Press, Oxford
Helleringer G, Sibony A-L (2016–2017) European consumer protection through the behavioral lens. Columbia J Eur Law 23:607–646
Howells G (2014) Product safety—a model for EU legislation and reform. In: Purnhagen K, Rott P (eds) Varieties of European economic law and regulation: Liber Amicorum for Hans Micklitz. Springer, Cham, pp 525–538
Howells G, Micklitz H-W, Wilhelmsson T (2006) European fair trading law. Ashgate, Aldershot
Howse R, Regan D (2000) The product/process distinction—an illusory basis for disciplining 'Unilateralism' in trade policy. Eur J Int Law 11(2):249–289
Jackson JH, Davey WJ, Sykes AO (2013) International economic relations: cases, materials, and text on the national and international regulation of transnational economic relations, Sixth edn. West, St. Paul, MN
Kysar DA (2004) Preferences for processes: the process/product distinction and the regulation of consumer choice. Harv Law Rev 118:526–642
Lewis KE, Grebitus C (2016) Why U.S. consumers support country of origin labeling: examining the impact of ethnocentrism and food safety. J Int Food Agribusiness Mark 28(3):254–270
Mavroidis PC (2012) Trade in goods, 2nd edn. Oxford University Press, Oxford
Mavroidis PC (2013) Driftin' too far from shore–why the test for compliance with the TBT Agreement developed by the WTO Appellate Body is wrong, and what should the AB have done instead'. World Trade Rev 12(3):509–531
OECD (2002) Towards sustainable consumption: an economic conceptual framework (No. NV/EPOC/WPNEP(2001)12/FINAL)

Pew Research Centre (2015) 'Global Publics back U.S. on Fighting ISIS, but Are Critical of Post-9/11 Torture', p. 31, http://www.pewglobal.org/files/2015/06/Balance-of-Power-Report-FINAL-June-23-20151.pdf. Accessed 15 May 2018

Proposal for a Council Regulation on the indication of the country of origin of certain products imported from third countries (COM(2005) 661 final)

Proposal for a Regulation of the European Parliament and of the Council laying down rules and procedures for compliance with and enforcement of Union harmonization legislation on products and amending Regulations (EU) No 305/2011, (EU) No 528/2012, (EU) 2016/424, (EU) 2016/425, (EU) 2016/426 and (EU) 2017/1369 of the European Parliament and of the Council, and Directives 2004/42/EC, 2009/48/EC, 2010/35/EU, 2013/29/EU, 2013/53/EU, 2014/28/EU, 2014/29/EU, 2014/30/EU, 2014/31/EU, 2014/32/EU, 2014/33/EU, 2014/34/EU, 2014/35/EU, 2014/53/EU, 2014/68/EU and 2014/90/EU of the European Parliament and of the Council (COM (2017) 795 final)

Ricardo D (1821) On the principles of political economy and taxation. John Murray, London

Rigod B (2015) Optimal regulation and the law of international trade: the interface between the right to regulate and WTO law. Cambridge University Press, Cambridge

Sajn N (2018) EPRS briefing: EU legislation in progress–strengthening market surveillance of harmonised industrial products. http://www.europarl.europa.eu/RegData/etudes/BRIE/2018/614696/EPRS_BRI(2018)614696_EN.pdf. Accessed 15 May 2018

Sankari S (2013) European Court of justice legal reasoning in context. Europa Law Publishing, Groningen

Saydé A (2017) Freedom as a source of constraint. Expanding market discipline through free movement. In: Koutrakos P, Snell J (eds) Research handbook on the law of the EU's internal market. Edward Elgar, Cheltenham, pp 29–53

Schaefer A, Crane A (2005) Addressing sustainability and consumption. J Macromarket 25 (1):76–92

Shuibhne NN (2017) Exceptions to the free movement rules. In: Barnard C, Peers S (eds) European Union Law, 2nd edn. Oxford University Press, Oxford, pp 471–503

Smith A (1776) Wealth of Nations

Snell J (2010) The notion of market access: a concept or a slogan? Common Market Law Rev 47:437–472

Van den Bossche P, Schrijver N, Faber G (2007) Unilateral Measures addressing Non-Trade Concerns. A study on WTO consistency, relevance of other international agreements, economic effectiveness and impact on developing countries of measures concerning non-product-related processes and production methods. The Ministry of Foreign Affairs of The Netherlands, The Hague

Chapter 14
The Importance of Labelling Food Items: Information, Food Security and Sustainable Consumption

Fabiana D'Andrea Ramos and Tatiana Cardoso Squeff

Contents

Abstract Food security encompasses access to nutritious and safe food and sufficient calories, these being the key elements of the human right to adequate nutrition. However, not every person has the same dietary needs, despite the 2200-kcal minimum intake suggested by the Food and Agriculture Organization (FAO). After all, food intake depends on region, culture, religion, work and education—to name but a few of the many salient factors. It is thus imperative that consumers are aware of the specifics of what is in the food they consume, so they can make informed decisions. The 'right to information' and its concomitant 'obligation to inform' thus play an important role in the effectiveness of the general human right to adequate food. This chapter therefore details the interrelations between food security, consumption and the right to information, addressing in particular the labelling of food items and its role in helping consumers choose their groceries properly, as well as how it may assist sustainable consumption patterns more generally.

F. D'. Ramos
Universidade Federal do Rio Grande do Sul (UFRGS), Porto Alegre, Brazil

T. C. Squeff (✉)
Universidade Federal de Uberlândia (UFU), Uberlândia, Brazil
e-mail: tatiana.squeff@ufu.br

A. d. Amaral Junior et al. (eds.), *Sustainable Consumption*,
https://doi.org/10.1007/978-3-030-16985-5_14

14.1 Introduction

Food security encompasses access to nutritious and safe food and sufficient calories, these being key elements of the human right to adequate nutrition. However, not every person has the same dietary needs, despite the 2200-kcal minimum intake suggested by the Food and Agriculture Organization (FAO). After all, food intake depends on region, culture, religion, work and education—to name but a few of the many salient factors.

It is thus imperative that consumers are aware of the specifics of what is in the food they consume, so they can make informed decisions. The 'obligation to inform' plays an important role in the effectiveness of the general human right to adequate food, especially in a context in which the McDonalidization of food consumption habits worldwide continues apace. Conversely, the problem of food insecurity is not just about not having enough food—'it also involves the quality, origin and elaboration of that food. It's not just about eating, but about eating well.'[1]

This chapter discusses the connections among food security, consumption and the right to information, addressing the importance of proper labelling of food items, so consumers can choose their food items properly, i.e. in the most rational way possible. This also implies meeting the paradigm of sustainability that surrounds every segment of life, including—and especially—the market, which is one social sphere that must be held accountable for the great degradation we are experiencing worldwide today.

14.2 The Links Between Sustainable Consumption and Food Security

Sustainable consumption has many definitions. It may refer to 'a level of consumption which causes a level of environmental impact over time that does not degrade basic ecosystem services, such as the provision of fresh water, fertile soil and a protective ozone layer.'[2] It can also relate to selection of 'products that use less natural resources in their production, which guarantee decent employment to those who manufactured them, and that will be easily reused or recycled.'[3] However, despite many dimensions that sustainable consumption might have, it is possible to draw a clear through-line that connects them all: *sustainability*. More precisely, sustainability entails activities that are concurrently socially relevant, ecologically prudent and economically viable.[4] It follows that sustainable consumption

---

[1]Cerri (2013).

[2]Salzman (1997), p. 1246.

[3]Brazilian Ministry of Environment (n/d).

[4]Sachs (2009), p. 35.

encompasses the utilization of goods and services that respects these three dimensions, be it in production, usage or eventual disposal.

Today, sustainable consumption is a paradigm of world society. As established by Kuhn, the term paradigm is used in two different senses:

> [O]n the one hand it indicates the whole constellation of beliefs, values, techniques, etc., shared by the members of a particular community. On the other hand, it denotes a type of element of such constellation: the concrete pieces of puzzles that, framed as models or examples, can substitute explicit rules as a basis for solving the remaining puzzles of normal science.[5]

This means that sustainable consumption is both a standard to be respected/achieved, and a foundation of society more generally, one that cannot be disrupted without disturbing the entire planet, especially as environmental issues do not 'stop at the border'.

Nonetheless, since the industrial revolution, our society has witnessed the acceleration of production volumes, the exploitation of global resources on a massive scale, and the enlargement of environmental and human health problems, harming not only the present generations, but also putting at risk those to come.[6] The 1992 Rio Earth Summit explicitly pointed to this exact scenario, where it was stated that 'the major cause of the continued deterioration of the global environmental are the sustainable patterns of consumption and production'[7] and all nations represented at the conference were pressed to start moving towards the sustainable consumption paradigm.

Certainly, it is simply impossible to sustain Western levels of consumption on a global scale indefinitely. This fact was mentioned by the then prime minister of Norway, Gro Harlem Brundtland[8]—known for producing the widely recognized report *Our Common Future* in 1987[9]—in 1994. Brundtland noted that ten earths would be needed to satisfy everyone's desires if all were to adopt Western consumption patterns, it thus being imperative to refrain from assuming that as a development goal. Instead, a more sustainable set of objectives are needed, which shall be applicable to all products brought to market. Food also clearly comes within this framework.

Hence, the basic connection among sustainable consumption and food is that those items placed in the market to be purchased by people shall also follow the sustainable paradigm, meaning they should be socially relevant, ecologically prudent and economically viable. All food items that do not reflect this triad cannot be considered sustainable, and thus, should not be produced, traded or consumed, as they may negatively impact the world patterns, leading to more degradation.

---

[5]Kuhn (1970), p. 146.

[6]Squeff (2016), pp. 11–25.

[7]United Nations (1992a).

[8]Brundtland (1994), p. 62.

[9]United Nations (1987).

Nevertheless, this view may sound problematic at first sight when considering the current situation of food insecurity that troubles the world. The FAO estimates that 815 million people currently experience chronic hunger globally—a decline from the 963 million in this situation in 2008, but well above the 777 million in chronic hunger 2016.[10] Moreover, the FAO estimates that in 2017 food uncertainty issues affected a further '124 million people in 51 different countries [who] went hungry or faced a food insecurity...11 million more people than the previous year', in which 'this tragic scenario was present in ['only'] 48 nations.'[11]

In the context of global food insecurity, one could be forgiven for arguing that food must be provisioned to the poor at all costs and it does not matter how this is achieved. To do so is simply imperative, given the crisis of chronic hunger and the human right to food prescribed under Article 25 of the 1948 Universal Declaration of Human Rights and Article 11 of the 1966 International Covenant of Social, Economic and Cultural Rights. Yet, feeding the world's most hungry people is not a matter of simple math, as many variables are present, especially the ones regarding sustainability. After all, food requirements should be met in the short term without 'sacrificing renewable and non-renewable natural resources, making it possible to preserve the conditions guaranteeing long-term food availability.'[12] In this context, therefore, the role of natural resources in providing nutritious food for present and future generations stands out, which includes not only the existence of arable land, but also a balanced ecosystem and, above all, the adoption of sustainable practices, ensuring full integration between the economic, social and environmental dimensions of sustainable development.[13]

More specifically, there are a core set of concerns that are deemed very important to guarantee the sustainability paradigm and to achieve food security. These include: *how* food is produced, as the level of land degradation and soil pollution, salinization and desertification, the use of Genetically Modified Organisms (GMOs) that may reduce biodiversity and respect for labor rights, which can impact the food production costs; *where* food is produced, which relates to the economic viability of producing certain food items, respect for the local environment, and the preference for local production; *who* is making the food, so rural and traditional populations, and women are empowered, and; *what* kind of food is produced, respecting the social, historic, cultural and religious backgrounds of a certain community.[14]

Food security is not just about the availability of food—meaning, the quantity. It encompasses three spheres that, as asserted by Valente,[15] must be considered if the

---

[10]FAO (2017); Ziegler et al. (2012), pp. 215–232.

[11]Sanz (2018).

[12]Maluf et al. (1996), p. 74.

[13]United Nations (1992a, b), Principle 4: 'In order to achieve sustainable development, environmental protection shall constitute an integral part of the development process and cannot be considered in isolation from it'.

[14]Thrupp (2000), p. 275.

[15]Valente (2003), p. 53.

right to food is to be fully respected: the biological dimension (nutritional quality), the material (the availability of food itself) and the economic (the ability to access it). Thus, despite hunger being 'mainly physiological', Valente brings our attention to the fact that these dimensions are not independent, and it is impossible to analyze food (in)security in isolation, given the likelihood that doing so would be highly incomplete and reductionist.

The indivisibility of the three dimensions can be understood through the concept of hunger itself. Hunger implies the 'deterioration of health and/or social and productive performance resulting from ingestion either of poor quality food or the wrong kind of food'—or both when it is impossible to access better food items.[16] Simply put, an individual may also be considered to be hungry when he or she lacks the income needed to access a sufficient volume of nutritious food in close proximity at a reasonable price. The result is that nutritional status is impaired, meaning the individual cannot acquire the energy level needed to enjoy a healthy and active life.

Therefore, to have a 'dignified life' or even to survive, it is not enough for a person to attain the daily calorie *quantum* needed to sustain a normal, active life. It is also necessary to consume food items that are sufficiently nutritious to eliminate the odds of 'hidden hunger'—a term coined by Josué de Castro[17] (the renowned Brazilian physician and politician who participated in the foundation of FAO in the 1940s) to define the type of hunger caused by 'food monotony'. Thus, 'in addition to the quantity of food ingested (to be transformed into energy), the human organism needs that the food ingested contain specific nutrients', which require a diversity of food sources and depends on the individual's intrinsic (sex, age, race, cardio, etc.) and extrinsic (profession, religion, social class, etc.) particularities.[18] Conversely, consumption of food items with a high caloric load but a poor nutritional content may allow an individual to meet the recommended daily intake but at the risk of creating a situation of severe food insecurity.[19]

Consequently, to ward off the threat of hunger, *awareness* of the role and influence of individuals and their practices in ensuring their own food and of the community they are part of is essential (especially towards the realization of the pillars of sustainability),[20] but also consciousness regarding the nature of the food being ingested as well. These are both vital to guarantee food security, a topic that we take up further in the remainder of the chapter.

[16] Jonsson (1986), p. 50.

[17] Castro (1957), pp. 90–91.

[18] Beurlen (2009), p. 23.

[19] Hernandez et al. (2012), p. 439; Mcdonald (2010), p. 4.

[20] UNESCO (1976), cited in Teitelbaum (1978), p. 51.

14.3 Food Labelling As the Basis for Rational, Conscious Decision-Making

Sustainable consumption is a key factor in guaranteeing food security. Without conscious consumption choices, some regions of the world where access to high-quality food items because of socioeconomics, culture, gender, age, etc. is limited may be directly impacted. To guarantee sustainable consumption and food security, consumers need to be properly informed, so they can make the best food choices possible given what is available and avoid problems regarding food production/consumption in the future.

As MacKerron and Sow observe, 'Extended Producer Responsibility' (EPR) is emerging as the framework for moving towards sustainable consumption. According to the authors:

> [EPR] is a policy that shifts responsibility for collection and recycling of post-consumer goods from governments to producers. Since companies have the greatest influence over product design and marketing decisions, they are the ones who have the greatest ability to reduce the overall environmental impact of their products.[21]

The public policy is dedicated to promoting corporate social responsibility, whereby firms develop sustainable options so the environment will not be negatively affected by newer technologies. However, such proposals can also promote a more general social consciousness regarding products that are brought to market because they serve as the direct contact point among consumers and sellers/producers. If properly informed, consumers will choose what they deem best for them and their surroundings, because they are fully informed about what they are eating, and the potential impacts of their purchase/consumption choices.

The policy, therefore, may have other uses, going further than initially thought and creating a third pillar, which is consumers themselves. Thus, by stimulating consumer awareness of the choices they make, food consumption responsibility is then split between the State—the typical guarantor of social rights—food companies (and all of those who are part the food chain; that is, from producers to sellers) and consumers. Above all, responsibility should not be restricted only to the first two. Governments and companies do not have complete power over individuals and their (ir/rational) choices. What they can do is stimulate better choices, not only to fulfil the human right of adequate food, but also to underpin sustainable consumption patterns, which may stimulate the private sector to keep making smarter/better product design choices.

The proposal, thus, draws heavily on rational choice theories (RCT) of human behavior, which exploits the insights of behavioral economics,[22] incorporating notions of the rational limits to human preferences, motivations and choices, to

[21] Mackerron and Sow (2011), p. 117.

[22] On the different approaches to behavioral economics, see Korobkin and Ulen (2000), pp. 1071–1074.

make their actions predictable.[23] RCT is composed of three elements: (a) the *perception* of the world by the social agent, interpreting and assessing different circumstances; (b) the emergence of *options*, which are always measurable and stable, making it possible to classify them; (c) the *rational choice* made by individuals based on their intentions, even if variable, increasing their usefulness.[24]

Thus, when acting rationally, an individual makes a particular choice because it is considered the most advantageous of all the options available. Rationality can be defined as what seems to be reasonable and logical, where a person consciously opts for a certain path after assessing or weighing the implications of the options presented.[25] Simply put, RCT departs from the deliberation between external costs and the benefits to an individual, eliminating any ethical or moral choices,[26] adopting at the end a perspective full of rationality that values the maximization of utility.[27]

There is no such a thing as a neutral choice.[28] The choices of rational individuals always seek to maximize their own interests and well-being, which is called utility. Conversely, utility translates the satisfaction of individuals associated with the potential results of their choices,[29] which, in the food scenario, can be defined as avoiding food insecurity both in the present and in the future. This perception, mainly argued by Gary Becker,[30] advocates for conduct that maximizes individual benefits, enabling them to make their choices rationally at a specific time, in accordance to their current preferences and conditions/surroundings.

Proper food labelling thus represents a core aid for consumers in their pursuit of maximal utility, in so far as it helps them select the most rational option considering all available variables. By being properly informed by producers/sellers, consumers can choose what is better for their own nourishment. Producers and sellers, in their turn, can also be stimulated through policies set by the government under EPR to improve their own products and overall actions to meet consumer demands, and thus, keep selling/making food items, and receiving government benefits for doing so. Nonetheless, it is the responsibility of governments to turn the human right to adequate food into effect within national boundaries.

In Brazil, as in other 40 nations around the world, food labelling has developed a lot in the past few years through discussions regarding GMOs and the need of duly informing the consumer he is buying a transgenic item, despite the great resistance of the biotechnology industry.[31] The discussion within Brazil concerns the mandatory

[23]Roberts (1997), p. 161.

[24]Ricker (1990), pp. 172–174; Flanagan and Montagnani (2010), pp. 46–47.

[25]Pinheiro and Saddi (2005), p. 168.

[26]For a critical approach, see: Sen (1977).

[27]Friedman (1953), pp. 15, 22, 31.

[28]Thaler and Sustein (2009), p. 3.

[29]Posner (1973), p. 3; Cooter and Ulen (1998), pp. 119–120.

[30]Becker (1968, 1976).

[31]According to Grère and Rao (2007), pp. 51–52: '[d]uring the last ten years, more than 40 countries have adopted labelling regulations, but the characteristics of the regulations and their degree of

introduction of a 'T' symbol on the labels of every product directed to human or animal consumption, if 1% or more of its content consists of GMOs,[32] as established by Decree No. 4680 of 24 April 2003.[33]

Such a law was introduced into the Brazilian legal system to promote more conscious decision-making by consumers, so they may know what they are eating, especially in a highly technological society, where risks[34] and uncertainties[35] are prevalent. Considering that society is deemed anthropogenic, human beings are the only animals capable of avoiding, reducing and addressing the consequences of human activity, which has been particularly impactful since the industrial revolution.[36] Thus, Decree No. 4680 was enacted to assist human beings, at least, to be more conscious with such a ('post-modern') grey area concern the impacts of the ingestion of GMOs.[37]

Further, it should also be recalled that the decree came into existence 'as a consequence of state intervention to ensure, before market failure, that consumers receive adequate information to enable them to consciously and freely exercise their consumption choices'. This was in accordance with Brazilian constitutional prescriptions under Articles 5/XXXII and 170/V,[38] and especially the need to regulate the 'right to information' prescribed within the Brazilian Consumer Code, Articles 6 and 31 of Law No. 8078 of 11 September 1990.

Therefore, such law balances the most common pillars mentioned before (the State plus all the food companies involved in the supply chain) by not completely prohibiting the usage of transgenic items in food items, and for caring for human well-being,[39] but it also introduces the third pillar as it includes the consumer in this

---

implementation vary greatly. While a large majority of countries belonging to the Organization for Economic Cooperation and Development (OECD) have implemented some type of labelling policy, only a few developing countries have introduced labelling laws, and even fewer have implemented them.'

[32]Brazil (2003), Art. 2: 'In the marketing of food items and food ingredients intended for human or animal consumption containing or produced from genetically modified organisms with a presence above the limit of 1% of the product, the consumer shall be informed of the transgenic nature of that product.'

[33]It is imperative to highlight that this order altered the prescription established by Decree No. 3871 of 18 July 2001, which prescribed that consumers be alerted to the existence of GMOs only if the content exceeded 4% of the product. See Brazil (2001), Art. 1.

[34]For bioethics, 'risks relate to future damage and are estimated by a scale of probability by which they succeed' (Junges 1999), p. 48.

[35]Uncertainties are related to the risk society, where there is an increased increase in technical options entailing the incalculability of its consequences so that the effects are unknown and unexpected (Beck 1992, p. 22).

[36]Beck et al. (1997).

[37]In this context, one can quote Warat (1994), who asserts that 'with modernity a cycle of questions and answers was finalized. And what is called postmodernity is a time interval in which the questions are not known, but in which, on the other hand, we will have to be prepared to construct answers' (p. 84).

[38]Benjamin (2012), p. 235.

[39]Currently, as noted by Efing et al. (2008), the collectivity of individuals is day after day more 'unconscious[ly] manipulated by forces like the market and the big corporations, that in search of

circle, allowing someone to introduce GMOs into her diet if she so wishes. However, it should be stressed that such a framework can only be effective if the consumer is properly informed and can duly ponder the options (and choose what suits her best regarding her well-being[40]).

If there is a gap in the information prescribed, one cannot blame consumers for their (bad) choice, and, consequently, say that it is solely the consumers' behavior that leads to a food insecurity scenario. Conversely, if accurately implemented, such law is a great example of both RCT and EPR because it relates to the right that individuals have to be correctly informed of what they are eating for them to freely choose among the options in the market, fomenting the food chain to improve even more its products (what could even include measures directed to fighting food insecurity) if supervized and stimulated by the State.

It has now become commonplace to place a certain product in the spotlight such that:

> the distribution of products in a supermarket gondola can facilitate the exploitation of the power of certain food item brands, maximize the profit of the commercial establishment itself, value the consumption of healthy and organic products or even promote the valorization of local products to the detriment of multinationals.[41]

Now imagine what proper labelling may do to promote food security and sustainable consumption by emphasizing the product itself wherever it may be placed. For this reason, we argue that the right to information is a pillar to both food security and sustainable consumption, as will be clarified in the final part of the present chapter.

14.4 The Obligation to Inform As a Pillar of Food Security and Sustainable Consumption

As discussed above, information allows for a more rational decision-making process in consumer behavior and thus leads to a more sustainable consumption pattern. The right to information is the very ground of Consumer Law, as highlighted by US President John F. Kennedy in his famous speech before the Congress on 15 March 1962.[42] As Kennedy pointed out,[43] consumer choice is strongly influenced by mass

---

the profit guide the wills of the people' (p. 10). Therefore, the State must act to impede such obscure future that tends to violate human dignity.

[40]For a discussion regarding choices and human dignity, especially addressing the idea that choices are not directed to gain advantage over the others, but to preserve one's well-being, see Nussbaum (2013), p. 248.

[41]Braga and Piovesan (2016), p. 319.

[42]Other important prescriptions in comparative law can also be pointed out, such as the inclusion of the right to information in the 'EEC Preliminary Program for a Consumer Protection and Information Policy' as one of the rights consumers in the EU have since 1975. See Fellous (2003), pp. 102–106.

[43]Kennedy (1963), pp. 235–243.

propaganda, which is developed through the art of persuasion. In the absence of effective labelling, it is impossible for a consumer to know what an item is made of, whether it follows particular standards or—in the specific case of food items—whether it is higher in nutrients than another. For these reasons, product information is deeply important. Consumers are at a disadvantage to producers in this regard, in the sense that only the supplier/manufacturer can know for sure what the contents of a certain item are.[44] This specific (and presumed) vulnerability of consumers in this regard may even be thought of as what distinguishes them from the supplier/manufacturer of goods, as the former have no control over the production process (manufacture, distribution, commercialization), but only the final step (consumption).[45] This situation makes consumers especially fragile and in a passive position, as they do not have conditions *a priori* to attest to the truth of the information provided.[46] In relations based on the inequality of the parties, especially those involving consumers, the weaker party is essentially beholden to expert systems.[47] By being unaware and having no influence over them, it is necessary to act exclusively based on the efficiency of the system.[48]

Thus, the way to protect them is to inform them properly, giving rise to the 'right to information' every consumer has, it being considered an intrinsic right of the human person within the 'society of consumption' where we are currently entrenched. This right means that any person has the right to be duly informed of what he consumes, especially regarding item publicity and labelling, so one may establish his own judgement over the item,[49] even providing for fair competition[50] (linked to good quality and fair price).

---

[44]Cavalieri Filho (2016), p. 6.

[45]Miragem (2010), p. 65.

[46]Marques (2006), p. 147.

[47]See the excellent study by Lorenzetti (2000), who advocates for the existence of a new *fattispecie*, characterized by an increase in social complexity. This would give rise to a new conceptual and legal technique for legal relations that are different from consent—as for personal ties—and adhesion—to mass relations; namely, an offer based on appearance and acceptance based on trust. He formulates a new form of binding for the parties—in this case, a contractual one, applicable only to what he calls 'complex contracts'—based on their factual behavior, generating a situation of appearance and trust, which must be protected by law. This new form of linkage would be grounded in what the author calls 'expert systems', into which contemporary society would be inserted. On expert systems, or what he also refers to as 'abstract systems', see Giddens (1991). The author states: '[b]y expert systems I mean systems of technical excellence or professional competence that organize large areas of the material and social environments in which we live today' (Giddens 1991, pp. 35–37, 87–91).

[48]Trust acts as a true mechanism for reducing social complexity. On the function of trust, see Luhmann (2000), pp. 27–38.

[49]Lapeña (2005), p. 158.

[50]It is important to highlight that on 2 July 1890, the 'Sherman Act' was enacted in the United States, which not only sought to promote free competition and restrict market monopoly, but also set the basis for other laws implemented worldwide regarding fair market competition. See Gloria (2003), pp. 61–62.

Brazilian law incorporated the right to information in consumer relations, especially through the provisions of the Consumer Protection Code (CPC), Law No. 8090/1990. In this context, the right to information has an important rebalancing function. Today, science proves that the decision-making process is rarely rational; emotions play a very important role in the brain processes during the development of conviction and the materialization of consent. Today, these emotions are generally produced, stimulated, and encouraged by suppliers. Thus, nothing feels more natural than the CPC, as it protects consumers by rebalancing consumption relations, transferring the risk of business to those who enjoy a privileged position in these transactions–namely, the supplier.

Such transfer of risk finds its foundation in 'strict good faith'.[51] Inserted into the CDC as a principle of the national policy of consumer relations, it acts as a true general clause. Therefore, it plays a fundamental role in restoring the balance deemed necessary for such disproportionate relations as consumer transactions. It assumes, thus, an important ethical function, with a specific teleological application of protecting the weaker party, namely the consumer.

One of the ways to achieve such this rebalancing, which is an expression of 'strict good faith', is through the transparency regime[52] within consumer relations. In this context, transparency can be understood as 'the practical result that law substantially pursues through what can be called the information principle'.[53] Consequently, information acts as an important factor to rationalize consumer options and it is an instrument through which the desired and necessary transparency is achieved. It is not surprising, thus, that in the normative scope it appears as one of the basic consumer rights (Article 6, CDC).

It is important to highlight that the Code has referred to a 'right' to information. This happens because, despite the natural connection it might imply, the differences between the *right to information*, the *obligation* it creates and the *duty* to inform must be known. In a general outline, the right to information is 'understood as the right to seek, receive, share and publish information'[54] and is a corollary of the fundamental right to freedom of expression.

Hence, it is more of a guarantee of accessing and transmitting general information related to the right to be informed than of the requirement of a legal information provision.[55] It is not a matter of requiring information to be formulated, but of having

[51]See the now classic description of the meaning of the 'strict good faith' made by Claudia Lima Marques: 'Strict good faith means, therefore, a 'reflected' performance, an action of consideration, thinking about the other, the contractual partner, respecting her legitimate interests, their reasonable expectations, her rights, acting with loyalty, without abuse, without obstruction, without injuring or causing excessive disadvantage, cooperating to achieve the good end of the obligations: the fulfillment of the contractual objective and the realization of the interests of the parties' See: Marques (2006), pp. 181–182.

[52]Expressly provided for in Article 4 of the CPC.

[53]Tomasetti Jr. (1992), p. 53.

[54]Lôbo (2001), p. 61.

[55]In a complementary sense, Barbosa understands that the right to information (in the sense of enforcing a provision) is a fundamental (constitutional) consumer right, implicitly foreseen in the

access to existing information. It is also the freedom to spread information—a constitutionally guaranteed right (so, a fundamental right) manifested through the freedom of expression (set out under Article 5, IX, of the Brazilian Constitution) and the specific right of access to information (set out under Article 5, XIV), whose instrument of effectiveness—*habeas data*—is also prescribed at the constitutional level (under Article 5, LXXII).

The right to information set out in the CPC, even if deriving from the Constitution, has certainly a more objective conception, resembling the duty and obligation to inform. In the context of Consumer Law, specifically, the right to information 'is the right to a positive provision to be delivered by all of those who offer products and services in the consumer market'.[56] To this right there is a corresponding duty or obligation of the supplier to establish or formulate the information (positive provision).

Indeed, the distinction between duty and obligation is subtle. In a general sense, duty relates to a rule of conduct of legal origin and of permanent character, to which all individuals are subject; while obligation is the 'passive expression of personal law', a legal tie through which one or more individuals (debtors) are subject to a provision to be delivered by another (creditors) because of a contract, a *quasi*-contract, an offense, a *quasi*-offence or even a statute.[57]

Concerning information in consumer relations, the 'duty to inform' is often referred to as the *passive* aspect of the right to information,[58] although it is, as seen above, an obligation. What happens is that the supplier's obligation to inform is confused with the theory of duties that derive from the conduct under obligation relations, which, by its turn, was built from the application of 'strict good faith'.[59]

Under this theory, all obligations are constituted by 'duties of performance' and other duties of conduct.[60] From this origin, the duty to inform arises as a corollary of

Federal Constitution, deriving from the principle of human dignity, and from systematic interpretation of the Constitution itself in harmony with the CPC. See: Barbosa (2008), pp. 45–54.

[56]Lôbo (2001), p. 62.

[57]Fabre-Magnan (1992), pp. 4–5.

[58]See Bruno Miragem's statement on the matter: 'It should be noted that [...] the right of information presents its co-respective effectiveness when imposing on suppliers in general the duty to inform' (Miragem 2010, p. 121).

[59]The application of the principle of good faith in obligation transactions has transformed its own conception, and is no longer considered as a simple static relation, represented only by the duties of provision and consideration, but one that should be conceived as a complex (made of other duties of conduct, besides those duties of provision) and variable (from which its conception as a process derives) structure. Regarding the transformation of the obligation relation through the influence of good faith, see: Larenz (1987, p. 6 et seq). On Brazil, see Silva (1976); Stiglitz (1997), pp. 11–12. Stiglitz specifically refers to the new duties of the obligation relations, which will be called as 'secondary rules of conduct', namely '[those] manifestations of contractual good faith, which translate into generic categories such as cooperation and loyalty, and specific directives that operate as detachments from the previous ones, such as information, fidelity, commitment, capacity of sacrifice, aid to the other part, etc.'

[60]According to Larenz, obligation relations (mainly those of contractual origin) consist of performance duties (giving, doing or not doing) and other duties of conduct. Performance duties have their

the general principle of law of 'strict good faith', which becomes effective because of a general duty of cooperation.[61] In this context, the expression 'duty' has a different meaning when compared to the one explained above, which is more generic.

Duty, in this strict sense, means a part of the obligation. This obligation, in its turn, may have a contractual or an extra-contractual origin. In the case of the duty to inform in consumer relations regulated by the CPC, it is a *legal* obligation.[62] It is the prescription of a duty of conduct, implicit in any obligation. It is, therefore, much more of an obligation, as its direct source is the law rather than a contractual obligation. What was just a duty (a part of the obligation), became the obligation itself.[63] Thus, what is often called a 'duty to inform', whether in the generic sense (legal and permanent rule of conduct) or in the strict sense (part of an obligation), it is, in fact, not a duty, but simply an obligation whose source is the law.[64] For this reason, the term adopted hereafter will be 'obligation to inform'.

Having resolved the terminological question, we can move on to the study of the obligation to inform in two important aspects. The first one concerns its content (requirements); and the second one, the timing of compliance. Regarding content,

content determined from the beginning of the relation, and the provision should result in the achievement of a specific outcome. They are divided into primary provision obligations (which determine the type of the obligation relation) or secondary provision obligations (that derive from default). Those would even include a further subdivision into a primary type (which determine the features of the provisions) and a secondary type (regarding other agreed duties). The other duties of conduct, in their turn, reflect, in a general sense, the duties of protection and loyalty; they are not related to a specific type of obligation, being content determined and developed throughout the relation, and which may vary in its intensity. They mean that the parties must behave as the situation requires to avoid harm and to the extent of achieving the purpose of the obligation; thus, it does not seek an exact result, but it only requires a certain behavior (Larenz 1987, pp. 6–15).

[61]Rubén S. Stiglitz says that '[t]he pre-contractual duty of information seems to integrate a broader one, that of cooperation, even if one and another are necessary derivatives of good faith' (Stiglitz 1997, p. 13). On the matter of cooperation, Wilhelmsson warns that it is now a new way of approaching the contractual relation, replacing the traditional antagonistic approach. According to him, 'the contract is seen and evaluated more as a form and tool of cooperation, with the objective of achieving results according to the purposes of the contract'. see: Wilhelmsson (1996), p. 11.

[62]Stiglitz states that '[t]he *duty of information constitutes a legal obligation*, based on an accessory rule of conduct, whose content consists of cooperating, from the negotiations stage, with who is in a weakened position in relation to the person who provides the information'. Stiglitz (1997), p. 12 (emphasis added).

[63]Lôbo states that '[T]he duty of information is not only the realization of the principle of good faith. In the evolution of consumer law, it took on an increasingly 'strict' nature in relation to the licit activity of supplying products and services. ... The development of consumer law went even beyond, *transforming it into the co-respective of the right of information, as a fundamental right, and ascending it to the level of a legal constraint and of a decisive guide to the content of the supplier's main provision. It is not just an attached duty*' (emphasis added) (Lôbo 2001, p. 67).

[64]Likewise, in Canada, where, according to Nicole L'Hereux, both federal and provincial legislators have imposed on the supplier an obligation of contractual information. See: L'Hereux (2000), pp. 273–275.

the obligation to inform will only be effectively fulfilled when it meets the requirements of adequacy, sufficiency and veracity.[65]

Suitability is a requirement prescribed by Article 6, III, of the Consumer Code, which sets forth the need for 'adequate and clear information'. This prescription regards to the use of means compatible to the product or service, and of clear and ostensive language (in Portuguese).[66] Moreover, the information must be transmitted in a way that allows and facilitates the consumer's comprehension and understanding[67] over a certain product or service.[68] In the case of packages and labels, the information contained therein is considered an offer under Article 30 of the CPC, and it is binding, that is, it obliges the supplier to its veracity and compliance.

Information should also be sufficient/complete. There should be no gaps or omissions in its fulfilment. It concerns not only the characteristics and qualities of the product or service, but also their inherent risks[69] that may affect both consumers' health and physical integrity, as well as those that may influence the rational decision to conclude the action of consumption.[70] In this case, the correct information about

[65]Tomasetti refers to sufficient information, and states that '[t]he transparency model implies not only the diffusion of information, but also the efficiency of the informational message. Efficient information can be said to appropriately (wholly or highly satisfactory) achieve the objective of transparency, enabling the consumer to act in accordance to the weighting and ordering of the five behavioral rationality variables considered.' The five variables proposed by the author are: '(a) critical awareness of the desires of consumption and the prioritization of the preferences that concern them; (b) the possibility of being ascertained, in accordance to technical and economic criteria accessible to the layman, the qualities and price of each product or service; (c) creation and multiplication of opportunities to compare the variety products and services; (d) knowledge of the subjective and proper legal positions that manifest themselves in the contextuality of the endless series of consumer situations and; (e) expedite and effective preventive presence of the State, as in mediating or deciding over the conflicts in the consumer market' (Tomasetti 1992, pp. 57–58).

[66]See Brazil (1990), Art. 31: 'The offer and the presentation of products and services must include *correct, clear, precise and ostensive information that is published in the Portuguese language* [...]' (emphasis added).

[67]See Brazil (1990), Art. 46: 'Contracts governing consumer relations shall not oblige consumers if they are not afforded the opportunity to have prior knowledge of their content or *if their instruments are worded in such a way as to make it difficult to understand their meaning and scope*' (emphasis added). Furthermore, Article 54, paragraph 3, stipulates that: 'Written adhesion contracts shall be written in clear terms and in clear and legible characters, *in order to facilitate its understanding by the consumer*' (emphasis added). Besides, Article 54, paragraph 4, establishes that: 'Clauses that imply limitation of consumer rights should be drafted with prominence, allowing their *immediate and easy understanding*' (emphasis added).

[68]Lôbo (2001), p. 74.

[69]The following conclusion was unanimously approved at the Fifth Brazilian Congress of Consumer Law promoted by BRASILCON: 'It is the duty of the supplier in the consumer relation contracts to keep the consumer adequately and permanently informed of all aspects of the contractual relationship, especially those related to risks, quality of the product or service or any other relevant circumstance to its consumption decision throughout the period in which the contractual relation lasts' (Benjamin et al. 2010, p. 69).

[70]Macedo Jr. (2000), p. 120.

the presence of allergenics or concerning the origin and the way the food item was made may be decisive for consumer health and decision-making.

Finally, the information must be truthful/honest. This means that the supplier must refrain from asserting anything, unless he is fully aware of the content of the information to be provided, otherwise he could mislead the consumer.[71] It is clearly a derivation of the moral rule of not lying. Hence, information corresponding to 'the actual characteristics of the product and service, as well as the correct data about its composition, content, price, deadlines, guarantees and risks' must be truthful.[72]

In Brazil there have been repercussions in this matter, regarding fruit juice and ice cream suppliers that used marketing strategy to basically lie on the labels of their products. More specifically, the products were associated with a touching story about their origin and recipe, whose facts were not exactly true. In the case of juice, the manufacturer made it clear that only raw materials from local producers were used, which may be considered an important piece of information for conscious and sustainable consumer decision-making. In fact, the manufacturer also used large suppliers, who did not necessarily practice sustainable agriculture, deceiving consumers who wished to follow a more environment-concerned path of life and who were worried about their health (regarding the ingestion of 'fresh' food items).

The last essential point concerns the moment of fulfilling the obligation to inform. As it is an autonomous legal obligation, its incidence occurs throughout the consumer relation/transaction—from the simplest social contact (that has the consumption as its main purpose), to the conclusion of the contract, its execution and even the post-contractual moment.[73] However, it is correct to assert that there are times when the observance of the obligation to inform will be particularly relevant. This is the case of the pre-contractual stage, as the information transmitted therein is essential for the materialization of consumer consent. That is why all the information contained onlabels, as those relating to the quantity and quality of the product, until those regarding the predominant colors, nutritional values and origin of the raw material, for example, are fundamental for the consumer to be able to participate in an effectively way to building a more sustainable consumption.

Considering that such protective norms have among their goals the reintegration of rationality[74] in consumer relations, it is natural to expect greater effort from the suppliers so the choices of consumers are reflected and self-determined.

---

[71] See Brazil (1990), Art. 37: 'All misleading or abusive advertising is prohibited. Paragraph 1: Any type of information or communication of an advertising character, totally or partially false, or in any other way, even by omission, that is capable of misleading the consumer regarding the nature, characteristics, quality, quantity, properties, origin, price and any other data on products and services is misleading.'

[72] Lôbo (2001), p. 70.

[73] Marques states that the transparency of the market carried out through information 'governs the pre-contractual moment, the eventual conclusion of the contract, the contract itself and the post-contractual moment' Benjamin et al. (2010), p. 68.

[74] On re-insertion of rationality in the process of consumer decision-making and the autonomy of the rational will, see: Chardin (1988).

Rationalization of consumer choices is certainly the most important function of the obligation to inform.[75]

14.5 Conclusion

The present chapter has sought to demonstrate that consumer decision-making is an essential and important part of building more sustainable consumption. Moreover, to accomplish such an ambition, consumers must be able to make conscious, and as far as it is possible, rational decisions. This type of decision-making, however, is only possible if consumers are correctly and sufficiently informed about the product they are purchasing within the market. Thus, although consumers are a central part of post-modern society, where risks to life and integrity are prevalent on a daily basis, sellers/suppliers must make up the most ground. After all, they bring products to the market, meaning they have the actual power to indirectly shift consumption patterns—either in a more sustainable direction, or in a more degraded one.

Further, when it comes to food items, it is possible to say that producers have a double responsibility, as they are responsible not only for the products (and services) they bring to the market, but also for the safety of human health, whether at the point of consumption or in terms of the waste produced by it. Therefore, in the case of food, the veracity and accuracy of the information contained in labels and packaging can ensure that consumers make safe and healthy decisions for themselves based on their habits and social patterns, and that they make choices that are safe and beneficial to the environment as a whole.

This is why we had argued that food security must be understood not only in relation to health and care with consumers as individuals, but also (and especially) regarding the collectivity and the environment where they live. This is also why we assert that it is sellers/producers who must bear prime responsibility for leading consumers towards such a path. This mission should be carried out in good faith, with information that is proper and accurate, and does not hinder consumer welfare in any way.

Such an approach can contribute both to a more competitive and sustainable market but also to the improvement in overall consumer and environmental welfare through the promotion of more effective choices, helping to reduce the numbers of people vulnerable to food insecurity in the world, and aiming at better days to come. Hence, the words of Braga and Piovesan[76] capture beautifully our thoughts on this matter: 'The choice of consuming a certain item means freedom to define what is best, individually and collectively; dignity to live in order to allow the broad

[75]This statement was made by Tomasetti Jr. (1992): 'Information has a functional sense of rationalizing consumer's options'. He continues: 'The rationalization of business declarations and legal consumer relations is a result that is not obtained by the suggestion (advertising, in particular), which primarily acts by encouraging human senses' (p. 53).

[76]Braga and Piovesan (2016), p. 321.

flowering of human capacities and the future, so that humankind preserves an environment ecologically balanced for all who are yet to be born.'

References

Barbosa FN (2008) Informação: direito e dever nas relações de consume. RT, São Paulo
Beck U (1992) Risk society: toward a new modernity. Sage, London
Beck U, Giddens A, Lash S (1997) Modernização reflexiva: política, tradição e estética na ordem social moderna. Unesp, São Paulo
Becker G (1968) Crime and punishment: an economic approach. J Polit Econ 76(2):169–217. https://doi.org/10.1086/259394
Becker G (1976) The economic approach to human behavior. University of Chicago Press, Chicago
Benjamin A (2012) Práticas Abusivas. In: Benjamin A, Marques CL, Bessa L (eds) Manual de Direito do Consumidor, 4th edn. RT, São Paulo
Benjamin A, Marques CL, Bessa L (2010) Manual de Direito do Consumidor, 3rd edn. RT, São Paulo
Beurlen A (2009) Direito Humano à Alimentação Adequada no Brasil. Juruá, Curitiba
Braga F, Piovesan F (2016) O consumo consciente e solidário: direitos humanos, movimentos ecológico-sociais e a promoção do desenvolvimento sustentável. Veredas do Direito 13 (25):307–330. https://doi.org/10.18623/rvd.v13i25.600
Brazil (1990) Federal Law n. 8.078
Brazil (2001) Decree n. 3.871
Brazil (2003) Decree n. 4.680
Brazilian Ministry of Environment (n/d) Consumo Sustentável, http://www.mma.gov.br/responsabilidade-socioambiental/producao-e-consumo-sustentavel/conceitos/consumo-sustentavel. Accessed Mar 26 2018
Brundtland G (1994) What is world prosperity? Bus Strategy Rev 5(2):57–69
Castro J (1957) Geopolítica da Fome, vol 1. Brasiliense, São Paulo
Cavalieri Filho S (2016) Programa de Direito do Consumidor, 4th edn. Atlas, São Paulo
Cerri A (2013) Fome e obesidade: dois fatores sociais decorrentes da insegurança alimentar. In: eCycle. https://www.ecycle.com.br/component/content/article/62-alimentos/4084-fome-e-obesidade-dois-fatores-sociais-decorrentes-da-inseguranca-alimentar.html. Accessed 26 Mar 2018
Chardin N (1988) Le contrat de consommation de crédit et l'atonomie de lavolonté. Librairie Générale de Droit et de Jurisprudence, Paris
Cooter R, Ulen T (1998) Law and economics. Scoot, Foresman and Co., Glenview
Efing A, Baggio A, Mancia K (2008) A informação e a segurança no consumo de alimentos transgênicos. Revista de Direito do Consumidor 68:9–27
Fabre-Magnan M (1992) De l'obligation d'information dans lès contrats: essai d'une théorie. Librairie Générale de Droit et de Jurisprudence, Paris
FAO (2017) How close are we to #ZeroHunger? The state of food security and nutrition in the World, Rome. www.fao.org/state-of-food-security-nutrition/en/. Accessed 26 Mar 2018
Fellous B (2003) Proteção do Consumidor no Mercosul e na União Europeia. RT, São Paulo
Flanagan A, Montagnani M (2010) Intellectual property law. Edward Elgar Publishing, Chelteham
Friedman M (1953) The methodology of positive economics. In: Friedman M (ed) Essays in positive economics. University of Chicago Press, Chicago
Giddens A (1991) As Conseqüências da Modernidade. UNESP, São Paulo
Gloria D (2003) A livre Concorrência como garantia do consumidor. Del Rey, Belo Horizonte
Grère G, Rao S (2007) A review of international labeling policies of genetically modified food to evaluate India's proposed rule. AgBioForum 10(1):51–64

Hernandez L, Magallanes M, Quinonez H (2012) Obesidad, conducta alimentaria e inseguridad alimentaria en adolescentes de la Ciudad de México. Bol Med Hosp Infant Mex 69(6):431–441
Jonsson U (1986) As Causas da Fome. In: Valente F (ed) Fome e Desnutrição: determinantes sociais. Cortez, São Paulo
Junges J (1999) Bioética: perspectivas e desafios. Ed. Unisinos, São Leopoldo
Kennedy JF (1963) Special message to the congress on protecting consumer interests. In: Public Papers of the Presidents of the United States: John F. Kennedy, 1962. Office of the Federal Register National Archives and Records Service General Services Administration, Washington
Korobkin R, Ulen T (2000) Law and behavioral science: removing the rationality assumption from law and economics. Calif Law Rev 88(4):1051–1144
Kuhn T (1970) The structure of scientific revolutions. University of Chicago Press, Chicago
L'Hereux N (2000) Droit de la consommation, 5th edn. Éditions Yvon Blais, Québec
Lapeña I (2005) Da rotulagem de produtos transgênicos. In: Varella MD, Barros-Platiau A (eds) Organismos Geneticamente Modificados. Belo Horizonte, Del Rey
Larenz K (1987) Lehrbuch des Schuldrechts, Band I, Allgemeiner Teil, 14edl. Beck, München
Lôbo P (2001) A informação como direito fundamental do consumidor. Revista Direito do Consumidor 37:59–76
Lorenzetti R (2000) La oferta como aparencia y la aceptación basada en la confianza. Revista Direito do Consumidor 35:9–38
Luhmann N (2000) Vertrauen, 4th edn. Lucius und Lucius, Stuttgart
Macedo R Jr (2000) Direito à informação nos contratos relacionais de consumo. Revista Direito do Consumidor 35:113–122
Mackerron C, Sow A (2011) Moving toward sustainable consumption in electronics design, production, and recycling. Utah Environ Law Rev 31(1):117–127
Maluf R, Menezes F, Valente F (1996) Contribuição ao Tema da Segurança Alimentar no Brasil. Revista Cadernos de Debate 4(4):66–88
Marques CL (2006) Contratos no código de defesa do consumidor: o novo regime das relações contratuais, 4th edn. RT, São Paulo
Mcdonald B (2010) Food security. Polity Press, Cambridge
Miragem B (2010) Curso de Direito do Consumidor, 2nd edn. RT, São Paulo
Nussbaum M (2013) Fronteiras da Justiça: Deficiência, Nacionalidade, Pertencimento à Espécie. Martins Fontes, São Paulo
Pinheiro A, Saddi J (2005) Direito, Economia e Mercados. Elsevier, Rio de Janeiro
Posner R (1973) Economic analysis of law. Aspen, New York
Ricker W (1990) Political science and rational choice. In: Alt J, Shepsle K (eds) Perspectives on positive political economy. Cambridge University Press, New York
Roberts J (1997) The rational constitution of agents and structures. In: Burch K, Denmark R (eds) International political economy yearbook: constituting international political economy, vol 10. Lynne Rienner Publisher Inc, Boulder
Sachs I (2009) Caminhos para o desenvolvimento Sustentável. Garamond, Rio de Janeiro
Salzman J (1997) Sustainable consumption and the law. Environ Law 27:1243–1293
Sanz B (2018) Fome e insegurança alimentar aumenta no mundo, diz ONU. In: Portal R7. https://noticias.r7.com/internacional/fome-e-inseguranca-alimentar-aumentam-no-mundo-diz-onu-22032018. Accessed 26 Mar 2018
Sen A (1977) Rational fools: a critique of the behavioral foundations of economic theory. Philos Public Aff 6(4):317–344
Silva CC (1976) A obrigação como processo. Bushatsky, São Paulo
Squeff TC (2016) Análise Econômica do Direito Ambiental. Lumen Juris, Rio de Janeiro
Stiglitz R (1997) La obligación precontractual y contractual de información: el deber de consejo. Revista Direito do Consumidor 22:9–25
Teitelbaum A (1978) El papel de la educación ambiental en América Latina. UNESCO, Paris
Thaler R, Sustein C (2009) Nudge: o empurrão para a escolha certa. Elsevier, Rio de Janeiro

Thrupp L (2000) Linking agricultural biodiversity and food security: the valuable role of sustainable agriculture. Int Aff 76(2):265–281

Tomasetti A Jr (1992) O objetivo de transparência e o regime jurídico dos deveres e riscos de informação nas declarações negociais para consumo. Revista Direito do Consumidor 4:52–90

UNESCO (1976) Taller Subregional de Educación Ambiental para la Enseñanza Secundaria. Chosica, Peru, 1–19 March 1976

United Nations (1987) Our common future, New York. http://www.un.org/documents/ga/conf151/aconf15126-1.htm. Accessed 26 Mar 2018

United Nations (1992a) Report of the United Nations Conference on environment and development, New York. http://www.un.org/documents/ga/conf151/aconf15126-1.htm. Accessed 26 Mar 2018

United Nations (1992b) Rio Declaration on environment and development

Valente F (2003) Fome, desnutrição e cidadania: inclusão social e direitos humanos. Saúde e sociedade 12(1):51–60. https://doi.org/10.1590/S0104-12902003000100008

Warat L (1994) O outro lado da dogmática. In: Rocha L (ed) Teoria do Direito e do Estado. SAFE, Porto Alegre

Wilhelmsson T (1996) Regulação de Cláusulas Contratuais. Revista Direito do Consumidor 9:11–51

Ziegler J, Golay C, Mahon C, Way S (2012) The right for the right to food. Palgrave Macmillan, Geneva

Chapter 15
Tobacco Packaging As a Contribution for Promotion of a Healthy Environment in Brazil

Luís Renato Vedovato and Cristiane G. F. Vianna

Contents

Abstract The chapter aims to demonstrate that plain packaging of tobacco products does not violate commercial freedom of expression. It argues, rather, that plain packaging is entirely consistent with the World Health Organization (WHO) Framework Convention on Tobacco Control (FCTC) itself and the guidelines that consist in rules in international administrative law about packaging and labelling of tobacco products. These guidelines are intended to assist Parties in meeting their obligations under Article 11 of the Convention, and to propose measures that Parties can use to increase the effectiveness of their packaging and labelling. Article 11 stipulates that each Party shall adopt and implement effective packaging and labelling measures within a period of 3 years after entry into force of the Convention for that Party. The chapter discusses how the implementation of tobacco plain packaging could help countries improve the general health of their populations. It argues that to do so, the FCTC Parties should adopt measures to restrict or prohibit the use of logos, colours, brand images and other promotional information on packaging other than brand names and product names displayed in a standard colour and font style (i.e., plain packaging). The chapter highlights how the FCTC comports with World Trade

L. R. Vedovato (✉)
State University of Campinas (UNICAMP), Campinas, São Paulo, Brazil
e-mail: lrvedova@unicamp.br

C. G. F. Vianna
Law and Civil Procedure and Public Health from the Oswaldo Cruz Foundation, Rio de Janeiro, Brazil
e-mail: cvianna@theunion.org

A. d. Amaral Junior et al. (eds.), *Sustainable Consumption*,
https://doi.org/10.1007/978-3-030-16985-5_15

Organization regulations, as has been found in recent Australian tobacco packaging disputes, in the context of tobacco control and human rights. The chapter also discusses the Parties that have already adopted FCTC measures and how they are entirely compatible with Brazilian domestic law.

15.1 Introduction

The Brazilian constitutional norms concerning freedom of expression, inscribed in Article 5, IV and IX, must be considered as fundamentally connected to political discourse and, indeed, to broader intellectual, artistic, and scientific concerns. Indeed, freedom of expression in these areas is a prerequisite for the functioning of democracy. At the same time, it remains quite challenging to establish the limits of political discourse because they tend to expand over time. However, even in the face of such difficulties, the stated constitutional objectives of the republic (Brazilian Federal Constitution, Article 3) constitute the basis for the norms that determine the conditions and the extension of constitutional freedoms of political discourse and other forms of protected expression. At the same time, free enterprise and free commerce is guaranteed in the Brazilian Federal Constitution (Title VII, Chapter I), which addresses the general principles of economic activity. Free enterprise is one of the foundations of the economic order.

Tobacco plain packaging does not fit in any of these places. Freedom of expression does not breathe the same atmosphere as tobacco packaging, so to speak, which obviously enough is concerned neither with democracy nor republican values. At the same time, it is not synonymous with free enterprise, as long as tobacco is an important public health concern. Nor is it related in any meaningful way with freedom of speech; a public health issue like tobacco assumes a different hermeneutic role. According to the World Health Organization's (WHO) Framework Convention on Tobacco Control (FCTC), plain packaging strengthens public health and helps control the global epidemic of tobacco use.

It would therefore be reasonable to conclude that there can be no confusion: packaging of a commercially marketed product is a question neither of freedom of expression or free enterprise, in which freedom of commercial expression is embedded. They consist, in fact, in different interpretive fields, which make it even more surprising that it is so common to find people confused in assuming the two rights—to freedom of expression and to freedom to trade—as one and the same.

In Brazil, we find this confusion alive and well, most notably in the case of advertising of alcoholic beverages, especially for beer. On 3 January 2014, in his regular weekly column in *Folha de S. Paulo*, Ruy Castro stated that beer advertising should be required to be more realistic. For him, every day and hour, the TV hammers the message that drinking beer guarantees us eternal youth and a euphoric life by the sea, surrounded by friends, fun, and romance. Additionally, he added for good measure that in such advertising, 'no one gets drunk or hungover, much less, gets sick and throws up'.[1]

One can fairly easilty picture the response from industry—most notably from the marketing side. Indeed, on the site of Meio & Mensagem online on 8 January 2014, an article by André Porto Alegre was published in which the latter argued that Castro was basically a hypocrite, in advocating complete freedom of expression in the case of unauthorised biographies of notable people, and at the same time suggesting it be restricted in the case of beer advertising. But there is no contradiction; freedom of speech is not unlimited and it certainly does not extend to advertising and marketing—especially of products that can cause serious damage to health. The regulation of beer advertising is about public health, so there is another fundamental right on the table, which cannot be found in the case of unauthorised biographies. This analogy extends certainly to tobacco plain packaging, which does not restrict freedom of expression either. Tobacco plain packaging promotes health and helps to build a healthier environment. For precisely this reason, it falls outside the borders of freedom of expression.

15.2 The Borders of Freedom of Expression

As the foregoing discussion suggests, the question we are concerned with in this chapter is whether tobacco plain packaging really does violate freedom of expression. As we shall detail further below, freedom of expression must be connected with the dissemination of information and ideas, and with democracy. It is thus, as we have already intimated, in a different field from free enterprise and freedom of commercial expression. So, regulation to achieve promotion of public health does not belong within the universe of freedom of expression. The rules about tobacco plain packaging concern public health.

Because of this, it is possible to affirm that plain packaging does not violate freedom of expression, which means that the FTCT and its guidelines can be applied inclusively in relation to plain packaging, and do not imply a violation of fundamental rights. But it is important to explain the reasons that this is in fact the case.

The Inter-American Court of Human Rights has declared freedom of expression to be the cornerstone on which the very existence of democratic society rests. Indeed, it has a special significance in the Americas, where most countries have built democracy after a long period of authoritarianism, which has had a profound institutional and cultural influence in the region.[2]

According to Grossman, the approach to freedom of expression adopted by a country serves as a social barometer of its democratic health.[3] Traditionally, serious

[1]Castro (2014).

[2]Grossman (2012).

[3]Grossman (2012).

problems of political, social and economic security have been used to justify authoritarianism and the denial of the value of pluralistic view of society. Unlike absolute governments, made up of political or military groups that intend to lead their nations, a human rights framework offers an alternative, aiming at the rule of law, including the right to free speech, to represent the values of human dignity and action to face social problems. Thus, to be understood as freedom of expression, it must be linked to founding elements of democracy.

From the human rights point of view, freedom of expression is seen as a necessary limit on the State, its rulers and social elites that seek to monopolise information. This allows for the free and plural flow of information, which is essential for democracy, facilitates individual participation in the democratic process, and strengthens civil society. The adoption of the Declaration of the Rights and Duties of American Men in 1948, and the American Convention on Human Rights (Pact of San José, Costa Rica) in 1969, created the normative framework of human rights in the Americas and were freely adopted by the Member States of the Organization of American States (OAS). Despite normative statements, with human rights obligations and standards set out in such documents, there is still a long way to go. Indeed, the region as a whole, which includes Brazil, still needs to achieve the implementation of such norms and freedom of expression, including the adoption of a normative framework that guarantees a plural society, so that the press is not an exclusive prerogative of governmental power or private monopolies.

It is not uncommon in the Americas to find that the right to freedom of expression inadequately protected by national law or the rules designed to protect it disregarded. In general, according to Advisory Opinion 5/85 of the Inter-American Court of Human Rights, measures that seriously violate the right to freedom of expression include: prior censorship; seizure or restriction of publications; absence of an effective regulatory framework concerning information monopolies and guaranteeing pluralism; as well as procedures that subject freedom of expression or disclosure of information to governmental control.

In exercising censorship, bureaucracies decide what people can see, read, write and produce, invoking justifications such as national security, public order, protection of morals, truth in information and personal reputation. However, as the potential for abuse is great, the constant promotion of free debate is preferable to censorship. Barriers to freedom of expression are laws that penalise scornful speech or expressions that are offensive to public officials or individuals on issues relevant to society in general. Such conduct, however, is included in the penal codes of various countries of the world, in effect criminalising them. Unjust punishment has always been understood as a logical corollary for authoritarian governments, but it is incompatible with the development of full democracy. On one hand, allowing criticism, free from fear of punishment, especially when directed at authority or matters of public concern, strengthens egalitarian principles and ensures that public officials perform their duties transparently and accountably. On the other hand, the threat or imposition of criminal sanctions stifles democracy and reflects an authoritarian logic incompatible with the principles of pluralistic society.

Subsequent and disproportionate liability in cases of slander and defamation is often imposed under the pretext of the defence of individual reputation. Such abuse severely limits the right to free expression of ideas. The monopolisation of information, by governments or private entities, restricts and ignores the pluralism of democratic societies. However, attacks on freedom of expression are not limited to regulatory deficiencies. Unfortunately, many crimes committed against journalists, including homicides, go unpunished. An attack on journalists is a threat to the population as a whole and is a complicated reminder that anyone, including the most visible, can be eliminated if they dare to fully exercise their right to free speech.

Other actions that seriously violate freedom of expression are economic measures to punish or reward the press for their ideas, as well as toleration of public and private monopolies in the media. These serious normative and concrete deficiencies in legal protection for freedom of expression still exist in many national legal systems, reaffirming the need for international laws and, in this case, regional protections of this fundamental freedom.

In addition to what is specified in the Brazilian Federal Constitution, the right to freedom of expression is also regulated in the inter-American system by Articles 13 and 14 of the American Convention, as well as in Article 4 of the American Declaration. Analysis of the norms of the inter-American system, especially those of the Convention, is useful for understanding the general regime of protection of this right, preventing exclusive focus on domestic legislation, which may distract the interpreter from the universality of human rights.

The Court of San José interpreted freedom of expression in cases such as: *Usón-Ramírez v. Venezuela*; *Canese v. Paraguay*; *Baruch Ivcher-Bronstein v. Peru*; *Juan Pablo Olmedo v. Chile (The Last Temptation of Christ)*; *Palamara-Iribarne v. Chile*; *Ríos, et al. v. Venezuela*; *Herrera Ulloa v. Costa Rica*; *Claude Reyes v. Chile*; *Kimel v. Argentina*; *Valle-Jaramillo, et al. v. Colombia*; *Perozo, et al. v. Venezuela*; *Tristán Donoso v. Panama*; and *López Álvarez v. Honduras*.

For its part, the Inter-American Commission on Human Rights has interpreted the different aspects of the right to freedom of expression in cases such as: *Martorell v. Chile*; *Hector Felix Miranda v. Mexico*; *Horacio Verbitsky v. Republic of Argentina*; *Alejandra Matus Acuña and others v. Chile*; *Jehovah's Witnesses v. Argentina*; *Victor Manuel Oropeza v. Mexico*; *Dudley Stokes v. Jamaica*, and *Hugo Bustíos Saavedra v. Peru*, among others.

To illustrate the debate, it is worth outlining Article 13 of the American Convention in full:

Article 13. Freedom of Thought and Expression

1. Everyone has the right to freedom of thought and expression. This right includes freedom to seek, receive, and impart information and ideas of all kinds, regardless of frontiers, either orally, in writing, in print, in the form of art, or through any other medium of one's choice.
2. The exercise of the right provided for in the foregoing paragraph shall not be subject to prior censorship but shall be subject to subsequent imposition of liability, which shall be expressly established by law to the extent necessary to ensure:

   a. respect for the rights or reputations of others; or
   b. the protection of national security, public order, or public health or morals.

3. The right of expression may not be restricted by indirect methods or means, such as the abuse of government or private controls over newsprint, radio broadcasting frequencies, or equipment used in the dissemination of information, or by any other means tending to impede the communication and circulation of ideas and opinions.
4. Notwithstanding the provisions of paragraph 2 above, public entertainments may be subject by law to prior censorship for the sole purpose of regulating access to them for the moral protection of childhood and adolescence.
5. Any propaganda for war and any advocacy of national, racial, or religious hatred that constitute incitements to lawless violence or to any other similar action against any person or group of persons on any grounds including those of race, color, religion, language, or national origin shall be considered as offenses punishable by law.

Article 13 (1) of the American Convention establishes the right to think and express oneself freely. It also details what freedom of expression means to seek, receive and impart information and ideas of all kinds, regardless of frontiers, and emphasises that the medium used is irrelevant, as the expression can be communicated orally, in writing, in the form of art, or by any other chosen way.

The decisions of the Inter-American Commission on Human Rights and the Inter-American Court on Human Rights interpret the right to freedom of expression as a prohibition of prior censorship, accepting only the subsequent discussion of responsibility, but in exceptional situations provided for in Article 13 (4) for the 'moral protection' of young people.

The inter-American system has also identified three distinct types of discourse that are especially protected, which include political discourse and discourse involving matters of public interest; the speech of public officials in the exercise of their functions and candidates for public office; and speech that is an element of the identity or personal dignity of the person who expresses himself.

All these specially protected forms of expression demonstrate the connection between discourse and democracy. Candidates and civil servants should be subjected to an electoral and public electoral review, which clearly links to their choice to enter the public field. The effort to protect identity-related discourse is designed to protect vulnerable groups, especially as democracy strengthens when minority protections are built. Such categories of discourse are particularly relevant in balancing different factors for assessing the responsibilities of the case. Both the Commission and the Court have repeatedly stated that in the inter-American system there is a strong connection between the right to freedom of expression and democracy.

From the interpretative work of the Commission and the Court, according to Grossman the following characteristics of freedom of expression can be extracted in the context of the inter-American system: (a) the so-called 'dual special character' (b) indivisibility of expression and diffusion of thought; (c) multiplicity of forms of expression; (d) protection of the means necessary for the dissemination of ideas; (e) protection of reproduction of expression; (f) exclusion of direct and indirect restrictions on speech; and (g) incompatibility of public and private monopolies in the media with the right to freedom of expression.[4]

As intimated, freedom of expression has a 'dual special character'—namely, the right of individuals to express themselves, but also their right to receive information and ideas. Thus, as the Court explains in its Advisory Opinion OC-05/85, a violation of the right to freedom of expression infringes not only an individual right, but also a collective right to receive any information. The Court elaborated on this interpretation in *Juan Pablo Olmedo v. Chile (The Last Temptation of Christ)*, in which it stated that freedom of expression is a way of exchanging ideas and information between people, which includes the right to communicate their point of view to others, but also implies the right of everyone to know opinions, reports and news.

The Commission had several opportunities to discuss this further. In the case *Martorell v. Chile*, where the censorship of the book *Impunidad Diplomatica* was in question, the Commission stated that arbitrary interference violating this right not only affects the individual right to express information and ideas, but also the right of the community as a whole to receive information and ideas of all kinds.

Baruch Ivcher-Bronstein v. Peru expands this dual character. The case was initiated when the Peruvian government deprived the majority shareholder and director of the Peruvian television channel Frecuencia Latina-Canal 2 of his Peruvian nationality. The government action was in retaliation for the transmission of the channel of several reports of human rights violations committed by the Fujimori regime in Peru during 1990–2000. Because foreigners could not own television or radio stations in Peru, the removal of Peruvian nationality from Ivcher-Bronstein resulted in his forced withdrawal from the channel's management. The new owners dismissed the journalists who had produced the programs deemed problematic to the Fujimori regime and ceased broadcasting negative news about it. When it decided on this case, the Commission stated that the social character of the right to freedom of expression was much broader than its individual aspects, which protects all those seeking and receiving information or opinions emanating from the media. In this case, the Commission alleged, and the Court confirmed, that every society is a victim when the freedom of an individual of expression is violated.

In Advisory Opinion OC-05/85, the Court defined the scope of the indivisibility of expression and disclosure, stating that restrictions that are imposed on the dissemination of information represent, in equal measure, a direct limitation on the right to express oneself freely. Furthermore, he said that the importance of the norms applicable to the press derives from this concept. Finally, he added that for the average citizen, it is as important to know the opinions of others or to have access to information in general as it is the right to give their own opinions.

The Court had an opportunity to deepen its position on this issue in *Palamara Iribarne v. Chile*. In this case, the Chilean government had seized and destroyed all hard and electronic copies of the book *Ethics and Intelligence Services* and prohibited their distribution. The Court considered that, because of the guarantee

---

[4]Grossman (2012), p. 380.

of the right to freedom of expression, the State should not only protect individual expression, but also its dissemination by any appropriate means.

The Commission followed a similar course in the case of *Martorell v. Chile*, in which it stated that the decision to prohibit the entry, circulation and distribution of the book *Impunidad Diplomatica* in Chile violates the right to transmit information and ideas of all kinds protected under the right to freedom of expression. As the cases demonstrate, the jurisprudence of the inter-American system has consistently confirmed the indivisibility of freedom of expression and the dissemination of ideas.

It should also be said that the right to freedom of expression is not limited to verbal expression; all types of expression are protected, including silence. The case of *Jehovah's Witnesses v. Argentina* is an example of the broad scope of the right to freedom of expression developed by inter-American jurisprudence. In 1976, the Argentine military dictatorship promulgated Decree No. 1867/76, which prohibited the public exercise of the religion by Jehovah's Witnesses in Argentina. The government claimed that this religion was based on principles contrary to Argentine nationality and basic state institutions. Followers of the religion were persecuted because of the decree. More than 300 children were expelled from school after refusing to swear allegiance to the country or singing the Argentine national anthem. The students opted for silence because their religion forbade them from engaging in such veneration of national symbols. Under Resolution No. 02/79, the Commission condemned the Argentine government's action, which it considered responsible for the alleged violations. As exemplified by the Commission's decision in the Jehovah's Witness case, all forms of expression, including silence, are protected under the right to freedom of expression.

Having affirmed the right to disseminate opinions and ideas, both the Commission and the Court have determined that the American Convention provides that freedom of thought and expression encompasses the right to disseminate information and ideas by any means. In Advisory Opinion OC-05/85, the Court stated that freedom of expression cannot be separated from the right to use any appropriate means to disseminate ideas and to bring them to as wide an audience as possible. The Commission stated in the *Baruch Ivcher-Bronstein v. Peru*, that the American Convention enshrines the right to disseminate information and ideas in an artistic way, or by any other means.

The right to freedom of expression includes the right of reproduction of expression from others. In the case of *Herrera Ulloa v. Costa Rica*, the State of Costa Rica had condemned the petitioner on the criminal defamation charge, based on the content of several articles published by the newspaper *La Nación*. Such articles, which had already been disclosed in the Belgian press, attributed illegal acts to the representative of Costa Rica at the International Atomic Energy Agency in Austria. The law of Costa Rica determines that a person who expresses himself must prove the truth of the facts reported. The Court considered that this norm is incompatible with Article 13 of the Convention and has dissuasive and inhibiting effects on all those who practice journalism, preventing public debate on issues of interest to society.

In the case of *Dudley Stokes v. Jamaica*, the petitioner, Dudley Stokes, editor of several prominent Jamaican newspapers, republished an Associated Press report regarding US investigation into tips allegedly given to former Jamaican Tourism Minister Eric Abrahams. The Associated Press withdrew the report the day after its publication but did not advise Jamaican newspapers that it had done so. Three days later, the story was published in Jamaica, the petitioner issued a statement based on a denial issued by Abrahams. However, a court found him guilty of defamation and assured Abrahams of a high conviction in pecuniary reparations. Stokes appealed the damages award originally granted and obtained a reduction in value. The case then went to the Commission. Although the petition was specifically accepted in 2004, specifically to deal with issues of proportionality of damages, the Commission ultimately denied the petitioner's claim, believing that the State did not violate the petitioner's right to freedom of expression. The Commission's decision raised serious procedural and substantive issues, as it held that proportionality issues should be decided by national legal systems without giving any guidance in that regard. As Jamaica did not accept the mandatory jurisdiction of the Court, the decision was final. However, in *Kimel v. Argentina*, the Court had the opportunity to do what the Commission could not, namely develop the notion of proportionality.

By criminalising the reproduction of information from third parties, in the absence of gross negligence or gross negligence, the Commission may have generated a restriction on the free flow of ideas in an increasingly complex global reality, where information flows from multiple and often distant sources. The exclusion of responsibility for reproducing this type of information does not mean, however, excluding the responsibility of those from which the information originated or the responsibility of those who reproduce such information with intent or guilt, without necessarily tracking the information.

Article 13 (3) of the American Convention prohibits restrictions on freedom of expression which are carried out by indirect means designed to prevent communication. The Special Rapporteur on Freedom of Expression has drawn up indirect measures such as those which, although not structured specifically to restrict freedom of expression, have in practice a negative impact on the free movement of ideas. Unlike direct constraints, these are harder to detect and therefore rarely investigated.

Baruch Ivcher Bronstein v. Peru is an example of an indirect restriction on freedom of expression. In this case, the government did not use traditional restrictions to silence Bronstein. Instead, the regime decided to withdraw its nationality. In this way, the decisions of the inter-American system broadened the concept of indirect restrictions. In the case *Canese v. Paraguay*, the Commission recognised punitive measures as an indirect restriction on freedom of expression. Canese was a journalist who wrote about allegations of corruption against a powerful presidential candidate, Juan Carlos Wasmosy. Canese was dismissed from the newspaper where he worked and saw criminal proceedings opened against him. He was sentenced to 4 months in prison and then denied permission to leave Paraguay. The Commission stated that the effect of applying such a punitive measure can generate self-censorship of a person who wishes to speak, which produces almost the same effect as direct censorship. Declaring that this way, opinions do not circulate. In that case, the Court ruled that the criminal proceedings and the consequent sentence imposed on Canese, including the restrictions on leaving the country for almost 8 years and 4 months, were an indirect means of restricting his freedom of thought and expression.

In the case of *Perozo and others v. Venezuela*, the Court held that there must be an effective restriction of freedom of expression for a violation to occur. In this case, the Court understood that the Venezuelan government did not violate the rights of the petitioner alone but would have failed in its obligation to prevent indirect restrictions by private actors.

Moreover, the Special Rapporteur for Freedom of Expression has been investigating new indirect means employed to restrict this right in the region. In its 2003 annual report, the Rapporteur assessed how irregular use of official advertising can be turned into a restriction. For example, abuse of funding distribution policies to benefit those who are in favour of the government or its agents, punishing media agencies seeking a more critical approach. Saying that while there is no inherent right to receive government advertising revenue, advertising income cannot be denied only for specific outputs based on discriminatory criteria.

Both the Court and the Commission have also taken the view that the existence of public or private monopolies prevents the dissemination of individual ideas and the receipt of opinions by the public. Consequently, the existence of monopolies in the media sector is incompatible with freedom of expression. In Advisory Opinion OC-05/85, the Inter-American Court affirmed that it is the mass media that makes the exercise of freedom of expression a reality. To ensure that this medium is not restricted, it has determined that there must be a plurality of means of communication, the abolition of all monopolies and guarantees for the protection of freedom and independence of journalists.

One of the main characteristics of the right to freedom of expression, as interpreted by the inter-American system, is that it only permits prior censorship in an exceptional situation, in specific cases to regulate public entertainment, specifically for safeguarding the morals of children and adolescents. Article 13 (2) of the Convention states that freedom of expression cannot be subject to prior censorship but does not exclude subsequent liability. This prohibition attempts to avoid the danger of creating filters capable of determining what people can hear, see or read. Therefore, the American Convention rejects justifications such as national security, morality or morality that could easily be used as pretexts to eliminate or overlap with the free expression of ideas.

In the Western Hemisphere both the Court and the Commission have interpreted the prohibition of prior censorship as the exclusion of the defence of reputation as the basis for prior censorship and the identification of the scope authorising exceptions. Some Member States have used the right to privacy, inscribed in Article 11 of the Convention as a basis for restricting the right to freedom of expression protected in Article 13. The basis of his argument is that the defence of reputation should be excluded from the prohibition of prior censorship.

Chile, for example, defended this argument in *Martorell v. Chile*. The Chilean government argued that in the event of a conflict between Articles 11 (right to privacy) and 13 (right to freedom of expression) of the American Convention, the former should prevail. In deciding the case, the Commission rejected that position and went further, arguing that the rules included in those two articles do not represent a collision of principles. Thus, the Commission cited the European Court of Human Rights, which, in *Sunday Times v. The United Kingdom* (1979), where a similar issue was discussed, considered that it was not faced with a choice between conflicting principles, one of which was freedom of expression, but the principle of freedom of expression which was subject to a series of exceptions should be interpreted narrowly.

The American Convention authorises, as an exception, prior censorship of public entertainment, always with the sole purpose of regulating access to such events, with the purpose of protecting the morals of children and adolescents. This exception, however, is only permitted within the inter-American system if it complies with the requirements of legality, necessity, reality or imminence or valid purpose.

In fact, freedom of expression, as said at the beginning of the section, must be connected with the dissemination of information and ideas, and with democracy. It is thus in a field different from free enterprise and freedom of commercial expression.

That leads the interpreter to guarantee that plain packaging does not violate freedom of expression, which means that the Framework Convention on Tobacco Control, and its guidelines, can be applied inclusively in relation to plain packaging, and do not imply a violation of fundamental rights.

15.3 Plain Packaging and the World Health Organization

The WHO affirms that plain packaging is a key part of successful implementation of the UN tobacco control treaty, the FCTC.[5] Article 11 of the FCTC, which relates to packaging and labelling of tobacco products, states:

Article 11 Packaging and Labelling of Tobacco Products

1. Each Party shall, within a period of three years after entry into force of this Convention for that Party, adopt and implement, in accordance with its national law, effective measures to ensure that:

   a. tobacco product packaging and labelling do not promote a tobacco product by any means that are false, misleading, deceptive or likely to create an erroneous impression about its characteristics, health effects, hazards or emissions, including any term, descriptor, trademark, figurative or any other sign that directly or indirectly creates the false impression that a particular tobacco product is less harmful than other tobacco products. These may include terms such as 'low tar', 'light', 'ultra-light', or 'mild'; and
   b. each unit packet and package of tobacco products and any outside packaging and labelling of such products also carry health warnings describing the harmful effects of tobacco use, and may include other appropriate messages. These warnings and

[5]World Health Organization (2003).

messages: (i) shall be approved by the competent national authority, (ii) shall be rotating, (iii) shall be large, clear, visible and legible, (iv) should be 50% or more of the principal display areas but shall be no less than 30% of the principal display areas, (v) may be in the form of or include pictures or pictograms.

2. Each unit packet and package of tobacco products and any outside packaging and labelling of such products shall, in addition to the warnings specified in paragraph 1 (b) of this Article, contain information on relevant constituents and emissions of tobacco products as defined by national authorities.
3. Each Party shall require that the warnings and other textual information specified in paragraphs 1(b) and paragraph 2 of this Article will appear on each unit packet and package of tobacco products and any outside packaging and labelling of such products in its principal language or languages.
4. For the purposes of this Article, the term 'outside packaging and labelling' in relation to tobacco products applies to any packaging and labelling used in the retail sale of the product.

Article 13 of the same convention, concerning tobacco advertising, promotion and sponsorship, states:

1. Parties recognise that a comprehensive ban on advertising, promotion and sponsorship would reduce the consumption of tobacco products.
2. Each Party shall, in accordance with its constitution or constitutional principles, undertake a comprehensive ban of all tobacco advertising, promotion and sponsorship. This shall include, subject to the legal environment and technical means available to that Party, a comprehensive ban on cross-border advertising, promotion and sponsorship originating from its territory. In this respect, within the period of five years after entry into force of this Convention for that Party, each Party shall undertake appropriate legislative, executive, administrative and/or other measures and report accordingly in conformity with Article 21.
3. A Party that is not in a position to undertake a comprehensive ban due to its constitution or constitutional principles shall apply restrictions on all tobacco advertising, promotion and sponsorship. This shall include, subject to the legal environment and technical means available to that Party, restrictions or a comprehensive ban on advertising, promotion and sponsorship originating from its territory with cross-border effects. In this respect, each Party shall undertake appropriate legislative, executive, administrative and/or other measures and report accordingly in conformity with Article 21.
4. As a minimum, and in accordance with its constitution or constitutional principles, each Party shall:

   prohibit all forms of tobacco advertising, promotion and sponsorship that promote a tobacco product by any means that are false, misleading or deceptive or likely to create an erroneous impression about its characteristics, health effects, hazards or emissions;

   a. require that health or other appropriate warnings or messages accompany all tobacco advertising and, as appropriate, promotion and sponsorship;
   b. restrict the use of direct or indirect incentives that encourage the purchase of tobacco products by the public;
   c. require, if it does not have a comprehensive ban, the disclosure to relevant governmental authorities of expenditures by the tobacco industry on advertising, promotion and sponsorship not yet prohibited. Those authorities may decide to make those figures available, subject to national law, to the public and to the Conference of the Parties, pursuant to Article 21;
   d. undertake a comprehensive ban or, in the case of a Party that is not in a position to undertake a comprehensive ban due to its constitution or constitutional principles,

restrict tobacco advertising, promotion and sponsorship on radio, television, print media and, as appropriate, other media, such as the internet, within a period of five years; and

e. prohibit, or in the case of a Party that is not in a position to prohibit due to its constitution or constitutional principles restrict, tobacco sponsorship of international events, activities and/or participants therein.

5. Parties are encouraged to implement measures beyond the obligations set out in paragraph 4.
6. Parties shall cooperate in the development of technologies and other means necessary to facilitate the elimination of cross-border advertising.
7. Parties which have a ban on certain forms of tobacco advertising, promotion and sponsorship have the sovereign right to ban those forms of cross-border tobacco advertising, promotion and sponsorship entering their territory and to impose equal penalties as those applicable to domestic advertising, promotion and sponsorship originating from their territory in accordance with their national law. This paragraph does not endorse or approve of any particular penalty.
8. Parties shall consider the elaboration of a protocol setting out appropriate measures that require international collaboration for a comprehensive ban on cross-border advertising, promotion and sponsorship.

The Conference of the Parties has approved guidelines about tobacco packaging that intend to increase health promotion. The guidelines are as follows:

> About Article 11: Plain packaging 46. Parties should consider adopting measures to restrict or prohibit the use of logos, colours, brand images or promotional information on packaging other than brand names and product names displayed in a standard colour and font style (plain packaging). This may increase the noticeability and effectiveness of health warnings and messages, prevent the package from detracting attention from them, and address industry package design techniques that may suggest that some products are less harmful than others.

> About Article 13: Packaging and product features. 15. Packaging is an important element of advertising and promotion. Tobacco pack or product features are used in various ways to attract consumers, to promote products and to cultivate and promote brand identity, for example by using logos, colours, fonts, pictures, shapes and materials on or in packs or on individual cigarettes or other tobacco products. 16. The effect of advertising or promotion on packaging can be eliminated by requiring plain packaging: black and white or two other contrasting colours, as prescribed by national authorities; nothing other than a brand name, a product name and/or manufacturer's name, contact details and the quantity of product in the packaging, without any logos or other features apart from health warnings, tax stamps and other government-mandated information or markings; prescribed font style and size; and standardised shape, size and materials. There should be no advertising or promotion inside or attached to the package or on individual cigarettes or other tobacco products. 17. If plain packaging is not yet mandated, the restriction should cover as many as possible of the design features that make tobacco products more attractive to consumers such as animal or other figures, 'fun' phrases, coloured cigarette papers, attractive smells, novelty or seasonal packs.

If we were to summarise these points, we would contend that WHO FCTC Article 11 and its Guidelines describe plain packaging as one of the 'measures to restrict or prohibit the use of logos, colours, brand images or promotional information on packaging other than brand names and product names displayed in a standard colour and font style'. The WHO FCTC Article 13 Guideline defines that FCTC parts must turn tobacco packaging in:

> black and white or two other contrasting colours, as prescribed by national authorities; nothing other than a brand name, a product name and/or manufacturer's name, contact details and the quantity of product in the packaging, without any logos or other features apart from health warnings, tax stamps and other government-mandated information or markings; prescribed font style and size; and standardized shape, size and materials. There should be no advertising or promotion inside or attached to the package or on individual cigarettes or other tobacco products.

Following these rules, the FCTC Parties are adopting plain packaging policy. The guidelines contain an important recommendation stressing 'packaging and product design are important elements of advertising and promotion. Parties should consider adopting plain packaging requirements to eliminate the effects of advertising or promotion on packaging. Packaging, individual cigarettes or other tobacco products should carry no advertising or promotion, including design features that make products attractive'.

The terms 'plain' and 'standardized' to describe packaging are used interchangeably. Some countries prefer the term standardised, as packaging with large, graphic health warnings does not look plain. As laid out above, the Guidelines to Articles 11 and 13 refer to 'plain' packaging and the WHO also uses this terminology, and the measure applies to packaging and to tobacco products.

At this point, it is possible to say that plain packaging promotes health and it is not a violation of freedom of expression, because it is not included in spreading ideas and thoughts. The World Trade Organization (WTO) has come to the same conclusions.

Australian plain packaging regulations are based on the FCTC Guidelines and they open doors to other FCTC Parties, despite many claims against it by the tobacco corporations. As Kennedy has written:

> It is widely known that tobacco is both a deadly product and a profitable industry. On November 21, 2011, Australia passed the world's toughest anti-tobacco law. Since its enactment, this law has sparked several national and international lawsuits and has put spotlight and increased pressure on the inherent tension between public health and global trade.[6]

The pressure on Australian regulators (by the tobacco companies) has become an important case, and subsequently went to the dispute settlement system of the World Trade Organization (WTO). As McGrady and Jones note, this case has turned into a clash between trade, health and human rights: 'As implementation of the WHO FCTC accelerates and states seek to address risk factors for non-communicable disease more broadly, tension has increased between the law of the World Trade Organization (WTO) and public health.'[7] Moreover, as Hammond has noted, working with the increasingly tense relationship between public health and international trade laws is one of the biggest challenge in the present and promises to continue to be in the near future.[8]

[6]Kennedy (2014), p. 599.

[7]McGrady and Jones (2013), p. 271.

[8]Hammond (2012).

15.4 Plain Packaging and the World Trade Organization

The WTO is about to decide the panel concerning the Australia–Tobacco Plain Packaging (Indonesia) case. Additionally, the decision could be fundamental to the global progress on plain packaging implementation. The final WTO panel decision was expected by the end of 2017, but it has yet to be finalised. In any event, in May 2017 Bloomberg news reported, that the 'World Trade Organization upheld Australia's right to impose plain-package label restrictions on the sale of tobacco products, dealing a blow to the cigarette industry, according to two people close to the situation.'[9] The international trade and investment laws have already indicated that public health comes first to any other remarks about trade, investment or intellectual property. So, it is reasonable to suppose that Bloomberg is correct in its assessment.

Public health was considered by the International Centre for Settlement of Investment Disputes (ICSID), in the Uruguay case about tobacco control measures. Human rights—as a public health concern—have thus reached the WTO system, which it now seems cannot but avoid weighing in for some time to come. By these precedents, public health will inform the WTO decision system, even if it is about freedom of initiative, or commerce, or advertising and marketing. It is fairly sure, then, that plain packaging does not violate WTO regulations.

The decision by the High Court of Australia regarding plain packaging regulation was sensible. The Court ruled[10] that:

1. The relevant rights of the tobacco companies were 'negative rights' (i.e., concerning trademarks);
2. The tobacco companies may have lost something of commercial value, but commercial value is not the object of constitutional protection;
3. The regulatory scheme is no different in kind from other legislation requiring health or safety warnings;
4. The requirements of the scheme are conditions on the sale of tobacco products—the commonwealth does not use tobacco packaging or products;
5. The scheme allows the continued use of brand names (including trademarked brand names)—the ability to use such names is valuable;
6. Intellectual property rights are created to serve public purposes, but they are not sacrosanct, and they do not operate above or in isolation from other laws created to serve other public purposes.

The High Court of Australia supported the plain packaging regulations and decided that public health has an important role in the 'human rights play'. Moreover, International Law upholds the domestic court's decision.[11]

[9]https://www.bloomberg.com/news/articles/2017-05-04/wto-said-to-uphold-australia-s-ban-on-cigarette-logos. Accessed 23 February 2019.

[10]Liberman (2013).

[11]Amaral Jr (2012).

15.5 Plain Packaging in the World

As we it unfold today, the global progress of tobacco plain packaging concerns the implementation of FCTC guidelines. Such a move is important for public health and for sustainable consumption. Up to the time of writing, the following countries had implemented plain packaging measures of some kind[12]:

- Australia: Sell-through period ended 1 December 2012.
- France: Sell-through period ended 1 January 2017.
- United Kingdom: Sell-through period ended 20 May 2017.
- Ireland: Law and regulations passed (sell-through period 30 September 2017–30 September 2018).
- New Zealand: Law passed, and regulations made (sell-through period 14 March 2018–6 June 2018).
- Hungary: All new brands or variants in plain packaging from 19 August 2016 and existing brands from 20 May 2019.
- Slovenia: Act passed in February 2017 requiring plain packaging within 2 years.
- Norway, Romania, Thailand, Georgia: Acts passed but awaiting regulations.
- Chile, Ecuador, Panama, Brazil: Legislative proposals have been tabled.
- Canada, Singapore, Sri Lanka, Uruguay: Among the group of countries with publicly announced policy processes.

Moreover, Brazilian draft legislation on plain packaging began to circulate in 2016 and represents a large step in promoting a healthy environment in the country, especially to inform consumers more effectively.

15.6 Conclusion

Freedom of expression cannot be understood as synonymous with freedom of trade. Public health is more important than trade or intellectual property. Precisely for this reason, plain packaging is considered an important measure in tobacco control. Soon, the WTO will decide about the Australian plain packaging case, but precedent tells us that the WTO panel has a tendency to rule in favour of Australia laws, which is considered a very important step for the interests of public health. The FCTC and its guidelines have already indicated plain packaging as a recommendation to promote health and, to implement these measures, a list of countries has already approved plain packaging regulations. Brazil can implement such a project in the coming years and this can be seen as a key means to create a healthy environment for future generations of Brazilian consumers.

[12]World Health Organization (2017).

References

Amaral A Jr (2012) Curso de Direito Internacional Público, 3a edn. Atlas, São Paulo

Castro R (2014) País das libações. Folha de S. Paulo, 2. http://www1.folha.uol.com.br/colunas/ruycastro/2014/01/1392576-pais-das-libacoes.shtml. Accessed 27 Jan 2018

Grossman C (2012) Challenges to freedom of expression within the inter-American system: a jurisprudential analysis. Hum Rights Q 34(2):361–403

Hammond D (2012) Plain packaging regulations for tobacco products: the impact of standardizing the color and design of cigarette packs. Salud Pública de México 52(2):226–232

Kennedy MS (2014) Australia's tobacco plain packaging act: convergence of public health and global trade. NCJ Int Law Com Reg 36(3):591–620

Liberman J (2013) Plainly constitutional: the upholding of plain tobacco packaging by the High Court of Australia. Am J Law Med 9(2):1–27. https://ssrn.com/abstract=2215346. Accessed 27 Jan 2018

McGrady B, Jones A (2013) Tobacco control and beyond: the broader implications of United States-Clove Cigarettes for non-communicable diseases. Am J Law Med 39(2–3):13–35. https://ssrn.com/abstract=2263270. Accessed 27 Jan 2018

World Health Organization (2003) Framework convention on tobacco control. http://www.who.int/fctc/en/. Accessed 27 Jan 2018

World Health Organization (2017) Plain packaging: a key part to successful implementation of the UN tobacco control treaty. http://www.who.int/fctc/mediacentre/news/2016/FCTC-WNTD-2016/en/. Accessed 27 Jan 2018

Part IV
Sector-Specific Approaches I: Transnational and International Law

Chapter 16
The International Regulation of Living Modified Organisms

Alberto do Amaral Junior and Luciane Klein Vieira

Contents

Abstract This chapter focuses on the contradictory regulation of living modified organisms in contemporary international law. Specifically, it analyses the consequences that emerge from this collision of rules. Our conclusion mainly involves two assertions. The first is that this situation results, on the one hand, from the fragmentation of International Law and, on the other, from opposing state interests in different regulatory domains. The second section discusses the Agreement on Sanitary and Phytosanitary Measures, whose peculiarity is the requirement of scientific evidence as a condition for unilateral restrictions on international trade by WTO Member States. The most salient features of the Cartagena Protocol on Biosafety, which covers transboundary movements of living modified organisms—largely founded on the precautionary principle—are also brought to bear. Opposing logics and rules govern the two treaties and reveal a dilemma at the heart of modern international law.

A. d. Amaral Junior
University of São Paulo (USP), São Paulo, Brazil

L. Klein Vieira (✉)
Universidade do Vale do Rio dos Sinos (UNISINOS), São Leopoldo, Brazil
e-mail: lucianevieira@unisinos.br

A. d. Amaral Junior et al. (eds.), *Sustainable Consumption*,
https://doi.org/10.1007/978-3-030-16985-5_16

16.1 Introduction

In early times, man discovered that seed selection improves plant quality and yield. Similarly, it is easy to see how one can accentuate or reduce physical traits in the animal world through selective breeding. That the use of such rudimentary techniques, which are the distant origin of modern biotechnology, would culminate in the spectacular progress observed in recent decades could not have been anticipated. None of it would be possible without extensive knowledge of the molecular structures necessary for the transfer of genetic information between different living organisms with the purpose of modifying them. This is the fundamental essence of genetic engineering: the modification of living organisms by exploiting scientific principles to obtain previously defined results.[1]

The proliferation of certain desirable characteristics in plants and animals at a much more intense pace than would have been expected a generation ago is one of the consequences of biotechnology, a feat that is unthinkable by using traditional methods of genetic transfer.[2] The many benefits that derive from living modified organisms include increased agricultural productivity and greater protection of human health and the environment. New species show greater resistance to insects and weeds, as well as permitting farmers to use herbicides that do not jeopardise crop yields. Adaptive seedlings from poor soils subject to long periods of dry weather can be sowed in inhospitable terrain where aridity has heretofore prevented agriculture.

The advantages associated with sustainable agriculture, food safety and quality now available to consumers at significantly reduced prices are also notable. Crops can now be grown to fight disease by producing vaccines and other useful medicines. Genetic alteration of rice was instrumental in addressing the vitamin A deficiency that leads to blindness for millions of children on several continents, especially Asia.

Genetic changes that delay ripening act to prolong the shelf life of fruits and vegetables. The sudden increase in agricultural productivity without a corresponding need to expand the amount of land under cultivation helps not only the maintenance of existing forests but also contributes to the creation of new reserves for environmental preservation.

Unfortunately, as we all know, the role of biotechnology in contemporary society does not only present benefits. One of the risks is harm to biological diversity by virtue of the proliferation of genetically modified crops that replace traditional strains on farms, including small ones. Rapidly declining genetic diversity is the primary reason behind the fear that even more dangerous pests, with immeasurable destructive potential, could emerge with catastrophic effects.[3] If this scenario occurs, developing countries will be most disadvantaged, and that is where most of the planet's biological wealth is hosted. Food security is a concern shared by significant portions of civil society and by nongovernmental organisations tying together

[1] Sampson (2005), p. 144.

[2] Schoenbaum (2001), p. 27.

[3] See Telesetsky (2011).

different parts of the world. Meanwhile the power of a few transnational corporations will strengthen and expand as more farmers at each planting period opt to sow new seeds produced in the laboratories of large corporations. We are currently in the midst of a profound change that will impact the nature, structure and ownership of the food production system.

The transposition of genetic material unbounded by the natural mechanisms that have hitherto prevented species from mixing—the transposition of animal genes into plants being a prime example—raises delicate ethical and religious problems. The debate over the manipulation of nature by the voluntary conduct of man has become more intense than ever as the perspectives and possible consequences of scientific progress expand beyond our ability to contain them. Both the cloning of human beings and the unrestricted patentability of living organisms, for example, run into central tenets of world religions[4] because issues pertinent to the integrity of articles of faith are at stake. Many effects inherent to biotechnology that are difficult to predict also threaten human, plant and animal life and health.

Recent discoveries have made it possible to alter the nutritional value, toxicity and allergic character of certain foods. Economic motivations, driven by anticipation of portentous profit, steer the research sponsored by multinational companies in the field of biotechnology. The trend towards market oligopolisation in this sector makes it possible for firms to price their products very high, ostensibly to recoup their colossal investments in applied research. In this scenario, farmers in developing countries tend to bear the burden of increasing costs. The largest corporations, impelled by the desire to reap the most benefits possible from their investments, devote their efforts to innovations that can be marketed in rich, developed countries to the detriment of the legitimate aspirations of the poor nations. For instance, there is a growing risk that injustices will be perpetuated because of the collection by private entities of genes extracted from tropical plants that improve varieties found in nature and are subsequently exported at high prices back to the developing countries from whence the original genetic material was obtained. The risks and rewards detailed above call for a flexible regulatory framework, one capable of dealing with the multiple facets of the issue.

This chapter focuses on the contradictory regulation of living modified organisms in the contemporary international law. Furthermore, it analyses consequences coming from that collision of rules. Our conclusion mainly involves two assertions. The first is that this situation results, on the one hand, from the fragmentation of International Law and, on the other, from opposing state interests in different regulatory domains. The second section discusses the Agreement on Sanitary and Phytosanitary Measures, also known as the SPS Agreement, whose peculiarity is the requirement of scientific evidence as a condition for unilateral restrictions on international trade by Member States of the World Trade Organization (WTO). The most salient features of the Biosafety Protocol on transboundary movements of living modified organisms, largely founded on the precautionary principle, are

[4]See Shaffer (2012).

also brought to bear. Opposing logics and rules govern the two treaties and reveal a dilemma at the heart of modern international law.

16.2 The Agreement on Sanitary and Phytosanitary Measures

The expansion in international trade over the last decades has been paralleled by a formidable growth in mandatory technical regulations,[5] voluntary standards and conformity assessment procedures designed to protect relevant social values, such as consumer safety, health and the environment.[6] After the Second World War, tariffs and quotas, the trade barriers traditionally employed, were lowered in successive trade rounds. Protectionism was seen as the primary obstacle for the objectives that would be served by the liberalisation of international trade. Therefore, the fear of a return to protectionism under subtle and disguised ways that are quite hard to identify is understandable.[7] Yet governments must have some flexibility to protect important collective values that do not involve disguised restrictions on trade.

Public sensitivity to the possibility of undesired effects is neither static nor uniform but rather is the product of specific circumstances that are to be analysed from the perspective of common values. The Agreement on Technical Barriers to Trade and the Agreement on Sanitary and Phytosanitary Measures regulate the setting of standards in the WTO. Both agreements ban disguised barriers to trade that prevent access to markets by erecting artificial conditions of competitiveness. Of note, the Agreement on Technical Barriers to Trade and the Sanitary and Phytosanitary Agreement were concluded in an environment much different from that of the mid-1940s when the General Agreement on Tariffs and Trade (GATT) was formulated. As parties of the 'single undertaking' that emerged from the Uruguay Round, all WTO members are bound by it, not just groups of states as occurred in the Tokyo Round. Further, certain behaviours are encouraged in line with the promotional function of law that Norberto Bobbio advocates. Rather than punish unwanted conduct, they promote values by allowing behaviours that serve valid objectives. The goal is to favour social change in accordance with national experiences while combating arbitrary interference.

The authors of the SPS Agreement deliberately invoked scientific knowledge to enshrine a notion of universal reason as a counterpoint to global conditions of stark diversity.[8] By building trust in a universal rationale, national decisions are less susceptible to governmental whims or intentionally masked protectionist measures. Domestic policies that reflect and express the dominant social convictions of a

---

[5]See Falkner and Jaspers (2012), pp. 30–55.

[6]See Trujillo (2013), pp. 12–15.

[7]Balassa (1978), p. 422.

[8]Scott (2007), p. 77.

society are permitted for the sake of attaining a reasonable degree of safety and eliminating adverse circumstances.

For example, the Agreement on Sanitary and Phytosanitary Measures allows WTO members to set their protection levels above international standards. Doing so is not mandatory, although it has been encouraged. International standards are guidelines and recommendations established by the Codex Alimentarius Commission (which covers food production and safety), the World Organization for Animal Health (*Office International des Epizooties*, OIE), the International Plant Protection Convention (IPPC). The voluntary nature of standards does not oblige governments to follow higher levels of protection than those considered to be appropriate. Very often international standards are more rigorous than rules enacted by domestic administrative authorities. There has been widespread commitment to harmonise standards because compliance with the SPS Agreement is presumed. Despite the possibility that domestic standards will be set higher than those existing internationally, balance is attained by the emphasis on the scientific foundation of state measures. Where international standards are absent or displaced by national rules, each state must provide scientific evidence for its position.

Once the standard level of protection is defined, a government is precluded from implementing a more severe or rigorous measure than necessary. In this respect, Article 2.2 establishes that sanitary or phytosanitary measures should only be applied if necessary for human, animal or plant life. For the WTO's Appellate Body, a measure must not only be indispensable to be necessary; it must also contribute towards accomplishing its objective. An indissoluble link is required between a measure and the elimination of the targeted risk for the scientific evidence requirement to be fully met. In the case *Japan Varietals*, the Appellate Body defined scientific process as follows: 'a process characterized by systemic, disciplined and objective enquiry and analysis, that is, a mode of studying and sorting out facts and opinions.'[9]

Fundamentally, as already observed, there must exist an objective relationship relating a measure to available scientific evidence, determined case by case in virtue of concrete circumstances. In the *Japan Varietals* dispute, the panel pointed out the lack of sufficient scientific evidence to justify quarantining various products. To be necessary for the protection against an identified risk, each element of a domestic measure must cite sufficient scientific evidence.

Sanitary and phytosanitary measures thus presuppose risk assessment by importing countries, other WTO members or international organisations. Accordingly, Article 5.1 proclaims: 'Members shall ensure that their sanitary and phytosanitary measures are based on an assessment, as appropriate to the circumstances, of the risks to human, animal or plant life or health, taking into account risk assessments techniques developed by the relevant international organizations.'

The first definition of such assessment regards risks related to plagues or diseases, while the second involves risks to human or animal health from substances found in

[9]World Trade Organization (1998b), par. 73.

foods, beverages or animal ratio. The assessment comprises three essential elements: (a) the identification of the disease or plague whose entry, establishment or dissemination it intends to ban, as well as biological and economic consequences derived from the entry, establishment or dissemination of such diseases or plagues; (b) an assessment of the likelihood of entry, establishment or dissemination of diseases or plagues, as well as economic and biological effects that may arise; and (c) an assessment of the probability of entry, establishment or dissemination of diseases or plagues in conformity with sanitary and phytosanitary measures susceptible to be applied.[10]

An extremely relevant evaluation of risk assessment appeared in the *EC–Hormones*[11] case:

> [G]eneral studies do not indeed show the existence of a general risk of cancer; but they do not focus on and do not address the particular kind of risk here at stake—the carcinogenic or genotoxic potential of the residues of those hormones found in meat derived from cattle to which the hormones had been administered for growth promotion purposes (…) Those general studies, are in other words, relevant but do not appear to be sufficiently specific to the case at hand.[12]

Similarly, in the *Japan–Apples* case, under the SPS Agreement the obligation to conduct an assessment of 'risk' is not satisfied merely by a general discussion of the disease sought to be avoided by the imposition of a phytosanitary measure. The Appellate Body found the risk assessment in the *EC–Hormones* not to be sufficiently specific although the scientific articles cited by the importing member had evaluated the 'carcinogenic potential of entire categories of hormones, or of the hormones at issue in general'. To constitute a 'risk assessment' as defined in the SPS Agreement, the WTO's Appellate Body concluded, the risk assessment should have reviewed the carcinogenic potential, not of the relevant hormones in general, but of residues of those hormones found in meat derived from cattle to which the hormones had been administered for growth promotion of purposes. Therefore, when discussing the risk to be specified in the risk assessment in *EC–Hormones*, the Appellate Body referred in general to the harm concerned (cancer or genetic damage), as well as to the precise agent that may possible cause the harm (that is, the specific hormones when used in a specific manner and for specific purposes).[13]

Both decisions highlight a common element: the conviction that the risk must be specific; that is, it must be the result of a product and not of a general nature, such as involving a combination of several different products. In the risk assessment of pests and diseases, each member will evaluate policy alternatives before making its final decision. Hence, Article 5.1 sets forth the obligation for sanitary and phytosanitary measures to be based on an appropriate assessment of the circumstances. The content

[10]Scott (2007), pp. 91–92.

[11]Pavoni (2010), p. 678.

[12]World Trade Organization (2008), par. 200.

[13]Amaral (2011), p. 239.

of this expression, which is eminently vague, becomes clearer in light of Articles 5.2 and 5.3, which facilitate the interpretive task. So, Article 5.2 provides:

> [I]n the assessment of risk, Members shall consider available scientific evidence; relevant processes and production methods; relevant inspection, sampling and testing methods; prevalence of specific diseases or pests; existence of pest—or disease—free areas; relevant ecological and environmental conditions; and quarantine or other treatment.

Article 5.3, in turn, reads:

> [I]n assessing the risk to animal or plant life or health and determining the measure to be applied for achieving the appropriate level of sanitary and phytosanitary protection from such risk, Members shall take into account as relevant economic factors: the potential damage in terms of loss of production or sales in the event of entry, establishment of spread of a pest or disease: the costs of control or eradication in the territory of the importing Member; and the relative cost-effectiveness of alternatives approaches to limiting risks.

Article 5.2 enumerates, for example, without any intent of exclusivity, the factors to be considered in the process of risk assessment. The economic factors[14] count only for the risks to life or health of animals and plants, as the panel in the *EC–Hormones* case explained in interpreting Article 5.3. However, the first definition of risk in Annex A.4 includes:

> [T]he evaluation of the likelihood of entry, establishment or spread of a pest or disease within the territory of an importing Member according to the sanitary or phytosanitary measures which might be applied, and of the associated potential biological and economic consequences; or the evaluation of the potential for adverse effects on human or animal health arising from the presence of additives, contaminants, toxins or disease-causing organisms in food, beverages or feedstuffs.[15]

In this context, a disturbing question arises: can a minority scientific opinion, under Article 5.1, justify the adoption of a sanitary or phytosanitary measure? In the case *EC–Hormones* the Appellate Body decided:

> [W]e do not believe that a risk assessment has to come to a monolithic conclusion that coincides with the scientific conclusion or view implicit in the SPS measure. The risk assessment could set out both the prevailing view representing the 'mainstream' of scientific opinion, as well as the opinions of scientists taking a divergent view. Article 5.1 does not require that the risk assessment must necessarily embody only the view of a majority of the relevant scientific community. In some cases, the very existence of divergent views presented by qualified scientists who have investigated the particular issue at hand may indicate a state of scientific uncertainty. Sometimes the divergence may indicate a roughly equal balance of scientific opinion, which may itself be a form of scientific uncertainty. In most cases, responsible and representative governments tend to base their legislative and administrative measures on 'mainstream' scientific opinion. In other cases, equally responsible and representative governments may act in good faith on the basis of what, at a given time, may be a divergent opinion coming from qualified and respected sources. By itself, this does not necessarily signal the absence of a reasonable relationship between the SPS measure and the risk assessment, especially where the risk involved is life-threatening in

[14] Pavoni (2010), p. 661.

[15] Annex A.4 of the WTO Agreement on the Application of Sanitary and Phytosanitary Measures, opened for signature 15 April 1994, 1867 UNTS 493 (entered into force 1 Jan 1995).

> character and is perceived to constitute a clear and imminent threat to public health and safety. Determination of the presence or absence of that relationship can only be done on a case-to-case basis, after account is taken of all considerations rationally bearing upon the issue of potential adverse health effects.[16]

Minority scientific opinions that support the affirmation that sufficient positive evidence for the purported risk will be deemed credible only if derived from truly qualified or respected sources. It is generally accepted that governments rely on minority opinions when the risk in question endangers life or poses a clear and imminent threat to public health. The reassessment of risk is appropriate in circumstances where members obtain new information capable of altering the earlier determination because of the discovery that the asset in question could cause harm to human health or the environment. The concept of risk assessment is relative and adapts to changing facts to preserve the autonomy of Member States to deal with new information, information that in principle could modify the dominant scientific conclusion.

Governments can, as a precaution, take provisional measures in the absence of established scientific certainty. The Appellate Body has made clear in the *EC–Hormones*[17] case that provisional measures should not be used broadly or without limits. In cases where relevant scientific evidence is insufficient, Article 5.7 provides that a member may provisionally adopt sanitary or phytosanitary measures based on the available relevant information, including information from relevant international organisations, as well as sanitary and phytosanitary measures applied by other members. In such circumstances, a WTO member shall seek the additional information required for a more objective risk assessment and shall review the sanitary and phytosanitary measure within a reasonable frame of time.

Joan Scott states that Article 5.7 could be divided into four parts, considered cumulatively: (a) a measure is applied in the event of insufficient relevant scientific information; (b) it should be adopted based on the relevant available information; (c) additional information necessary for the objective determination of the risk should be sought by the member; and (d) review of the measure should be undertaken within a reasonable period of time.[18] Despite the uncertainties that remain, Article 5.7 seems to confer an autonomous right, not an exception to Article 2.2. The panel in the *EC–Biotech* case considered that Article 5.7 codifies an exception to Article 5.1, so that a measure that is not inconsistent with Article 5.1 is thus not automatically presumed to be incompatible with Article 2.2. Defining Article 5.7 as a qualified exception rather than as a mere exception has, in the opinion of the *EC–Biotech*[19] panel, important consequences for the distribution of the burden of proof. It is for the complainant and not for the complained party to demonstrate that the contested measure is incompatible with requirements under Article 5.7.

---

[16]World Trade Organization (1998a), par. 194.

[17]Pavoni (2010), p. 677.

[18]Scott (2007), p. 111.

[19]See Trujillo (2013), p. 21.

In the *Japan–Apples* case, the Appellate Body explicitly correlated Article 5.7 with Article 5.1:

> [T]hese contextual elements militate in favour of a link or relationship between the first element of Article 5.7 and the obligation to perform a risk assessment under Article 5.1: 'relevant scientific evidence' will be 'insufficient' within the meaning of Article 5.7 if the body of available scientific evidence does not allow, in quantitative or qualitative terms, the performance of an adequate assessment of risks as required under Article 5.1 and as defined in Annex A to the SPS Agreement.[20]

Insufficiency has a qualitative dimension, one that involves a lack of confidence in the results because of inadequate methodology or inconclusive results, as well as a quantitative dimension, which generally reflects the volume or scarcity of relevant evidence available to experts. It is only natural that, in circumstances of insufficiency, the members should not have the right to take measures to address theoretical risks that are in fact yet to be established by reliable data.

Insufficient scientific evidence makes the adoption of sanitary or phytosanitary measures possible if adequate relevant information is available. We have every reason to believe that the predominant view at the WTO favours a strong, tight relationship between any provisional measure and the existing international standards, to the point that this relationship cannot be proven if a provisional measure challenges the standards in certain areas. There is no criterion guiding the search for additional information referred to in Article 5.7 or for the review of the measure in a reasonable period of time, which are dependent on the specific circumstances of each case, according to the *Japan Varietals* decision.

The *EC–Hormones* case provoked debate over the nature of the precautionary principle in public international law and the role it plays in the SPS Agreement. The Appellate Body rejected the motion challenging it to express itself on the status of the precautionary principle in contemporary international law, as it found doing so unnecessary and even reckless. It highlighted, however, some aspects involving the relationship of the precautionary principle to the SPS Agreement. First, the principle was not written into the SPS Agreement as a way to justify measures incompatible with the obligations of the members contemplated in particular provisions of that Agreement. Second, the precautionary principle appears in Article 5.7, but this article does not exhaust its relevance. It also finds expression in the sixth paragraph of the preamble and in Article 3.3, which explicitly recognises the right of members to establish their own level of sanitary protection, which may be higher than established in current international standards, guides and recommendations. Third, a panel asked to determine, for instance, whether there is 'sufficient scientific evidence' to authorise the maintenance by a member of a sanitary or phytosanitary measure, should bear in mind that representative and responsible governments commonly act with prudence and caution where they incur irreversible risks to life or damage to human health. Finally, the precautionary principle does not diminish the duty of the panel to apply the principles of customary international law

[20]World Trade Organization (2008), par. 179.

concerning the interpretation of treaties when reading the provisions of the agreement.[21]

16.3 The Biosafety Protocol

The Cartagena Protocol on Biosafety,[22] addresses the transboundary movement of living modified organisms produced with modern biotechnology. The Protocol was born under the auspices of the Convention on Biological Diversity, which saw the light at the 1992 Rio Conference, and was adopted on 29 January 2000, at the conclusion of negotiations that spanned four years. It was opened to signature at the Fifth Meeting of the Conference of the Parties to the Convention on Biological Diversity, held in Nairobi from 15 to 26 May of that year. The protection of biological diversity and human health, the main goals of the new Protocol, have been embodied in standards relating to the safe transfer, use, treatment and disposal of living modified organisms. Contrary to wishes of certain governments, who preferred to refer to genetically modified organisms, the US delegation succeeded, in the end, in imposing the use of their preferred term, living modified organisms, meaning 'any living organism that possesses a new combination of genetic material obtained through the use of modern biotechnology' (Article 7).

Underlying the Washington proposal was the explicit intention to emphasise that products made from genetic engineering are no more dangerous than those that modified using other processes. Different configurations of countries came together according to the nature of their interests, shaping the negotiation[23] of the Cartagena Protocol on Biosafety.[24] Mostly represented by developing countries, the Like-Minded Group[25] was committed to the defence of positions that lacked the necessary conditions to assess the risks peculiar to genetically modified products because of the absence of qualified professionals and specialised national agencies. The suggestions it put forward underscored the need for the Protocol to be broad in scope, to resist limitations on the use of the precautionary principle, and to allow for decisions to consider social and economic repercussions in addition to adverse effects on human health.

Composed of the United States, Argentina, Australia, Canada, Chile and Uruguay—all large producers of artificially developed seeds—the Miami Group advocated policies diametrically opposed to those of the Like-Minded Group, namely restricting the scope of the protocol so that it not include genetically modified products. The Miami Group also recommended that the use of the precautionary

[21]World Trade Organization (1998a), par. 123–124; European Communities (2006).

[22]See Lefeber (2012).

[23]See Gupta (2010), pp. 34–38.

[24]See Kelemen and Knievel (2015), p. 955.

[25]See Gayathri and Kurup (2009), p. 239.

principle would only be allowed on rigidly circumscribed occasions, including those concerning social and economic conditions. Scientific criteria, it was held, should be the only parameters guiding the decisions based on the Protocol. The precautionary principle would lead, from their perspective, to inevitable abuses when dealing with hypothetical risks about which science had not provided assertive conclusions.

Analogously to the Like-Minded Group, but in response to specific factors stemming from consumer pressure, the European Union generally aligned itself around the Like-Minded Group's positions, noting its fear of unfamiliar threats. Considering the drawbacks of protectionism, it suggested the importance of the non-discrimination principle between imported goods and those produced domestically would prevent abuse of the precautionary principle by importing countries. It warned, as did the Like-Minded Group, that irreversible damage could understandably be caused by the ongoing ignorance of the potential harms of modified. The bloc of Eastern and Central European countries, for their part, sought solutions that could be accepted and applied by all states.

The Compromise Group, made up of Japan, Mexico, the Republic of Korea, Singapore, Switzerland and News Zealand, struggled to overcome the antagonism to reach a mutually acceptable agreement. Social groups mobilised in protest during the negotiations phase against the positions taken by the Global Industry Coalition, herald of business interests—chiefly pharmaceutical and food companies. That coalition supported the Miami Group, while environmental associations and consumer movements allied with the Like-Minded Group.

Because of the long-lasting negotiations, the final text of the Protocol embodies the desire harmonise antagonistic positions despite a clear predilection for the arguments advanced by the Like-Minded Group and the European Union. Prior informed consent, a fundamental procedure in the legal architecture of the Protocol, is intended to provide the authorities of the importing country, through notification, with detailed information on products entering the natural environment, such as seeds, before the first shipment takes place. It is essential that the importer ascertain and communicate the characteristics of the imported good, the changes it has undergone and the rules governing it, based on which commerce is authorised or prohibited. The risk assessment report by the exporter, which completes the list of mandatory information, must be added to those requirements.

Within two hundred and 70 days from the date of receipt of the notification, the importing Party must communicate to the notifier and the Biosafety Clearing-House[26] of its decision whether to approve the importation, which would include determination of how the decision would be applied to subsequent imports of the same living modified organism (Article 10.3). The same must apply in cases where it was considered appropriate to prohibit importation or to request relevant or

[26]'The Biosafety Clearing-House (BCH) is a mechanism set up by the Cartagena Protocol on Biosafety to facilitate the exchange of information on Living Modified Organisms (LMOs) and assist the Parties to better comply with their obligations under the Protocol. Global access to a variety of scientific, technical, environmental, legal and capacity building information is provided in the six official languages of the UN.' Biosafety Clearing-House (2018) par. 1.

additional information under the domestic legal system.[27] With a view to reducing risks for biological diversity and human health, the importing state is authorised to take decisions on the importation of a living modified organism if there is no scientific certainty as to the adverse effects it causes. Documentation specifying the characteristics and providing information on handling, storage, transport and use, and the name and address of the importer and exporter of living modified organisms entering the environmental shall be required. Only part of the trade in these goods occurs under the rules concerning the prior informed procedure.

This procedure also excludes medicines, goods in transit, 'contained use' defined by Article 3 (b) of the Protocol, products controlled by specific measures that limit contract and impact on the environment, and the goods declared safe by the Meeting of the Parties to the Protocol on Biosafety. Moreover, this list includes consumer products derived from living modified organisms. Different and far less stringent requirements govern the notification of living modified organisms intended for direct use as food or feed or for processing, including most of the grains that represent more than 90% of the trade in living modified organisms. The documentation accompanying them, to clarify the nature of future uses, must require that the importer does not intend to introduce them into the environment.

Highly relevant is the extreme difficulty of including living modified organisms in the category of goods released into the environment by voluntary act or between those intended for human or animal consumption or processing. Lower costs explain the behaviour of farmers who, instead of using imported grains for feed or processing, prefer to use them as seeds for planting new crops. Meeting in the city of Curitiba in March 2006, the Contracting Parties decided by consensus to keep the words 'may contain living modified organisms' in Article 18.2 (a) of the Cartagena Protocol. It is incumbent upon states to enact domestic laws requiring disclosure about crops of products altered by biotechnology only when the identity is known to involve means such as the identity preservation system.[28] Similarly, this happens with modified organisms that are introduced into the environment. The Protocol recognises, based on the precautionary principle,[29] the validity of domestic measures that forbid or interrupt the flow of imported products that cause risks to human health and biological diversity.

Forged in the arena of International Law during the negotiations of various environmental conventions in the 1980s and 1990s, the precautionary principle influenced the Cartagena Protocol as genetically modified products were developed in the wake of advances in biotechnology, sparking fear of potential harm to human health and the environment by the cross-border movement of these goods. The importance of precautionary principle is reflected in the frequent reference to it throughout the Protocol, in particular in Articles 1 and 17. A point of contention in the negotiations, the precautionary principle is an undeniable achievement of the

[27]See Kelemen and Knievel (2015), pp. 945–965.

[28]Redick (2007), pp. 51–116.

[29]Kelemen and Knievel (2015), p. 961; Trujillo (2013), p. 22.

Protocol, as it authorises signatories to prevent imports when scientific certainty regarding the likelihood of future damage is lacking. Once implemented, the ban will likely be extended indefinitely because the importing State will not feel compelled to seek information that resolves the prevailing scientific uncertainty. The exporter shall be entitled to apply for review of the restrictive measures if the circumstances justifying it change or previously unknown relevant information is found. Domestic authorities may, if they wish, maintain the initial decision as long as it communicates the grounds of the administrative act exercised. Yet this rule only applies to modified organisms that are introduced into the environment by voluntary act. It differs in this respect from the stricter limitation on governments established by the SPS Agreement, which obliges the competent governmental body to seek additional information and to review the measure within a reasonable period of time.

The flexibility to adopt measures against unknown risks, a central element of the precautionary principle, has been a demand of developing countries since the Cartagena Protocol negotiations began. It was given priority because those countries lacked the necessary scientific expertise, border security and internal laws to anticipate or prevent unwanted consequences. Within the WTO, however, developing countries have advocated a strict interpretation of the Agreement on Technical Barriers and the SPS Agreement together with international standards and scientific evidence. In the former case, almost all developing countries import living modified products, while in the latter case they are also usually exporters of agricultural product who are afraid of undue restrictions on their exports. This duality shows that nations with scarce technical and financial resources are uncomfortable taking unequivocal positions under conditions of increasingly commercial complexity.[30]

Phillips and Kerr, however, point out that the Cartagena Protocol runs counter to the guiding philosophy of the WTO by allowing trade restrictions based on manufacturing processes and methods, rather than exclusively focusing on product characteristics.[31] They assert that: 'The BSP is clearly inconsistent with the WTO in a number of areas (. . .) it is imperative that jurisdictional issues should be sorted out quickly.'[32] Matthee and Vermersch share the same view, arguing: 'It can be concluded that under the current WTO Agreement and with the current definitions of the precautionary principle, no full reconciliation of the precautionary principle and trade liberalization is possible.'[33]

The reason for this is simple: because the lack of scientific evidence justifies exercise of the precautionary principle, the need for the measures in question cannot be proven, nor can the country which deploys them demonstrate that they are not a disguised restriction on trade.[34] The possibility of conflict[35] between the

[30]Zarrilli (2001), pp. 614–621; Hagen and Weiner (2000), pp. 697–716.

[31]Winham (2003), p. 146; Redick (2007); Jacob (2001), pp. 79–90; Glasse (2001), pp. 491–518.

[32]Phillips and Kerr (2000), p. 74.

[33]Ibid, p. 69.

[34]Winham (2003), p. 146.

[35]See Pavoni (2010), pp. 653–657.

precautionary principle envisaged in the Cartagena Protocol and the requirement for scientific risk assessment, as set out in Article 5.1 of the SPS Agreement, has already been raised. This is certainly a major challenge for the WTO Dispute Settlement System. Notwithstanding the rule in Article 1 (1) of the Dispute Settlement Understanding that the decisions rendered rely on the agreements set out in Appendix 1 of same, panels and the Appellate Body may dilute the meaning of sufficient scientific evidence to be compatible with the precautionary principle.

A pressing problem concerns the application of agreements under the multilateral trading system to living modified organisms. There seems to be no doubt that Articles III, XI and XX of the GATT apply to hypotheses that are not fully supported. The principle of national treatment in Article III of the GATT, when transposed to the reality of living modified organisms, prohibits the importing country from imposing measures that are costlier for foreign products than they are for similar domestic goods. Article XI, whose intention is eliminating quantitative restrictions, stipulates that:

> [N]o Contracting Party shall establish or maintain, for the importation of a product originating in the territory of another Contracting Party or for export or sale for export of a product to prohibit or restrict, other than customs duties, taxes or other charges, whether they application is by quotas, import or export licences or by any other processes.

Article XX, however, provides for exceptions to free trade that have greatly contributed to the acceptance and effectiveness of the commitments that the signatories undertook. In the judgement of the *US–Shrimp* case,[36] the Appellate Body considered that a balance must be struck between the right to invoke one of the exceptions in Article XX and respect for the other members provided for in the treaty. A measure relied upon must be among the exceptions in Article XX and satisfy the requirements of the caput, that is, it must constitute unjustified or arbitrary discrimination between countries where the same conditions prevail, or arbitrary discrimination against international trade. If a violation of the SPS Agreement is alleged, a WTO Member wishing to suspend imports of living modified organisms shall be required to demonstrate that the decision is necessary to protect human, plant or animal life or health. The success of an attempt to resort to the SPS Agreement is uncertain, as clear scientific evidence on the risks brought by living modified organisms is lacking. The difficulties are not reduced with respect to the Agreement on Technical Barriers to Trade: the doctrine examines whether the import ban carried out by a government, characterises a technical regulation and how different the genetically modified organisms are from the conventional species and whether they exhibit a similar nature.

The 'substantial equivalence test' has been recommended as a means of resolving divergences. This test operates on the premise that identical legal treatment should be applied to genetically modified products and conventional products that possess the same external characteristics, for it is assumed that having identical external characteristics implies that the products are equally safe or dangerous. The European

[36]See Trujillo (2013), p. 7; Pavoni (2010), pp. 661–666.

Union and certain developing countries have accused the 'substantial equivalence test' of lacking any scientific basis and insist that genetically modified foods and conventional foodstuffs are different because the former contain DNA or proteins that are absent in the latter.

16.4 Conclusion

Several observations can be drawn from the foregoing discussion:

1. The rapid acceleration and proliferation of living modified organisms have raised doubts as to their effects as to human, animal and plant life and health. Possible harms, at present, cannot be scientifically proven or disproven.
2. Fundamentally, two treaties regulate living modified organisms: the SPS Agreement and the Biosafety Protocol.
3. While the former establishes the need for scientific evidence to ban an imported good, the latter allows prohibition on the grounds of the precautionary principle.
4. This contradiction might be attenuated, eventually, by the Appellate Body under the jurisdiction of the WTO. If the contradiction is not resolved, however, problems of legitimacy of International Law could arise.
5. Given the great expansion of International Law witnessed over the last decades, the creation of incompatible rules is likely. The case in question is a clear-cut indication of what experts have called the fragmentation of International Law, whose outcome is the existence of contradictory rules.

One cause of fragmentation of International Law is the nature of state behaviour in international affairs, which varies in multiple arenas in following different configurations of interests at the domestic level. This explains why developing countries supported scientific evidence in the SPS Agreement yet backed the precautionary principle in the Biosafety Protocol.

References

Amaral A Jr (2011) Comércio internacional e a Proteção do Meio Ambiente. Atlas, São Paulo

Balassa B (1978) The new protectionism and the international economy. J World Trade Law 12 (5):409–436

Biosafety Clearing-House (2018) Welcome to the BCH central portal. https://www.bch.cbd.int/. Accessed 10 May 2018

European Communities (2006) Measures affecting the approval and marketing of biotech products: reports of the panel. WT/DS291/R, WT/DS292/R, WT/DS293/R. http://docsonline.wto.org/imrd/gen_searchResult.asp?RN=0&searchtype=browse&q1=%28%40meta%5FSymbol+WT%FCDS293%FCR%2A+and+not+RW%2A%29&language=1. Accessed 12 Dec 2018

Falkner R, Jaspers N (2012) Regulating nanotechnologies: risk, uncertainty and the global governance gap. Global Environ Polit 12(1):30–55. https://doi.org/10.1162/GLEP_a_00096

Gayathri PG, Kurup RR (2009) Reconciling the bio safety protocol and the WTO regime: problems, perspectives and possibilities. Am J Econ Bus Admin 1(3):236–242
Glasse JA (2001) The merits of ratifying and implementing the Cartagena Protocol on biosafety. Northwestern J Int Law Bus 21(2):491–518
Gupta A (2010) Transparency as contested political terrain: who knows what about the Global GMO Trade and why does it matter? Global Environ Polit 10(3):34–38
Hagen PE, Weiner JB (2000) The Cartagena Protocol on Biosafety: new rules for international trade in living modified organisms. Georgetown Int Environ Law Rev 12(3):697–716
Jacob T (2001) The Cartagena protocol—a first step to a global biosafety structure? Trans Lawyer 14(1):79–90
Kelemen R, Knievel T (2015) The United States, the European Union, and international environmental law: the domestic dimensions of green diplomacy. I CON 13(4):945–965. https://doi.org/10.1093/icon/mov057
Lefeber R (2012) The legal significance of the Nagoya-Kuala Lumpur supplementary protocol: the result of a paradigm evolution. In: Amsterdam Law School. Legal studies research paper no 2012–87. http://ssrn.com/abstract=2151282. Accessed 14 Feb 2018
Pavoni R (2010) Mutual supportiveness as a principle of interpretation and law-making: a watershed for the 'WTO-and-Competing-Regimes' debate? Eur J Int Law 21(3):649–679
Phillips PWB, Kerr WA (2000) Alternative paradigms: the WTO versus the biosafety protocol for trade in genetically modified organisms. J World Trade 34(4):63–75
Redick TP (2007) The Cartagena protocol on biosafety: precautionary priority in biotech crop approvals and containment of commodities shipments. Colorado J Int Environ Law Policy 18 (1):51–116
Sampson G (2005) The WTO and sustainable development. United Nations University Press, New York
Schoenbaum TJ (2001) International trade in living modified organisms. In: Francioni F (ed) Environment, human rights and international trade. Hart, Portland
Scott J (2007) The WTO agreement on sanitary and phytosanitary measures: a commentary. Oxford University Press, New York
Shaffer G (2012) International law and global public goods in a legal pluralist world. Eur J Int Law 23. http://ssrn.com/abstract=2243063. Accessed 10 Jan 2018
Telesetsky A (2011) The 2010 Nagoya-Kuala Lumper supplementary protocol: a new treaty assigning transboundary liability and redress for biodiversity damage caused by genetically modified organisms. Insights 15(1):1–11
Trujillo E (2013) A dialogical approach to trade and environment. Legal studies research paper series 13–6. http://ssrn.com/abstract=2235071. Accessed 5 Jan 2018
Winham GR (2003) International regime conflict in trade and environment: the biosafety protocol and the WTO. World Trade Rev 2(2):131–155
World Trade Organization (1998a) EC measures concerning meat and meat products (hormones) AB-1997-4: report of the Appellate Body. WT/DS26/AB/R, WT/DS48/AB/R. http://www.worldtradelaw.net/reports/wtoab/ec-hormones(ab).pdf. Accessed 25 Nov 2017
World Trade Organization (1998b) Japan—measures affecting agricultural products: report of the panel. WT/DS76/R. https://www.wto.org/english/tratop_e/dispu_e/76r.pdf. Accessed 10 Jan 2018
World Trade Organization (2008) Japan—measures affecting the importation of apples—AB-2003-4—report of the Appellate Body WT/DS245/AB/R. https://docs.wto.org/dol2fe/Pages/FE_Search/FE_S_S009-DP.aspx?language=E&CatalogueIdList=36202&CurrentCatalogueIdIndex=0&FullTextHash=&HasEnglishRecord=True&HasFrenchRecord=True&HasSpanishRecord=True. Accessed 25 May 2019
Zarrilli S (2001) International trade in genetically modified organisms and multilateral negotiations: a new dilemma for developing countries. In: Francioni F (ed) Environment, human rights and international trade. Hart, Portland

Chapter 17
The Effects of International Agreements on Water Security: A Critical Study of the EU and MERCOSUR Approaches

Kleverton Melo de Carvalho and Clara María Minaverry

Contents

Abstract The consequences of overuse of the planet's water supplies are the subject of emerging concern. This chapter aims to critically discuss the effects of different regulatory approaches to water security, comparing the European Union (EU) and the Southern Common Market (MERCOSUR) and paying close attention to industrial activity. Water security can be understood as the adequate availability of water to meet the needs of society, the economy and the ecosystem. Thus, the study intends to: (i) question the legal base, its coercive power and extent; (ii) explore effects of agreements on internal water regulation, focusing especially on the industrial sector; (iii) identify hydro-sustainable policies adopted by governments to control industrial activity considering the domestic and international laws. The chapter employs the critical hermeneutics methodology, examining publications of governments, international organisations, NGOs, academic articles, and international newspapers and magazines. The chapter notes that the most effective treaties regarding water security are those that have economic coercion clauses, such as the Basel and Stockholm

K. M. de Carvalho (✉)
Federal University of Sergipe, Aracaju, Brazil

C. M. Minaverry
Faculty of Law, University of Buenos Aires, Buenos Aires, Argentina
e-mail: cminaverry@derecho.uba.ar

A. d. Amaral Junior et al. (eds.), *Sustainable Consumption*,
https://doi.org/10.1007/978-3-030-16985-5_17

Conventions. It concludes that current regulation and policies, especially in MERCOSUR, need to be urgently revised to ensure the quantity, quality and accessibility of water reserves are to be maintained for present and future generations.

17.1 Introduction

The excessive exploitation of the planet's natural resources and our present way of life are unsustainable. Humanity has been warned through various indicators of the urgent need to review the format of our society's relationship with nature. Humanity has been eminently unsustainable and has lived in a scenario of complete insecurity at multiple risks. Can we live much longer on the planet? We do not know. One of the key resources and indicators of life—safe water—is seriously compromised. Between 1970 and 2012, humanity experienced a 58% decrease in freshwater habitats because of the fragmentation of rivers and streams and excessive water withdrawal.[1] Industrial activity emerged as a core factor in (over) use and contamination of water.

This chapter is structured around a comparison of a selection of international instruments and public policies in two integration blocs that are fundamental at the international level: the European Union (EU) and MERCOSUR. In pursuing this analysis, it adopts the critical hermeneutics methodology and focuses on the multidisciplinary concept of 'water security', which basically considers water availability for different users integrating politics among peace and cooperation. This is a preliminary study, as we anticipate the publication of more in-depth research, and it has the same objectives concerning to water highlighted in the United Nations Sustainable Development Goals (SGDs), that are based on conserving fresh water habitats in the face of industrial pollution and unsustainable water withdrawals.[2]

The EU has high level development of policies and environmental laws, as each country in the bloc must necessarily include in its internal legal system everything regulated at the supranational level. MERCOSUR, in contrast, has yet to develop any relevant supranational authority. Moreover, it has not (so far) had the proper policy to include environmental and hydric concerns, and has struggled to address the context of droughts and contamination which its members are experiencing individually. However, we stress that water use is a key factor for integration in the South American bloc because of the giant water resources shared by the countries of the region.

The chapter proceeds as follows. The first section describes the conceptual construction of water security focusing on the industrial sector. The second

---

[1]World Wildlife Fund (2016).

[2]UN (2015).

examines existing multilateral international instruments that have an impact on water security in the industrial sector. Keeping with the theme of industrial activity and water, the third section compares regional regulation and public policies in MERCOSUR and the EU.

17.2 Towards a Consensus Vision of 'Water Security': Discourse, International Cooperation, and the Regulation of Industrial Activity

To date no scientific consensus on the definition of 'water security' has emerged, because of the diverse epistemologies that underlie this actual approach. In this section, we will briefly describe some of its main guidelines and links between them and industrial activity.

The idea of 'securitization', as developed by the Copenhagen School, is grounded in an internationalist perspective. In the post-Cold War period, this theoretical current has risen to prominence through analysis of perceived threats to a range of values (including ones not traditionally captured by security analysis) and how they manifest in the collective imagination. Indeed, 'securitization' has been a bedrock of cooperation between the EU's Member States, through the development of laws and policies to remedy all manner of non-traditional security challenges. This orientation has blossomed—especially in the defence area—since the 9/11 attacks. The basic ground of this approach is the requirement to manage certain issues without necessarily having conditions of objective threat present—what is central here is the social construction of perceived risk. This implies that immediate actions must be taken by public decision makers if they see them as appropriate while facing a certain threat or risk of damage to acquired values.[3]

This notion has been expanded to the water arena, bringing together political, environmental, economic and social considerations. From the political perspective, beyond the fantasy that a world war over water will erupt, we highlight efforts to construct multilateral legal regimes for the reduction of risks, especially concerning water scarcity and pollution. This action has influenced cooperation to tamp down conflicts between transboundary countries. The EU is arguably a good example of water cooperation. In contrast, the dispute between India and Pakistan over the Indus River and contestation over the River Jordan constitute signal examples of transboundary water conflict.[4]

Furthermore, awareness has grown of the crucial role played by ecosystems in guaranteeing the quantity and quality of water for various users. The environmental approach outlines the threats suffered by human communities and nature. Based on this understanding, the vision of sustainability has been overcoming the orientation

[3]Stritzel (2007), Spring (2010) and Balzacq and Guzzini (2015).

[4]See Allenby (2000) and Magsig (2012).

of water security, which could be defined as an expression of the risk society in different scenarios to establish actions and policies. Achieving water security requires decision-making based on rational risk calculation that must be transparent (i.e., open to scrutiny and testing). Risk analysis provides a framework for the exploration of costs, benefits and compensation associated with various policies to guarantee water security. It could be seen as a process of social construction which aims to reach a rational prevention approach against unsustainable effects. The new conceptual paradigm of 'water security' has therefore superseded 'sustainable water'—which dominated in the 2000s—as a central organising framework, because it is more practical and focuses on access to water by all users, including non-human ones.[5]

To ensure the social perspective of water is heard in the international debate, the United Nations Human Rights Council recognised that access to drinking water and sanitation are fundamental human rights. Countries and international organisations have also been lobbied to provide financial resources, enhance capacity and transfer technology through international assistance and cooperation on water security. This UNGA Resolution (64/292/2010) calls on countries to commit themselves to:

1. elaborate laws, plans and strategies;
2. monitor the planning and implementation service of providing safe drinking water, considering also the involvement of affected local communities (with special attention to vulnerable and marginalised groups);
3. observe human rights in the impact assessments of delivering water services; and
4. ensure that non-State service providers accomplish their human rights responsibilities, through a good quality and sufficient quantity of safe drinking water, including effective complaints mechanisms which could be implemented by users.

Considering the industrial sector, the economic element is crucial if we interrelate it with water security, which observes environmental, social and political factors. This relationship is crucial, as international geopolitics has redesigned the industrial production of the planet, especially through consolidation of interregional economic blocs. The trade of industrial goods depends and has a significant impact on water resources because of the pressure exerted from its use and contamination. This tension is increasing with the expansion of global production chains, especially those of industries whose manufacturing processes span different countries. The virtual water trade (i.e., the water resource embedded in products through the value chain) today represents approximately 15% of total world water use.[6]

Industrial expansion has seen contamination of freshwater resources on an unsustainable scale, warranting urgent attention. Today, between 300 and 500 million tonnes of heavy metals, solvents, toxic sludge and other waste is dumped by industry into waterways annually, making it the largest source of water pollution on

[5]See Beck (1995), Goldblat (1996), Hall and Borgomeo (2013) and Lupton (2013).

[6]See Dalin et al. (2012), Hoekstra (2015), Carvalho and Santos (2017) and Hoekstra (2017).

the planet. These pollutants make water unsafe while contaminating fish stocks, which are an important source of protein for a large part of the world's population, especially in developing countries. Contaminated water also flows through the food chain through agriculture or livestock, another underappreciated risk to end users. In developing countries, 70% of industrial waste is dumped without treatment.[7]

Moreover, industrial activity itself consumes vast quantities of water, although less than in agriculture. The various industrial uses of water include hydropower, nuclear and thermoelectric power generation, as well as in the manufacturing processes. The annual total water used by this sector can be expected continue to increase significantly, from 752 km^3 per year in 1995 to an estimated 1170 km^3 per year in 2025. This covers 24% of total freshwater withdrawals on the planet. To address this problematic scenario, the Sustainable Development Goals (2015–2030) presented by the UN in the Agenda 2030 declaration, incorporated various challenges and commitments about water. For example, SDG 6 establishes the obligation to 'ensure water availability and its sustainable management and sanitation for all', clearly stressing the importance of the industrial sector regarding water security and demonstrating a systemic view, considering the ecological, economic and social aspects of water resources.[8] The sub-goals are as follows:

- By 2030, improve water quality by reducing pollution, eliminating dumping and minimizing release of hazardous chemicals and materials, halving the proportion of untreated wastewater and substantially increasing recycling and safe reuse globally.
- By 2030, substantially increase water-use efficiency across all sectors and ensure sustainable withdrawals and supply of freshwater to address water scarcity and substantially reduce the number of people suffering from water scarcity.
- By 2030, implement integrated water resources management at all levels, including through transboundary cooperation as appropriate.
- By 2020, protect and restore water-related ecosystems, including mountains, forests, wetlands, rivers, aquifers and lakes.
- By 2030, expand international cooperation and capacity-building support to developing countries in water and sanitation related activities and programs, including water harvesting, desalination, water efficiency, wastewater treatment, recycling and reuse technologies. Support and strengthen the participation of local communities in improving water and sanitation management.[9]

To achieve those objectives, it is fundamental to develop policies from a multilevel and complex perspective, considering global, regional and local actions for water security within the industrial sector. Thus, we must consider as the concept of water security the integration of aspects such as:

---

[7]Hoekstra (2015) and Hoekstra (2017).

[8]UN (2011, 2015).

[9]UN (2015).

1. human access to clean and potable water;
2. the protection of livelihoods, human rights and cultural and recreational values;
3. the preservation and protection of ecosystems;
4. water availability;
5. the ability to cope with water-related hazards such as floods, droughts and pollution, among others; and
6. good governance, accountability, adequate and effective legal regimes, transparent, participatory and accountable institutions, infrastructure and capacity building.[10]

Globally, the challenge is to build this orientation within an international regulatory structure that is still dispersed and ineffective.

17.3 Multilateral Regulatory Instruments Related to Water

There is not yet a specific, integrative and multilateral document regulating freshwater protection, unlike other environmental issues, such as climate change. The excessive number of possibilities of uses of this resource and the legal fragmentation are two barriers to the creation of a global approach to fresh water. Furthermore, the increasing pressure on resources is exacerbated by the marketing of bottled water, domestic consumption, industrial activities, energy production, mining and oil extraction, irrigation and transportation among others.

After examining some international legal instruments related to water protection and management, we propose a specific focus on the industrial sector. Initially, we could say that there are only a few coercive international instruments. This section seeks to evaluate some of the international agreements concerning freshwater and industry, observing two categories of documents: (a) those that consider water as a central object, not necessarily related to industrial activity, but which in some way has an impact on it and; (b) those that are related to industrial activity and have greater power of imperium internally in the countries. In the first case, we will analyse the UN Convention on the Law of the Non-Navigational Uses of International Watercourses or the New York Convention (1997) and the Helsinki Convention (1992). In the second case we decided to examine specifically water protection instruments existing in the Conventions of Basel and of Stockholm.

The New York Convention incorporated some fundamental principles that were already applied through customary law. It entered into force on the 17 August 2014, despite its adoption by the General Assembly of the United Nations in 1997. This delay was due to the resistance of the transboundary countries which had relevant water resources and, in fact, only 36 countries (largely Arab ones) have ratified it. This instrument establishes that all Member States should make an equitable and

[10]UN (2013).

reasonable use of water, to prevent them from being harmed. The agreement is based on one of the fundamental principles of International Environmental Law, which is 'prevention of transboundary environmental damage'. This principle is directly linked to other principles such as solidarity, cooperation and sustainable development.[11]

The Helsinki Convention of 1992 (also known as the UNECE Convention) is very similar to the New York Convention. This treaty counts 56 signatories—especially in the EU but also on other continents—and defines aquifers and surface water as hydrologically connected watercourses, and occurs within an extensive part of water resource around the world. The convention has a more proactive approach (practical), because it considers the consequences of water uses and their environmental impacts. It is inspired by the precautionary principle, which is based on the application of actions against evidence of serious and irreversible damages in nature, even facing a lack of absolute scientific certainty. These ideas are associated with the implementation of specific plans, programs and activities allied to the management of transboundary watercourses, to provide rapid access to water for as many people as possible.[12]

This agreement seeks to achieve reductions in chemical emissions discharged in water by industry. Some of its dispositions are predominantly imperatives, functioning as coercion mechanisms. The Helsinki Convention states that to prevent, control and reduce transboundary impact, Parties shall develop, adopt, implement and, as far as possible, harmonise relevant legal, administrative, economic, financial and technical measures to ensure, *inter alia*, that: (i) the emission of pollutants is prevented, controlled and reduced at its origin through the application of, among other things, low pollution technology or pollution without residues; (ii) transboundary waters are protected against pollution from point sources by prior authorisation granted by national authorities, besides monitoring and controlling those discharges; (iii) the permissible wastewater discharge limits are based on the best available technology; (iv) more severe requirements are imposed, considering the water quality needed for ecosystems.

We note a clear link between the Helsinki Convention and the Basel and Stockholm Conventions. These documents establish lists of the industrial sectors and those linked to the handling of dangerous substances that appear in the applicable international regulations or conventions within its scope. Furthermore, the provisions shall not affect the rights or obligations of the Parties under their national legal systems and applicable supranational provisions to protect information relating to industrial and commercial secrecy, including intellectual property or national security.

The Basel Convention on the Control of Transboundary Movements of Hazardous Wastes and their Disposal was adopted in 1989 and entered into force in 1992 and has now been signed by 186 States Parties. This convention maintains a more

[11]Soares (2015).

[12]See Mccaffrey (2009) and Minaverry (2016).

substantive relationship with the restrictive clauses of the Helsinki Convention. The statute is strictly in control, from the process of generating hazardous wastes to their storage, transportation, treatment, reuse, recycling, recovery and disposal. As the most important aspect in the perspective of this chapter, we could mention the compulsory actions for the disposal and treatment of waste at the place of origin. The impact of the agreement on policies and legislation are visible, especially in relation to industrial and health activity. The Convention applies the 'prior informed consent' procedure, which further obliges Parties to ensure that hazardous and other wastes are environmentally handled and disposed of properly, mainly considering: (i) reduction of marketed quantities; (ii) treatment and disposal of waste as close as possible to where it is created and; (iii) reduction of waste generation at source. Practically, it means that the discharge of hazardous wastes must cease.[13]

The term 'waste' can be understood as substances or objects whose disposal procedures comply with the provisions of national legislation, such as: biomedical and health care waste; used lead oils and accumulators; waste with persistent organic pollutants; chemical products and pesticides that remain in nature for many years; bio accumulate that cause diseases and; compounds used in industry. Regarding the measures of coercion, the Basel Convention prohibits in its Annex 4 a series of operations, such as the elimination of waste or of their treatment in land areas (these two possibilities have a direct impact on aquifers), deep injection, poured into wells, lagoons or extension of continental waters. We note that the document has one of the most complete mechanisms for cases of non-compliance, formalised in the Basel Protocol on Liability and Compensation for Damage Resulting from Transboundary Movements of Hazardous Wastes and their Disposal.[14]

The Stockholm Convention on Persistent Organic Pollutants (POPs) of 2001 came into force in 2004. It seeks to restrict POPs, detailing the affected substances, as well as the rules of production, import and export of the same. The agreement has a significant scope and has 179 members. POPs are chemical products that have toxic characteristics and that, unlike other chemicals, resist degradation. POPs are very harmful to human health and to the environment because they accumulate in nature. They are transferred by air, water and migratory species, accumulating in terrestrial and aquatic ecosystems. Pollution caused by POPs is a local and transboundary problem, requiring the implementation of essential actions. The agreement establishes coercive measures to reduce or eliminate releases derived from stocks and waste like that of Basel.

[13]Stockholm Convention (2004).

[14]Secretariat of the Basel Convention (2014).

17.4 Regional Regulatory Aspects and Public Policies: The Cases of MERCOSUR and the European Union

This section aims to establish a cross-sectional analysis between regulatory aspects, public policies and their possible results for water security in the industrial sector, considering the EU and MERCOSUR.

17.4.1 MERCOSUR

In MERCOSUR[15] we do not find specific norms for water governance, although the issue has been considered, in some way, in its program 'the environment support for deepening the process of economic integration and sustainable development of MERCOSUR (ECONORMS)'. This regional bloc was created in 1991 by the Treaty of Asunción, expressing in its preamble that States Members should achieve, through the idea of the most effective use of available resources, the preservation of the environment in connection with different sectors. This idea has grounded attempts to unify the legal framework of environmental protection in the region. Thus, Resolution No. 22/92 of the Common Market Group of MERCOSUR establishes the following objectives:

1. To analyse the current legislation of each State Member and draw conclusions about its political and legislative asymmetries in environmental matters;
2. To prepare a proposal for basic guidelines on environmental policy. Resolution No. 38/95 of the MERCOSUR Common Market Group, concerning work methods, specialized meetings and ad hoc groups, refers to Sub-Working Group 6 'Environment'. The objectives of this group are to achieve aspects such as non-tariff restrictions, competitiveness and environment, international and voluntary standards, and the creation of an environmental legal instrument, green labels and environmental information systems.[16]

The MERCOSUR Framework Agreement on the Environment in its preamble sets out an approach linked to the principle of sustainable development. It stresses 'States Members emphasize the need to cooperate for the protection of the environment and the sustainable use of natural resources'. However, although the issue is formalised in this document the bloc has done little progresses in practice, concerning to achieve environmental goals. Even considering actions of the Working Subgroup No. 6 on the Environment,[17] its general objective of formulating

[15]This regional bloc is formed by the following countries: Argentina, Brazil, Paraguay, Uruguay and Venezuela.

[16]MERCOSUR (1992).

[17]GMC Resolution No. 20/95, CMC No. 59/00.

normative proposals and implementing programs, projects, strategies and guidelines to ensure the protection and integrity of the environment, the results were not extremely positive. As an isolated example, we must highlight the Technical Cooperation between MERCOSUR and Germany to promote cleaner production in the industries of the bloc to reduce environmental pollution by this economic sector.

We note that this regional bloc has applied the doctrine of 'weak sustainability', whereby:

> the obligation of present generations is to act in such a way that future generations have the same level of wealth as the present [. . .] all types of capital can be exploited, as long as it is replaced by an equivalent amount of it.[18]

Likewise, the ecosystem approach is not present here, as it signals the management of natural resources as an object in itself, not oriented to the integral management of ecosystems—normally based on the interrelationship between natural environment and its services. Such a disconnect is usually internally visualised in the countries, as in the case of in Brazil.

Additionally, in the case of the riparian states: Brazil, Bolivia, Paraguay, Argentina and Uruguay of the River Plate Basin, through its consuetudinary practices and its local and international regulations, they consolidated a solid and coherent legal body developing a *corpus aquarum ambientalis,* which is applicable to multiple uses in the basin, and to achieve a sustainable water management.[19]

The Tietê River Basin in São Paulo (Brazil) Only three agreements with express international coercive power connected with industrial activity and water security have become local norms in Brazil. The Basel Convention is related to industrial activity to the extent that it takes care of wastes, many of them of industrial origin. The convention was internalised in Brazil through Decree No. 875/93, Decree No. 4581/2003 and Resolution CONAMA No. 452/2012. These documents have emerged as the normative bases of Brazil's National Solid Waste Policy, through Law No. 12305/2010.

Perhaps the Stockholm Convention, which deals with POPs, is the most important to inhibit emissions of pollutants by industry in rivers and aquifers. As mentioned, the objective of the Convention is the elimination and/or restriction of POPs, their stocks and residues, the reduction of the release of their unintentional emissions into the environment, as well as the identification and management of areas contaminated by these substances. In a preventive position, the treaty determines that governments promote the best technologies and practices in their technological field and prevent the development of new POPs. Brazil approved the text of the Convention through Decree No. 204/2004 and Decree No. 5472/2005. The Secretariat of Climate Change and Environmental Quality of the Ministry of Environment of Brazil plays the role of Technical Focal Point of the Convention, together with the

[18]See Penna and Cristeche (2008).

[19]See Capaldo (2010).

Division of Environmental Policy and Sustainable Development of the Ministry of Foreign Affairs, which acts as an Official Focal Point.

Preventive efforts have been made through the implementation of regulatory policy. At the federal level, the framing of water bodies establishes that the level of quality must be reached or maintained over time. Law No. 9433/1997 aims to ensure the quality of water to reduce the costs of combating pollution. This rule has been reflected in public policies for the reduction of solid and liquid waste, in addition to its reuse.[20] Brazil has also been developing an internal waste management policy based on Law No. 12305/2010 (enacted in light of the Basel Convention), which provides tools to deal with the main environmental, social and economic problems arising from inadequate management of solid wastes. The regulation establishes the prevention and reduction of waste, the increase in recycling, the reuse and adequate management of waste that cannot be recycled or reused. Moreover, based on goals at the national, State, micro regional and municipal levels, it establishes shared responsibility among waste generators: industries, importers, distributors, merchants, citizens and responsible for the reverse logistics of waste and post-consumer packaging.[21]

The reuse of water is one of the main possibilities for dealing with waste generated by industries. Article 2 of Resolution No. 54/2008 of the National Council of Water Resources of Brazil, classifies sewage, discarded water/waste and liquid effluents from industries as wastewater. In Brazil, the Project of Law No. 183/2015 is currently under analysis. It exempts taxes on industrialised products and imports for companies that decide to acquire equipment and machines for water reuse, also providing credit lines for this purpose.[22]

Regarding discharges, in the State of São Paulo, Articles 18 and 19A of State Decree No. 8468/76 establish the parameters for the release of treated effluent into rivers or sewage networks. These regulations establish certain limits, such as the maximum amount of pollutants allowed in the effluent. Another criterion evaluated is water quality in the receiving body; quality is monitored closely to ensure the discarding of pollutants is not resulting in excessively diminished water.[23]

São Paulo State represents the most industrialised area of Brazil, with 449 industries operating in the Tietê River Basin, predominantly belonging to the metal and mechanical sectors. This industrial activity has the greatest impact on the contamination of aquifers in the Tietê River Basin, followed by 38 landfills, 52 districts with partial sanitation and 134 service stations with fuel leaks.[24]

Some projects have been implemented to reduce the use of water in the industrial sector, drawing heavily on the public-private partnership model. One example is the program for water reuse derived from the sewer system, implemented between a

---

[20]Government of Brazil (1997).

[21]Government of Brazil (2010).

[22]Chamber of Deputies of Brazil (2015).

[23]Government of São Paulo State (1976).

[24]Hirata and Ferreira (2017).

private company and the Sanitation Company of the State of São Paulo (SABESP), in which the State holds a majority stake. This represents the largest enterprise for the production of water for industrial reuse in South America, and the fifth largest on the planet. The result of this association provides by contract 650 litres/second (l/s) of water for reuse for the Petrochemical Plant of the well-known ABC Paulista Region (encompassing the cities of Santo André, São Bernardo do Campo and São Caetano do Sul). This is equivalent to supplying a city of 500,000 inhabitants. The plant is capable of producing 1000 l/s of water for reuse.[25]

The public-private partnership model has also progressed in cleaner production projects. In 2017, the Environmental Company of the State of São Paulo (CETESB) and the Federation of Industries and the Ministry of Environment signed a memorandum of understanding to promote sustainable solid management in the State. The cooperation agreement has among its objectives to encourage practices of circular economy, cleaner production and the adoption of environmental management systems in the area of waste, such as recycling programs for industry and reverse logistics, involving municipal mayors for the implementation of regional dumps.[26]

17.4.2 The European Union

The EU[27] has had a significant policy development in connection with the protection and sustainable management of water resources, especially because the bloc has established a coherent framework at the supranational level. We highlight the central aspects of Directive 2000/60/EC of the European Parliament and the Council of the European Union of October 23, 2000, which establishes a community framework for action in the field of water policy and with the purpose of unification. The Directive has long-term planning objectives and states that watersheds function as a management unit, applying an ecosystem approach that also provides joint treatment to the entire water system in general (considering rivers and aquifers, among others).

However, one of the main difficulties is that it has a great complexity because of its technical contents, to achieve an adequate state to all water bodies. To fulfil this strategic goal, the community has adopted some intermediate objectives: (i) to establish concrete measures for the reduction of discharges, emissions and losses of priority substances and the interruption or gradual suppression of discharges, emissions and losses of priority hazardous substances; (ii) to ensure the sustainable use of groundwater and progressive reduction in pollution of it and; (iii) to contribute

[25]National Association of Research and Development of Innovative Companies of Brazil (2010).

[26]Federation of Industries of São Paulo (2017).

[27]This regional bloc is formed by the following countries: Austria, Belgium, Bulgaria, Croatia, Cyprus, the Czech Republic, Denmark, Estonia, Finland, France, Germany, Greece, Hungary, Ireland, Italy, Latvia, Lithuania, Luxembourg, Malta, The Netherlands, Poland, Portugal, Romania, Slovakia, Slovenia, Spain, Sweden and the United Kingdom.

to mitigating the effects of floods and droughts. It also incorporated various fundamental social aspects, such as public participation and the drafting and approval of hydrological plans (which are regulated by the same directive). In all the Member States of the bloc, public consultation processes must be carried out to detect problems, their costs and to include them in their respective basin hydrological plans.

The EU has adopted the concept of 'ecological state' and it represents the degree of alteration of the (superficial) water mass compared with its basic reference conditions. To classify them, the indicators of biological, chemical and physico-chemical quality elements are normally applied. It is considered that social support is the key to making the normative application a success. Likewise, in the directive's text, the paradigm of integrated water resources management (IWRM) is applied, in which water masses (and also their sources and tributaries) are treated as a unit.

IWRM is grounded on an ecosystem-based approach and was incorporated in the Convention on Biological Diversity (1992) and the Helsinki Convention (1992). Finally, the normative base that guides IWRM is complemented by other specific laws:

- The Directive on Environmental Quality Standards (2008);
- The Framework Directive on the Marine Strategy (2008);
- The Floods Directive (2007);
- The Surface Water Directive (2006);
- The Bathing Water Directive (2006);
- The Drinking Water Directive (1998);
- The Urban Wastewater Directive (1991);
- The Nitrates Directive (1991).

As one of the main objectives of this chapter, we will examine whether the normative base in the EU has been reflected in practical actions to increase water security.

The Rhine Transboundary River This watercourse crosses several countries, namely Switzerland, France, Germany, Luxembourg and The Netherlands. The quality of Rhine water declined during the 1850s because of discharges due to industrialisation and urbanisation, which continued until the 1970s. Between 1970 and 1980, the Rhine Commission developed concrete strategies to fight pollution of the river, through the reduction of industrial waste emissions, the implementation of purification plants and public health policies. However, in 1986 there was an accident caused by a pharmaceutical company in Switzerland (in Basel, the city by which the Treaty was named)—20 tonnes of insecticides, fungicides and herbicides reached the river and pollution reached up to 250 km away, killing various aquatic species and altering water quality.[28] This incident accelerated projects to clean up the river. The program of recuperation proposed joint action across environmental

[28] Swissinfo (2016).

protection agencies, universities and research institutes, in addition to the obligation to compensate those affected by the company to start the program. The first objective was to reduce by 50% the hazardous substances discharged into the river (70% of some heavy metals).[29]

A plan was developed to restore ecological habitats and guarantee the quality of water in the basin, and was very successful in reducing phosphorus concentrations and recovering aquatic ecosystems. Further, actions related to the EU Water Directive were adopted to continue quality control and ecological rehabilitation. In 2003 the European Commission requested the incorporation of the European Union in the Rhine Commission, and its main attributions were: (i) the adoption of common regulations necessary for the safety of navigation; (ii) solutions based on research of complaints or of failures in the river's activities and; (iii) the decision-making in connection with the appeals filed against judgements related to navigation in the river. It has a large structure for monitoring and following up on activities by the Member States, conducts constant plenary meetings, the functioning of Committees and Working Groups, and other scheduled activities.

The Central Commission for the Navigation of the Rhine River is an international organisation, based in Strasbourg and created in 1815, whose mission is to promote the prosperity of river navigation, ensuring its safety and respect for its environment. This commission is regulated by the Convention for the Navigation of the Rhine of 1868 and includes the countries that the river passes through: Germany, Belgium, France, The Netherlands and Switzerland. The main objective of this convention is to protect and improve the river's ecosystem and to consolidate cooperation between the Community and the riparian States of the Rhine.[30]

Furthermore, it stands out for three core principles, which are fundamental, and that could serve as a model to be followed in application to other watercourses:

- Freedom of navigation—whenever security measures are adopted in general;
- The unity in the scheme and equal treatment—traffic regulations concerning the Rhine River are applicable to its entire length;
- Maintenance and improvement of navigable channels—conservation and improvement of navigable waterways.

Currently the Rhine River supplies drinking water to 20 million people and there are large salmon communities living there. The work has been going on constantly for more than 25 years, and various contaminations have been controlled (such as zinc, copper, mercury and other).[31]

[29]Swissinfo (2006).

[30]Central Commission for the Navigation of the Rhine (2017).

[31]La Nación (2017).

17.4.3 Beyond Norms and Policies: A Critical Discussion

From all that was highlighted in this chapter, it is possible to establish a critical discussion. The situation in the two blocs indicates South America is in a scenario of water insecurity that should not be underestimated. The cases selected show that the process of water integration began in the EU even before the formation of the bloc, and that in the Southern Cone there are some conquests and conflicts which show that cooperation is still a pending issue. As an example of a successful legal framework, but not necessarily under the auspices of MERCOSUR, the Guaraní Aquifer Treaty can be highlighted. This aquifer crosses the bloc in its most industrialised areas, from Buenos Aires in Argentina to São Paulo in Brazil, and industrial contamination is a recurring issue.[32]

Although there is an effort to introduce regulatory policies, we note that the relationship between 'water and industry' in Brazil is one of the Achilles' heels in the environmental arena. In Brazil, the industrial sector is responsible of 17% of the withdrawal flow (313 m^3/s) and 7% of the water consumption flow (69 m^3/s). Water is used in different activities: as an input in the productive process, for water cooling machines and for sanitary purposes, among others. The lack of water recycling in industry is the factor most salient in waste of the resource. The challenges for sustainable use of water in this sector include recycling and reuse of water, the treatment and proper disposal of industrial waste, an adequate transport and energy consumption levels in their processes. The manufacturing of food, beverages, pulp, paper, metallurgy and petroleum products account for 85% of national water demand.[33]

In the case of the Tietê River system in São Paulo, the region combines very high population density, low water availability and an index of contamination with prohibited chemical materials that indicates that discharges continue in high proportions among the almost 413 industries operating in the region of Alto Tietê. Leaks that directly impact the basin are another key source. Moreover, there is a history of pollutant emissions, a remedy for which would require not only a ban on chemical emissions and polluting waste, but a gigantic public policy of decontamination, as occurred in the case of the Rhine River, something that is still pending.[34]

Additionally, we could add to this the effects of the extreme changes in water pressure. São Paulo is the State that uses the highest quantity of water for industrial activities, especially through the Tietê River system (Alto Tietê, Tietê-Jacaré, Tietê-Sorocaba, Baixo Tietê) and Paraíba do Sul. Droughts expose the water insecurity of the State. The Metropolitan Region of São Paulo is located in Alto Tietê, with 39 Municipalities and a population of almost 20 million people, with seven times less water per capita than the United Nations considers as the minimum acceptable—50–100 litters of water per person daily.[35]

---

[32]Kuhn and Roisenberg (2017).

[33]National Water Agency of Brazil (2017).

[34]Mortatti et al. (2010) and Hirata and Ferreira (2017).

[35]UN (2017) and National Water Agency of Brazil (2017).

This picture of water insecurity in São Paulo is exacerbated by successive water crises. The collapse that occurred between 2014 and 2015 had a direct impact on the local industry, which saw a drop in production and had to reduce investments and increase layoffs. Out of 413 industries, 67% faced rationing and 54% reported not having alternative sources of water. Among the more affected sectors, the food, pharmaceutical and chemical industries were the most prejudiced, generally those that use more water as an input. Further, some industries moved their production plants to other provinces, specifically the beverage, paper and pulp sectors, or in a less drastic manner, granted collective vacations to employees.[36]

The most industrialised region in Latin America offers a lesson on how water security and industry are inherently linked variables. São Paulo lived its worst water supply crisis in history. To guarantee water for the industrial sector, under the argument of avoiding emptying and saving water, SABESP reduced water pressure at night. With that, water stopped arriving adequately around the periphery of the capital, which is where the poorest regions of the city are. Therefore, it is a priority to understand the social perspective of this picture of insecurity, considering the complexity of the water issue. It is remarkable that the ascension of the human variable in considerations about the water security and industry, as much in the subject of the contamination as in situations of decrease in the water availability.[37]

Other areas of MERCOSUR reproduce the scenario of the environmental chaos of the Tietê River. In the Matanza Riachuelo Basin (the industrial area of the Buenos Aires metropolitan region) there are several environmental problems caused by deficits in urban infrastructure, unplanned urban growth, lack of an effective system to control productive activities, poor management of urban solid wastes, proliferation of open pit dumps, a lack of compliance of environmental actors in watersheds and a lack of urgency from public policy managers regarding the anomic behaviours of urban dwellers. The aggravating circumstance is that in this river substances prohibited by the Stockholm Convention continue to be released, with the authorisation of the Matanza Riachuelo Watershed Authority (ACUMAR), which was created through Law No. 26168/2006. It was identified that the human right to water resources under adequate conditions had been poorly respected, which was resolved thorough a judicial action filed at the Argentine National Supreme Court of Justice, with the case *Beatriz Mendoza and others against the National State on damages*.[38]

Especially relating to the River Plate Basin, recent history brings us an emblematic case involving industry, pollution and water. The International Court of Justice has made express reference in its jurisprudence to the principles contained in the New York Convention (1997). One of the precedents of international jurisprudence was issued in 1997, in relation to the dam Gabcikovo Nagymaros (*Hungary*

[36]Environmental Portal (2015) and Carta Capital Magazine (2015).

[37]Exame Magazine (2015).

[38]See Supreme Court of Argentina (2006).

v. Slovakia). In the MERCOSUR area, the case of the Uruguay River pulp mills (*Argentina v. Uruguay*) was highlighted, and it was issued on 20 April 2010. The two countries share the Uruguay River and signed an international treaty on its management in 1975. The Court declared that Uruguay did not comply with the Uruguay River Statute signed by both countries, because it did not inform and ask Argentina about the implementation of an activity that could affect water quality and navigation.

Moreover, the EU has historically developed the basis for an orientation focused on precaution and safety in connection with the water issue. As discussed earlier, the security approach has been one of the foundations for cooperation between Member States of this bloc, which develops laws and policies fighting different threats. In addition to the Rhine River, the case of the River Thames in the United Kingdom (which is still a member of the bloc) is also worth underlining. The rehabilitation of the river occurred on all fronts, such as industrial effluents, sewage systems and education actions addressed to the population. A mathematical model was elaborated and allowed to fix with precision the proportion of oxygen necessary in the estuary, to eliminate fetid odours and allow the migration of fish. New methods were used for the treatment of wastewater, and large purification stations were constructed. The River Sanitation Program lasted more than 20 years and is currently ongoing (with goals and objectives still to be met).[39]

To achieve water security, we note that public policies in the EU transcend the multilateral and regional regulatory base to carry out deep and shared structural actions among many actors, involving control over factories (relocation in related industrial parks), treatment of sewage effluents, water and soil cleaning, population relocation and granting of health plans, joint work among several countries or jurisdictions (effective coordination) and more demanding and effective environmental audits. The vision of water security is clearly consolidated in the bloc, which is something that MERCOSUR is still unfortunately far from achieving. Although there is a normative and institutional effort on water governance that precedes MERCOSUR itself (the Rio de la Plata Treaty is an example), the region is characterised by a lack of coordination in water management and instances of previous consultations before the decision-making by coastal countries. There is also a notable lack of exchange of information and coercive power for the application of sanctions to offenders at the regional level or at the level of shared basin committees.[40]

[39]The Telegraph (2013).

[40]Capaldo (2010).

17.5 Conclusions

The European continent, which was the cradle of the Industrial Revolution, has learned through an intense history of conflicts and health crises—moreover the systematic destruction of natural environments by population pressure and industrial and economic development—that cooperation and integration might be the best way to recover and preserve natural resources for future generations. The achievements over the last decades in optimised water management, the reconstruction of ecosystems and the reduction of emission of industrial contaminants, signal a confluence of complex factors that seem to explain successful cases such as those of the Rhine and Thames Rivers. We could mention as important factors the precautionary principle as a regulatory and public management base, many transboundary rivers, a culture based on innovation and human rights and on civil participation. Furthermore, we can say that the supranational integration, the ecosystem approach, the common political concern for the environment and for the scarcity of water resources, the systematic application of coercive treaties (the Helsinki/Basel/Stockholm Conventions and others), the Directive 2000/60/EC and the internal laws of Member States are all aspects facilitating the management of water security in the region.

MERCOSUR has the largest volume of fresh water available on the planet. This bloc has a less conflictive history of frontiers than the European continent, immense water availability and a process of integration far from the historical construction that helps explain the supranationality of the EU. Additionally, the history of the two biggest industrial areas of the South American bloc—São Paulo and Buenos Aires—was built with an almost complete disregard for water. As an example, the Metropolitan Region of São Paulo was built on top of more than 300 rivers, many of which became underground sewers.[41] The scenario constituted by fragmented norms and public policies allows us to say that there is a discourse and a practice in MERCOSUR, which has formed a regional environmental normative framework guided by 'weak sustainability'. The environmental rules of the bloc work more like 'soft law', reflecting in countries that do not establish dialogue between their environmental and water legislation, and internal policies. One positive point that we observed in our analysis might be the application of the Basel and Stockholm Conventions internally in the countries, albeit in a disconnected regional perspective.

Finally, beyond ecological and human aspects—considering that the water security approach is also based on political and economic aspects—it is important to comment that there are risks of potential setbacks for the EU and even for the non-environmentally oriented MERCOSUR. The advances of a new model of integration—more inter- than intra-regional—has weakened the debate on regional cooperation and strengthened one more based on the interests of the investment financial sector and a legal approach extremely dangerous for the environment in the investor–State relationship. In this model contracts dictate rules to the detriment of countries' environmental legislation. We have seen an ascendant pressure from

[41] Carvalho and Santos (2016).

economic groups for the EU to reduce its environmental requirements because of the Transatlantic Trade and Investment Partnership, which has been negotiated with the United States.[42] For MERCOSUR, the biggest threat is the almost complete abandonment of the environmental debate in the bloc, given the economic priorities of the new profile of government established in the two largest countries in this market. It is necessary to reflect that without clean, available and sufficient quantity of water there is no economic development and there is no well-being. To put it bluntly, there is no future. The lesson from the EU shows us (or, at least, should show us) that cooperation is the best road to progress further in securing hydric resources for current and future generations.

References

Allenby BR (2000) Environmental security: concept and implementation. Int Polit Sci Rev 21 (1):5–21

Balzacq T, Guzzini S (2015) Introduction: what kind of theory–if any–is securitization? Int Relat 29 (1):97–102

Beck U (1995) Ecological politics in an age of risk. Polity, Cambridge

Capaldo G (2010) Internacionalización del derecho constitucional— constitucionalización del derecho internacional. 2010. Available via http://www.derecho.uba.ar/investigacion/investigadores/publicaciones/capaldo-cuencas-hidrograficas-transfronterizas-y-procesos-de-integracion.pdf. Accessed 4 Jan 2018

Carta Capital Magazine (2015) A seca já começou a afetar a economia. Available via http://www.cartacapital.com.br/revista/835/a-seca-da-economia-4105.html. Accessed 10 Nov 2017

Carvalho KM, Santos MEP (2016) Riesgo, conflictos y gestión pública del agua en São Paulo: un estudio crítico. XL ENANPAD. Encontro da Associação Nacional de Pós-Graduação e Pesquisa em Administração. Costa do Sauípe

Carvalho KM, Santos MEP (2017) Global value chains and international water security: a critical study. Available via http://wwc2017.iwra.org/congress/resource/ABSID431_ABSID431_Global_value_chains_and_international_water_security.pdf. Accessed 21 Dec 2017

Central Commission for the Navigation of the Rhine (2017) Available via http://www.ccr-zkr.org. Accessed 1 Nov 2017

Chamber of Deputies of Brazil (2015) Project of Law No. 183/2015. Available via http://www.camara.gov.br/proposicoesWeb/fichadetramitacao?idProposicao=945876. Accessed 28 Nov 2017

Dalin C et al (2012) Evolution of the global virtual water trade network. Proc Natl Acad Sci 109 (16):5989–5994

Environmental Portal (2015) Crise da água faz indústrias cogitarem férias coletivas em SP. Available via http://www.portaldomeioambiente.org.br/agua/10062-crise-da-agua-faz-industrias-cogitarem-ferias-coletivas-em-sp. Accessed 21 Nov 2017

Exame Magazine (2015) Crise da água em SP mostra disparidade entre pobres e ricos. Available via http://exame.abril.com.br/brasil/noticias/crise-da-agua-em-sp-mostra-disparidade-entre-pobres-e-ricos. Accessed 12 Nov 2017

[42]The Guardian (2013).

Federation of Industries of São Paulo (2017) FIESP e CIESP assinam protocolos para iniciativas de logística reversa. Available via http://www.fiesp.com.br/noticias/fiesp-e-ciesp-assinam-protocolos-para-iniciativas-de-logistica-reversa/. Accessed 13 Dec 2017

Goldblat D (1996) Social theory and the environment. Polity, Cambridge

Government of Brazil (1997) Law number 9433. Available via http://www.planalto.gov.br/ccivil_03/leis/L9433.html. Accessed 25 Oct 2017

Government of Brazil (2010) Law number 12305. Available via http://www.planalto.gov.br/ccivil_03/_ato2007-2010/2010/lei/l12305.htm. Accessed 24 Oct 2017

Government of São Paulo State (1976) Decree number 8468. Available via https://www.al.sp.gov.br/repositorio/legislacao/decreto/1976/decreto-8468-08.09.1976.html. Accessed 1 Oct 2017

Hall J, Borgomeo E (2013) Risk-based principles for defining and managing water security. Philos Trans R Soc Lond 371(2002):407–430

Hirata RCA, Ferreira LMR (2017) Os aquíferos da Bacia Hidrográfica do Alto Tietê: disponibilidade hídrica e vulnerabilidade à poluição. Revista Brasileira de Geociências 31 (1):43–50

Hoekstra AY (2015) The water footprint of industry. In: Klemes JJ (ed) Assessing and measuring environmental impact and sustainability. Elsevier, Oxford

Hoekstra AY (2017) Water footprint assessment in supply chains. In: Bouchery Y et al (eds) Sustainable supply chains–a research-based textbook on operations and strategy. Springer, New York

Kuhn IA, Roisenberg A (2017) Combining lead isotopes and cluster analysis to distinguish the Guarani and Serra Geral Aquifer Systems and contaminated waters in a highly industrialized area in Southern Brazil. J Environ Radioact 177:24–31

La Nación (2017) El ejemplo del río Rin, en Europa. 2006. Available via http://www.lanacion.com.ar/835664-el-ejemplo-del-rio-rin-en-europa. Accessed 12 Dec 2017

Lupton D (2013) Risk and sociocultural theory: new directions and perspectives. Cambridge University Press, London

Magsig BO (2012) Rising to the challenge of water security: international (water) law in need of refinement. Int J Sustain Soc 4(1–2):28–44

Mccaffrey S (2009) Convención de las Naciones Unidas sobre el derecho de los usos de los cursos de agua internacionales para fines distintos de la navegación. 2009. Available via http://legal.un.org/avl/pdf/ha/clnuiw/clnuiw_s.pdf. Accessed 18 Dec 2017

Mercosur (1992) Mercosul/GMC/res. N° 22/92. Available via http://www.mercosur.int/msweb/portal%20intermediario/Normas/normas_web/Resoluciones/PT/GMC_RES_1992_022_PT_RMedio_Ambien.PDF. Accessed 15 Dec 2017

Minaverry C (2016) La protección jurídica ambiental del agua para consumo humano. Panorama de dos provincias argentinas y del ámbito internacional. Revista Civilizar Ciencias Humanas y Sociales 31:17–40

Mortatti J, Hissler C, Probst JL (2010) Distribuição de metais pesados nos sedimentos de fundo ao longo da bacia do Rio Tietê. Geologia USP. Série Científica 10(2):3–10

National Association of Research and Development of Innovative Companies of Brazil (2010) Sabesp e Foz do Brasil fazem projeto de água de reuso. Available via http://anpei.org.br/anpeinews/sabesp-e-foz-do-brasil-fazem-projeto-de-agua-de-reuso-2/. Accessed 12 Dec 2017

National Water Agency of Brazil (2017) Água na indústria: uso e coeficientes técnicos. ANA, Brasília

Penna JA, Cristeche E (2008) La valoración de servicios ambientales: diferentes paradigmas. Instituto Nacional de Tecnologia Agropecuaria, Buenos Aires

Secretariat of the Basel Convention (2014) Basel Convention on the control of transboundary movements of hazardous wastes and their disposal: Texts and annexes. Available via http://www.basel.int/Portals/4/Basel%20Convention/docs/text/BaselConventionText-e.pdf. Accessed 21 Dec 2017

Soares APM (2015) A Convenção de Nova York de 1997: notas sobre a água doce nas relações internacionais em face da sua (não) implementação. Espaço Jurídico J Law 6(1):21–32

Spring UO (2010) Ampliando la agenda de seguridad y seguritizarla. Una propuesta de políticas públicas. Available via http://www.afes-press.de/html/pdf/uos/29_UOS_Ampliando_agenda_seguridad_seguritizarla_101005.pdf. Accessed 22 Nov 2017

Stockholm Convention (2004) Convention text. Available via http://chm.pops.int/default.aspx. Accessed 21 Dec 2017

Stritzel H (2007) Towards a theory of securitization: Copenhagen and beyond. Eur J Int Relat 13 (3):357–383

Supreme Court of Argentina (2006) CSJN, M. 1569. XL, Judgement 20/06/2006. Available via http://old.csjn.gov.ar/data/sda.pdf. Accessed 15 Nov 2017

Swissinfo (2006) El Rin se ha recuperado del desastre de 'Schweizerhalle'. Available via https://www.swissinfo.ch/spa/el-rin-se-ha-recuperado-del-desastre-de%2D%2Dschweizerhalle-/5534222. Accessed 22 Nov 2017

Swissinfo (2016) Chemical disaster still burns in Swiss memory. Available via https://www.swissinfo.ch/eng/schweizerhalle-fire_chemical-disaster-still-burns-in-swiss-memory/42559954. Accessed 7 Dec 2017

The Guardian (2013) This transatlantic trade deal is a full-frontal assault on democracy. Available via https://www.theguardian.com/commentisfree/2013/nov/04/us-trade-deal-full-frontal-assault-on-democracy. Accessed 15 Nov 2017

The Telegraph (2013) The clean up of the River Thames. 2010. Available via http://www.telegraph.co.uk/news/earth/wildlife/8059970/The-clean-up-of-the-River-Thames.html. Accessed 3 Nov 2017

UE (2000) Directive 2000/60/EC of the European Parliament and of the council of 23 October 2000–establishing a framework for Community action in the field of water policy. Available via http://eur-lex.europa.eu/resource.html?uri=cellar:5c835afb-2ec6-4577-bdf8-756d3d694eeb.0004.02/DOC_1&format=PDF. Accessed 2 Nov 2017

UN (2011) Agua y industria en la economía verde. Available via http://www.un.org/spanish/waterforlifedecade/green_economy_2011/pdf/info_brief_water_and_industry_spa.pdf. Accessed 27 Oct 2017

UN (2013) Water security and the global water. Available via http://www.unwater.org/fileadmin/user_upload/unwater_new/docs/analytical_brief_oct2013_web.pdf. Accessed 24 Oct 2017

UN (2015) Sustainable development goals. Available via http://www.un.org/sustainabledevelopment/. Accessed 22 Oct 2017

UN (2017) The human right to water and sanitation. Available via http://www.un.org/waterforlifedecade/pdf/human_right_to_water_and_sanitation_media_brief.pdf. Accessed 17 Dec 2017

World Wildlife Fund (2016) Planeta vivo informe 2016–riesgo y resiliencia en una nueva era. WWF, Gland Suiza

Chapter 18
Sustainable Water Consumption, Foreign Direct Investment and the Human Right to Water

Andreia Costa Vieira

Contents

Abstract This chapter sketches a picture of the entry of the private sector into the water business through foreign direct investment. It does so against the backdrop of the worldwide demand for sustainable water consumption and the human right to water. A region-by-region inventory of the global water crisis is detailed to show the full extent of 'peak water'. The water privatization processes undertaken in different countries is detailed, with the aim of identifying the extent to which the 'business of water privatization' represents sound policy in the face of this world water crisis and the demand for sustainable water consumption. This discussion supports a broader claim: that of the human right to access water. Private sector involvement is necessary in some countries and regions of the world to promote 'water social inclusion' given the lack of public resources for investment. However, the success of such policy, in most places, depends on previous public regulation that considers sustainable development and human rights.

A. C. Vieira (✉)
Catholic University of Santos, São Paulo, Brazil
e-mail: andreia.vieira@unisantos.br

A. d. Amaral Junior et al. (eds.), *Sustainable Consumption*,
https://doi.org/10.1007/978-3-030-16985-5_18

18.1 Introduction

Today the world confronts a dilemma in addressing the cross-cutting issues of fresh water scarcity and the crisis of supply, increasing contamination of water sources, suboptimal water management and the eagerness of the private sector to capture the water market, which has now become 'big business'. This present chapter thus draws a picture of the entry of the private sector into the water business through foreign direct investment against the backdrop of the worldwide demand for sustainable water consumption and the human right to water.

The chapter begins with a region-by-region inventory of the global water crisis to show the full extent of 'peak water'. Next, an overview of private investments in water services is provided, assessing degree to which the 'business of water privatization' represents sound policy against the backdrop of a worldwide water crisis and the need to ensure sustainable water consumption. To this end, the privatization processes undertaken in different countries, including France, the United Kingdom, Argentina, Bolivia, Uruguay, Chile and Brazil, will be analyzed.

Opponents of some of the more suboptimal water privatizations undertaken worldwide have raised a distinct claim: the human right to water. This study will detail the evolution and consolidation of this basic rights claim from its first documentation in the United Nations General Comment No. 15 (2002) through to the most recent publication of the General Assembly (2010), the reports of the UN *Special Rapporteur* (2013) and the 2015 Sustainable Development Goals (SDGs). Finally, the content of the human right to water will be scrutinized to show how it might be raised in international and domestic *fora* to promote and support SDG 6, which covers access to water and sustainable water consumption.

18.2 The Global Water Crisis

After raising many environmental concerns related to water, the United Nations World Water Development Report,[1] makes it clear that growing water consumption in the most industrialized countries (those with high per capita incomes), combined with demographic concerns (such as increasing urbanization and population growth in general), is the core of the global water crisis. The key issues, then, combine questions of economics and sustainability, wherein ensuring sustainable sources of water for household and industry against a backdrop of rising incomes and a growing world population is thrown into sharp relief.

The International Water Management Institute presented a report on water resource management for the years 2010–2012, highlighting global water scarcity, which affects one-third of the world population. The report identifies two kinds of scarcity: economic (where a lack of investment and infrastructure in the water sector

[1]United Nations (2009), p. 14.

limits effective supply) and physical (where the limited quantum of water is itself the key issue).[2] Hundreds of millions of people around the world are affected by water scarcity.[3] Studies show declining volumes of water in many reservoirs around the world, which according to scholars, has raised the risk of international disputes (possibly even leading to war) in water scarce regions.[4] Some have pictured worrying panoramas for the water supply crisis in the world.[5] We can detail this crisis on a regional basis.

The Middle East has lacked sufficient water for human consumption since 1970.[6] It is the first region in the world to fall subject to chronic water scarcity, which means that there is not enough water to satisfy basic needs of its population and industry. Some have pointed out that the region ran out of water and that such information was not widespread because of lack of political interest as political costs of such an information would have serious consequences.[7] In many parts of Africa, the situation is no different, but is particularly bad in Libya, Egypt, Sudan, Kenya and South Africa.[8] As part of a Libyan water supply project, an 1860 km aqueduct has been built to transport fresh water from the Kufra basin into the Saharan desert region.[9]

Fresh water has also been at the top of the agenda of the Chinese government. In a biennial report published by the Pacific Institute, China's limited water supply was raised as an issue of serious concern.[10] The situation in China has become worse with climate change. Having the highest population in the world (around a billion inhabitants), only 6% of the water in China is freshwater.[11] More than half of the population has had to deal with water scarcity and shortages. India has exploited its underground water potential currently to exhaustion and a large part of rivers and lakes have dried up.[12]

The crisis is not limited to developing countries either. Australia's so-called 'Millenium drought' lasted from 1996–2010, threatening agriculture in the Murray-Darling river region, known as the country's 'food bowl' for its high concentration of farming industries.[13] Australia is already known for long periods of water scarcity. In California, USA, the last decades have witnessed the worst drought in history, and the effects of such climatic changes have been felt not just in

---

[2]IWMI (2009).

[3]Allan (1997).

[4]Vieira (2016b).

[5]Barlow and Clarke (2002).

[6]Allan (1997), p. 3.

[7]Barlow and Clarke (2002).

[8]Ibid.

[9]Ibid.

[10]Palaniappan and Gleik (2008), p. 79.

[11]Ibid.

[12]Barlow and Clarke (2002).

[13]Palaniappan and Gleik (2008), p. 79.

the large urban centers but in agriculture as well.[14] Water for irrigation has been brought in from Canada, leading a former Californian Governor to publicly declare that 'water is [now] more precious than gold'.[15]

In Latin America, concerns about water issues are also very serious. In the Mexican border cities, published reports suggest that the population, in certain regions, consumes Pepsi-Co and Coca-Cola instead of water because of scarcity and the higher prices of bottled water. In Lima, Peru, the poorest pay around US$3 for each cubic metre of water and must fetch it in buckets.[16] Brazil has one of the largest freshwater reservoirs in the world—estimates range from 12 to 20% of the planet's entire freshwater reserves.[17] Nevertheless, despite all this potential, Brazil is not immune to water supply crises, as access to water across the country's vast territory varies significantly. According to research published by Tucci and colleagues, 68% of all Brazilian freshwater is located in the north of the country, where just 7% of the population live; the northeast, in contrast, has 29% of the population, but only 3% of freshwater reservoirs. The southeast, meanwhile, has just 6% of Brazilian freshwater for 43% of the country's residents.[18]

Desalinization has been presented as a solution for this water crisis. However, generating water in this way is tremendously energy intensive (adding to fossil fuel use) and inefficient—each gallon of sea water processed produces just one-third of a gallon of potable water. The remainder is brine containing concentrated toxic residues.[19] When this brine returns to the sea, it can produce environmental damage, contributing to the death of many marine species. It is, besides, very expensive technology and economically unfeasible in poorer countries. Desalinization is thus a suboptimal way to source water for sustainable consumption, especially if there are alternatives.

This global water crisis underpins what has come to be known as 'peak water', referencing the more well-known idea of 'peak oil'.[20] While reservoirs are not running out of water, as such, what is suggested is that the planet's capacity to sustain all the water demand is reaching its limit, as water is a renewable resource. After use, water typically returns to the cycle (evaporation, condensation, soil absorption, underground formation or ground waters) in a continuous process. However, in many regions, water use has exceeded this rechargeable capacity and the world now witnesses intensifying competition for the so-called 'blue gold'.[21]

[14]Caubet (2006), p. 178.

[15]Barlow and Clarke (2002), p. 94.

[16]Ibid.

[17]Ibid, p. 95.

[18]Tucci (2000).

[19]Barlow and Clarke (2002), p. 213.

[20]Palaniappan and Gleik (2008), p. 3.

[21]Barlow and Clarke (2002).

In 2003, the United Nations proclaimed the International Decade Water for Life (2005–2015), coordinated by UN Water.[22] Definitive goals were established for water supply and basic sanitation: half of the world population should have full access to clean potable water by the year 2015.[23] Despite good achievements registered by the UN in 2015, the right of access to water and basic sanitation together with the claim for sustainable water consumption is still a fight for many across the world. A report presented by the United Nations Special Rapporteur on the Human Right to Water (2013) was centered on sustainability of the accomplishment of the right of access to water and basic sanitation. The Rapporteur raised her concern about the fact that the 2008–2009 global financial crisis has exacerbated long-running shortfalls in investment in the water sector, not to mention raised tariffs for water as governments scramble to plug growing public deficits.[24] She concludes that better public policies to guarantee the right of access to water and sustainable water consumption on a global scale are required.

18.3 'Big Water': Private Investment in Water Services

By the nineteenth century, water supply services and basic sanitation emerged as matters of core public concern in the United States and Europe. The first water supply systems were privately managed and consisted of supplying water for the upper classes. States took water supply services into public hands, mainly in Europe, when they realized that provision of clean water was a matter of public health and economic development. At the beginning of the twentieth century, public management of water sources had already been adopted in most European countries and in the United States.[25]

The neoliberal ideology that has come to dominate, mainly in Europe and in the United States, since the 1970s, largely influenced the water sector in developing countries. Public sector inefficiency in providing a proper supply of fresh water and basic sanitation triggered suggestions that the private sector should become involved. By the end of the 1980s, in many countries, privatization of water services meant a solution for that financial issue because this was a more politically palatable option for plugging public finances than raising taxes. The idea of privatizing the water sector was 'sold' to many countries as a way of promoting sustainability in the sector. The International Monetary Fund (IMF) and the World Bank—as the leading sources of international loans and financing—were the major preachers of that neoliberal policy, supporting investments by private companies in the water sector

[22]UN Water (2015a).

[23]Palaniappan and Gleik (2008).

[24]United Nations (2013), p. 12.

[25]Budds and McGranaham (2003), p. 90.

in developing countries.[26] These two institutions advocated a policy posture of minimum State intervention in the economy and the entry of private providers into those parts of the public sector deemed inefficient and/or uncompetitive.

At the United Nations Conference on Water and Environment (1992) in Dublin, a Declaration on Water Principles was made. Among the many principles enshrined therein, was the contentious notion that water has an economic value (Principle n. 4), which has been interpreted by many as a door through which the private sector may enter the domain of water management. Whether the principle has this effect, it remains true that loan conditionality for developing countries has included a core demand that they commit to the privatization of state assets.[27]

The term 'privatization of the water sector' captures a range of different management models adopted worldwide. Budds and MacGranaham define it as private commercial participation (making investments and assuming the follow-on risks) in the supply of water and basic sanitation services but not necessarily with a transfer of property to private hands.[28] Different contract types are in use in the water sector, which range from agreements to provide specific services (for a short period) to Build–Operate–Transfer (BOT) models and concession contracts. There are also joint ventures and cooperatives. In all the contract models implemented in Latin America, little competition was seen in the public procurement procedures.

Many have defended privatization of water services based on the claim that the private sector is best suited to assure sustainable water consumption and broader access to water. However, as Budds and McGranham have remarked:

> A number of multinational water companies have asserted that low income populations do not represent an attractive market because they are too poor to be profitable and represent too great a financial risk. . . . Representatives of Veolia stated that profits depend on 'sufficient and assured revenues from the users of the service', which are unlikely to include poor groups. Biwater's general manager, referring to Zimbabwe, also claimed that: 'From a social point of view these kinds of projects are viable but, unfortunately, from a private sector point of view they are not'.[29]

Leaving such environmental and social concerns to the private sector might not be the right political decision either. A 2008 study collected data covering four decades of water services management from different parts of the world, concluding that efficiency in the water sector is not related to the public or private nature of that service or its management. This study also clarifies that whenever the private sector enters the business, it only remains if there is profit, which has not complemented the public interest or the goal of sustainability. Even in countries where water services were almost totally privatized, the State has had to regulate to assure environmental and consumer protection.[30] As is well remarked by Rogers and Hall:

---

[26]Budds and McGranaham (2003), p. 91.

[27]Ibid.

[28]Ibid, p. 89.

[29]Ibid, p. 109.

[30]González-Gomez and García-Rubio (2008).

> There is a growing perception that the governance of water resources and water services functions more effectively with an open social structure which enables broader participation by civil society, private enterprises and the media, all networking to support and influence government. Moreover, examining the role of networks or distributed governance helps to overcome the sterile debate about private versus public water service delivery and the role of the community. (...)There is no single model of effective water governance; indeed, to be effective governance systems must fit the social, economic and cultural particularities of each country.[31]

More than 90% of all water supply and sewage services in the world are provided by the public sector. This is likely to remain the case, although the public sector has been accused of inefficiency and of not having enough resources to make the needed investments.[32] Transfer of property to the private sector might not be the right policy to be adopted in most places whenever the issue is sustainability, but allowing the entrance of the private sector into the water business when the public sector lacks sufficient revenue to make the necessary investments is sometimes the only option left. Proper regulation—established in advance—has been the key for assuring sustainability in the water sector.

The next sections of the chapter will detail select cases of water privatization across the planet in recent decades, to highlight the concerns and issues that arise in private sector involvement, particularly concerning sustainability.

18.3.1 France

In the nineteenth century, water services in France were supplied by the private sector, whose main concern was to take water to the upper classes. The success attributed to the private business of water services in France was because of the incapacity of the *communes françaises* to adopt a public municipal model.[33] Implementation of a liberal State, in France, had already left the *communes* with large deficits and so there was not enough public revenue to invest in the water sector. This explains the large specialization of French private companies in water services today. Two French water companies—Vivendi and Suez—are the largest 'water businesses' in the world and as publicly traded companies are ranked 91st and 118th globally in terms of market capitalization as they supply water services for more than 100 million people spread over 130 different countries worldwide.[34]

Moreover, in France, *La Déclaration des droits de l'homme et du citoyen*—which remains a powerful source of the French understanding of basic human and civil rights—provides for a robust protection of the right to private property. As water resources in France have historically been included in conceptions of private

[31] Rogers and Hall (2003), p. 26.

[32] Ibid, p. 32.

[33] Barraqué (1995), p. 427.

[34] Barlow and Clarke (2002), p. 85.

property, it follows that water developed into a good to be marketed for consumption. That was a historic achievement for the *citoyen français*. Water in France has historically been a matter of a right to private property and, as such, it developed into a good to be marketed for consumption.[35]

In France, private sector dominance came under challenge with the introduction of environmental protection, sustainability and the polluter-pays principle. In the second half of the twentieth century, the river basins committees were established and ever since regulatory agencies in each water basin have assumed principal responsibility for the management of water services for the population.[36] Recently, continuous increases in the tariffs charged for water supply services have triggered social movements—mainly in Paris and larger French cities—that claim water should be classed as a public asset in France. In Paris, water supply services were into public hands in January 2010. A public company was created (l'Eau de Paris) for the management of water supply services. Tariff reductions have followed as l'Eau de Paris implements a policy of narrowing profit margins and boosting sustainability,[37] as affordability is one elements of universal access, itself a central component of sustainable consumption.

18.3.2 The United Kingdom

In the United Kingdom at the end of the nineteenth century, water services were largely taken over by local public authorities, with very few entities remaining in private hands. In 1987, the Thatcher government undertook a neoliberal restructuring of the sector, initiating a process of privatization. The government's contention was that the private sector would be more efficient, private firms had the capital to finance the large investments necessary to maintain and develop the sector and that privatization would boost competition in the market.[38]

The *Water Act 1988* transferred the property of the UK water management system to ten private companies. Moreover, it granted concessions in water services to these companies for a period of 25 years, free from any competition, thus creating a system of temporary regional monopolies. At the same time, state subsidies were introduced to assure the survival of those ten companies, financing these with direct consumer payments or by public levies.[39]

In the following years, very high tariffs for water became commonplace and quality showed no noticeable improvement while the annual profits of these companies soared as the sector's labor force shrunk. High prices in the system caused

[35] Barraqué (1995).

[36] Ibid, p. 441.

[37] Eau De Paris (2017).

[38] Lobina and Hall (2001), p. 6.

[39] González-Gomez and García-Rubio (2008), p. 53.

social exclusion in many UK regions and water-borne diseases increased because of attenuated access to potable water.[40] The entrance of private actors into the water business in the UK therefore did not mean sustainability of the sector. Quite the opposite. Indeed, privatization in the UK was responsible for higher prices—around 39% higher—which rose so that the companies' profits margins would remain far above inflation over the years analyzed.[41] At the same time, there was no appreciable improvement in efficiency in the sector, as measured by industry productivity.[42]

The years 2007 and 2008 saw the highest figures for complaints about water services to the UK Consumer's Council for Water since privatization was put into effect. In 2011, the Council again reported a spike in consumer complaints.[43] This relation between consumer rights and efficiency is also a key factor in assessment of the sustainability of the sector.

18.3.3 Argentina

In the 1990s, Argentina privatized public water services, adopting the neoliberal approach recommended by the IMF and the World Bank, which recommended this liberalization approach so that the country could raise the necessary funds to manage its international debt payments. A regional organization for the private companies entering the Argentine market was established. Most Argentine provinces privatized their water companies and by the end of the decade, 65% of all Argentina water companies were in private hands.[44]

In 1993, the Buenos Aires municipal government launched a public tender for the public water company. A consortium named Aguas Argentinas, led by the French firm Suez, won the bid. Under contract, the company proposed to reduce tariffs and improve water quality. A regulatory agency—ETOSS (Entre Tripartito de Obras y Servicios Sanitarios)—was established to oversee the concession contract and to represent Buenos Aires' water consumers. Eight months after the contract came into force, ETOSS authorized Aguas Argentinas to raise tariffs, based on a contractual clause that tied local rates to the US dollar. The same contract called on the company to make further capital investments in the water system. Three years into the contract, the level of new investment was found to have fallen well short of what had been set under the terms of the agreement. By the end of that decade, the Argentinian Ministry of Natural Resources and Sustainable Development took control of the concession from ETOSS, which had been privileging the water

[40]Lobina and Hall (2001), p. 8.

[41]Ibid.

[42]Ibid.

[43]UK Consumer's Council for Water (2010).

[44]Azpiazu (2005), p. 45.

company with continuous authorizations for higher tariffs and no penalties for breach of contract.[45]

In 2002—at the height of the Argentinian debt economic crisis—the Argentine government adopted a price-freeze provision for all tariffs across public services. Aguas Argentinas raised a claim before the International Centre for Settlement of Investment Disputes (ICSID) for breach of contract. The IMF and the World Bank sent representatives to Buenos Aires. The Kirchner government decided to terminate the contract with Aguas Argentinas, based on a claim of public necessity. The water crisis in Argentina had spread across the country and underinvestment and the poor quality of the water supplied was commonplace in many other Argentine regions.[46]

The poor quality of supplied water and the decrease in accessibility and affordability for the Argentine consumer are factors that support the conclusion that privatization undermined sustainability in water services in Argentina.

18.3.4 Bolivia

The privatization of water services in Bolivia began at the end of the 1990s, when the Bolivian government—like that of Argentina—was pressed by the IMF and the World Bank to introduce liberalization policies in public services to guarantee payment of the country's external debt.[47] In 1999, a concession contract was awarded to an international consortium—Aguas del Tunari, led by International Water Ltd.[48] Aguas del Tunari took over the debts of the previous water management entity, but transferred them, through tariff adjustments, to consumers. These tariff adjustments saw an increase of about 200%, thus shutting off a large part of the Bolivian population from access to water services.[49]

In 2001, the residents of Cochabamba organized a social movement and many went into the streets to protest against Aguas del Tunari's water management. A general strike followed, and eight people died in the ensuing protests.[50] After direct involvement by human rights groups, the law on water services privatization was repealed—the concession contract was cancelled in 2000. Aguas del Tunari raised a claim before the ICSID.[51] Since 2000, water services in Bolivia have stayed in public hands and a public agency was created to regulate and supervise the water sector and since this time public management of water services has been associated with sustainable water consumption.

---

[45]Celli (2009), p. 219.

[46]Budds and McGranaham (2003), p. 100.

[47]Celli (2009), p. 217.

[48]Ibid.

[49]Ibid.

[50]Caubet (2006), p. 173.

[51]Ibid.

18.3.5 Chile

At the end of the 1990s, Chilean public water services were privatized and a regulatory agency was created to oversee the sector. Tariffs were to be readjusted every 5 years to cover investments. A commitment to transparency was made under the concession contract and allowances for tariff adjustments were part of the many government incentives for the new private water company.[52] In the wake of privatization, public demand for water has generally been met. Tariffs adjustments have been accompanied by high subsidies benefitting the consumer—these cover around 60 to 80% of the water bill in registered households.[53] The private companies controlling the Chilean water market have often increased tariffs, mainly on the eve of periods of drought. However, it must be said that water quality and water accessibility in Chile have improved considerably since the private companies entered the Chilean market.[54]

The Chilean experience of water privatization has been considered successful, mainly because of the public policies adopted to smooth the entry of private actors into the water sector. The Chilean Water Regulatory Agency has been called a 'model agency'[55] and has gained respect from the Chilean population for its attention to the environment and human rights.

Thus, in Chile, sustainable water consumption has been met through the entry of the private sector but having a previous and fair public regulation together with adequate adoption of public policies that supported sustainable development in the sector.

18.3.6 Uruguay

In 1992, the Uruguayan people, in a referendum, voted against privatization of public services. Nevertheless, in 2000—under pressure from the IMF and the World Bank—water services in Uruguay were privatized and Aguas de Barcelona won the bid in a public tender. Aguas de Barcelona entered into a concession contract of 30 years for water management and sanitation services in Montevideo and other major Uruguayan cities. Another subsidiary of Suez also won the bid to exploit underground waters in Uruguay.[56] Tariffs increased in the privatized regions but there was no improvement in quality of the water services in return. Sustainability in the sector was therefore far from being achieved with the privatization policy adopted in Uruguay.

---

[52]Celli (2009), p. 220.

[53]Ibid.

[54]Foster (2005).

[55]Ibid, p. 21.

[56]Barlow and Clarke (2002), p. 189.

In 2004, the Uruguayan population voted in another referendum against privatization of water services. In this referendum, the right of access to water was put forward by the opponents of privatization as a human right and it was determined that water management should return to public hands. The Uruguayan parliament duly voted an amendment to the Uruguayan Constitution to include a human right of access to water in the constitutional text. Uruguay was the first country in the world to make privatization of water services illegal and the first in Latin America to have a human right of access to water enshrined in the Constitution. The water concession contracts were terminated on the ground that they were no longer constitutional. Ever since, under Uruguayan law, sustainable water consumption has been closely linked to constitutional human rights.

18.3.7 Brazil

In Brazil, water supply and basic sanitation are predominantly in public hands. However, many regions and large cities have seen large private sector investment in water since 1996, according to the Associação Brasileira das Concessionárias Privadas de Serviços Públicos de Água e Esgoto.[57] Such investments are based on concession contracts and it has been estimated that around 4% of the urban population in Brazil has had water supply and basic sanitation services from private companies.

The concession contracts have established a regime in which the water services infrastructure remains public property but operates under private management. Although the Brazilian Federal Constitution (Articles 20 and 26) enshrines water is a public asset this does not preclude the possibility of water services being privatized under a concessionary management model like the one that has been introduced. In fact, there are 65 concession contracts for different cities/regions and each State in Brazil.[58] One concession contract that has raised concerns is the one granted to a subsidiary of Suez for a period of 30 years in the city of Manaus from the year 2000. Some studies have pointed out to a considerable increase in tariffs combined with a decrease in access to potable water in Manaus since privatization was launched.[59] In the State of Parana, there are studies that show water exclusion for a large part of the population (mostly low-income earners) since the concession

[57]Associação Brasileira das Concessionárias Privadas de Serviços Públicos de Água e Esgoto (2017).

[58]These are located in the states of São Paulo (the cities of São Carlos and Ribeirão Preto), Rio de Janeiro (Búzios, Cabo Frio, São Pedro da Aldeia, Iguaba, Petrópolis and 90 other cities), Espírito Santo (Cachoeiro do Itapemirim), Mato Grosso do Sul (Campo Grande), Mato Grosso (some small cities), Santa Catarina (some small cities), Minas Gerais (some small cities), Paraná (the entire State), Pará (some small cities) and Amazonas (the capital city Manaus).

[59]Olivier (2006).

contract was signed with a subsidiary of Vivendi after which tariffs increased.[60] In 2017, the State of Rio de Janeiro announced its intention to privatize water services to plug a gap in the public finances. This is another privatization that has raised much concern in Brazil since the State of Rio de Janeiro has been in the headlines for bribery scandals and all its deficits are associated with corruption in its institutions. Users of water services are therefore about to pay the proverbial Pied Piper for the mendacity of the State's political elite.

Brazil's National System for Information on Basic Sanitation (SNIS) reports that in 2011 there were 23 state-owned enterprises, 95 municipally run operators and 27 private companies operating water services.[61] On close inspection, the privatization of water services in Brazil has generally reflected a lack of efficiency in public institutions and has been on occasion closely linked to bribery, but a desire to address a shortfall in public finances has not necessarily been a key factor here. Thus, sustainable water consumption in Brazil is much more related with sustainability of the public institutions involved, encompassing a fight against corruption.

The studies that have been published on water privatization policies in different countries around the world show how problematic twenty-first century water governance issues in fact are. On the one hand, the rise of a range of 'water conflicts' is undisputable; on the other hand, it is quite uncertain what the best policy to deal with them is. That said, studies related to water governance have offered some pointers as to what really matters in the sector. Instead of drawing standard models that all countries should follow in a one-size-fits-all approach, they contend that tailored solutions are optimal.[62] For these studies the water crisis is in fact a 'governance crisis' and that the governability of the water sector is closely linked to the historic needs of each country. All of them show the importance and necessity of regulation from government before any sort of privatization is carried out and of strengthening public institutions against corruption.

18.4 Conclusion: Sustainable Water Consumption as a Claim for a Human Right to Access Water

The human right to access water has evolved historically, having been developed at the global level, affirmed internationally and then implemented domestically. The crisis in the water sector worsened by an objective water supply crisis in different parts of the world has given rise to a claim for a human right to water. This, however, has not been guaranteed in most countries, sometimes because of inappropriate

---

[60] Lobina and Hall (2001).

[61] Sistema Nacional de Informações sobre Saneamento (SNIS) (2016) available at http://www.snis.gov.br/. Accessed 20 January 2017.

[62] Rogers and Hall (2003), Castro (2007), Ribeiro (2008), and Vieira (2016b).

privatization policies undertaken in the water sector, sometimes because of marginalizing trade policies, and sometimes because of inefficient public management.[63]

Three distinct social movements have been at the forefront of the claim to a right to water: environmentalists, those advancing social justice and groups focusing on sustainable development.[64] For the environmentalists, all water sources must be preserved for present and future generations. For the social rights movement, water must be accessible to all and minorities should not be excluded. For the sustainable development movement, balanced and equitable access to water of sufficient quantity and quality are the ground for an adequate standard of living and sustainable water consumption.

But what is the meaning of a 'human right to water'? Conceptually, it rests first on the ground that water is an essential facet of the right to life, because water is crucial for human survival and a core element of a decent standard of living, thus encompassing two dimensions (civil and economic) of human rights. In the third dimension, the human right to water might be affirmed within the right to a balanced environment. The World Water Council report explains the broad meaning of the human right to water, affirming that:

> The obligation to respect requires that States Parties ... refrain from interfering directly or indirectly with the enjoyment of the right to water ... The obligation to protect requires that States Parties prevent third parties such as corporations from interfering in any way with the enjoyment of the right to water (...). The obligation to fulfil requires that States Parties adopt the necessary measures to achieve the full realization of the right to water.[65]

The human right to water was first developed in opposition to the neoliberal economic tendency of privatization of water services. From such opposition, protection of the right to water was launched, seeing it as an asset that is essential to life, just as clean air is. Concerns related to water privatization have driven internationalists to build up principles on the right of access to water. In 2002, General Comment No. 15 on the Right to Water (GC15/2002) was adopted as an authoritative interpretation of Articles 11 and 12 of the United Nations Covenant on Economic, Social and Cultural Rights (UNCESCR), disposing of the right to water and its content, as well as Member States' duties to implement it. This document has been used to monitor the implementation of the right to water on a global scale. The International Law Association introduced the Berlin Rules on Water Resources in 2004, which state, in Article 17, the right of access to sufficient, safe, acceptable, physically accessible and affordable water to meet all vital human needs. This is a summary of the meaning of sustainable water consumption.

In 2010, the United Nations General Assembly recognized that the right of access to potable water as a human right.[66] In 2015, Member States adopted a set of goals to end poverty, protect the planet and ensure prosperity for all as part of a new

---

[63]Caubet (2006), Mirandola and Sampaio (2005), Petrella (2009), and Vieira (2016b).

[64]Mirandola and Sampaio (2005), p. 4.

[65]World Water Council (2003), p. 7.

[66]GA/10967 Document A/64/L.63/REV.1.

sustainable development agenda.[67] Each goal has targets to be achieved over the next 15 years. SDG 6 is related to water and sanitation for all. This formal recognition of the human right to water has certainly placed the issue squarely onto the international agenda but, unfortunately, has not answered the many questions posed about the precise content of this right.[68]

GC15/2002 classifies water, in paragraph 1, as a 'limited natural resource and a public good fundamental for life and health' and disposes that the right to water should be a concern for developed, as well as for developing countries. Given that GC15/2002 is, in fact, an authoritative interpretation of Articles 11 and 12 of the UNCESCR it can be said to constitute the normative content and legal basis for the right to water.

Article 11, paragraph 1 states that:

> The States Parties to the present Covenant recognize the right of everyone to an adequate standard of living for himself and his family, including adequate food, clothing and housing, and to the continuous improvement of living conditions. The States Parties will take appropriate steps to ensure the realization of this right, recognizing to this effect the essential importance of international co-operation based on free consent.

In turn, Article 12 maintains that

> 1. The States Parties to the present Covenant recognize the right of everyone to the enjoyment of the highest attainable standard of physical and mental health.
> 2. The steps to be taken by the States Parties to the present Covenant to achieve the full realization of this right shall include those necessary for:
>
>   (a) The provision for the reduction of the stillbirth-rate and of infant mortality and for the healthy development of the child;
>   (b) The improvement of all aspects of environmental and industrial hygiene;
>   (c) The prevention, treatment and control of epidemic, endemic, occupational and other diseases;
>   (d) The creation of conditions which would assure to all medical service and medical attention in the event of sickness.

Bringing about an interpretation of these two clauses, the GC15/2002 sets the legal basis and a broad meaning of the right to water, in paragraph 2, as follows:

> 2. The human right to water entitles everyone to sufficient, safe, acceptable, physically accessible and affordable water for personal and domestic uses. An adequate amount of safe water is necessary to prevent death from dehydration, to reduce the risk of water-related disease and to provide for consumption, cooking, personal and domestic hygienic requirements.

From this paragraph, some main elements of the content of sustainable water consumption can be found: (a) water availability (sufficient and continuous water); (b) water quality; and (c) water accessibility (economically, physically, in a non-discriminatory way and making information available).

Paragraph 6 of the GC15/2002 broadens the content to include different purposes:

---

[67] UN SDGs (2015b).

[68] Winkler (2012), and Vieira (2016a).

> Water is required for a range of different purposes, besides personal and domestic uses, to realize many of the Covenant rights. For instance, water is necessary to produce food (right to adequate food) and ensure environmental hygiene (right to health). Water is essential for securing livelihoods (right to gain a living by work) and enjoying certain cultural practices (right to take part in cultural life). Nevertheless, priority in the allocation of water must be given to the right to water for personal and domestic uses. Priority should also be given to the water resources required to prevent starvation and disease, as well as water required to meet the core obligations of each of the Covenant rights.

Moreover, in a subchapter termed 'Normative content of the right to water', the CG15/2002, the UNCESCR expresses that the right to water has freedoms and entitlements: 'The freedoms include the right to maintain access to existing water supplies necessary for the right to water, and the right to be free from interference, such as the right to be free from arbitrary disconnections or contamination of water supplies'. Meanwhile, the entitlements 'include the right to a system of water supply and management that provides equality of opportunity for people to enjoy the right to water' (paragraph 10).

The UNCESCR suggests that 'the adequacy of water should not be interpreted narrowly' (paragraph 11), which means by mere reference to volumetric quantities and technologies. Thus, water should be treated as a cultural and social asset, and not primarily as an economic good. Besides, realization of the right to water should be sustainable, considering it for present and future generations.

Water availability, under the terms of paragraph 12(a), must be sufficient and continuous for domestic and personal use. That paragraph recommends taking World Health Organization Guidelines as the standard for the minimum quantity of water that is necessary for each person (40 liters per person, per day), which might vary because of climate, health, cultural and working conditions.[69] However, the amount itself is less important than the set of goals to be achieved. Billions of people do not have access to the minimum amount of 40 liters per person per day, but that is not due necessarily to water scarcity; sometimes, inadequate public policies are to blame. Many water scarce countries have been able to provide a minimum amount of water to their population through different and efficient public policies.

Quality is important as well. Potable water, under paragraph 12(b) of CG15/2002 is water 'free from micro-organisms, chemical substances and radiological hazards that constitute a threat to a person's health [with] an acceptable color, odor and taste for each personal or domestic use'. The World Health Organization (WHO) defines potable water as a supply that does not represent a risk to human health and that has an acceptable color and odor, as well as a palatable taste for human beings.[70]

GC15/2002 separates accessibility into distinct aspects: physical, economic and informational accessibility. Non-discrimination is a through-line across these. It states, in general, that 'water, and adequate water facilities and services, must be within safe physical reach for all sections of the population. Sufficient, safe and acceptable water must be accessible within, or in the immediate vicinity, of each

[69]WHO (2011).

[70]Ibid.

household, educational institution and workplace'. Moreover, 'water, and water facilities and services, must be affordable for all', including 'the most vulnerable or marginalized sections of the population, in law and in fact'.

Therefore, accessibility to water services must include fair tariffs and a continuous service to all, which means the adoption of public policies that: (i) cover all costs in such a way that continuity is not interrupted and (ii) assure an affordable price for the service, adopting policies of social inclusion to attend to those that cannot afford it.[71] Assuming the classification of the right to water as a public right (GC15/2002, paragraph 1), it is important to bear in mind that the responsibility of the public authority for water supply is a broad one because the service is essential and public. No matter what kind of water management the State chooses—private or public—public authorities must have overall control over the water supply and basic sanitation services. Such control must include, in particular, investments in the water sector, monitoring of water quality and water tariffs, continuity of the water supply services, regulation of the water sector, as well as participation of the consumer in decision-making procedures as a way of guaranteeing efficiency.

Moreover, priorities should not be forgotten. Around 70% of all available water for consumption is used in agriculture, mainly in irrigation.[72] Paradoxically, the overall increase in production of food as a direct effect of improvements in irrigation techniques is not proportional to the decline in figures of starvation and sub-nutrition in the world. Large scale agriculture and industry should not be priority in the management of water resources.

The right to water is part of the right to health, and must encompass not just the right to sufficient quantity of water and acceptable quality but also the right to basic hygiene. Therefore, from an environmental hygiene perspective, governments should consider the need to protect water sources from pollution and contamination as well their responsibility for water supply within a minimum standard of public health (related to installations and sewage treatment). Protection of the right to water as part of the right to health must inform public policies related to basic sanitation. The progress of a nation's development is also measured in terms of access to basic sanitation, which directly affects public health outcomes. According to the WHO and UNICEF, diarrhoea—a disease largely caused by poor basic sanitation—is the main cause of child mortality in the world.[73]

The World Water Council report calls attention to the fact that protection of the right to water is also part of a right to culture in many regions of the world.[74] In many indigenous cultures water is part of religious ceremonies and other cultural practices. In India, for instance, bathing rituals in the Ganges river are considered sacred. In many South African communities, fishing in fresh water is part of food culture. Thus,

[71]Smets (2000).

[72]WWC (2003), p. 18.

[73]WHO/UNICEF (2000).

[74]WWC (2003).

public authorities should consider cultural issues—including those of ethnic minorities—in planning and managing sustainable water policies.

Water access is also directly related to housing quality. It is estimated that around 600 million people in urban areas and around 1 billion people in rural areas live in substandard housing.[75] These figures are because of poor conditions of water supply and basic sanitation. In 2002, the UNCESCR produced a special report on the right to housing, wherein it asserted that '[n]o dwelling should be deprived of water because such deprivation would render it unliveable'.[76]

Furthermore, protection of the right to water and sustainable water consumption are also part of the right to development. Cançado Trindade understands the right to life and the right to development as intrinsically linked—because satisfaction of those human needs without which a dignified life cannot be lived can only be met if adequate levels of economic development are achieved.[77] It is a matter of having the human being as a core subject of development, and not merely as a factor of production.

The human right to water has distinct implications related to water supply itself. Examining such implications, it is important to set the right to water in a broader context and to have it analyzed differently from the other kinds of water uses, but together with other human rights that also depend on water for their realization.[78] Hence, GC15/2002 promotes the right to water and sustainable water consumption, encouraging Member States to develop public policies within the domestic sphere and establish local instruments for protection and management of water resources, within a context of sustainable development. This is nothing less than a claim for implementation of the human right to water.

Guarantees of the right to water require state action. The GC15/2002 lists, in paragraphs 21 to 36, the three main obligations of the State in relation to implementation of the human right to water: to respect, protect and fulfil water access. States should, therefore refrain 'from interfering directly or indirectly with the enjoyment of the right to water' (paragraph 21). This is a negative duty for the States. The note also demands Member States refrain from, *inter alia*:

> [E]ngaging in any practice or activity that denies or limits equal access to adequate water; arbitrarily interfering with customary or traditional arrangements for water allocation; unlawfully diminishing or polluting water, for example through waste from State-owned facilities or through use and testing of weapons; and limiting access to, or destroying, water services and infrastructure as a punitive measure, for example, during armed conflicts in violation of international humanitarian law (GC15/2002, paragraph 21).

States should also protect the right to water, preventing 'third parties from interfering in any way with the enjoyment of the right to water' (paragraph 23). This is an obligation that includes, *inter alia*:

---

[75] Ibid.

[76] E/CN.4/2002/59, para. 56.

[77] Cançado Trindade (1993), pp. 81, 179.

[78] Winkler (2012).

> [A]dopting the necessary and effective legislative and other measures to restrain, for example, third parties from denying equal access to adequate water; and polluting and inequitably extracting from water resources, including natural sources, wells and other water distribution systems.

The 'obligation to fulfil' further divides into the obligations to facilitate, promote and provide (paragraph 25) water—as such, they are positive obligations. The obligation to fulfil, requires Member States to, *inter alia*:

> [Accord] sufficient recognition of this right within the national political and legal systems, preferably by way of legislative implementation; [adopt] a national water strategy and plan of action to realize this right; [ensure] that water is affordable for everyone; and [facilitate] improved and sustainable access to water, particularly in rural and deprived urban areas'.

Moreover, each State also has an international obligation in relation to the right to water (paragraph 30), respecting the enjoyment of such a right in other countries, adopting policies for international cooperation and assistance and refraining from any activity that would deprive another country of the ability to realize the right to water for persons in its jurisdiction (paragraph 31). As a matter of international politics, 'water should never be used as an instrument of political and economic pressure' (paragraph 32). Paragraph 33 of GC15/2002 sets up a bold clause that calls for 'steps [to] be taken by States Parties to prevent their own citizens and companies from violating the right to water of individuals and communities in other countries', which could be done through legal or political means. As members of international organizations, Member States should ensure that their actions 'take due account of the right to water' (paragraph 36). This paragraph is emphatic in highlighting some of the main institutions that should be considered: the IMF, the World Bank and regional development banks. It requires Member States to 'take steps to ensure that the right to water is considered in their lending policies, credit agreements and other international measures'.

Including the right to water within a list of universal human rights constitutes an important step to make it reality for many and, at the same time, to guarantee sustainable water consumption. In summary, the World Water Council report concludes that recognition of a human right to water implies:

1. Potable water is an individual right that can, consequently, be claimed before courts and tribunals—water is not a mere commodity nor are water services supplied by the State merely out of good will or charity;
2. Recognition of the right to water as a human right will accelerate the process of guaranteeing access to water for all;
3. Social inequalities diminish, and marginalized classes are included in the policies of access to water;
4. The most vulnerable groups and communities must have access to the decision-making in public policies toward access to water;

5. United Nations human rights monitoring mechanisms (and those of other organizations) are to be used to monitor the enforcement of the right to water within the jurisdiction of States Parties.[79]

The report concludes that:

> Approaching development from a rights perspective informs people of their legal rights and entitlements, and empowers them to achieve those rights. Rather than seeing people as passive recipients of aid, the rights-based approach puts the individual at the centre of development.[80]

Thus, a right to water requires action first, from public authorities, which oversee water supply and sanitation services. If public authorities do not provide such services, then individuals may seek by right a remedy and bring an action against the State, either in administrative or judicial tribunals.

Public authorities have the obligation to plan public policy around a set of water use priorities. Dubreuil identifies a hierarchy of priorities in this context.[81] First, the State must recognize that water is a matter of survival—'water for life'. Public authorities should therefore accord the highest priority to guaranteeing sufficient quantities of high-quality water in quality for human and animal consumption and for personal hygiene. Second, authorities should prioritize water in the public interest—'water for citizens'—such as for health, education and leisure. Last, public authorities should pay attention to 'water for development', which incorporates agriculture, industry, electricity and other sectors. In this way, sustainable water consumption becomes a reality as a policy to be adopted and a right to be claimed.

18.5 Conclusion

Establishing the link between the human right to access water and sustainable water consumption is a matter of understanding that these are simply two faces of the same coin. The United Nations GC15/2002 calls on Member States to take due steps to prevent their own citizens and companies from violating the right to water. Thus, public authorities should adopt distinct measures to guarantee the human right to water for all and to facilitate full enjoyment of this right. This may in some cases mean a role for the private sector in water supply services through FDI, but only to the extent that such services cannot be managed publicly or funded by public revenue. Cases across the globe in which privatization of water services has gone awry have triggered a public claim for a human right to water. This makes it clear enough that privatization *per se* is not necessarily the optimal policy choice where

[79] WWC (2003), p. 9.

[80] Ibid.

[81] Dubreuil (2006), p. 4.

access to water is concerned. Indeed, if the goal of water privatization is simply to raise funds to plug public sector deficits, then it is certainly suboptimal.

The present study looked at water privatizations in the United Kingdom, Argentina, Bolivia, Uruguay and Brazil and found evidence of this conclusion. For instance, while sustainable water consumption in Uruguay is a matter of a constitutional human right to be claimed, in Brazil this is a matter of promoting efficiency in public institutions and developing policies that drive away corruption.

Where water privatization has produced good outcomes, such as in Chile and France, government regulation that considered real environmental and human rights concerns were in place beforehand. Even so, in some French cities such as Paris, water services returned to public hands after the private companies put in charge of providing them raised tariffs beyond acceptable levels. Thus, in France one might say that sustainable water consumption is closely related to fair and affordable water services.

A number of distinct ways of assisting public authorities implement the right of access to water[82] have been proposed:

1. Legislation and integrated policies entitling public authorities to act whenever possible;
2. Creation or destination of judicial mechanisms for monitoring the right of access to water;
3. Encouragement of local action, such as river basin committees; and
4. Encouragement of local or regional regulation of the right to water, mainly when a privatization policy takes place. All these means could also be a way of leading to a sustainable water consumption.

To promote so-called 'water social inclusion', performance indicators for monitoring the implementation of the human right to water must treat 'access' in terms of 'affordability' as paramount. In this context, one of the most efficient mechanisms is progressive tariffs, based on the economic capacity of consumers.[83] Nevertheless, every country must define its own 'poverty line' (and in many, such as in Brazil, the poverty line will vary from region to region), to guarantee the human right to water for all, as well as sustainable water consumption, either under public or private management of water supply services.

References

Allan T (1997) Virtual Water: a long term solution for water short Middle Eastern economies? In: Water and development session. British Association Festival Science. University of Leeds. TUE.51, 14.45, September, 1997

---

[82] Dubreuil (2006), p. 15.

[83] Roaf (2005).

Associação Brasileira das Concessionárias Privadas de Água e Esgoto., Available at: http://www.abcon.com.br/ (Accessed em 10 January 2017)
Azpiazu D (2005) Água potable y saneamiento en Argentina. Privatizaciones, crisis, inequidades y incertidumbe futura. Cuaduernos del Cendes, Mayo, Caracas 22(59):45–68
Barlow M, Clarke T (2002) Blue Gold. The battle against corporate theft of the World's Water. Earthscan Publications, London
Barraqué B (1995[2005]) Politiques de l'eau em Europe. Revue Française de Science Politique 45 (3):420–453
Brasil, Ministério Das Cidades, Sistema Nacional de Informações sobre Saneamento 2016. Available on http://www.snis.gov.br/. Accessed 10 Jan 2017
Budds J, McGranaham G (2003) Are the debates on water privatization missing the point? Experiences from Africa, Asia and Latin America. Environment and Urbanization. International Institute for Environment and Development, SAGE, pp 87–114
CanÇado Trindade (1993) A. A. Direitos Humanos e Meio Ambiente. Paralelo dos Sistemas de Proteção Internacional. Sergio Antonio Fabris, Porto Alegre
Castro JE (2007) Water Governance in the twentieth-first century. Ambiente e Sociedade Campinas 10(2):97–118
Caubet CG (2006) Domínio da água ou direito à água? Rivalidades nas relações internacionais do século XXI. In: Nasser SH, Rei F (orgs) Direito Internacional do Meio Ambiente. Atlas, São Paulo, pp 165–183
Celli Jr U (2009) Comércio de Serviços na OMC. Liberalização, Condições e Desafios. Juruá, Curitiba
Dubreuil C (2006) The right to water: from concept to implementation. World Water Council
Eau De Paris (2017) L'Entreprise Publique. Available on http://www.eaudeparis.fr/. Accessed 10 Jan 2017
Foster V (2005) Ten years of water service reform in Latin America: toward an Anglo-French model. Water Suply and Sanitation Sector Board Discussion Paper Series, n. 3. The World Bank Group, Washington DC
González-Gomez F, García-Rubio M (2008) Efficiency in the management of urban water services. What have we learned after four decades of research? Revista de Economía Pública de la Hacienda Pública Española. Instituto de Estudios Fiscales 185(2):39–67
International Water Management Institute (IWMI) (2009) IWMI Annual report 2008. Colombo, Sri Lanka: International Water Management Institute (IWMI) 57p
Lobina E, Hall D (2001) UK Water Privatization—a briefing. Public Services International Research Unit, February, 2001. Available on: http://www.psiru.org. Accessed 10 Jan 2017
Mirandola CMS, Sampaio LS (2005) Universalizing the Right to Water in Brazil. Liberalization, Regulation and Human Rights Public Policies. IDCID Working Paper Series. Working Paper n. 0001. December 2005
Olivier A (2006) Water Tariff Increase in Manaus (Brazil): an evaluation of the impact of households. Document de Travail, Paris, DT/2006-10, 2006.
Palaniappan M, Gleik PH (2008) Peak Water. In: Gleik (org) The World's Water 2008–2009. The biennial report on freshwater resources. Pacific Institute, Island Press, Washington, pp 1–16
Petrella R (2009) Le Manifeste de l'eau pour le XXIe siécle. Pour un pacte social de l'eau. Fides, Montréal
Ribeiro WC (2008) Geografia Política da Água. Annablume, São Paulo
Roaf V (2005) Monitoring implementation of the right to water: a framework for developing indicators. Global Issue Papers, n. 14, March, 2005
Rogers P, Hall A (2003) Effective Water Governance. The Background Papers. Global Water Partnership. n. 07. Suécia
Smets H (2000) The right to water as a human right. Environmental Policy and Law Report. 30/5, IOS Press, pp 248–250
Tucci CEM (2000) Relatório Nacional sobre o Gerenciamento da Água no Brasil. Instituto de Pesquisas Hidráulicas. Available at: http:www.bvsde.paho.org. Accessed 1 Mar 2017

United Kingdom Consumer's Council for Water (2010) Company Performance Report, 2010, Available on https://www.ccwater.org.uk/publications/companies-performance-report/. Accessed 10 Jan 2017

United Nations (2009) UNESCO. The United Nations World Water Development. Report Water in a Changing World, n. 3

United Nations (2013) Report of the Special Rapporteur on the human right to safe drinking water and sanitation, Catarina de Albuquerque (A/HRC/24/44), July 2013

United Nations (2015a) UN Water, Report on Water and Sustainable Development, 2015. Available at http://www.unwater.org/fileadmin/user_upload/unwater_new/docs/WaterandSD_Vision_to_Action-2.pdf. Accessed 10 Jan 2017

United Nations (2015b) Sustainable Development Goals. Available http://www.un.org/sustainabledevelopment/sustainable-development-goals/. Accessed 10 Jan 2017

Vieira AC (2016a) O Direito Humano à Água. Arraes, Belo Horizonte

Vieira AC (2016b) International law, governance and trade of water services. In: Parzatzis P, Gavouneli M (Org) Reconceptualising the rule of law in global governance, resources, investment and trade, vol 1, 1st edn. Hart, Oxford, pp 277–288

Winkler IT (2012) The human right to water. Hart, Oxford

World Health Organization (2011) Guidelines for drinking water quality, 4th edn

World Health Organization/UNICEF (2000) Joint Monitoring Programme on Water Supply and Sanitation, Assessment Report

World Water Council (2003) Report: Financing Water for All

Chapter 19
Building Upon Sustainable Consumption and Production for Food and Apparel

Rodrigo Carvalho de Abreu Lima and Josiane Godoy Lima

Contents

Abstract This chapter aims to address inherent challenges in characterizing sustainable consumption and production (SCP) as a tangible reality when it comes to the food and apparel sectors. Addressing how SDG 12 is adopted and managed in different sectors is fundamentally important to measure if, and to what extent, environmental and social concerns are fulfilled, considering, in a broader view, their interdependence with other Sustainable Development Goals (SDGs), such as poverty eradication, zero hunger and food security, energy security, decent work and economic growth, climate action and partnerships for the goals. There are critical environmental agendas related to food and apparel production that deserves attention when it comes to sustainable production. Deforestation, excessive use of agrochemicals, biodiversity loss, soil degradation and climate change are key issues related to food production. From a social perspective, it is relevant to quote labor rights— and, more broadly, human rights—as a baseline scenario to be considered when addressing SDG 12. For the apparel sector, social issues are critical, especially considering big trademarks and the vast, decentralized network of suppliers and manufacturers concentrated in developing countries. Slave and child labor, poor wages and lack of basic labor rights are at the core of the challenges when it comes to sustainable production and consumption at the apparel sector. Regarding

R. Carvalho de Abreu Lima (✉)
Agroicone Ltd., São Paulo, Brazil

J. G. Lima
University of São Paulo, São Paulo, Brazil

A. d. Amaral Junior et al. (eds.), *Sustainable Consumption*,
https://doi.org/10.1007/978-3-030-16985-5_19

consumption, it is relevant to analyze the level of consumption in different countries, food and apparel loss and wastage, access to nutritious food and overconsumption.

19.1 Introduction

The UN's 2030 Agenda for sustainable development, encompassing 169 targets within 17 Sustainable Development Goals (SDGs), constitutes a decisive plan of action for the long-term prosperity of people and the planet. SDG 12—ensuring sustainable consumption and production (SCP) patterns to reduce environmental and social impacts, manage risks and allow the adoption of good practices on the production and consumption—is a fundamental goal within the overall agenda.

SDG 12 sets a broad range of targets to address management and use of natural resources and of chemicals, reduce waste and loss of food and other products, reduce fossil-fuel use and phase out harmful subsidies. Moreover, it works within both public and private governance structures to shift to sustainable patterns, as well as promoting awareness among consumers. It enshrines an overarching goal that calls on the State, the private sector and consumers to take concrete steps towards SCP.

This chapter therefore addresses the main issues around SCP in two specific sectors: food and apparel. Moreover, it seeks to lay down minimum grounds for measuring outcomes on SDG 12 concerning SCP in those two sectors. The first section will outline the debate around SDG 12 with a view to framing its content, highlighting especially its specific targets, and shedding light on what is at stake in achieving the goal overall. It will also discuss the strengths and weaknesses of SDG 12, considering how it professes to put forward a broad and complex agenda, thus putting in perspective the interdependencies between the SDGs. The second and the third sections will deal with key environmental and social issues at the heart of the food and apparel sectors, respectively. A core contention of the chapter is that there is no single, easy and stepwise approach to achieving SCP when it comes to these two sectors. The fourth section will address partnerships that are working to implement the SDGs, especially SDG 12. The final section will reflect on the discussion of the food and apparel sectors when it comes to SDG 12, arguing that the SDGs are in fact mutually reinforcing and that all must be addressed in conjunction to reach the targets of the 2030 Agenda.

19.2 Sustainable Consumption and Production: What Is at Stake with SDG 12?

The debates around SCP are connected to the emergence of the environmental global agenda during the 1980s, the key milestones being the World Commission on Environment and Development and the publication of the Brundtland Report, known also as *Our Common Future*.[1] The establishment of the Commission on Sustainable Development in 1992 launched an intense process to build upon this agenda at the UN General Assembly. Paragraph 42 of the Report of the General Assembly's Sixth Session acknowledged the need to promote sustainable production and consumption as a key challenge. The process launched at Rio+20 in 2012 draws upon the accumulated experience gained under the UN system towards building sustainable development and triggered the development of a broader, complementary, collaborative and ambitious universal agenda.

The UN resolution that launched the SDGs in 2015[2] forms the bedrock of international efforts to strengthen the conceptualization and practice of sustainable development across the world, regardless of varying levels of development.[3] The United Nations High-Level Political Forum on Sustainable Development (HLPF) arose from the United Nations Conference on Sustainable Development as the forum to follow up and review the 2030 Agenda and implementation of the SDGs. The HLPF has the following in-depth review schedule, having reviewed goals 1, 2, 3, 5, 9 and 14 in 2017. In 2018, it is scheduled to review goals 6, 7, 11, 12 and 15 and goals 4, 8, 10, 13 and 16 in 2019.[4] Along with the Forum, countries should prepare voluntary national reviews, addressing progress towards meeting the SDGs and sharing progress, experiences, lessons learned and challenges towards it. At the HLPF meeting that took place in July 2017, 43 countries presented their voluntary national reviews, adding to the 22 countries that presented reports in 2016.[5]

It is important to underline that the SCP concept is predominantly connected to environmental concerns. In 1994, the Oslo Symposium proposed a working definition of sustainable consumption as 'the use of goods and services that respond to basic needs and bring a better quality of life, while minimizing the use of natural resources, toxic materials and emissions of waste and pollutants over the life cycle, so as not to jeopardize the needs of future generations'.[6] There are intrinsic environmental principles of SCP, regardless of the fact that the definition is relatively open and has evolved over time:

1. Improving the quality of life without increasing environmental degradation and without compromising the resource needs of future generations.
2. Decoupling economic growth from environmental degradation by:
   (a) Reducing material/energy intensity of current economic activities and reducing emissions and waste from extraction, production, consumption and disposal, and;

[1]United Nations (1998).

[2]United Nations (2015a, b).

[3]FAO and the SDGs Indicators (n.d.-b), p. 4.

[4]United Nations (2016).

[5]United Nations (2017a, b, c).

[6]Oslo Roundtable on Sustainable Production and Consumption (1994).

(b) Promoting a shift of consumption patterns towards groups of goods and services with lower energy and material intensity without compromising quality of life.[7]

As stated by Jeffrey Barber,[8] the concept of SCP is somewhat vague and open:

> Much of the fuzziness of the concept lies in the varied meanings of 'sustainability', which like other important concepts, such as love, justice, truth and freedom, have various interpretations, the dominance of which often depends on the interpreter in power. Thus, like other words, the specific meaning will depend on the context, the stakeholders and the stakes or interests involved. ... Thus, we can expect a number of different answers to the question what makes a particular production and consumption system or process 'sustainable'.

A specific analysis of SDG 12 is essential to frame the main objectives to be pursued when dealing with the characterization of SCP. Table 19.1 presents the targets and indicators as agreed within the 2030 Agenda.

It is important to highlight that SDG 12 calls on countries to establish programs on SCP, as well as to enact sustainable public procurement policies. Moreover, it calls on private companies to adopt sustainable practices and integrate sustainability information into their reporting. Awareness for sustainable development and lifestyles in harmony with nature are also a core target, with indicators basically attached to education.

The nature of the objectives agreed under SDG 12 puts forward a general framework aiming to guide countries towards reaching the goal, without necessarily establishing specific criteria on key social concerns attached to production and consumption. As the HLPF report from 2017 notes, it is possible to point to the vague nature of SDG 12 when it comes to an objective definition or assessment of SCP.[9] For some countries, the approach is to adopt the 'green economy' or 'circular economy' objective, highlighting the economic benefits that SCP can bring. Sustainability reporting could be one way to manage and promote SDG 12, as well as the adoption of public procurement policies, green taxation, awareness raising and stakeholder engagement.[10]

Considering that the focus of this chapter is SCP in the food and apparel sectors, the sections below will put forward concrete examples of principles and criteria that should be in place to measure SCP outcomes effectively. The authors do not intend to prescribe any closed set of criteria, but rather to establish a debate about key concerns in implementing SDG 12 within these sectors.

---

[7]UNEP (2015a, b).

[8]Barber (2018), p. 2.

[9]United Nations (2017a, b, c).

[10]Ibid.

Table 19.1 SDG 12–Ensure sustainable consumption and production patterns

Targets	Indicators
12.1 Implement the 10-year Framework of Programmes on Sustainable consumption and Production, all countries taking action, with developed countries taking the lead, considering the development and capabilities of developing countries	**12.1.1** Number of countries with SCP national action plans or SCP mainstreamed as a priority or a target into national policies
12.2 By 2030, achieve the sustainable management and efficient use of natural resources	**12.2.1** Material footprint, material footprint *per capita* and material footprint as a share of GDP **12.2.2** Domestic material consumption, domestic material consumption per capita, and domestic material consumption as a share of GDP
12.3 By 2030, halve per capita global food waste at the retail and consumer levels and reduce food losses along production and supply chains, including post-harvest losses	**12.3.1** Global food loss index
12.4 By 2020, achieve the environmentally sound management of chemicals and all wastes throughout their life cycle, under the agreed international frameworks, and significantly reduce their release to air, water and soil to minimize their adverse impacts on human health and the environment	**12.4.1** Number of parties to international multilateral environmental agreements on hazardous waste, and other chemicals that meet their commitments and obligations in transmitting information as required by each relevant agreement **12.4.2** Hazardous waste generated *per capita* and proportion of hazardous waste treated, by type of treatment
12.5 By 2030, substantially reduce waste generation through prevention, reduction, recycling and reuse	**12.5.1** National recycling rate, tonnes of material recycled
12.6 Encourage companies, especially large and transnational companies, to adopt sustainable practices and to integrate sustainability information into their reporting cycle	**12.6.1** Number of companies publishing sustainability reports
12.7 Promote public procurement practices that are sustainable under national policies and priorities	**12.7.1** Number of countries implementing sustainable public procurement policies and action plans
12.8 By 2030, ensure that people everywhere have the relevant information and awareness for sustainable development and lifestyles in harmony with nature	**12.8.1** Extent to which (i) global citizenship education and (ii) education for sustainable development (including climate change education) are mainstreamed in (a) national education policies; (b) curricula; (c) teacher education; and (d) student assessment
12.A Support developing countries to strengthen their scientific and technological capacity to move towards more sustainable patterns of consumption and production	**12.A.1** Amount of support to developing countries on research and development for sustainable consumption and production and environmentally sound technologies

(continued)

Table 19.1 (continued)

Targets	Indicators
12.B Develop and implement tools to monitor sustainable development impacts for sustainable tourism that creates jobs and promotes local culture and products	**12.B.1** Number of sustainable tourism strategies or policies and implemented action plans with agreed monitoring and evaluation tools
12.C Rationalize inefficient fossil-fuel subsidies that encourage wasteful consumption by removing market distortions, under national circumstances, including by restructuring taxation and phasing out those harmful subsidies, where they exist, to reflect their environmental impacts, fully considering the specific needs and conditions of developing countries and minimizing the possible adverse impacts on their development in a manner that protects the poor and the affected communities	**12.C.1** Amount of fossil-fuel subsidies per unit of GDP (production and consumption) and as a proportion of total national expenditure on fossil fuels

19.3 Approaches to Sustainable Consumption and Production in the Food Sector

The availability of safe, nutritious and sufficient food is at the core of SDG 2. The *State of Food Security and Nutrition in the World Report* from 2017,[11] presents an alarming figure, pointing to 815 million people undernourished in 2016, compared to 777 million in 2015. Sub-Saharan Africa, Central and Southern Asia are the most critical regional affected with food insecurity prevalence. SDG2 details the obstacles that must be overcome to guarantee food security and access to nutritious food for all of humanity. This goal covers sustainable agriculture, tackling food wastage, ensuring minimum environmental protection while increasing resilient food production systems, enhancing productivity and preventing trade distortions.[12]

From an environmental perspective, food production has explicit challenges when it comes to natural resources conservation and biodiversity loss, deforestation, soil degradation, greenhouse gas (GHG) emissions, use of agrochemicals and fertilizers, adoption of technologies and other aspects. Moreover, it is important to consider socio-economic issues like access to technology, capacity building and rural extension, land tenure rights, the strengths and weaknesses of agricultural policies, the existence of cooperatives or other organizational frameworks, gender issues in agricultural work and labor rights among others.

It is possible to frame the debate about sustainable food within a systems approach, by which the interactions of the components of a specific agricultural operation, such as the scale of production, natural resources, labor, nutrition, genetic

[11]FAO (2017).

[12]United Nations (2015a, b).

stock, equipment, knowledge and others, are considered systemically and holistically.[13] According to the FAO:

> Within a system, value chains are composed of the full range of farms and enterprises and their value-adding activities, which produce agricultural raw materials and transform them into food products that are sold to final consumers and disposed of after use.[14]

Regarding food production and what drives sustainability, it is generally believed that industrial production of food, usually based on monocultures and large areas under cultivation, use of inputs and biotech, would not be considered sustainable compared to a small-scale, diverse, organic agriculture and other systems integrating agriculture, forests and livestock. Agricultural systems and patterns are diverse and have evolved in distinct ways over a long period of history. Countries with a vast extension of arable land usually have different systems mixing monocultures, pastures and small-scale models. In small countries, the size of the farms and the model of agricultural systems are generally on the small end, privileging more diversified food systems.

Jonathan Mockshell and Josey Kamanda propose a blended sustainability approach to achieving sustainable agriculture, avoiding the choice of one system over another and capturing continuous improvements and benefits from different models:

> The debate has centred on two sustainable agricultural approaches: AEI [agro-ecological intensification] and SAI [sustainable agricultural intensification]. The findings from this analysis suggest that the actors differ in their concept and vision of farming, as well as along the economic, ecological and social dimensions of agricultural sustainability. These fundamentally different perceptions make the question of which pathway to take to feed the growing population even more pertinent. At the same time, it is important to accept that there are synergies and tradeoffs between the AEI and SAI approaches. Acknowledgment of the tradeoffs implies that one pathway cannot be considered a panacea to achieving sustainable agriculture and food security. Rather, location-specific situations should determine the best set of practices by considering the strengths, opportunities, weaknesses, and threats of AEI and SAI and moving towards an integrated approach.[15]

A range of elements is connected to defining SCP in food, usually driven by consumer demands and trends. Consumer preferences have a strong role to play in shaping demand and creating new standards for the whole food system. Awareness of the possibility to acquire fresh, diverse and minimally industrialized food and the role of local and regional food production as a driver of optimal social and health outcomes therefore has an important function when it comes to consumer behavior. For instance, pressure to reduce meat consumption based on health, GHGs emissions, animal welfare and other environmental impacts has become a trend. Science must be foregrounded in any discussions here, however, including dialogue about

---

[13]USDA (n.d.).

[14]FAO (2017).

[15]Mockshell and Kamanda (2018), p. 18.

the health benefits from the consumption of meat, especially for infant children, and the adoption of good practices.

Conversely, the argument against meat consumption should be driven by improving production standards, reducing impacts on the environment and allowing the diversification of agricultural outputs from the same area, considering rational and diversified food production and consumption. It does not seem reasonable to condemn meat consumption *per se* as a way to reduce environmental and health impacts, especially given that in most developing countries the volume consumed, measured in kg/capita/year, is extremely low.

Data from the FAO for 2013 for a range of countries is instructive: Mozambique (1.04 kg per capita per year); Congo (4.35 kg); Ethiopia (3.61 kg); China (5.23 kg). This compares with 21.26 kg per capita per year in Switzerland; 36.24 in the United States, 29.1 in Uruguay, 23.81 in France, 28.46 in Denmark, 30.25 in Canada, 33.86 in Australia, 39.25 in Brazil and 55.48 in Argentina.[16] This prejudice-against-meat-consumption approach can undermine or destabilize cultural or personal choices. Instead, it is preferable to push for the adoption of best practices that will allow reduced impacts on the production side, as well as the application of key internationally recognized principles, such as for soil and native vegetation restoration and conservation.

The argument in favor of climate smart agriculture (CSA), for instance, is a key driver towards promoting a continuous approach to sustainable consumption of food. Although there is no single methodology of CSA, it is important to recognize the FAO approach:

> [CSA] is an approach to help the people who manage agricultural systems respond effectively to climate change. The CSA approach pursues the triple objectives of sustainably increasing productivity and incomes, adapting to climate change and reducing greenhouse gas emissions where possible. This does not imply that every practice applied in every location should produce 'triple wins'. Rather the CSA approach seeks to reduce trade-offs and promote synergies by taking these objectives into consideration to inform decisions from the local to the global scales and over short and long time horizons, to derive locally-acceptable solutions.

CSA comprises the adoption of good agricultural practices, restoring soil productivity, reducing nitrogen emissions, improving productivity, reducing deforestation while promoting forest restoration, fomenting integration of agriculture and livestock and promoting the resilience of the agricultural systems. Moreover, the FAO has developed a common vision and an integrated approach to manage sustainability across agriculture, forestry and fisheries, based on five principles:

1. Improving efficiency in the use of resources
2. Conserve, protect and enhance natural resources
3. Protect and improve rural livelihoods, equity and social well-being
4. Enhance resilience of people, communities and ecosystems

[16]FAO (n.d.-a).

5. Responsible and effective governance mechanisms.[17]

FAO is pivotal to measure the achievement of the SDGs, taking as the core goal SDG2, but considering its links with several different SDGs. Some of the key indicators from the FAO perspective are:

- Prevalence of undernourishment;
- Prevalence of moderate or severe food insecurity in the population, based on the Food Insecurity Experience Scale;
- Average income of small-scale food producers, by sex and indigenous status;
- Proportion of agricultural area under productive and sustainable agriculture;
- Number of plant and animal genetic resources for food and agriculture secured in medium- or long-term conservation facilities;
- Change in water use efficiency over time;
- Global food loss index; and
- Percentage of land that is degraded over total land area.

One of the challenging and critical issues to be addressed under SDG2 and SDG12 is food loss and waste. According to FAO, every year approximately 1.3 billion tonnes of food is lost or wasted, representing one-third of total food output. Reducing consumption cannot be considered, *per se*, a strong indicator towards building upon SCP. As stated by Lewis Akenji and Magnus Bengtsson, there is clearly a need to balance consumption needs based on the elimination of extreme poverty as a key driver and the adoption of new patterns of producing and meeting society needs that are less ecologically damaging and social balanced:

> Reduced levels of consumption by developed countries. For the SDGs to be seen as applicable to all, industrialized countries need to, while ensuring the wellbeing of their citizens, commit to reducing their level of material consumption. This is as much desired as it is imperative in order to give the Post-2015 Development Agenda fairness and legitimacy.
>
> Increased quality of consumption by developing countries. Poverty is unsustainable; developing countries need to increase the level of consumption of low-income groups, especially of basic necessities, to at least meet minimum requirements for health and dignity, while adopting the most sustainable methods available to avoid causing the same levels of ecological harm as has been seen with developed countries.[18]

The ability to address loss and waste of food is critical to improve SCP patterns. Wasting food during production, during transportation, at the industry level, at the retail level and at the household level requires different strategies, policies and actions. This implies country-level strategies, from best production practices, to adequate logistics and sanitary regulations, promotion of the cold supply chain,

---

[17]FAO (2018).

[18]Akenji and Bengtsson (2018), p. 524.

Table 19.2 Sustainable consumption and production indicators relating to food

• Reduce deforestation rates • Restoration of native vegetation Projects aimed at conserving and restoring native vegetation • Native vegetation areas set aside for conservation on private lands • Native vegetation set aside for protection on public lands • Land tenure rights • Biodiversity conservation actions • Conservation of genetic resources • A percentage of native vegetation protected by innovative projects • Raise awareness about biodiversity conservation • Soil management and restoration activities • Increase the wood and coal produced using planted forests • Promote livestock intensification aimed at reducing GHG emissions	• Productivity increases because of capacity building and technology access • Policies to promote climate smart agriculture • Reduced level of government subsidies • Reduce and manage energy and water use to produce food • Reduce and make a responsible use of inputs • Increase the use of biological control of pests • Reduce trade restriction of food products to developing countries • Food security and nutrition levels • Private sector initiatives aiming to improve food production and promote awareness about responsible consumption • A percentage of small farmers adopting technology and improving productivity • Policies towards promoting gender balance on agriculture	• Increase of area allocated to low carbon and diversified agriculture • Promote agriculture expansion over converted areas • Increase the area under organic agriculture • Increase the area under agroforestry systems • Increase area under crop-pasture-forest integration • Reduced prices for organic food • Increase demand for local, regional and organic food • Increase household level access to quality food • A percentage of food loss and waste reduction • Improve nutrition patterns • Raise awareness about consumption behavior (meal patterns, eating away from home, snacking, highly processed foods, activity levels) • Labor in agriculture follows countries regulations and ILO Conventions

reducing informality in food retailing and promoting awareness on food use and consumption at the household level.

It is reasonable to argue that each country has its own scope concerning SCP, and that each model has its own indicators and good practices aiming to build upon desirable practices related to sustainable food production and consumption. Table 19.2 presents possible substantive indicators that countries, companies and governments can consider in defining and measuring SCP for food.

The list above is not meant to be prescriptive; it is not a closed set of indicators. It is merely a compilation of success drivers that can have positive impacts towards sustainable food production and consumption.

19.4 Approaches Towards SCP in the Apparel Sector

When it comes to the apparel sector, the challenges of promoting sustainable production and consumption patterns are closely related to social issues, including some key items on the environmental agendas as well. This section analyzes the main concerns associated with the SCP in textiles, garment, shoes, accessories and any other products based on fabrics.

The apparel production chain is highly segmented, starting with the production of natural fibres and filaments or chemical fibres, through to spinning, weaving, knitting and finishing either at the artisanal or industrial scales, through to distribution and the sale of clothes and accessories in retail stores around the world. The apparel sector has, indeed, a long production chain, highly dependent on labor and the use of chemical inputs and water as key components during processing. Moreover, the possibility of geographically locating parts of the chain where costs are lower gives rise to a decentralized production network, ranging from fibre production, to basic manufacturing, finishing and, finally, retailing worldwide. The countries of Southeast Asia, where labor costs are lower, feature prominently upstream in the production chain.

The production of clothes, footwear and accessories creates the opportunity to generate low-skilled jobs in extremely populous regions that suffer from poverty and limited opportunities. But it also can promote the exploitation of workers without safeguarding fundamental rights—such as minimum wages, limitations on working hours, the right to form unions, paid weekly rest—and encourages the use of child and slave labor.

There are several cases of threats to social and labor rights in the apparel sector, such as in Cambodia, China, Indonesia, India, Pakistan and Bangladesh. Moreover, recent cases of exploitation in Brazil have been reported, involving migrants from Bolivia, South Korea and Haiti. The Nike case in the 1990s is a paradigm that illustrates this concern. The denunciation of employees earning a dollar, with journeys longer than 10 h to produce a pair of US$80 sneakers, drew attention to the precariousness of the clothing chain. In 1995, Nike had no direct claims related to labor issues, but the campaign that began in Asian countries generated more than 300 articles linking Nike to conditions of exploitation, slave labor, child labor and abuse until 1997.[19] Because of the negative repercussions—damage to its reputation, financial impacts and the pressure from several countries where there were workers producing sneakers and clothes for Nike—the company started to adopt best practices and mechanisms to ensure that its products did not were made disregarding basic labor rules. The Report *Work Faster or Get Out*, published by the Human Rights Watch, provides another example of a reported denunciation of numerous violations of social rights in garment factories in Cambodia. Brands such as H&M, Marks and Spencer, GAP and Joe Fresh were among the purchasers of the products made by workers subject to strenuous working journeys, low wages, sexual abuse and child labor in conditions analogous to slavery.[20]

When it comes to environmental concerns, the apparel sector faces challenges related to water use in the production of fibres through the processing cycle. Cotton growing, textile production (dyeing/finishing/treating) and consumer laundering all

[19]Conroy (2007), pp. 11–12.

[20]Human Rights Watch (2015).

consume large quantities of water. China, India, USA, Pakistan, Bangladesh, Turkey and Brazil are textile supply chain countries that already have regions with high water stress.[21] The use of chemicals at the apparel sector relates to an increasing number of purposes. They can offer certain benefits, as the case of fabrics that reflect the sun, flameproof furniture fabrics, durable airbags for vehicles, color-coordinated bath towels, hardwearing sweaters or workwear that effectively withstands moisture and dirt. However, the use of chemicals has a downside, posing risks to human health and the environment. Persistent substances that are difficult to break down, bio-accumulative substances that accumulate in or toxic to living organisms are particularly problematic. Chemicals therefore, need to be used responsibly and according to the best available science.

The use of energy is another aspect that deserves attention when it comes to apparel production. The sector is highly energy-intensive and associated GHGs emissions are high. Better management of energy resources is a key aspect of progress towards SDGs 13 and 12. Recycling and managing solid waste is another trending topic related to apparel, that goes much beyond the responsibility of producers to deal with waste at the processing level and after consumption. In fact, recycling may become a challenge for some type of garments, depending upon the materials used. Consumers have a strong role in putting pressure on manufacturers and retailers regarding the impacts of apparel production on the environment.

The increasing demand for apparel, especially in developed countries, raises demand for fibres, energy, water, chemical inputs and labor. Responsible consumption can be a subjective concept, but a rational approach to apparel consumption should integrate awareness initiatives from consumer groups, companies, schools, NGOs and the State. Certainly, defining what is reasonable depends on cultural, educational, economic, personal and even health attributes. Despite this, it is important to stress that SCP of apparel lies in changing consumer behavior and the way brands approach the issue. Overconsumption clearly puts pressure on resources, impacting the environment and society alike.

The export of recycled clothing is also a trend. In one way, this is because it allows developed countries to dispose of millions of tonnes of used clothes that will not necessarily be worn in developing countries. As a recent Greenpeace report notes:

> Up to date and comprehensive figures on the volumes of clothes waste and used clothing globally are not compiled. This lack of standardised information shows limited interest by policymakers and a lack of transparency from the fashion industry about its use of resources and the amounts wasted. In the EU 1.5–2 million tons of used clothing is generated annually, only 10–12% of the best quality clothes are re-sold locally and much of the rest is likely to be exported to countries in the Global South. Figures from the EU as a whole are not available but in the UK 70% of the 540,000 tonnes of clothes that are collected for reuse are exported; in the USA 53% (800,000 tonnes) is exported. The export of used clothing has risen dramatically since the year 2000, with 4.3 million tonnes traded in 2014. The leading exporters are the USA, Germany, the UK, South Korea, Japan, Netherlands, Malaysia,

[21]GLASA (2015).

> Belgium, China and France and the main importers include Pakistan, Malaysia, Russia, and India, although these may not be the final destinations. For example, large amounts of used clothes are reprocessed in India and Pakistan and re-exported to Africa.[22]

In some cases, these imports damage local apparel industries in developing countries, causing economic and social losses beyond the burden of managing huge amounts of used clothing.

Switching now to social issues it is essential to discuss that the 1998 ILO Declaration on Fundamental Principles and Rights at Work establishes the fundamental principles and rights to be met and promoted by its members, independently of the ratification of the respective Conventions, which are:

1. Freedom of association and the effective recognition of the right to collective bargaining;
2. The elimination of all forms of forced or compulsory labor;
3. The effective abolition of child labor; and
4. The elimination of discrimination in respect of employment and occupation.

In relation to each of these principles, some ILO Conventions are fundamental:

- Convention concerning Freedom of Association and Protection of the Right to Organize–C087;
- Convention concerning the Application of the Principles of the Right to Organize and to Bargain Collectively–C098;
- Convention concerning Forced or Compulsory Labour–C029;
- Convention concerning the Abolition of Forced Labour–C105;
- Convention concerning Minimum Age for Admission to Employment–C138;
- Convention concerning the Prohibition and Immediate Action for the Elimination of the Worst Forms of Child Labour–C182;
- Equal Remuneration Convention–C100;
- Convention concerning Discrimination in Respect of Employment and Occupation–C111.

The effective application of the ILO guiding principles and Conventions are at the core at the debate about SCP in the apparel sector. The innate feature of being a long and widespread chain, covering several countries with particular labor conditions and regulations means the social agenda for apparel is particularly important. For that reason, it is essential to read SDG 12 together with SDG 8, aimed at promoting sustained, inclusive and sustainable economic growth, full and productive employment and decent work for all. It is reasonable to argue that the targets agreed in SDG 12 lack minimum criteria related to social and labor goals, as it does for environmental concerns.

Child and slave labor in poor countries with weak labor regulations and/or enforcement and deteriorated social conditions, lack of a proper education system, informality and unemployment are critical challenges towards achieving decent

[22]Greenpeace (n.d.), p. 5.

work. According to the *World Employment Social Outlook Trends 2018*, global unemployment is estimated at more than 190 million people, with vulnerable employment achieving the worrying figure of 1.4 billion people during 2017. Extreme working poverty accounted for more than 300 million workers in developing countries living with less than US$1.90 per day.[23]

It is extremely helpful to put in perspective some of the specific targets of SDG 8, which include the need to:

> 8.3 Promote development-oriented policies that support productive activities, decent job creation, entrepreneurship, creativity and innovation, and encourage the formalization and growth of micro-small and medium-sized enterprises, including through access to financial services;
>
> 8.4 Improve progressively, through 2030, global resource efficiency in consumption and production and endeavour to decouple economic growth from environmental degradation, in accordance with the 10-year framework of programmes on sustainable consumption and production, with developed countries taking the lead;
>
> 8.5 By 2030, achieve full and productive employment and decent work for all women and men, including for young people and persons with disabilities, and equal pay for work of equal value;
>
> 8.7 Take immediate and effective measures to eradicate forced labour, end modern slavery and human trafficking and secure the prohibition and elimination of the worst forms of child labour, including recruitment and use of child soldiers, and by 2025 end child labour in all its forms;
>
> 8.8 Protect labour rights and promote safe and secure working environments for all workers, including migrant workers, in particular women migrants, and those in precarious employment.

SDG 8 will be reviewed in-depth at the HLPF in 2019 and should be read together with SDG 12 when it comes to defining proper indicators to measure what progress has been made towards SCP in the apparel production chain. Table 19.3 presents possible indicators for this.

This section does not pretend to exhaust the debate about what can or not be considered under the SCP umbrella when it comes to apparel. In fact, the idea is to put this concern under discussion considering the multiple challenges that affects countries and the private sector ability to achieve the SDGs.

19.5 Partnerships to Promote SDG12

As SDG 12 relies on countries' ability to approve and implement the 10-Year Framework of Programmes on Sustainable Consumption and Production, it is also quite relevant to connect SDG 17 and the role of partnerships towards SCP patterns as a way to foster implementation. The aim of this part of the chapter is not to detail

[23] ILO (2018).

Table 19.3 SCP indicators relating to apparel

• Reduce water use and promote water recycling • Optimize energy use and reduce emissions • Increase the use of renewable energy at the apparel sector • Rate of recycling and proper disposal of solid waste against used apparel exports • A certain percentage of used apparel recycled • Effective compliance with local and/or regional labor regulations	• Comply with the ILO Conventions • Adopt basic labor regulations in countries with weak rules • Follow the ILO Declaration on Fundamental Principles and Rights at Work • Countries regulations prohibiting imports of used clothes without a minimum standard • Improve labor conditions, especially in least-developing countries	• Effective ratification and adoption of ILO Conventions • Eradicate child and slave labor • Raise awareness about modern slavery • Raise awareness about consumption behavior in the apparel sector • Engagement of the private sector on initiatives to promote worker inclusion, considering social and labor rights

specific initiatives but rather to point to the importance of the engagement of the private sector using two examples, The Global Compact and The Consumer Goods Forum.

The United Nations Global Compact has brought together around 9000 companies plus 4000 non-businesses members, including academics, foundations, NGOs, public sector organizations and cities.[24] It is relevant to mention some shared values of the Global Compact members regarding SDG 12 based on the organizations own reports:

- Strive for sustainably sourced key commodities including palm oil, soy, paper and pulp and beef;
- Reduce food and solid waste along production and supply chains, in collaboration with other stakeholders including suppliers, consumers, retailers and governments;
- Reduce packaging and increase recycling of end products and by-products of the production process;
- Reduce water consumption in the production process (especially for beverages);
- Raise consumer awareness of the importance of sustainable consumption and practical steps they can take to live more sustainably; and
- Develop and apply common standards and methodologies for sustainability across the life cycle of a product.[25]

The Global Compact's role engaging and working with capacity building and awareness related to multiple environmental and social challenges in over 160 countries allows a cooperative and a continuous approach towards building upon the SDGs.

[24]UNGC (n.d.).

[25]UNGC (n.d.).

The Consumer Goods Forum (CGF) was established in 2009. It is a coalition of more than 410 companies around the world, including agricultural and beverage companies, industries, retailers, wholesalers, service and education businesses, as well as organizations such as the World Economic Forum and the International Monetary Fund.[26] CGF's mission is to bring together manufacturers of consumer goods and retailers to pursue efficient business practices to promote positive changes in the industry, benefitting buyers, consumers and the world, without impeding competition.

When it comes to the SDG agenda and the challenges towards SCP patterns in food, it is important to mention the Food Waste Resolution approved in 2015, recognizing that food waste is a major social, environmental and economic challenge that stands in the way of food security, contributes to climate change, consumes natural resources unnecessarily and costs money. CGF members are committed to helping avoid food waste and maximize their recovery with the goal of halving waste within the retail and manufacturing operations of CGF members by 2025, considering 2016 as a baseline. Further, as a means of contributing to the achievement of SDGs by 2030, CGF members are committed to halve per capita food waste at the consumer level and reduce food losses along supply and production chains, including post-harvest losses and working to maximize the value of the remaining food waste.

It is also worth noting the Deforestation Resolution, adopted in 2010 by the CGF, which established a commitment to mobilize resources within the business of its members to help achieve zero net deforestation within production chains by 2020. This objective will be pursued individually by companies and through partnerships with governments and NGOs. Regarding the broader social agenda that encompasses the food and the apparel sectors, it is crucial to name the Social Resolution on Forced Labour approved in 2016, whereby CGF members recognize their role as companies responsible for respecting and promoting human rights and decent working conditions around the world. Moreover, members recognize the broad social problem of modern slavery and the need to eradicate forced labor from value chains. With a view to implementing these commitments, and considering the power of collective action as an industry group to identify and resolve issues and geographies of common concerns, specific action plans will be developed to support the eradication of forced labor, drawing on the principles of the ILO. To this end, the CGF has committed with governments and civil society aiming to deal with specific themes and geographic areas, to adopt joint actions to avoid violations of rules, to promote the harmonization of industry supply chain standards and systems and to share best practices.

In sum, the Global Compact and the CGF are leading examples of initiatives seeking to drive private sector engagement to push for continuous improvement on several agendas related to the SDGs. Their broad membership, together with their

[26] CGF (n.d.).

focus on key environmental and social issues, is pivotal to the achievement of the goals.

19.6 Reflections

There is no single and straightforward path to implement SDG12. The challenges from the food and apparel sectors must be read together with the contrasts between countries and their varying capabilities to manage minimum criteria and indicators about what constitutes acceptable patterns of SCP. There is no shortcut to SDG12. The vagueness of the concept also requires inclusion of the social and labor agendas as baseline criteria when measuring and assessing SCP. Without this, it is not reasonable or fair to claim that sustainability is genuinely enshrined at the heart of the 2030 Agenda.

The weakness of social criteria within SDG12, even as it is picked up in SDG8, can be explained by the fact that sustainable development as a concept developed within the environmental agenda that emerged only during the 1980s and 1990s. Throughout the process carried out by the HLPF, as well as the other UN fora and processes and countries activities towards the SDGs, it is expected that the social agenda may be consolidated leading to a proper assessment of minimum patterns of SCP. The interdependency between the SDGs is a reality that requires a cross-sectional approach. Therefore, it is not feasible to consider SDG 12 as a stand-alone goal. On the contrary, countries, industries, consumers, NGOs, academia and other interested stakeholders will only be successful in defining SCP criteria for it if they can frame and measure concrete steps towards it.

The fact that SDG 12 relies mostly on the ability of individual countries to implement the 10-Year Framework of Programmes on Sustainable Consumption and Production, its implementation may fall short. It may not be possible to accurately measure progress towards SDG 12 solely with the framework programs. Rather, it requires a full and effective participation of the private sector and society in a broader view. It seems critical to stress the interlinkages of SDG 12 and SDG 17, highlighting the role of the private sector and incentives to promote and disseminate awareness about how to improve SCP patterns.

Looking to the key issues around food production and consumption, it seems reasonable to say that environmental concerns and impacts must be managed in every production system. It is not feasible to simply choose a specific agricultural system as the core indicator of SCP. A blended approach can deliver better results, considering countries needs and abilities. Continuous improvement should be a baseline criterion, encompassing issues like tackling deforestation, giving value to forests, improving biodiversity while mainstreaming its value within society, adopting best production practices, restoring soil and forests, using inputs responsibly, among others. Tackling food loss and waste is critical at all levels, from government programs, to private sector initiatives and strong awareness and behavior campaigns. For the apparel sector, it is critical to stress the social agenda as a huge gap that must be fulfilled to be able to achieve SCP. No matter the improvements that the environmental concerns could achieve, related to water, energy, chemical and recycling aspects, sustainability will not exist for this sector if social and labor requisites are not met.

Contrasts between countries and cultures will always exist. However, the achievement of SCP will only be tangible if all the three pillars of sustainable development are considered and, more importantly, if countries can work cooperatively and respect differences. Achieving a level playing field on the baseline requirements to improve social and environmental outcomes in food and apparel is not an easy task. However, the more countries can do that without prejudging different approaches, the better the results will be. The recommended indicators to achieve food and apparel SCP merely list important targets, and do not constitute closed packages. They are, simply put, an indicative group that has strong connection to the debates regarding sustainability at the global level.

References

Akenji L, Bengtsson M (2018) Making sustainable consumption and production the core of sustainable development goals. Institute for Global Environmental Strategies, 2008–11. Sustainability 2014, 6, 513–529. Available via www.mdpi.com/journal/sustainability. Accessed 29 Jan 2018

Barber J (2018) Mapping the movement towards sustainable production and consumption in North America Integrative Strategies Forum. Available via http://isforum.org/pubs/documents/copyofmappingthemovement.pdf. Accessed 1 Feb 2018

Chemicals in Textiles: A Practical Advice for Companies in these Sectors. Source: A New Cloth Market. Fire2Fashion. Available via https://images.fibre2fashion.com/ArticleResources/PdfFiles/56/5503.pdf. Accessed 30 Jan 2018

Conroy ME (2007) Branded how the certification revolution is transforming global corporations. New Society Publishers, Philadelphia

Food and Agricultural Organization (2017) The state of food security and nutrition in the World 2017. In: Building resilience for peace and justice. Rome. Available via http://www.fao.org/3/a-I7695e.pdf. Accessed 25 Jan 2018

Food and Agricultural Organization (2018) Transition towards sustainable food and agriculture. 2018

Food and Agricultural Organization (n.d.-a) FAO STAT. Food Balance. Available at http://www.fao.org/faostat/en/. Accessed 12 Feb 2018

GLASA (2015) State of the apparel sector special report. Water. In: The sustainable business group. Available via https://glasaaward.org/wp-content/uploads/2015/05/GLASA_2015_StateofApparelSector_SpecialReport_Water_150624.pdf. Accessed 30 Jan 2018

Greenpeace (n.d.) Time out for fast fashion. Available via https://storage.googleapis.com/p4-production-content/international/wp-content/uploads/2018/01/6c356f9a-fact-sheet-timeout-for-fast-fashion.pdf. Accessed 31 Jan 2018

Human Rights Watch (2015) Work faster or get out. Labor rights abuses in Cambodia's Garment Industry. ISBN: 978-1-6231-32330. Available via https://www.hrw.org/sites/default/files/reports/cambodia0315_ForUpload.pdf. Accessed 30 Jan 2018

Instituto Pacto Nacional pela Erradicação do Trabalho do Trabalho Escravo. Available via http://www.inpacto.org.br/pb/2015/03/trabalhadoras-da-industria-de-vestuario-no-comboja-realizam-trabalho-forcado. Accessed 24 Apr 2017

International Labour Organization (2018) World employment social outlook trends 2018. International Labour Office – Geneva: ILO. Available via http://www.ilo.org/wcmsp5/groups/public/%2D%2D-dgreports/%2D%2D-dcomm/%2D%2D-publ/documents/publication/wcms_615594.pdf. Accessed 30 Jan 2018

International Labour Organization. Convention concerning Freedom of Association and Protection of the Right to Organize. Available via http://www.ilo.org/dyn/normlex/en/f?p=NORMLEXPUB:12100:0::NO::P12100_INSTRUMENT_ID:312232. Accessed 29 Jan 2018

International Labour Organization. Convention concerning the Application of the Principles of the Right to Organise and to Bargain Collectively. Available via: http://www.ilo.org/dyn/normlex/en/f?p=NORMLEXPUB:12100:0::NO::P12100_INSTRUMENT_ID:312243. Accessed 29 Jan 2018

International Labour Organization. Convention concerning Forced or Compulsory Labour. Available via http://www.ilo.org/dyn/normlex/en/f?p=1000:12100:0::NO::P12100_ILO_CODE:C029. Accessed 29 Jan 2018

International Labour Organization. Convention concerning the Abolition of Forced Labour. Available via http://www.ilo.org/dyn/normlex/en/f?p=1000:12100:0::NO::P12100_ILO_CODE:C105. Accessed 29 Jan 2018

International Labour Organization. Convention concerning Minimum Age for Admission to Employment. Available via http://www.ilo.org/dyn/normlex/en/f?p=NORMLEXPUB:12100:0::NO::P12100_ILO_CODE:C138. Accessed 29 Jan 2018

International Labour Organization. Convention concerning the Prohibition and Immediate Action for the Elimination of the Worst Forms of Child Labour. Available via http://www.ilo.org/dyn/normlex/en/f?p=NORMLEXPUB:12100:0::NO::P12100_ILO_CODE:C182. Accessed 29 Jan 2018

International Labour Organization. Equal Remuneration Convention. Available via http://www.ilo.org/dyn/normlex/en/f?p=NORMLEXPUB:12100:0::NO::P12100_ILO_CODE:C100. Accessed 29 Jan 2018

International Labour Organization. Convention concerning Discrimination in Respect of Employment and Occupation. Available via http://www.ilo.org/dyn/normlex/en/f?p=NORMLEXPUB:12100:0::NO::P12100_ILO_CODE:C11. Accessed 29 Jan 2018

International Labour Organization. Convention concerning Annual Holidays with Pay. Available via http://www.ilo.org/dyn/normlex/en/f?p=NORMLEXPUB:12100:0::NO::P12100_ILO_CODE:C132. Accessed 29 Jan 2018

International Labour Organization. Convention Concerning Night Work. Available via http://www.ilo.org/dyn/normlex/en/f?p=NORMLEXPUB:12100:0::NO::P12100_ILO_CODE:C171. Accessed 29 Jan 2018

Mockshell J, Kamanda J (2018) Beyond the agroecological and sustainable agricultural intensification debate: is blended sustainability the way forward? German Development Institute DIE. Discussion paper 2017. Available via https://www.die-gdi.de/uploads/media/DP_16.2017.pdf. Accessed 30 Jan 2018. ISSN 1860-0441

Organization for Economic Co-operation and Development. Towards Greener Households. Policy Highlights. Food. Available via https://issuu.com/oecd.publishing/docs/food_-_greening_household_behaviour/2. Accessed 28 Jan 2018

Oslo Roundtable on Sustainable Production and Consumption (1994) Available via http://enb.iisd.org/consume/oslo004.html#top. Accessed 28 Jan 2018

The Consumer Goods Forum. Our members. Available via https://www.theconsumergoodsforum.com/who-we-are/our-members/. Accessed 5 Sept 2017

Trabalho escravo nas oficinas de costura. In: Repórter Brasil. Available via http://reporterbrasil.org.br/wpcontent/uploads/2016/06/Fasc%C3%ADculoConfec%C3%A7%C3%A3o-Textil_Final_Web_21.01.16.pdf. Accessed 25 Apr 2017

United Nations (1998) Commission on Sustainable Development. In: Report on the Sixth Session (22 December 1997 and 20 April–1 May 1998). E/1998/29–E/CN.17/1998/20. Available via http://www.un.org/ga/search/view_doc.asp?symbol=E/CN.17/1998/20&Lang=E. Accessed 6 Apr 2016

United Nations (2015a) Transforming our World: the 2030 Agenda for Sustainable Development. In: Resolution adopted by the General Assembly on 25 September 2015. A/RES/70/1. Available via http://www.un.org/ga/search/view_doc.asp?symbol=A/RES/70/1&Lang=E. Accessed 2 Feb 2016

United Nations (2015b) Addis Ababa Action agenda of the third international conference on financing for development. New York 2015. Available via http://www.un.org/esa/ffd/wp-content/uploads/2015/08/AAAA_Outcome.pdf. Accessed 15 Jan 2018

United Nations (2016) Follow-up and review of the 2030 agenda for sustainable. Development at the global level. In: Resolution adopted by the General Assembly on 29 July 2016. A/RES/70/299. Available via http://www.un.org/en/ga/search/view_doc.asp?symbol=A/RES/70/299. Accessed 10 Jan 2018

United Nations (2017a) Format and organizational aspects of the high-level political forum on sustainable development. In: Resolution adopted by the General Assembly on 9 July 2013. A/RES/67/290. Available via http://www.un.org/ga/search/view_doc.asp?symbol=A/RES/67/290&Lang=E. Accessed 2 Feb 2017

United Nations (2017b) High-level political forum on sustainable development. In: 2017 voluntary national reviews. Synthesis Report. Available via https://sustainabledevelopment.un.org/content/documents/17109Synthesis_Report_VNRs_2017.pdf. Accessed 20 Jan 2018

United Nations (2017c) High-level political forum on sustainable development. Synthesis of voluntary national reviews 2017. Available via https://sustainabledevelopment.un.org/content/documents/17109Synthesis_Report_VNRs_2017.pdf. Accessed 1 Feb 2018

United Nations Environmental Programme (2015a) Sustainable consumption and production. A handbook for policymakers. 2015. Available via https://sustainabledevelopment.un.org/content/documents/1951Sustainable%20Consumption.pdf. Accessed 12 Jan 2018

United Nations Environmental Programme (2015b) Sustainable consumption and production indicators for the future SDGs. In: UNEP Discussion Paper – March 2015. Available via https://sustainabledevelopment.un.org/content/documents/2301SCP%20indicators.pdf. Accessed 28 Jan 2018

United Nations Global Compact (n.d.) SDG industry matrix. Available via https://www.unglobalcompact.org/docs/issues_doc/development/SDGMatrix-ConsumerGoods.pdf. Accessed 29 Jan 2018

United States Department of Agriculture (n.d.) National Institute of Food and Agriculture. Agricultural Systems, Available via https://nifa.usda.gov/topic/agricultural-systems. Accessed 28 Jan 2018

Chapter 20
Supermarkets and Private Standards of Sustainability: The Responsibility to Protect Without Protectionism

Tiago Matsuoka Megale

Contents

Abstract The current international economic scenario is characterised by the presence of an increasing number of multinational enterprises, the spread of global value chains and the creation of transnational regulatory networks. These phenomena contribute to the erosion of the regulatory capacity of the State and generate questions about how to regulate the structures created by private actors. This chapter aims to analyse the extent to which private standards of sustainability imposed by supermarkets protect common values, such as human life or health rather than constituting disguised restrictions to trade. Sustainability standards are initiatives that could be welcome on corporate social responsibility (CSR) grounds, but can be protectionist from a trade regulation perspective. These standards exist in a legal vacuum, given the multilateral World Trade Organization (WTO)-centred trade system does regulate global trade, but with rules that are binding only on Member States. This regulatory gap is generally filled by private standards of sustainability that regulate the production framework and the quality of goods sold in supermarket chains. These must be coherent with international standards on CSR. This chapter calls for the deepening of the international sustainability agenda to go beyond the

T. M. Megale (✉)
Escola de Administração de Empresas de São Paulo of Fundação Getúlio Vargas, São Paulo, Brazil

A. d. Amaral Junior et al. (eds.), *Sustainable Consumption*,
https://doi.org/10.1007/978-3-030-16985-5_20

classical economic, environmental and social pillars to encompass principles of good governance.

20.1 Introduction

Private standards introduced by non-governmental entities typically take the form of documents, namely a set of written requirements, prepared by private companies related to food safety or human, animal or plant life or health for common and repeated use by that private-sector entity.[1] They are increasingly prevalent in the global marketplace. They are different from private labels, which are constant indicators in the packaging of goods that are produced under the brand of a supermarket to be sold exclusively in its stores. Private standards thus constitute a broader category than private labels, which comprise different and non-binding standards.

Retailers and producers introduce private standards in the same domains where the government imposes public standards, such as safety and quality and the social and environmental aspects of production, retailing and consumption.[2] The standards arise from a demand from the market to establish minimum requirements of quality, safety and sustainability that will extend the standardisation of products and, in principle, facilitate trade.

Private standards are relevant and are continuously being implemented in the grocery retail market because they can reduce consumer uncertainty and information asymmetry about product characteristics[3] and thus guarantee the safety and quality of products and their conformity with environmental sustainability. These standards provide information regarding the characteristics of the final product (e.g., maximum residue levels), production practices in the food supply chain, traceability within the supply chain and the legal liability of the supply chain.[4] Private standards of sustainability are voluntary; meaning, they are non-governmental initiatives that aim to stimulate sustainable production and consumption by simultaneously creating demand for sustainable products and the supply of output to satisfy it.[5]

The proliferation of mandatory technical regulations, voluntary standards and procedures of conformity assessment that aim to protect relevant social values—such as the environment and the health and the safety of consumers—has been accompanied by a mounting concern that such regulations hide protectionist objectives that contravene the goals that have guided the liberalisation of international trade since the formation of the General Agreement on Tariffs and Trade (GATT) in

[1]SPS Committee (2014).

[2]Vandemoortele and Deconinck (2013).

[3]Ibid.

[4]Hammoudi et al. (2009).

[5]Jackson and Komives (2014).

the 1940s.[6] This fear that protectionism—albeit in veiled and subtle forms—is on the rise is quite legitimate. As far as intent is concerned, it is simply not possible to affirm the protectionist nature of all decisions that prevent access of imported products to the internal market. Certainly, it is difficult to deny the legitimacy of efforts to safeguard widely shared values—such as health and safety—in ways that do not signal privileged treatment to a particular group. In practice, however, there has been a restriction on market access as importers have been required to comply with private standards. There is thus no clarity on what to do about measures imposed on markets by non-state actors.

My working hypothesis is that private standards reflected in product packaging do not foster the promotion of sustainable development and sustainable consumption and are distant from the real complexity of sustainability challenges at the global level. Many private standards imposed by supermarkets aim to promote legitimate objectives, but there is frequently a lack of correspondence between the objective and the written description of the standard on product packaging. Private standards, when legitimate and coherent with the values being promoted, can make world trade easier and reduce the costs that are currently produced through multiple registrations, inspections, certifications and labelling processes.

The main goal of the present chapter is to compare the content of private standards of sustainability created by multinational retail companies and the sustainability policies created by these companies where private standards of sustainability are advanced. The comparison will involve the sectors where the standards are created and the extent to which they constitute trade barriers and will allow the identification of standards that protect common values. The question that permeates this research is the following: to what extent do private standards of sustainability imposed by different multinational retail corporations protect human life or health rather than constituting a new form of protectionism?

20.2 The Internationalisation of the Retail Sector and Private Standards of Sustainability

The internationalisation of the grocery retail sector is a product of the limited possibilities of expansion inside domestic markets and from the new and increasing opportunities in developing markets. Emerging countries, in many cases, have been creating new promises for retail sector expansion through changes in the regulatory system, in the conditions of market demand and in the supply structure.

The regulatory system is undergoing a liberalisation and standardisation process through multilateral agreements that have opened the possibility of international expansion, mainly in goods and services. The GATT was the watershed and

[6] Amaral (2011).

contributed towards coherent regulations for international trade in goods. The work of the GATT/WTO has not only triggered multilateral reductions in tariffs on products in international trade, but increasingly with non-tariff measures (NTMs) that also constitute impediments to trade.[7] WTO agreements have created international standards for all kinds of NTMs so that trade concessions negotiated between WTO members will not be undermined.[8] Within the framework of the General Agreement on Trade in Services (GATS), for instance, WTO Member States have agreed transparent national regulations that partly allow foreign competitors access to national markets, while leaving each Member State the right to define sub-sectors of services excluded from this liberalisation.[9] In this scenario, state services and the private retail sector may be protected from international competition. In practice, a high number of Member States follow or apply GATS regulations and the number of developing countries ready to open their markets to foreign capital has increased in recent years.

Plurilateral and bilateral trade agreements have also increasingly allowed retailers to expand to neighbouring countries under competitive conditions. The expansion of EU grocery retailers in Latin America is an example of internationalisation based on these liberalised regulation.[10] In recent years, hundreds of bilateral and regional free trade agreements have substituted and overcome the multilateral GATS settings, which has led to more favourable agreements concerning food import/export and a diversity of service-based activities inside the grocery sector.

Changes on the demand side in emerging markets have also stimulated internationalisation among grocery retailers. The most important conditions identified for the establishment of foreign super- and hypermarkets are the following: expanding market volumes based on increasing household income and a growing middle class; upheavals in social structures and modes of behaviour; increased ownership of refrigerators and vehicles, which open possibilities both for larger shopping volumes and storage; and rising urbanisation, which leads to falls in subsistence production and local market systems being replaced by modern retail formats.[11]

Finally, the change from a fragmented to a more cohesive supply system and the introduction of new grocery retail formats have seen upstream value chains emerge that underpin retail networks on a regional, national and global scale. The prerequisites for these structures are high-quality transportation systems, modern procurement systems and improved domestic commodity supply systems that efficiently deliver goods to new retail formats. Moreover, deeper cooperation and an information transfer between stakeholders inside the chain fosters innovation and trust.

[7]Matsushita et al. (2015).

[8]Ibid.

[9]Kulke and Suwala (2016).

[10]Wrigley and Lowe (2010).

[11]Kulke and Suwala (2016).

Private labels are a core part of the current strategy of international supermarket chains to increase profits. Examples of private labels are Tesco's Nature's Choice, Carrefour's Filiéres Qualité, Marks & Spencer's Field-to-Fork and Auchan's Filiére Controllée. Private labels are increasingly becoming more important as their market share increases and they trigger the delisting of A-brands. Despite some ups and downs, in absolute terms the private label market experienced robust growth between 2007 and 2012, with value sales in markets growing overall by just under US$250 to over US$350 billion.[12] Private label growth has been fostered by the expansion of large grocery retailers and the tendency in the direction of more refined lines that command higher prices and margins. The increase of private label sales can encourage more shelf-space and customer loyalty.

Current private labels are carefully managed and advertised with the aim of improving the retailer's competitive edge. Indeed, some are now perceived as brands in their own right. Labels support product differentiation and marketing from a business perspective, as well as the ability of consumers to make informed purchasing decisions.[13] Retailers are upping the stakes with value-added innovations of their own to improve private label's traditional image of inferior quality and further erode consumer loyalty to established brands.[14] The focus is changing from aggressive promotion and discounting of private labels to offering better quality and a unique offer. Developed markets have recently seen increased private label prices as a consequence of quality improvements, as many retailers created premium ranges that competed directly with branded products.

As briefly explained in the introduction, private standards are documents prepared by non-governmental entities containing written requirements related to food safety, animal or plant life or health for common and repeated use. The expression 'for common and repeated use' implicitly excludes other types of documents for uses inside the non-governmental entity.

20.3 Justification for the Case Selection of International Retailers

The three international retailers chosen for the present research—Casino, Tesco and Walmart—were chosen because of the availability of transparent and credible information about their supply chains and their sustainability practices related to private standards. Casino's website carries comprehensive reports of each year's sales and information about its positions in international markets. Tesco publishes its sustainability policies relative to food waste, responsible sourcing and support of local communities online. Walmart, in the same sense, publishes measures to

[12]Euromonitor (2014).

[13]Meliado (2017).

[14]Euromonitor (2010).

manage risk in its supply chains, its suppliers' standards manual and its responsible sourcing audit process.

Walmart adopts a comprehensive sustainability policy and led the 2008 OECD ranking of transnational retailers. The company's international sales were of US$113 million, a number that exceeded by more than US$20 million the value of international sales of the retailer ranked in second place. The company's approach aims to achieve the following objectives: reduce waste to zero, operate with 100% renewable energy, sell products that sustain resources and the environment and create a sustainable value chain. The main initiatives to attain these goals are support for the measurement and transparency in the supply chain, provision of affordable, healthier and safer food products, reduction in environmental impact and support for the dignity of workers.[15] All these initiatives concern the economic, environmental and social dimensions of sustainable development.

Inside the scope of this research, the two first initiatives deserve to be highlighted. Regarding the first initiative, the company is adopting issue-specific measurement instruments as the CDP (Carbon Disclosure Project) supply chain emissions disclosure program to verify climate and energy risks and to identify possibilities for improvement. The company is also making better product data publicly available to customers through, for instance, the Great for You icon for food in Walmart's US stores. Considering the second initiative, the company is promoting food safety through, for example, the reduction of salmonella in chicken and is stimulating animal welfare and responsible antibiotic use under the internationally recognised five freedoms of animal welfare: freedom from hunger or thirst, freedom from discomfort, freedom from pain, injury or disease, freedom to express normal behaviour and freedom from fear and distress.

Tesco does not occupy a leading position in the 2008 OECD ranking, but was selected for the analysis for two main reasons. The first is its poor and controversial corporate social responsibility record (which will be discussed in the following section) and the existence of its recognised private label, Nature's Choice, worldwide. Considering the first reason, Tesco was criticised for selling vegetables in China that contained pesticides that were either prohibited or exceeded the legal limit.[16] Considering the private label, it was based on the EN 45011 concept and is in conformity with the principles of continuous improvement. The label is inside a long and complex value chain that extends to 11,400 farms, 66 countries, 28 approved auditing companies and 204 approved auditors worldwide.[17]

Casino is also not a leader in the OECD rankings, but is included in this research because of its strong presence in many developing countries, especially in Brazil, where it controls two giant retailers—Extra and Pão de Açúcar—and to its ongoing efforts to consolidate private labels, such as Agriplus and Taeq. The analysis of Casino will show how standards have been received by consumers in emerging

[15]Walmart (2017).

[16]Greenpeace (2011).

[17]Cox (2007).

countries and if they constitute opportunities to optimise the system within a multinational enterprise. Casino's standards will not be analysed in isolation, but as a pillar of the company's policy designed to monitor and improve the social and environmental impacts of its supply chain.

20.4 The Plurality of Private Labels and Standards in Retail

From a theoretical perspective, three types of standards can be identified: food safety standards, food quality standards and social and environmental standards. Food safety standards aim to provide consumers with safe food and are the standards that matter to all consumers.[18] Examples of these standards are the limits imposed on the level of pesticide residues on vegetables and the prohibition of dangerous substances on food. Food quality standards ensure nutritional quality, taste, colour or size meet acceptable criteria of consumer preference or interest.[19] Standards linked to personal lifestyle choices are examples of this category of standards. Social and environmental food standards aim at regulating social and environmental issues and are linked with the ethical values of a society.[20] The prohibition of child labour and limits on carbon dioxide emissions in production process are examples of norms, provisions or measures that are found in certain types of private standards. Consumption and/or production externalities are, thus, present and can be either positive or negative.[21]

The provision to consumers of products that are in conformity with quality and safety standards that go beyond the minimum requirements is necessary to build a retailer's reputation, a key asset for present and future earning flows. Private standards have thus been seen as a way to maintain and even to add to a firm's reputation.[22] The recent advances in economic criminal law and the possibility of imposing public sanctions on corporations has also contributed to the growth and stringency of standards.

Food safety standards receive the highest priority in the sector. Food safety failures are seen to damage reputation and to generate significant negative effects on consumer confidence and thus future sales and earnings.[23] Moreover, food safety failures by one firm may have repercussions on others as consumers decrease trust in

---

[18]Swinnen and Vandemoortele (2011).

[19]Ibid.

[20]Ibid.

[21]Ibid.

[22]Fulponi (2006).

[23]Ibid.

retailers and the capacity of governments to ensure safe food.[24] All firms aim to achieve zero tolerance for food safety failure and, should failure occur, then rapid and complete recall must follow as soon as possible.

Considering food quality standards, taste, smell, texture and other gustative attributes as well appearance and product integrity are considered almost as important to reputation as food safety.[25] In order to increase expectations, a number of companies have created high quality product lines to which higher and stricter private standards are applied. Examples of include Loblaws' President's Choice, Tesco's Nature's Choice, Carrefour's Filiére Qualité, which are niche products with specific characteristics and lines of organic products and fair trade.[26]

The increasing relevance of social and environmental food standards is because of the current context in which consumers become more aware and interested in the way retailers conduct their business both at home and overseas and non-governmental organisations (NGOs) monitor firm behaviours at home and abroad.[27] Global chains with a significant market power have a role on the enforcement of labour codes of conduct outside the home country and need the cooperation of governments and NGOs. Environmental standards required by retailers are applied through farm practices schemes, such as SQF 1000, GlobalGap or the Good Agricultural Practice (GAP) requirement of the Global Food Safety Initiative (GFSI), a standard of food safety that increases transparency and allows a higher efficiency of costs on the food supply chain, and may include the environmental component of manufacturing processes and/or packaging, which are part of good manufacturing practices or GDP.[28]

The description above of the private standards reveals that supermarkets' sustainability standards are created in conformity with international economic, environmental and social standards and with stakeholders' interests. As the supermarkets sell products directly to consumers, their preferences will be highlighted. Five aspects that motivate the preference of the consumer are: good relations with the community, labels of environmental protection, good workplace relations, help in reducing energy consumption and avoiding ill treatment of animals.[29] Retailers use animal well-being as a niche marketing strategy and supermarkets offer products with better standards, such as the elimination of birth cells and battery cages for pigs and cattle.[30]

---

[24]Ibid.

[25]Ibid.

[26]Ibid.

[27]Ibid.

[28]Ibid.

[29]GV CES (2014).

[30]Ibid.

20.4.1 Walmart's Sustainability Policy and Private Standards

Walmart adopted a sustainability policy in its own interest and to meet society's expectations. Regarding the first aspect, the company aims to promote a positive self-image and to preserve or improve its own reputation and related profits in global markets. Regarding the second aspect, socially responsible performance has become an urgent need for businesses since reports of global warming and industrial disasters began to alarm the public about the detrimental and irreversible effects of industrialisation on human life and nature.[31] Socially responsible behaviour—understood as organisational performance in meeting society's expectations by contributing to economic growth and social development while acting within legal and ethical restraints—therefore indicates the growing influence of corporations and underlines increasing stakeholder pressure on business conduct.[32] The main initiatives adopted by Walmart are signing the Sustainability Pact and investment in sustainability products.

The Sustainability Pact is a recent initiative among the Brazilian retailers based on three axes: the Amazon river basin, responsible shopping and management of residues. Walmart Brazil signed the pact in 2009. The results are significant; in 5 years, the initiative has expanded, reaching more than 160 companies, which now commit to reporting their conduct in these three areas. In the Amazon region, the fight against the degrading work conditions in soy, timber and livestock production is a key aspect. Other important efforts are the training, orientation and fostering of certification of suppliers and the verification of origins of raw materials to avoid any link to deforestation of the Amazon. Responsible shopping concerns reviewing the life cycle of products involving the evaluation, on the chain of suppliers, of the development, the productive process, transport, sale and discard based on aspects of innovation, eco efficiency and social and environmental conformity. The management of residues aims to reduce the volume of packaging and single-use plastics, diminish the destination of waste to landfills and stimulate recycling.

Inside the Sustainability Pact, End-to-End Sustainability was an initiative created to expand the role of retail in pursuit of opportunities to reduce social and environmental impacts throughout the whole life cycle of products.[33] During the first 5 years of the Pact's existence, 29 companies participated in the initiative and introduced more than 40 innovative and socially and environmentally responsible products.

In 2015, all the partners that develop products with private labels entered the program of Qualification and Certification to Suppliers of Food. With this movement, the company seeks to obtain the GFSI. When dealing with suppliers, multinational corporations tend to focus regulatory efforts through contracts and

[31]Koca-Helvaci (2015).

[32]Ibid.

[33]Walmart (2015).

agreements with individual suppliers that incorporate the multinational corporation's framework for ethical standards.[34]

Walmart has, to date, provided only a rough outline of the environmental footprint of its suppliers, but it is creating a program that can be a game changer, because it will command important environmental data from some 100,000 global suppliers.[35] The program can lead to a new retail standard for the twenty-first century. The project, which is to be introduced in three phases over several years comprises the following steps: (a) a supplier sustainability assessment; (b) a lifecycle analysis database and; (c) a simple tool for customers.[36]

The topics below illustrate the process of creating and producing goods with the Walmart label:

1. **Inspiration**: specialists identify new products to better satisfy the needs of customers.
2. **Partnership**: suppliers through audits that follow guidelines and international criteria. The aim is to certify that they are in conformity with principles and the company's business strategy, besides guaranteeing quality and safety.
3. **Quality**: products are selected through frequent quality tests, including sensory evidence done by customers and laboratory and performance analysis.
4. **Results**: quality and safety are offered, with lower prices than similar products.

Everything guarantees satisfaction through an intelligent option, namely that Walmart products contain the best trade-off between cost and benefit.

Examples of Walmart brands are Grand Expedition, The Bakery, Ekonomico, Parent's Choice, Special Kitty, Equate, Great Value and Sentir Bem. The two last deserve to be highlighted. Great Value is an international brand that offers a variety of high quality products in the categories of food, house cleaning, clothing treatment and disposable. Sentir Bem reflects Walmart's sustainability strategy. It comprises healthy food available in the Light, Zero, Soy, Integral and Organic lines. These products are aligned to sustainability because of the low environmental impact of the packaging, with orientations to be discarded, information in braille and health tips and information about children rights. The relationship between the Sentir Bem private label and private standards on protection of the environment will be analysed below.

Walmart attained greater transparency in its seafood supply chain mainly because of the certification initiative. This initiative consisted in the adoption of the Marine Stewardship Council's (MSC) certification program for wild-caught fish that sets forth a broad set of certification standards based on the United Nations' Code of Conduct for Responsible Fishing.[37] The certification process is implemented by

[34]Backer (2007).

[35]Cutting et al. (2011).

[36]Ibid.

[37]Denend (2010).

impartial, MSC-accredited certifying bodies that audit fishery and processor compliance to guarantee that goods are managed sustainably from sea to the consumer.

Walmart sustainability policy has, however, come in for criticism recently because of violations of labour and social standards. Different media sources reported that a huge number of lawsuits were filed against Walmart denouncing wage violations, its anti-union stance, inappropriate health care and exploitation of workers. The company was also accused of working with suppliers in emerging markets where labour rights were commonly violated. Serious damage to Walmart's public reputation came from a fire in the Tazreen garment factory in 2012 where hundreds of Bangladeshi workers died while they were producing clothes for Walmart.

20.4.2 Tesco Sustainability Policy and Private Standards

The British supermarket chain Tesco has used private labels to innovate its business. The origins of premium private label brands stem from Europe, particularly the United Kingdom, which has a similarly concentrated grocery industry, where the key players utilise private label brands to exert significant amounts of power over their suppliers.[38] There is widespread consumer acceptance of such products owing to the wide range of offerings.[39]

The use of private labels, however, has not translated into sustainable practices. Recent disasters like the collapse of the Rana Plaza factory in Bangladesh and the 'human trafficking' scandal involving employers in the Thai shrimp industry who supply some of the world's largest retail chains (including Tesco) provoked a reaction from the company. Efforts were made to improve the monitoring and reporting of social capital in supply chains. In this respect, a review revealed that retailers were, in fact, auditing only a small portion of supply chains, omitting those segments where labour and environmental abuses were most likely to take place.[40]

Tesco sustainability policy covers several different areas, including: climate change, forests, the marine environment, farmland, fresh water and recycling. Concerning its environmental policy, Tesco established targets for carbon emissions reductions for 2020 and for 2030. These targets balance the need to produce more food to feed a growing global population and the need to embrace restorative farming practices and reduce field emissions. Tesco is working with its suppliers to achieve these targets through its dedicated Tesco Supplier Network where it can collaborate on carbon reduction measures.[41] Regarding forests, the company is working hard to regulate the four global drivers for deforestation that are relevant

---

[38] Sutton-Brady et al. (2017).

[39] Ibid.

[40] Lebaron and Lister (2015).

[41] Tesco (2017).

for its business: palm oil, cattle products, soy and timber. For each commodity, Tesco is mapping its supply chains to understand its exposure and putting in place sustainable procurement policies.[42] Analysing marine policy, to guarantee the future of the seas and the industries that depend on them, Tesco sources its seafood sustainably. On this issue, Tesco is working together with the MSC to increase its range of certified sustainable counter, pre-packed and frozen fish.[43] The company expects to have almost one hundred MSC-certified seafood products in its stores and is working to keep adding more MSC lines. Tesco also works with the Sustainable Fisheries Partnership (SFP) to map and risk assess all its seafood supply chains.[44] This collaboration with SFP enables the company to guarantee that all its wild-caught seafood is going through the process to be certified by MSC. The Aquaculture Requirements were created for farmed fish. These requirements outline standards that apply to all its own brand aquaculture producers to ensure good farming practices and address animal health and welfare, environment, ethics and feed.[45] The company also participates in sector-wide improvement initiatives as the work with fishermen, processors, brands and other retailers to take the iconic North Sea cod fishery to MSC certification and participates in the Shrimp Sustainable Supply Chain Taskforce in Thailand.[46]

Considering farmland, a huge variety of Tesco fresh food is cultivated and produced all over the world. The indirect impact derived from global agricultural supply chains is addressed through the stipulation of targets. In order to achieve a sustainable footprint on climate, water and biodiversity from its biggest agricultural products, by 2020 the company aims to: (a) reduce its greenhouse gas (GHG) emissions by 7%; (ii) reduce its water use including local reduction targets for water-stressed areas; and (iii) improve farmland biodiversity (i.e., soil health, pollinators and off-field biodiversity).[47] An essential part of these efforts is guaranteeing that the arable and livestock standards that apply throughout its supply chains contribute to meeting its environment goals and align with industry best practice. The company has also begun to work with some suppliers access tools and strategies that measure impact and lead to improvements in key agricultural products.

Moreover, in April 2016, Tesco joined the Sustainable Agriculture Initiative (SAI) Platform, a non-profit organisation that facilitates the sharing of knowledge and best practice in sustainable agriculture.[48] The participation in SAI Platform will allow the company to dialogue with like-minded corporations and help it define, conduct and implement environmental best practice in its supply chains. The main

[42]Ibid.

[43]Ibid.

[44]Ibid.

[45]Ibid.

[46]Tesco (2017).

[47]Ibid.

[48]Ibid.

sources of best practice are growing standards and guidance and on-the-ground collaboration.

Finally, there is the issue of fresh water. Tesco is monitoring its main supply chains to determine their exposure to water risk. This includes overlap with environmentally sensitive river basins.[49]

Tesco developed the standard Nature's Choice, which all its fresh produce growers around the world must comply with, to supply the supermarket with fresh fruit and vegetables. The standard aims to guarantee that its high-quality fresh products come from growers who apply sound agricultural practices, operate in an environmentally responsible manner and with adequate consideration for the health and wellbeing of their staff. The Nature's Choice code of practice lays down requirements that growers must fulfil. One of the main reasons for the creation of Nature's Choice was the expectation of Tesco's consumers that the company ensures that fresh products are grown safely and with reduced environmental impact. Tesco was the first supermarket to create a formal code of practice in 1991, which originally aimed to regulate chemical usage and to progressively develop environmentally sustainable production standards for its growers. Since then, the supermarket has revised it to become a standalone international framework. The standard is coherent and consistent for farmers all over the world, a feature that encourages continuous improvement. The standard encompasses all aspects of farming practice and aims to promote the best agricultural practice. A Technical Advisory Committee was created to guarantee that Nature's Choice is independent and remains current. CMi is the body responsible to operate as the registrar for this framework on behalf of Tesco and to manage the grower registration and certification process.

The standard comprises the following sections: rational use of crop inputs such as fertilisers and plant protection products through proper control of inputs, pollution prevention through the reduction of environmental pollutants, wildlife and landscape prevention through conservation plans, recycling, re-use and energy conservation through water and energy efficiency and protection of human health through the reduction of risks of accidents and ill health.

The following numbers reveal Tesco's sustainability policies and the impact of the company on the environment:

1. 93% of the whole palm oil used in all its own-brand UK products comes from a certified sustainable source
2. 99% of its own-brand food products is from a sustainable source
3. 60% of the supply chain carbon footprint and 97% of the water footprint have been reduced.

Tesco's operations, however, have suffered criticisms because of the violation of environmental and human rights norms. The horsemeat scandal, which saw reporting on the presence of horsemeat in burgers on shelves of some global retailers also reached Tesco. The scandal severely dented consumer trust in the food

[49]Ibid.

industry.[50] In order to recover reputation, Tesco assured the public that would enhance scientific assessment through the introduction of a world-class DNA testing system and would reformulate its food procurement strategies.

Furthermore, Human Rights Watch (HRW) found that many European companies that publicly embrace international labour standards in practice undermine workers' rights in their US operations. Tesco is one of the companies mentioned in HRW's report 'A Strange Case: Violations of Workers' Freedom of Association in the United States by European Multinational Corporations'. The report explains how some European multinational firms have carried out aggressive campaigns to prohibit workers in the United States to organise themselves and bargain, violating international standards and frequently US labour laws.[51] HRW found that Tesco management sought to muzzle workers' communication among themselves about unions.[52]

HRW had already expressed concern with working conditions in the facilities owned and operated by Tesco, which, together with four other large retailers, launched a new Global Social Compliance Program (GSCP). The omission of working conditions from the program created the situation in which GSCP member companies' direct employees enjoy fewer rights protections than those labouring in the supplier facilities.[53]

20.4.3 Casino Sustainability Policy and Private Standards

Since 2002, Casino has developed within the organisation a service of specialists that work on sustainable development. The company works on many projects, aiming to reduce its environmental impact and has created a charter of commitments. The measures foreseen in the charter favour products that respect the environment and are the developed through the offer of responsible products, such as Casino Bio and Casino Commerce Equitable. There firm committed to communicate the environmental impact of its products through the inclusion of the so-called index of Carbon Casino (discussed in detail below) on product packaging, improve the nutritional quality of its products through reductions in fat, sugar and salt, and reduce the amount of packaging of its products.[54]

The company's sustainability policy aims to ensure that each customer can choose products that are both affordable and friendly to people and the environment. Casino's major sustainable development challenges are the following: encouraging responsible forms of production and consumption, taking action to improve

[50]Tse et al. (2016).

[51]HRW (2010).

[52]Ibid.

[53]HRW (2007).

[54]Casino (2017).

nutrition, combating climate change, optimising waste management, meeting the challenges of sustainable construction and encouraging respect for human rights in the workplace.[55]

In 2008, Casino launched the index Carbon Casino, the first complete environmental label in France. The impact category that is clearly stated is the amount of greenhouse gases emitted from each product during five key life cycle stages of each Casino brand product: production, manufacture, transport from the field to the Casino warehouses, packaging from raw material extraction to recycling and distribution from the Casino warehouses to the consumer's home.[56]

The main objectives of the initiative are twofold. First, the provision to consumers of clear, relevant information about the environmental impact of the products they consume daily. Second, improvement of the environmental quality of Casino products through a global project actively involving all areas of the company—including purchasing, quality assurance and stores—and also in partnership with suppliers, major corporations and small- and medium-sized businesses.[57]

The main instrument to reach the reduction of carbon emissions is the influence of consumer conduct. Innovation is at the heart of the procedure of sustainable development of the group and represents the willingness of the company to respond as closely as possible to the expectations of its customers. The main impacts of the initiative are the use of the carbon index by the consumers and the increased competition between suppliers that can motivate them to further reduce their carbon footprints.[58] The index is fully financed by the Casino group so there are no direct costs or burdens of agricultural producers.[59]

The main benefits derived from the trade of products with the index are the following: beneficial relationships with stakeholders, consumers and French authorities, favourable coverage in numerous newspapers, on television and radio, the presentation of the initiative to the United Nations and the explanation of the new label at schools, famer's cooperatives and packaging or product manufacturers.[60]

Casino has also created two eco-citizen private labels: Casino Bio and Terre et Saveur. The products in the Casino Bio line contain at least 95% of ingredients that come from biological agriculture. The acquisition of products in the Casino Bio line contributes to the protection of the environment, because biological agriculture prohibits the use of chemical products, protects natural equilibria and guarantees animal well-being.[61] Terre et Saveur is a Casino label that was born through a partnership with farmers and quality-conscious seafood and fresh-water producers.

---

[55]CSR Europe (2013).

[56]Tzilivakis et al. (2012) and CSR Europe (2013).

[57]CSR Europe (2013).

[58]CSR Europe (2013).

[59]Tzilivakis et al. (2012).

[60]CSR Europe (2013).

[61]Casino (2017).

According to its label, Casino prefers modes of production that contribute to the protection of the environment and of the needs of plants and animals.[62]

The company also adheres to eco-citizen private standards. Casino has created a range of cleaning products certified with Ecolabel, the European Union ecological standard that guarantees reduced impact on the environment from the beginning to the end of the life cycle of the product.[63] The company was the first French distributor to trade frozen fish with the MSC standard.[64] The company offers pine garden furniture and eucalyptus with the Forest Stewardship Council FSC) standard.[65] Finally, Casino offers products with the standard FairTrade Max Havelaar. The association Max Havelaar France aims to develop and promote products that come from equitable trade.[66]

Besides the index, Casino launched a program known as Casino Avenir, which is a guide for the consumer to identify the company's responsible products, services and actions. The program helps consumers to identify the impact on ecology, citizenship or fair trade among the products of different areas of Casino.[67] Further, in 2016, Casino enlarged its group of agro-ecological products with the offer of pork meat guaranteed without treatment of the animal with antibiotics from birth. Finally, Easydis is Casino's logistics subsidiary group and employs sustainable modes of transport. Casino develops with specialised companies silent and less polluting refrigerated trucks to city centres.

20.5 The Search for Coherence Between Supermarkets' Sustainability Standards and Attainment of Common Values

Corporate social reporting refers to the issue of standalone reports that provide information regarding a company's economic, environmental and social performance.[68] These reports constitute a relevant source to identify CSR practices. However, they go beyond sustainability policies to indicate the results of the application of those policies.

Casino's 2016 annual report, Tesco's 2016 update on corporate responsibility commitments and Walmart's 2017 Global Responsibility Report unify their language on economic, environmental and social topics. The main economic issues identified are increased opportunities for small and medium business and stimulus

[62]Ibid.

[63]Casino (2017).

[64]Ibid.

[65]Ibid.

[66]Ibid.

[67]Casino (2017).

[68]Carroll and Shabana (2010).

for local farms. The main environmental issues are the increase of energy performance and the reduction of emissions of pollutants and of greenhouse gases. The main social issues are the protection of labour rights and the promotion of the company's accountability.

The identification of true sustainable elements in supermarkets' policies cannot be limited to a comparison of their reports on international standards of CSR, however. Such a methodology would risk hiding the dark sides of generic words that aim to reveal apparent protection of the environment and social and economic rights. Information collected on international independent databases is necessary for a full understanding of supermarkets' sustainability policies.

The empowerment of consumers by providing them information through mandatory reporting on social and environmental performance and the development of a social or CSR label is a recent phenomenon. It is true, as well, that an increasing number of consumers have become more sceptical of the environmental impact of the products they buy.[69] The main criticism concerns vague claims made by corporations. Consumers believe that companies portray themselves to be environmentally friendly and engage in green washing to increase sales and raise prices.[70] Companies are, thus, beginning to add more detailed performance information in their reports instead of only their mission and objectives.

Considering external transparency, business ethics are a key driver for all firms. At the same time, internal transparency through environmental management accounting has been on the rise, with a business rationale to improve efficiency and reduce costs, in relation to using natural resources, reducing emissions to air water and soil, as well as in reducing waste.[71] More recently, firms have come to see the relevance of sustainability for their long-term business continuity and identify sustainability challenges as a business growth driver.[72]

In recent years, in a context in which non-financial information is becoming more relevant in company valuation, international indexes have been created by independent organisations to evaluate the sustainable performance of different multinational corporations. These organisations require information from companies to produce rankings of the most sustainable companies encompassed by the index. These indexes are valuable as investors use sources that go beyond corporate reporting to make their investment decisions as publicly available information on the target companies, third-party databases and responses to questionnaires sent directly to companies.[73] Much of the data from data sets essentially reflect the state of the art rather than environmental policy outputs, policy regimes, institutions and

[69]Kamp-Roelands (2013).

[70]Ibid.

[71]Kamp-Roelands (2013).

[72]Ibid.

[73]Ibid.

administrations, but it might nevertheless be useful as indicators of environmental policy outcomes or environmental governance capacity.[74]

The absence of internationally-agreed standards for reporting on environmental performance at company level leads to variations in methodologies and indicators being used. Current measurement and reporting frameworks relevant to support sustainable growth include: the Sustainability Consortium; the International Integrated Reporting Council; the Guide to Corporate Ecosystem Valuation; the Greenhouse Gas Protocol; and the Climate Change Reporting Framework. The Sustainability Consortium is an effort of retailers, manufacturers and NGOs that have joined efforts to develop a sustainability measurement and reporting system for major product categories sold by retailers.[75] Sustainability profiles are attributed to each company, but, among the supermarkets analysed in the present research, only Walmart is under the scrutiny of Sustainability Consortium.

The Greenhouse Gas Protocol is an international accounting tool created by the World Business Council for Sustainable Development and the World Resources Institute that provides standards to measure and manage emissions. The corporate standard provides requirements and guidance for companies preparing a corporate-level GHG emissions inventory. The corporate value chain (Scope 3) standard allows companies to assess their entire value chain emissions impact and to identify where to focus reduction activities. The product standard is used to understand the full life cycle emissions of a product and to focus efforts on the greatest GHG reduction opportunities. The project protocol is a policy-neutral accounting tool for quantifying the greenhouse gas benefits of climate change mitigation projects.

Tesco has discussed the important role Scope 3 and Product Standards play in revealing new opportunities for GHG reductions. Walmart's representatives participated in a panel on smart thinking in delivering significant supply chain emissions reductions of greenhouse gases in 2012. Walmart Brazil was one of the first companies to voluntarily measure and publicly report its greenhouse gas emissions using the Brazil GHG Protocol Program.

It is also worth mentioning that Walmart created its own sustainability index with the objective of fostering progress on identified sustainability hot spots along global supply chains. After compiling the index results, scorecards that highlight key hot spots by category are produced and suppliers can see how they rank relative to the field and gain insight into improvement opportunities for each of the categories they supply.[76] Suppliers of laundry detergent, for instance, can see how they are progressing relative to all other suppliers on chemicals of concern; in beef, suppliers can see how they are doing relative to all other suppliers on animal welfare or emissions.[77] The index metrics are continuously refined by the Sustainability

[74]Duit (2016).

[75]Kamp-Roelands (2013).

[76]Walmart (2017).

[77]Ibid.

Consortium to align the index with those developed by organisations such as CDP and SAC.

20.6 Conclusion

Recent years have witnessed the rapid growth in the number of non-state, private legal regimes that create their own substantive law. In practice, multinational corporations focus regulatory efforts through contracts and agreements with individual suppliers that incorporate the multinational corporation's framework for ethical standards. The locus of authority is, thus, within the corporate sector and reflects a corporate–economic rationality. The strategy of many multinational corporations to create informal lawmaking forums to offer a low-cost path to create competing rules aims to undermine, change and reorient substantive legal provisions with which they disagree and advocate for legal norms that fit their substantive preferences. In terms of private standards of sustainability, codes of conduct, standards for social reporting and auditing and product labelling standards have been created and disseminated.

The regulatory vacuum in the regulation of NTMs fosters a debate in the governance field about the creation of institutions and formal mechanisms that can sanction abusive practices on economic, environmental and social issues disguised under standards with legitimate objectives. Examples of abusive practices are the aforementioned violations of human rights norms. Private regulation is present in the shadow of action by the State. In circumstances in which the State does not act, it creates a generic regulatory framework or general policy objectives and private actors regulate the implementation of measures themselves. Private standards have then an impact on the access to market of products that do not comply with pre-established requisites.

Private standards, however, cannot only be understood as potential regulatory barriers that reflect the new protectionism characterised by overlapping technical, sanitary and phytosanitary measures. They also constitute measures of corporate social responsibility. The supermarkets' sustainability reports reflect the increasing level of demand on matters of quality and safety and on protection of economic, environmental and social values.

The analysis of most recent Walmart sustainability report and its presence in the debate and the application of international indexes as Sustainability Consortium and Greenhouse Gas Protocol allows us to conclude that Walmart, through its private standards of sustainability, reveals a greater commitment to the protection of human life and health than Tesco and Casino. These two, on the contrary, tried to demonstrate the accomplishment of their sustainability policies mainly through sustainability reports. Tesco only reluctantly participated in a debate about the Greenhouse Gas Protocol's standards. These facts allow to us conclude that the sustainable function of Tesco's and Casino's private standards of sustainability is not clearly identifiable and justified and that these standards can constitute disguised restrictions to trade and create arbitrary and unjustified discrimination.

References

Amaral A Jr (2011) Curso de Direito Internacional Público. Atlas, São Paulo
Backer LC (2007) Economic globalization and the rise of efficient systems of global private law making: Wal-Mart as global legislator. Conn Law Rev 39(4):1739–1784
Carroll AB, Shabana KM (2010) The business case for corporate social responsibility: a review of concepts, research and practice. Int J Manag Rev:85–105
Casino (2016) 2016 Annual and Corporate Social Responsibility Performance Report. Available via: https://www.groupe-casino.fr/en/wp-content/uploads/sites/2/2017/06/RA-2016-GB.pdf. Accessed 15 July 2018
Casino (2017) 2017 Annual and Corporate Social Responsibility Performance Report. Available via: https://www.groupe-casino.fr/en/wp-content/uploads/sites/2/2018/05/2017-Annual-and-Corporate-Social-Responsibility-Performance-Report.pdf. Accessed 15 July 2018
Cox S (2007) Tesco Nature's Choice. An Example of a Private Scheme. Available via: https://www.wto.org/english/tratop_e/sps_e/...e/tesco_e.ppt. Accessed 15 July 2018
CSR Europe (2013) Casino—Casino Carbon Index. Available via: https://www.csreurope.org/casino-casino-carbon-index#.WpAlPa6nEdU. Accessed 15 July 2018
Cutting R, Cahoon L, Flood J, Horton L, Schramm M (2011) Spill the beans: good guide, Walmart and EPA use information as efficient, market-based environmental regulation. Tulane Environ Law J 24:291–334
Denend L (2010) Walmart's Sustainability Strategy. Available via: https://services.hbsp.harvard.edu/services/proxy/content/61806372/61806582/9c314cb833a47dc3000142beed0abc05. Accessed 15 July 2018
Duit A (2016) Environmental Politics among Advanced Industrial Democracies. Available via: http://www.oxfordbibliographies.com.sbproxy.fgv.br/abstract/document/obo-9780199756223/obo-9780199756223-0141.xml?rskey=C8jE7f&result=1&q=duit#firstMatch. Accessed 15 July 2018
Euromonitor (2010) Do Brands Still Matter? Focus on Private Label Packaged Food. Available via: http://www.euromonitor.com/do-brands-still-matter-focus-on-private-label-packaged-food/report. Accessed 15 July 2018
Euromonitor (2014) The New Face of Private Label: Global Market Trends to 2018. Available via: http://www.euromonitor.com/the-new-face-of-private-label-global-market-trends-to-2018/report. Accessed 15 July 2018
Fulponi L (2006) Private voluntary standards in the food system: the perspective of major food retailers in OECD countries. Food Policy 31:1–13
Greenpeace (2011) Tesco produce in China unsafe for three years now. Available via: http://www.greenpeace.org/eastasia/press/releases/food-agriculture/2011/tesco-pesticides-more-testing/. Accessed 15 July 2018
GV CES (2014) Comemos melhor ou pior? Página 22:81
Hammoudi A, Hoffman R, Surry Y (2009) Food safety standards and Agri-food supply chains: an introductory overview. Eur Rev Agric Econ 36(4):469–478
Human Rights Watch (2007) Eliminate Double Standard in Code. Available via: https://www.hrw.org/news/2007/07/16/eliminate-double-standard-code. Accessed 17 July 2018
Human Rights Watch (2010) A Strange Case: Violations of Workers' Freedom of Association in the United States by European Multinational Corporations. Available via: https://www.hrw.org/sites/default/files/reports/bhr0910web_0.pdf. Accessed 16 July 2018

Jackson A, Komives K (2014) Introduction to voluntary sustainability standard systems. In: Schmitz-Hoffmann C, Schmidt M, Hansmann B, Palekhov D (eds) Voluntary standard systems: a contribution to sustainable development. Springer, Berlin, pp 3–19

Kamp-Roelands N (2013) Private sector initiatives on measuring and reporting on green growth. Available via: http://www20.iadb.org/intal/catalogo/PE/2013/11878a05.pdf. Accessed 15 July 2018

Koca-Helvaci ZC (2015) Walmart and its employee relations: organizational stance-taking and legitimacy. Horizon 23(4):374–386

Kulke E, Suwala L (2016) Internationalisation of grocery retailing in the global south: general conditions, formats and spatial expansion patterns of selected MNEs. J Geogr Soc Berlin 147 (3):187–200

Lebaron G, Lister J (2015) Benchmarking global supply chains: the power of the 'ethical audit' regime. Rev Int Stud 41(5):905–924

Matsushita M, Schoenbaum T, Mavroidis P, Hahn M (2015) The World Trade Organization. Law, practice and policy. Oxford University Press, New York

Meliado F (2017) Private standards, trade and sustainable development: policy options for collective action. International Centre for Trade and Sustainable Development, Geneva

SPS Committee (2014) Report of the Co-stewards of the Private Standards E-Working Group on Action 1. Available via: https://docs.wto.org/dol2fe/Pages/FE_Search/FE_S_S009-DP.aspx?language=E&CatalogueIdList=130672,127571,125295,124840,123363,123250,122636,122660,121483,121398&CurrentCatalogueIdIndex=4&FullTextHash=1&HasEnglishRecord=True&HasFrenchRecord=True&HasSpanishRecord=True. Accessed 17 July 2018

Sutton-Brady C, Taylor T, Kamvounias P (2017) Private label brands: a relationship perspective. J Bus Ind Mark 32(8):1051–1061

Swinnen JFM, Vandemoortele T (2011) Trade and the political economy of food standards. J Agric Econ 62(2):259–280

Tesco (2016) Tesco 2016 update on corporate responsibility commitments. Available via: https://www.tescoplc.com/media/391787/corporate-responsibility-update_nov-2016-final.pdf. Accessed 16 July 2018

Tesco (2017) Reducing our impact on the environment. Available via: https://www.tescoplc.com/little-helps-plan/products-sourcing/reducing-our-impact-on-the-environment/. Accessed 15 July 2018

Tse YK, Zhang M, Doherty R, Chappell PJ, Garnett PR (2016) Exploring the hidden patterns in tweets: a study of product recall scandal. Paper presented at International Society for Business Innovation & Technology Management Conference, University of Michigan, Ann Arbor, United Kingdom

Tzilivakis J, Green A, Warner D, McGeevor K, Lewis K (2012) A framework for practical and effective eco-labelling of food products. Sustain Account Manage Policy J 3(1):50–73

Vandemoortele T, Deconinck K (2013) When are private standards more stringent than public standards? Am J Agric Econ 96(1):154–171

Walmart (2015) Sustainability Report 2015. Available via: https://www.walmartbrasil.com.br/wm/wp-content/uploads/2015/08/Relatorio-de-Sustentabilidade-2015-EN.pdf. Accessed 15 July 2018

Walmart (2017) Walmart 2017 global responsibility report. Available via: https://corporate.walmart.com/2017grr. Accessed 15 July 2018

Wrigley N, Lowe M (2010) The globalization of trade in retail services. Available via: http://www.oecd.org/tad/services-trade/46329746.pdf. Accessed 15 July 2018

Part V
Sector-Specific Approaches II: National Law

Chapter 21
Reasonable Credit in Canada: An Attempt to Avoid Over-Indebtedness

Marc Lacoursière

Contents

Abstract Consumer indebtedness may lead to meltdown, for the consumer and the financial system more generally. Overly generous access to credit and consumer indebtedness have been identified as one important reason for the 2007–2009 financial crisis. Admittedly, Canada passed through this crisis without any serious problems. However, growing indebtedness among Canadian households, fuelled mainly by increasing over-consumption and low interest rates in recent years, remains a continuing problem for the Canadian financial system. Facing the concerns of the Superintendent of Financial Institutions, recent steps have been taken by the Canadian and provincial regulators to address this issue. Unfortunately, despite these measures, reasonable credit, and hence, sustainable consumption of credit, has yet to emerge.

M. Lacoursière (✉)
Faculty of Law, Université Laval, Quebec City, QC, Canada
e-mail: Marc.Lacoursiere@fd.ulaval.ca

A. d. Amaral Junior et al. (eds.), *Sustainable Consumption*,
https://doi.org/10.1007/978-3-030-16985-5_21

21.1 Introduction

Consumer indebtedness may produce two negative outcomes: the meltdown of individual consumers, and/or the meltdown of the financial system itself. Overly generous access to credit, in particular for high-risk (subprime) borrowers—boosted by an expansionist monetary policy that led to a drastic drop in interest rates in 2000 following the burst of the dotcom bubble—was clearly one of the causes of the financial crisis of 2007–2009. However, this over-supply of credit would not have affected the US financial sector so significantly had it not been for a range of other factors, essentially stemming from the deregulation of financial markets in the 1980s and, especially, the 1990s. This trend, which is not without parallels to the 'free banking' episode that peaked in the nineteenth century, led to the development of financial innovations whose impact escaped the attention of the regulators and contributed to the most recent financial collapse. These include the securitisation of debt, where high-risk mortgage debt was used to back securities on monetary markets. Investors' trust in the quality of the securities was based on a sometimes overly accommodating valuation by the rating agencies. Last, the disregarding of rules of conduct and deficiencies in the oversight of market players also contributed to the financial chaos.

Canada came through this worldwide financial turmoil relatively unscathed, unlike other industrialised nations, including the United States, which was in the eye of the storm. Cultural, organisation and regulatory factors explain Canada's success. More specifically, these factors helped create a strong, effective regulatory environment, both in terms of regulations per se and in terms of supervision and control. Simply put, while the economic, political and regulatory situation of the 2000s encouraged US bankers to take enormous risks, their Canadian counterparts tended to be more prudent. Although Canada achieved a good overall outcome, it is important to mention the constant growth in the indebtedness of Canadians in recent years, due in particular to the ease of obtaining credit. This situation, which results from inadequate supervision of present-day financial practices, has become the Achilles heel of the Canadian financial sector.

Since the alarm was raised by the Superintendent of Financial Institutions[1] and by the Governor of the Bank of Canada,[2] a few measures have been implemented. They include actions by the federal parliament and the authorities that exercise supervision and control at the federal (Canada) and provincial (Quebec) levels. The authorities, aware of the consequences of the US financial crisis and, above all, of the level of consumer debt, are currently strengthening their oversight of credit operations.

This chapter presents an analysis of the legislative and regulatory approach to real estate credit in Canada and Quebec, designed to encourage a responsible approach to borrowing. First, we will look briefly at the background to the financial crisis, which has led to major legislative and regulatory reforms in Canada and Quebec (Sect.

[1]Dickson (2012), pp. 5–6.

[2]Governing Council of the Bank of Canada (2012), p. 27.

21.2). Second, we will look more specifically at the protection measures implemented by Canada and Quebec (Sect. 21.3).

21.2 Rationale for the Supervision of Real Estate Credit

The supervision of real estate credit in particular remains a key problem in Canada, although several deficiencies concern consumer credit in general.[3] Except for a recent intervention by the Superintendent of Financial Institutions, few legislative actions can be noted, at least until the onset of the financial crisis in 2008–2009. In this section, we will look briefly at the question of personal over-indebtedness (Sect. 21.2.1) and the impact of the crisis itself (Sect. 21.2.2).

21.2.1 Personal Over-Indebtedness

In Canada, consumer credit is distributed by four different intermediaries: deposit institutions (banks, credit unions and trust companies, sometimes referred to as 'banking intermediaries' or 'near banks'); other financial institutions specialising in consumer loans; commercial enterprises (superstores, furniture stores, automobile dealerships, in particular, known as 'non-banking intermediaries'); and alternative lenders (pawnbrokers, microcredit). These financial and commercial intermediaries are all involved in the move towards over-consumption and over-indebtedness. Consumers with little experience of the complex world of financial services are especially at risk.

Consumer credit is a relatively broad concept. Basically, and in general, consumer debts fall into two groups: personal debt—made up of personal loans such as lines of credit, credit cards and automobile loans—and debts secured on property, namely, mortgages (known as 'hypothecs' in Quebec). Consequently, it is important to distinguish between consumer credit (strictly defined) and real estate credit.

The problem facing the consumer is even more complex, for two reasons. First, there is leakage between consumer credit and real estate credit. For example, a consumer in debt because of a high credit card balance may find it hard to make mortgage payments, and vice versa. Next, Canadian financial institutions have, in recent years, offered home equity lines of credit, or lines of credit secured on immovable property. Once a mortgage has been partially repaid, the consumer can borrow again using a line of credit secured by the equity built up in the mortgaged property.[4] Originally, the concept was designed to give consumers easy access to a

[3]This chapter focuses on mortgage credit.

[4]The maximum amount available to be borrowed is a loan-to-value ratio of 65%: Office of the Superintendent of Financial Institutions [OSFI], Guideline B-20: Sound Business and financial

loan that would allow them to repair or renovate their home. However, consumers began to use their home equity line of credit for other, unrelated expenses, such as automobile purchases, vacations, credit card payments and even, in some cases, regular mortgage payments.[5] This link between the two forms of credit emerged in particular during the financial crisis of 2008–2009, as we will see below.

The solution to over-indebtedness is both straightforward and complex. In the final analysis, is the consumer or the lender responsible for the consumer's over-indebtedness? Each case is different, but it is apparent that, in general, responsibility must be shared. The average consumer needs to be aware of the risks associated with over-consumption and over-indebtedness,[6] and this awareness could be facilitated, in part, by better financial education. In keeping with this approach, the Task Force on Financial Literacy has made several appropriate suggestions to improve the knowledge and financial skills of consumers.[7] However, lenders still have an obligation to fully consider a borrower's ability to repay a loan when granting credit, and this obligation is not always respected in actual practice.

This is why stronger regulation of real estate credit is relevant today. Credit is not necessarily bad, but the risk of misuse means that legal boundaries must be set. Until recently, this obligation was met by the in-house practices of financial institutions, in particular because of the pressure exerted by the banking lobby, but the financial crisis of 2007–2009 drastically changed the situation.

21.2.2 The Financial Crisis of 2008–2009

Canadian banks sailed through the financial crisis of 2008–2009 in exemplary fashion, unlike most banks in other Western countries. According to the World Economic Forum, Canada ranked first worldwide for the solidity of its banking system in 2008–2009.[8] Canada's success, commended by many international authorities, can be traced to several factors. First, the structural organisation of the Canadian banking system makes it stronger than the US system, because despite the small number of Canadian banks overall, each institution can operate throughout the country. In contrast, the legal structure of the US banks is based on a system that encourages a large number of banks that, until recently, were prohibited from becoming established in more than one state. Second, the regulation of financial

---

Practices, October 2017, online: http://www.osfi-bsif.gc.ca/Eng/Docs/B20_dft.pdf [OSFI], "Guideline B-20". Practically speaking, the interest rate is usually above the mortgage rate.

[5]Bailliu et al. (2011–2012), p. 22.

[6]The Supreme Court of Canada has defined the 'average consumer' as a person who is vulnerable, credulous and inexperienced: *Richard* c. *Time Inc.*, [2012] 1 S.C.R. 265, 2012 SCC 8. For an analysis of this concept, see e.g., Lacoursière (2011).

[7]Task Force on Financial Literacy (2011).

[8]World Economic Forum (2008), p. 456.

institutions, combined with strong capitalisation and close supervision and control, made it possible to avoid some of the traps experienced in other countries. Last, the management culture of Canadian bankers is conservative by nature, meaning that decision makers are not inclined to take inordinate risks and generally comply with legislation. This contrasts with the culture of US bankers that is clearly more liberal, as reflected in the multitude of high-risk operations they undertake.[9]

Unlike the United States, where the crisis developed, erupted and caused widespread financial injury—not to mention the related economic, political and social damage—Canada came out relatively unscathed. It is generally accepted that the causes of the crisis were excessive consumer indebtedness, volatile interest rates in the 2000s (in particular following the bursting of the technology bubble), an expansionist monetary policy, the development of structured products, the actions of the rating agencies, a culture of risk-taking among bankers, the liberalisation of financial markets, and the shadow banking system. Conversely, a system of 'credit intermediation involving entities and activities outside the regular banking system'.[10] The latter example is unfortunately well known because of mortgage loans to high-risk borrowers and the securitisation of financial instruments on monetary markets backed by high-risk credit. Simply put, the impact was felt both in the shadow banking system (unregulated) and in the traditional banking system.

It is important to note a danger specific to traditional bank financing—household debt—that clearly played a role in the great financial crisis, although it was not extensively documented by the experts. The International Monetary Fund expressed its concern about household debt and its procyclical effect on macroeconomic variables, with the potential to cause economic shock.[11]

In both Canada and the United States, consumer over-indebtedness has been a cause of concern for several years. According to a recent survey by Statistics Canada, indebtedness in Canada, measured by the ratio between household debt and personal disposable income, has risen to over 150%, just above the rate in the United States.[12] Some people have claimed that borrowers are entirely to blame for their indebtedness—because they are unable to plan their budget and consume more than they can afford—and were in fact responsible for the financial crisis of the 2000s.[13] Given the unequal relationship between borrowers and lenders,[14] this minority position is questionable, to say the least. Instead, it is necessary to consider the complex set of reasons that underlie the problem: the excessive generosity of

---

[9]Ibid.

[10]European Commission (2012). See also: Financial Stability Board (2011), p. 1.

[11]International Monetary Fund (2012), p. 97.

[12]Statistics Canada (2012), p. 4. More precisely, the debt is distributed unequally among borrowing groups, whether it has to do with their income, their education or their region of living. See: Statistics Canada (2012), p. 6 (Table 1). In 2010, this indebtedness was 146.5% in Canada and 122.5% in the United States. See: OECD (2013), p. 28.

[13]On this topic, see e.g., Flávio de Oliveira and Moreira dos Santos Ferreira (2012), p. 31.

[14]French sociologist Bourdieu refers to capital inequality (*distribution inégalitaire du capital*). See: Bourdieu (2000), p. 244ff.

bankers in granting loans[15]; the excessive amounts borrowed by consumers, caused in particular by a society based on over-consumption[16]; job losses among consumers; and deficiencies in the legislative framework.

On the latter point, governments have been attempting since 2008 to correct the deficiencies by sporadic, targeted interventions, but these legislative approaches and the supervision system need more work, as we will see below.

21.3 Supervision of Real Estate Credit

The legal supervision of real estate credit, in contrast to consumer credit, took time to emerge in any effective form. The financial crisis, which we summarised briefly in the previous section, reversed this trend, by introducing a fairer balance between consumers and lenders. We will now look at the legislative measures proposed or adopted following the financial crisis (Sect. 21.3.1), along with the governance standards recently adopted by the supervisory authorities in Canada and Quebec (Sect. 21.3.2).

21.3.1 Legislative Measures

Both levels of government, federal and provincial, have taken steps to supervise consumer credit. Their reasons for taking action diverge but the consequences are convergent. In general, the federal government is concerned by the country's financial stability, and one of the variables it is attempting to control is the residential real estate market. Taking for granted that a real estate bubble was partly responsible for the crisis, its targeted, surgical interventions are designed to limit this concern, with a generally satisfactory result (See Federal legislative measures).

The Quebec government, meanwhile, has a more interventionist approach to the economy, and has taken action to attempt to limit consumer over-indebtedness. However commendable, the Quebec approach fails on two counts. First, because the Quebec National Assembly never passed the required legislative amendments, after a provincial election was called, and second because uncertainty about the proposed legislation did not provide a clear framework for the supervision of real estate credit, but simply of consumer credit (See Quebec legislative measures).

Federal Legislative Measures The regulation of credit remains insufficient at several levels. Although the federal and Quebec legislators have clearly regulated

[15]Lazarus (2009).

[16]On the rationales of over-indebtedness, see e.g., Plot (2009), p. 85; Duhaime (2003).

credit information, including interest rates,[17] the same does not apply to the assessment of credit applications, a crucial element in the process. Until recently, the legislators left financial institutions, and certain government bodies such as the Canada Mortgage and Housing Corporation (CMHC), to manage this aspect by applying the standards generally used in the Canadian financial services industry.

The financial crisis of 2007–2009 modified the landscape considerably. The triumvirate of the Canadian Minister of Finance, the Governor of the Bank of Canada and the Superintendent of Financial Institutions, worried about the consequences of a real estate bubble, have addressed the issue directly. One illustration of this was a public intervention by the Superintendent of Financial Institutions to announce an increase in vigilance and oversight over the quality of loans granted to consumers.[18] To ensure strong financial health, the federal rules for the granting of credit were amended between 2008 and 2012 to address the length of the amortisation period, mortgage insurance, compulsory down payments, mortgage ratios and, in part, a determination of the borrower's ability to pay. Next, the Office of the Superintendent of Financial Institutions (the OSFI) issued principles of governance for financial institutions issuing residential loans.[19] It is important to note that the government's actions affect the whole of the Canadian market although, in reality, some markets such as Toronto and especially Vancouver have specific problems because of extremely high property prices. Whereas the ideal housing cost/disposable income ratio is 30%,[20] it is higher in Canada and higher still in the markets of Toronto and Vancouver. However, it is important to remember that the level of mortgage debt, like the level of household debt, depends on several different variables, including interest rate, income and the age of the borrower.[21] In practice, consumers tend to borrow more than the minimum they need when a loan is granted, but also tend to pay back their loan before the scheduled end of the amortisation period, in particular because their income increases over time.[22]

The changes introduced were as follows. In 2012, the federal legislator codified the regulations applying to residential mortgage insurers (Canada Guaranty Mortgage Insurance Company, Genworth Financial Mortgage Insurance Company Canada, and PMI Mortgage Insurance Company Canada)[23] to harmonise the legislation,[24] especially with the regulation governing the CMHC.[25] Section 5, which is

[17]On the federal level: *Bank Act*, S.C. (1991), c. 46, s. 450; *Cost of Borrowing Regulations*, SOR/2001-101, ss. 5–6. In Quebec: *Protection Consumer Act*, CQLR, c. P-40.1, ss. 69–72 [CPA].

[18]See *supra* note 2 and accompanying text.

[19]*Supra* Sect. 20.2.2.

[20]The CMHC proposes a ceiling of 40% of gross monthly income: CMHC (2018a).

[21]See e.g.: Crawford and Faruqui (2011–2012).

[22]CMHC (2015).

[23]Protection of Residential Mortgage of Hypothecary Insurance Act, S.C. 2011, c. 15, s. 20.

[24]Eligible Mortgage Loan Regulations, SOR/2012-281.

[25]National Housing Act, R.S.C. 1985, c. N-11; Insurance Housing Loan Regulations, SOR/2012-282.

common to both regulations,[26] provides several key measures to lessen the effect of any real estate bubble likely to affect the economy and, more generally, to ensure better financial stability.

Under Section 5(1)(a), a high ratio loan (based on the loan-to-value ratio)[27] must meet several criteria, the first of which is that the amount borrowed must not represent more than 95% of the value of the property. Conversely, in Canada, nobody can borrow more than 95% of the value of a property and, if the down payment is less than 20%, the loan must be insured by the CMHC or a private mortgage insurer (those mentioned in the previous paragraph), because it is considered a high-risk loan. The compulsory down payment of 5% was introduced quickly in 2008, but was simply the codification of a banking practice. The Canadian approach presents a striking contrast with the United States, where loans can represent 100% of the value of a property, and sometimes even more to cover the associated legal costs.

Next, Section 5(1)(c) specifies that the maximum insurable period of amortisation, for a high ratio loan, is limited to 25 years,[28] after being set at 30 years in 2011,[29] 35 years in 2008[30] and 40 years before 2008.[31] In reality, the loan amortisation period is a variable occasionally used by the government to stimulate or dampen the mortgage market. In Canada, a loan is made for a total period of 25 years (or less in some cases), but the terms must be renegotiated on average every 5 years.[32] Borrowers have the option of reducing the length of the amortisation period by repaying their loan more quickly, something that occurs quite frequently.

---

[26]Insurance Housing Loan Regulations, SOR/2012-282; Eligible Mortgage Loan Regulations, SOR/2012-281.

[27]Section 1 of these regulations defines the high loan ratio as: 'a housing loan that is secured by an eligible residential property and whose principal amount, together with the outstanding balance of any loan having an equal or prior claim against the eligible residential property, is greater than 80% of the value of the eligible residential property at the time the loan is approved' (Insurance Housing Loan Regulations, SOR/2012-282; Eligible Mortgage Loan Regulations, SOR/2012-281). Conversely, a loan is considered of high ratio when the down payment is lower than 20% of the value of the property.

[28]On this topic, see: Department of Finance (Canada) (2012).

[29]Sec. 7(4)(b) of regulations Insurance Housing Loan Regulations, SOR/2012-282; Eligible Mortgage Loan Regulations, SOR/2012-281.

[30]Sec. 7(3)(b) of regulations: Insurance Housing Loan Regulations, SOR/2012-282; Eligible Mortgage Loan Regulations, SOR/2012-281.

[31]Department of Finance (Canada) (2008).

[32]Until the 1980s, financial institutions were giving on mortgage loans of an average 25 years (not divided into 5-year terms) a regular basis, for a fixed interest rate Following the economic crisis of the 1980s, which saw high rising interest rates, banks have encountered major loss in these operations. Since then, amortisation remains at 25 years, but terms are divided into 5-year periods, while financial conditions are renegotiated at these moments.

Moreover, to promote saving and, at the same time, discourage indebtedness, the maximum amount of a loan granted to refinance a mortgage is 80%,[33] a percentage previously set at 85% in 2011[34] and 90% in 2010.[35] Since 2012, the gross debt service ratio and the total debt service ratio cannot exceed 39% and 44%, respectively, when the loan is approved.[36] For a variable-rate loan, or a term below 5 years, the coefficients are calculated based on a fixed 5-year rate.[37] Moreover, since 2012, the rule only applies if the value of the property is below one million dollars.[38]

Last, although the federal legislator does not deal in detail with the question of the borrower's credit rating, its quality is considered, since a credit score of over 600, as established by a rating agency, is required.[39] In practice, however, the statistics show clearly that less than 20% of mortgage applicants, whether they were insured, had a credit score between 600 and 700 when their loan was granted; almost 80% had a credit score of over 700.[40]

Legislative Measures in Quebec The Quebec legislator, unlike its federal counterpart, has intervened extensively in the area of residential real estate law, in particular via several provisions of the Civil Code of Québec dealing with hypothecs, the examination of title and publication of rights, co-ownership, immovable property leases and dismemberment of the right of ownership.[41] Moreover, article 2332 deals with the revision of a rate of interest considered abusive by the court,[42] and other legislation establishes a framework for real estate brokerage.[43]

---

[33]Sec. 1 of regulations: Insurance Housing Loan Regulations, SOR/2012-282; Eligible Mortgage Loan Regulations, SOR/2012-281.

[34]Sec. 7(4)(a) of regulations: Insurance Housing Loan Regulations, SOR/2012-282; Eligible Mortgage Loan Regulations, SOR/2012-281.

[35]Sec. 7(3)(a) of regulations: Insurance Housing Loan Regulations, SOR/2012-282; Eligible Mortgage Loan Regulations, SOR/2012-281.

[36]Sec. 5(1)(h) of regulations: Insurance Housing Loan Regulations, SOR/2012-282; Eligible Mortgage Loan Regulations, SOR/2012-281.

[37]Sec. 5(3) of regulations: Insurance Housing Loan Regulations, SOR/2012-282; Eligible Mortgage Loan Regulations, SOR/2012-281.

[38]Sec. 5(1)(d) of regulations: Insurance Housing Loan Regulations, SOR/2012-282; Eligible Mortgage Loan Regulations, SOR/2012-281.

[39]Sec. 5(1)(g) of regulations: Insurance Housing Loan Regulations, SOR/2012-282; Eligible Mortgage Loan Regulations, SOR/2012-281. See Section 5(2) of these regulations for exceptions.

[40]Crawford et al. (2013), p. 69.

[41]These protective measures have been taken since the Civil Code of Lower Canada (CCLC) in 1866, a year after the foundation of the Canadian Confederation. It was replaced in 1994 by the Civil Code of Quebec.

[42]This section has replaced former Article 1040 (c) CCLC, in force in 1964, which reads as follows: 'The monetary obligations under a loan of money may be reduced or annulled by a court so far as it finds that, having regard to the risk and to all the circumstances, they make the cost of the loan excessive and the operation harsh and unconscionable.'

[43]*Real Estate Brokerage Act*, CQLR, c. c. C-73.2.

These general provisions, which concern both consumers and enterprises, are completed by the protection provided by the Consumer Protection Act. Paragraph b) of Section 6 of the Act states that the sale, lease or construction of an immovable are exempt from the application of the Act, but nevertheless subject to protection against unfair trading practices in the course of such activities.[44] Next, paragraph c) excludes 'credit secured by hypothec' and paragraph d) excludes 'the furnishing of services for the repair, maintenance or improvement of an immovable' from the application of the Act. However, because paragraphs c) and d) have never come into force, these matters are currently subject to the Consumer Protection Act.[45] With respect to real estate credit, it is important to read Sections 20 to 24 of the Regulation respecting the application of the Consumer Protection Act, which specifically exclude an immovable secured by an immovable hypothec ranking first from the provisions concerning consumer credit and, more generally, from almost all of the Consumer Protection Act. Simply put, residential real estate credit is guaranteed by the protection measures in the general law, and not the exceptional measures in the Consumer Protection Act.

In 2017, Quebec's consumer protection agency, the Office de la protection du consommateur (OPC), drew up a draft bill to combat consumer over-indebtedness. It was passed by the National Assembly in late 2017 but has not yet come into force.[46] To encourage lenders to lend more responsibly, and to make borrowers more aware of the dangers of over-indebtedness, the legislation makes it compulsory for lenders to assess a consumer's ability to pay.[47] During examination of the bill by a parliamentary committee, several stakeholders spoke in favour of making real estate credit subject to the Consumer Protection Act. In particular, the Quebec notaries' association, or Chambre des notaires du Québec, considered that 'Bill 134, if it is to truly modernize the field of consumer credit, ensure compliance with the Act for all loans and combat over-indebtedness, must necessarily redefine the rules applicable to immovable hypothecs.'[48]

By setting guidelines for the obligation to verify a borrower's ability to pay, the Quebec bill targeted consumer over-indebtedness and only hinted that another goal of the approach was to ensure the financial health of the economy in general. In fact,

[44] Art. 6.1 CPA.

[45] Article 1 of An Act mainly to modernise rules relating to consumer credit and to regulate debt settlement service contracts, high-cost credit contracts and loyalty programs, S.Q. 2017, c. 24 (assented to November 2017).

[46] An Act mainly to modernise rules relating to consumer credit and to regulate debt settlement service contracts, high-cost credit contracts and loyalty programs, ibid.

[47] Ibid, sec. 19 (proposed sec. 103.2 CPA).

[48] Chambre des notaires du Québec, Submission of the Chambre des notaires du Québec on Bill n° 134 titled Loi visant principalement à moderniser des règles relatives au crédit à la consommation et à encadrer les contrats de service de règlement de dettes, les contrats de crédit à coût élevé et les programmes de fidélisation, presented in October 2017 for the special consultations and public hearings on Bill 134 of the Committee on Citizen Relations, online: http://www.assnat.qc.ca/fr/travaux-parlementaires/commissions/CRC/mandats/Mandat-38403/memoires-deposes.html.

the Consumer Protection Act has been under the responsibility of Quebec's department of justice since 2005,[49] and the department's main focus is microeconomic. It is important to note that the federal government has chosen a regulatory, rather than a legislative, approach that gives it more scope to act and avoids the problems encountered by the Quebec government.

Last, some economists have alleged that reckless consumer behaviour, supported by easy access to credit, was in part to blame for the financial crisis of 2008.[50] The theories put forward about impulse buying patterns could, however, support the opposite position, that because of their vulnerability, some consumers have developed an addiction to credit. However, we should emphasise that the academic literature suggests that the number of consumers affected by this syndrome is not large.[51] It is still possible to suggest, though, that easy access to credit was created by expansionist US monetary policy, used by the government specifically to reinvigorate a sluggish economy following the bursting of the technology bubble in 2000.[52]

21.3.2 Interventions by Control and Supervision Authorities

The management of the biggest financial crisis in the last 80 years was, in fact, an attempt by the State to gain control of credit risk, liquidity risk and systemic risk. The adoption of the position of the Basel Committee on Banking Supervision by governments has helped increase the resilience of the national and international banking system. The central bank governors of most of the world's countries have expressed their confidence in the Basel Accord and the work of Basel Committee on Banking Supervision,[53] which was partly influenced by the recommendations made by the Financial Stability Board[54] and the G20. These approaches have been echoed in Canada, where the OSFI has proposed governance rules to rectify commercial and

[49]Sec. 1(i) CPA; Act Respecting the Ministère de l'Immigration et des Communautés Culturelles, CQLR, c. M-16.1, s. 48.

[50]*Supra* note 14 and accompanying text.

[51]There could be as many as 18 million Americans suffering from the compulsive buying disorder. See: Workman and Paper (2010).

[52]This point of view is shared among monetarists. See, e.g., Salin (2010).

[53]Members of the Basel Committee on banking supervision originate from various countries' central banks. See: Basel Committee on Banking Supervision (2013).

[54]The Bank of International Settlement is working along with the Financial Stability Board. Dated from 2009, and succeeding to the Financial Stability Forum (1999), the Financial Stability Board aims at coordinating national financial entities and at fostering prudential measures. It is chaired by Mark Carney, governor of the Bank of England and former governor of the Bank of Canada. The Board reports to the Group of Twenty (G20). For more information, see: Financial Stability Board, online: http://www.financialstabilityboard.org/about/overview.htm.

financial practices (Sound business and financial practices) using an approach based on the same principles (Governance principles of the OSFI).

Sound Business and Financial Practices In Canada, the policies of the Bank of Canada are strictly followed by the government and the OSFI, the organisation responsible for the control and supervision of financial institutions under federal jurisdiction, namely banks and 'near banks'.[55] The OSFI transfers the policies of the central bank into its own guidelines. Despite their name, these guidelines are standards adopted under the Office of the Superintendent of Financial Institutions Act,[56] which makes them enforceable. The unified and coherent approach in Canada is strikingly different to the approach in the United States. Canadian bankers sometimes disagree with some of the measures but comply with good grace, in particular because of their effectiveness.[57] This was not the case in the United States, where the implementation of the Basel II standards was an arduous task,[58] as will be, no doubt, the implementation of Basel III.

Consumer credit was one of the sources of the crisis, and is also a concern that regulators must monitor constantly to prevent real estate bubbles. In this connection, in June 2012 the OSFI issued a guideline to define its expectations for prudent residential mortgage underwriting.[59] Based on a guideline of the Financial Stability Board,[60] the guideline sets out five fundamental principles for the underwriting of mortgage loans, in keeping with the principle of responsible lending that must govern the control and supervision of consumer credit.[61]

In Quebec, the financial markets authority, the Autorité des marchés financiers or AMF, presented a guideline, based on the OSFI guideline,[62] in which it stated that it favoured a principles-based approach rather than a specific rules-based approach.[63] In addition to the policies imposed on the board of directors and senior management,

---

[55]Near banks include financial institutions other than banks, such as insurance entities, trust and loan companies, and credit union: *Insurance Companies Act*, S.C. 1991, c. 47; *Trust and Loan Companies Act*, S.C. 1991, c. 45; *Cooperative Credit Associations*, S.C. 1991, c. 48; *Act Respecting Financial Services Cooperatives*, CQLR, c. C-73.3 (Quebec). OSFI also supervises federal pension funds to comply with the statute: *Pension Benefits Standards Act, 1985*, R.S.C. 1985, c. 32 (2nd supp.), s. 4(2). OSFI's activities deal with two issues: regulation and supervision.

[56]R.S.C. (1985), c. 18 (3rd supp.), s. 6(4), 38.

[57]Canadian Bankers Association, *Review of the Federal Financial Sector Framework*, Submission to the Department of Finance Canada, November 15, 2016, online: https://www.cba.ca/Assets/CBA/Files/Article%20Category/PDF/sub-20161115-review-federal-financial-sector-framework-en.pdf.

[58]Blair (2013), p. 436; Bleier and Yoest (2013), p. 5; White and Case (2007); Bouton and Amadieu (2007), pp. 115–116.

[59]*Supra* note 5; Rolland (2011).

[60]Financial Stability Board (2012).

[61]Reifner (2009).

[62]*Supra* note 5.

[63]AMF (2013).

it is important to note the similarities with the OSFI concerning the assessment of the borrower's capacity to repay the loan, the loan-to-value ratio and hypothec insurance. The new AMF guideline, like its federal counterpart, expresses in clearer terms than Bill 134 the objective of assessing the capacity to pay compulsory to ensure the financial health of the economy. Further, we should note that Guideline B-20 is adopted by the OSFI, which is under the authority of the Department of Finance Canada[64] as are the Insurable Housing Loan Regulations[65] and the Eligible Mortgage Loan Regulations.[66] Financial services cooperatives, which are very popular in Quebec, are under the authority of the AMF, which has adopted standards similar to those issued by the OSFI and have coercive force.[67] In fact, the AMF states clearly in the preamble to the Residential Hypothecary Lending Guideline that it favours a principles-based approach rather than a specific rules-based approach and aims to ensure harmonisation with rules developed across Canada by placing 'more emphasis on the execution of the legal obligation than on its interpretation'.[68]

Governance Principles of the OSFI The five principles developed by the OSFI are also found in the Quebec standards, where they are subdivided into eight principles.[69] For our analysis, as the Quebec guideline is so similar to the OSFI guideline, we have chosen to look at the federal rules. Each principle is analysed separately below.

Principle 1 Under the first principle, financial institutions 'that are engaged in residential mortgage underwriting and/or the acquisition of residential mortgage loan assets in Canada should have a Residential Mortgage Underwriting Policy [RMUP]. Residential mortgage practices and procedures of FRFIs [federally-regulated financial institutions] should comply with their established RMUP.'[70] The financial institutions must, in particular, give consideration at the portfolio level, to risk management practices and processes with respect to residential mortgage loans and loans assets. They must also consider, at the individual residential mortgage loan level, acceptable underwriting and acquisition standards, criteria and limits for all residential mortgage products (e.g., credit scores, loan-to-value ratios, debt service coverage, amortisation period), as well as the complexity and financial institution's residential mortgage business.

The board of directors should draft and adopt policies on loans, understand the decisions made by senior management, and ensure that they are implemented as agreed. The board should also ensure the operational compliance of the policy by

[64] It is also the case for the AMF's *Residential Hypothecary Lending Guideline* (Ibid.)

[65] *Supra* note 26.

[66] *Supra* note 25.

[67] Act Respecting the Autorité des marchés financiers, CQLR, c. A-33.2, s. 8.

[68] AMF (2013), p. 3.

[69] Ibid.

[70] *Supra* note 5, p. 3.

exercising adequate control and supervision. According to the OSFI, financial institutions:

> [S]hould have adequate processes in place with respect to residential mortgages to independently and objectively:
>
> - Identify, assess and analyse the key risks;
> - Monitor risk exposures against the Board-approved risk appetite of the FRFI;
> - Ensure that risks are appropriately controlled and mitigated;
> - Ensure that risk management policies, processes and limits are being adhered to;
> - Provide exception reporting, including the identification of patterns, trends or systemic issues within the residential mortgage portfolio that may impair loan quality or risk mitigation factors; and
> - Report on the effectiveness of models.[71]

In this connection, it is important to understand that the principle must be read alongside the new updated version of the Corporate Governance Guideline.[72] This focuses on the need for better risk management and aims to ensure that FRFIs have prudent corporate governance practices and procedures that contribute to their safety and soundness; promote industry best practices in corporate governance; are consistent with OSFI's Supervisory Framework (2011)[73]; and address international standards, as articulated by organisations such as the Financial Stability Board (FSB), the Organization for Economic Cooperation and Development (OECD), the Basel Committee for Banking Supervision (BCBS) and the International Association of Insurance Supervisors (IAIS). Last, it is important to note that the Corporate Governance Guideline, and indirectly the first principle in the B-20 guideline,[74] are now harmonised with the OSFI Supervisory Framework.

Principle 2 Under the second principle, financial institutions 'should perform reasonable due diligence to record and assess the borrower's identity, background and demonstrated willingness to service his/her debt obligations on a timely basis.'[75] First, the OSFI indicates that financial institutions must 'make a reasonable enquiry' into the borrower's background, including the credit history and credit score. The OSFI adds that the credit score 'should not be solely relied upon to assess borrower qualification',[76] because it is not always up to date. Although the OSFI does not mention this directly, a credit file sometimes contains errors that can have a negative impact on the borrower. Next, the OSFI notes that 'maintaining sound loan documentation is an important administrative function for lenders.'[77] Financial institutions must keep the documents used to approve a loan, such as the documents

[71]Ibid. p. 5.

[72]BSIF (2013).

[73]BSIF (2010).

[74]*Supra* note 5.

[75]Ibid. p. 5.

[76]Ibid.

[77]Ibid. p. 6.

showing the borrower's employment status and verification of income.[78] Last, the OSFI is aware of the risk of hypothecary loans being used for money laundering, and has even issued a guideline on that topic.[79] If the financial institution knows or has reasonable grounds to believe that the loan will be used for illicit purposes, it must file a report with the centre that processes information about suspicious transactions, the Financial Transactions and Reports Analysis Centre of Canada. The declaration ensures compliance with the Proceeds of Crime (Money Laundering) and Terrorist Financing Act,[80] the relevant regulations,[81] and the standards enacted by the OSFI to mitigate the financial risks caused by the vulnerability of financial institutions to clients involved in money laundering transactions.[82]

Principle 3 The third principle, which stems directly from Principle 2, requires the financial institution to 'adequately assess the borrower's capacity to service his/her debt obligations on a timely basis.'[83] This is clearly a fundamental element that must be respected by lenders. The principle has two main focuses.

First, bankers must consider the borrower's income, in particular by verifying the borrower's employment status and income history. The same checks must be completed for the co-signor. Second, the debt serviceability ratios must be calculated. These are called the gross debt service ratio and total debt service ratio by the OSFI,[84] referring to the CMHC definition and criteria to determine the ratios used. The gross debt service ratio is calculated as the ratio between the borrower's gross income and the portion of the gross income especially dedicated to hypothecary payments, including the repayment of principal and the payment of interest, property taxes, heating and one-half of co-ownership charges. The total debt service ratio is the ratio between the borrower's gross income and all outgoings, including payments for housing, car loans and credit cards.[85] The CMHC suggests that the two

---

[78]Ibid, principle n° 3. The following documents are required: 'A description of the purpose of the loan; employment status and verification of income (see Principle 3); debt service ratio calculations, including verification documentation for key inputs (e.g., heating, taxes, and other debt obligations); LTV ratio, property valuation and appraisal documentation (see Principle 4); credit bureau reports and any other credit enquiries; documentation verifying the source of the down payment; purchase and sale agreements and other collateral supporting documents; an explanation of any mitigating criteria or other elements (e.g., 'soft' information) for higher credit risk factors; property insurance agreements; a clearly stated rationale for the decision (including exceptions); and a record from the mortgage insurer validating commitment to insure the mortgage, where applicable': ibid., principle n° 2.

[79]OSFI (2008).

[80]S.C. 2000, c. 17, sec. 5, 7.

[81]Proceeds of Crime (Money Laundering) and Terrorist Financing Regulations, SOR/2002-184, art. 11.2; Proceeds of Crime (Money Laundering) and Terrorist Financing Suspicious Transaction Reporting Regulations, SOR/2001-317.

[82]*Supra* note 80.

[83]*Supra* note 5, p. 8.

[84]Ibid.

[85]CMHC, *supra* note 21.

ratios should not exceed 32% and 40%,[86] basically matching market requirements, although the legislator indicated in 2012 that the rates should not exceed 39% and 44%.[87] The OSFI specifies that financial institutions should have clear policies with respect to the contributing factors for the calculation of gross debt service [GDS] and total debt service [TDS] ratios, including, but not limited to principal and interest; other sources of income; heating costs; property taxes; guarantor or co-signor income; and monthly payment amounts for other credit facilities.[88] Again, from a risk management perspective, the OSFI requires financial institutions to calculate these ratios conservatively. For example, it recommends that 'for insured mortgages, the Government of Canada's mortgage insurance guarantee framework requires that for all variable interest rate mortgages, regardless of the term and fixed rate mortgages with a term less than the standard 5-year term, lenders use the greater of the contractual mortgage rate or the 5-year benchmark rate published by the Bank of Canada.'[89] Last, the OSFI states that financial institutions must consider variations in income, whether because of seasonal work or changes in the general economic outlook.

Third, another essential element in assessing the capacity of the borrower to repay the debt concerns the amortisation period, for which financial institutions should have a policy. The OSFI recommends that the amortisation periods proposed should be shorter than the stated maximum. In practice, this element is often discussed by the lender, but is difficult to apply to a first-time purchaser, although over the years the amortisation period sometimes tends to decrease.[90]

Principle 4 The fourth principle requires financial institutions to 'have sound collateral management and appraisal processes for the underlying mortgage properties'.[91] Sound processes, based on risk assessment, must apply both for the granting of a loan and for the realising of security. In concrete terms, they target two aspects. First, the value of the property must be appraised by a bank employee or a third-party appraiser. The financial institution must ensure that the claim on the property can be realised in a reasonable period or ensure that title insurance from a third party is in place. Furthermore, the financial institutions should impose 'contractual terms and conditions that secure their full protection under the laws applicable in the relevant jurisdiction, and seek to preserve an appropriate variety of recourses',[92] such as a personal covenant. Next, the loan-to-value ratio must be evaluated, namely the percentage of the collateral value that can be used to guarantee the loan. Risk assessment is a fundamental element in the determination of the ratio. Although

---

[86]CMHC (2018b), p. 1.

[87]Sec. 5(1)(h) of regulations: *Insurance Housing Loan Regulations*, *supra* note 26; *Eligible Mortgage Loan Regulations*, *supra* note 25.

[88]*Supra* note 5, p. 9.

[89]Ibid.

[90]*Supra* note 33 and accompanying text.

[91]*Supra* note 5, p. 10.

[92]Ibid, p. 11.

the maximum ratio is 65%, the OSFI considers that it should be set at a lower value, depending on the risk presented by the borrower and market concerned. However, for residential hypothecary loans, the maximum ration of an uninsured loan is 80%, compared to 95% for insured loans.

It is interesting to note that the OSFI considers the origin of the money used as a down payment for a loan, and states that financial institutions 'should make reasonable efforts to determine if it is sourced from the borrower's own resources or savings'.[93] In particular, in the case of a gift, a letter from the donor, preferably notarised if the borrower is from Quebec, must be provided.

Last, the OSFI has the wisdom to address the question of home equity lines of credit.[94] These are lines of credit up to a fixed maximum, granted to a borrower and secured on equity in a property for an undetermined duration as revolving credit. The percentage available to the borrower is up to 65% of the equity in the property.[95] The original reason for these lines of credit, which appeared recently, was to help finance repairs or renovations to the property. In practice, however, financial institutions grant the lines of credit without making any serious checks on how the money is spent. Consequently, some borrowers used the line of credit to pay off the balance on a credit card, or to finance an automobile or vacation. Moreover, borrowers occasionally use the line of credit to make their regular mortgage payments, which substantially increases the credit risk and contributes to their over-indebtedness.[96]

Principle 5 The fifth and last principle states that financial institutions 'should have effective credit and counterparty risk management practices and procedures that support residential mortgage underwriting and loan asset portfolio management, including, as appropriate, mortgage insurance.'[97] In practice, mortgage insurance—in the form of life insurance or disability insurance—is systematically offered to borrowers, most of whom take it out.[98] Furthermore, the OSFI requires heightened prudence for higher-risk hypothecary loans, through greater board of directors and senior management oversight of the asset portfolio; increased reporting; stronger internal controls; and increased personal capital levels to reduce systemic risk.[99]

---

[93]Ibid, p. 13. This must be read in parallel with principle n° 2, *supra* notes 76–83 and accompanying text.

[94]It must be added that these are called 'home equity lines of credit' in Guideline B-20: *supra* note 5, p. 14.

[95]Ibid.

[96]*Supra* note 6 and accompanying text.

[97]*Supra* note 5, p. 15.

[98]OSFI (2014).

[99]*Supra* note 5, p. 3–5.

21.4 Conclusion

The financial crisis of 2008–2009 had the positive result of making the public authorities more aware of the need to provide a framework for real estate credit. The concerns of the Governor of the Bank of Canada, the Canadian Minister of Finance and the Superintendent of Financial Institutions led to the first serious attempts to structure real estate credit. The three authorities are complementary, given the accommodating monetary policy of the Bank of Canada, the measures proposed by the Department of Finance in the form of regulatory amendments—to define the minimum interest rate, term of amortisation, refinancing, down payments, assessment of the capacity to pay, and financial ratios—and the governance principles drawn up by the OSFI and adopted by the AMF.

These measures, acting as a counterweight to the banking lobby while making consumers more responsible, tend to converge with the ideas of responsible credit and even equality. The new legislative developments in Quebec from late 2017 focus only on consumer credit, rather than on real estate credit. This is regrettable and counter-productive in the movement towards responsible credit, despite the general governance principles previously adopted by the authorities responsible for control and supervision.

References

AMF (2013) Residential Hypothecary Lending Guideline, January, online: https://lautorite.qc.ca/fileadmin/lautorite/reglementation/lignes-directrices-assurance/g_res_hyp_2012_final.pdf

Bailliu, J, Kartashova K, Meh C (2011–2012) Household borrowing and spending in Canada. Bank Canada Rev: 16

Basel Committee on Banking Supervision (2013) A brief history of the Basel Committee. Bank of International Settlement, Available via http://www.bis.org/bcbs/history.pdf

Blair MM (2013) Making money: leverage and private sector money creation seattle. Univ Law Rev 36:417

Bleier ME, Yoest DP (2013) Federal banking agencies propose Basel III capital requirements for U.S. banking organizations: what it is and what it means. Bank Law J 130:3

Bourdieu P (2000) Les structures sociales de l'économie. Seuil, Paris

Bouton D, Amadieu D (2007) Les possibles conséquences d'une application différenciée de la réforme Bâle II aux États-Unis et en. Europe Revue d'économie financière 87(111):115–116

BSIF (2010) Supervisory Framework, December, online: http://www.osfi-bsif.gc.ca/Eng/Docs/sframew.pdf

BSIF (2013) Corporate Governance, January, online: http://www.osfi-bsif.gc.ca/Eng/Docs/CG_Guideline.pdf

CMHC (2015) Canadian Housing Observer 2014 (rev. 2015). Available via https://www03.cmhc-schl.gc.ca/catalog/productDetail.cfm?lang=en&cat=122&itm=27&fr=1517609366185

CMHC (2018a) Homebuying Step by Step. Available via https://www.cmhc-schl.gc.ca/en/co/buho/step-by-step/index.cfm

CMHC (2018b) Mortgage Planning Tips. Available via https://www.cmhc-schl.gc.ca/en/co/buho/plmayomo/plmayomo_001.cfm

Crawford A, Faruqui U (2011–2012) What Explains Trends in Household Debt in Canada? Bank Canada Rev: 3

Crawford A, Meh C, Zhou J (2013) The residential mortgage market in Canada: a primer. Financ Syst Rev 61:69

Department of Finance (Canada) (2008) News Release 2008-051, Government of Canada Moves to Protect, Strengthen Canadian Housing Market. Available via https://www.fin.gc.ca/n08/08-051-eng.asp

Department of Finance (Canada) (2012) News Release 2012-070, Harper Government Takes Further Action to Strengthen Canada's Housing Market. Available via https://www.fin.gc.ca/n12/12-070-eng.asp

Dickson J (2012) Boards of Directors and Risk Governance, Remarks to the Toronto Board of Trade, Toronto, Ontario. Available via http://www.osfi-bsif.gc.ca/Eng/Docs/jd20120405.pdf

Duhaime G (2003) La vie à crédit: consommation et crise. Les Presses de l'Université Laval, Québec

European Commission (2012) Green Paper: Shadow Banking, Brussels. COM (2012) 102 Final. Available via http://ec.europa.eu/internal_market/bank/docs/shadow/green-paper_en.pdf

Financial Stability Board (2011) Shadow Banking: Strengthening Oversight and Regulation—Recommendations of the Financial Stability Board. Available via http://www.financialstabilityboard.org/publications/r_111027a.pdf

Financial Stability board (2012) Principles for Sound Residential Mortgage Underwriting Practices. Available via http://www.financialstabilityboard.org/publications/r_120418.pdf

Flávio de Oliveira A, Moreira dos Santos Ferreira F (2012) Análise Econômica do Direito do Consumidor em Períodos de Recessão: Uma Abordagem a Partir da Economia Comportamental. Revista de Direito do Consumidor 21:13, p 31

Governing Council of the Bank of Canada (2012) Monetary Policy Report. Available via http://www.bankofcanada.ca/wp-content/uploads/2012/04/rpm-avril2012.pdf

International Monetary Fund (2012) World Economic Outlook: Growth Resuming, Dangers Remain, World Economic and Financial Surveys. Available via http://www.imf.org/external/pubs/ft/weo/2012/01/pdf/text.pdf

Lacoursière M (2011) Richard c. Time Inc. : à la recherche de la définition du 'consommateur moyen'! Can Bar Rev 90:495

Lazarus J (2009) L'épreuve du crédit. Sociétés contemporaines 76:17

Organisation for Economic Co-operation and Development (2013) OECD Factbook 2013: Economic, Environmental and Social Statistics—Household Debt. Available via http://www.oecd-ilibrary.org/economics/oecd-factbook-2013/household-debt_factbook-2013-28-en

OSFI (2008) Guideline B-8 – Deterring and Detecting Money Laundering and Terrorist Financing, December, online: http://www.osfi-bsif.gc.ca/Eng/Docs/b8.pdf

OSFI (2014) Guideline B-21 – Residential Mortgage Insurance Underwriting Practices and Procedures, November 2014, online: http://www.osfi-bsif.gc.ca/Eng/Docs/B21_GIAS.pdf

Plot S (2009) Du flambeur à la victime? Vers une problématisation consensuelle du surendettement. Sociétés Contemporaines 76:67

Reifner U (2009) 'A Call for Arms'—for regulation of consumer lending. In: Niemi J, Ramsay I, Withford WC (eds) Consumer credit, debt and bankruptcy. Hart, Oxford, p 105

Rolland S (2011) Banques canadiennes: la qualité des prêts sera plus surveillée, LesAffaires.com [de Montréal]

Salin P (2010) Revenir au capitalisme. Pour éviter les crises. Odide Jacob, Paris

Statistics Canada (2012) Household Debt in Canada. Available via http://www.statcan.gc.ca/pub/75-001-x/2012002/article/11636-eng.pdf

Task Force on Financial Literacy (2011) Canadian and Their Money: Building a Brighter Financial Future. Available via http://publications.gc.ca/collections/collection_2011/fin/F2-198-2011-eng.pdf

White & Case (2007) US Finally Adopts Basel II... So Now What? Financial Services Advisory Update 4(8). Available via http://www.whitecase.com/fsau_1207_2/#.UeqyquD5bqs

Workman L, Paper D (2010) Compulsive buying: a theoretical framework. J Bus Inq 9:89
World Economic Forum (2008) Global Competitiveness Report 2008–2009, Geneva. Available via http://www3.weforum.org/docs/WEF_GlobalCompetitivenessReport_2008-09.pdf

Chapter 22
Homes or iPhones? Diversion of Social Security Funds to Relieve Consumption-Fuelled Household Debt in Brazil

Maria Paula Bertran

Contents

Abstract This chapter seeks to shed light on the potential for well-meaning but poorly thought-out state interventions in Brazilian consumption indebtedness to produce sub-optimal outcomes in the last 15 years. The Brazilian population has for some time lived the paradox of taking on debt for common consumer goods when they could least afford it. This paradox was sharpened in 2016–2017 when the Brazilian government moved to draw on accumulated assets used to fund severance or sick pay (the so-called Fundo de Garantia por Tempo de Serviço–FGTS) as part of its efforts to address the effects of the 2014 economic crisis. Established in 1966, FGTS resources have traditionally been the most important and efficient means by which employees have been able to accumulate reserves to later use to help fund the purchase of a home. A significant share of FGTS funds was released to consumers in 2017, much of which was then used not to stimulate the economy but to pay down credit card debts. Foolishly, the resources historically earmarked for housing were thus reallocated to more prosaic—and arguably, unsustainable—consumption.

M. P. Bertran (✉)
University of São Paulo (USP), Ribeirão Preto, Brazil
e-mail: bertran@usp.br

A. d. Amaral Junior et al. (eds.), *Sustainable Consumption*,
https://doi.org/10.1007/978-3-030-16985-5_22

22.1 Introduction

I used to be poor. Nowadays, I'm still poor but I'm also in debt.—Anonymous Brazilian saying

Brazil went through a period of economic boom (from the early 2000s until 2014) followed by a deep crisis. During the period of expansion, there was a remarkable increase in money lending. The most common financial product in the period was revolving credit. Real estate financial products and payment-deducted loans to retired people and state employees were also two remarkable forms of credit that resulted in massive indebtedness within the Brazilian population.

Revolving credit is a renewable credit line for purchases that are not paid within the 30-day interest-free period after the purchase. Interest rates in Brazil for this kind of credit were extraordinarily high: 192.94% per year in 2013, rising after the 2014 economic crisis, whereby annual interest rates on credit cards reached 500% per year between 2016 and 2017.[1] As *The New York Times* put it in 2014, 'interest rates in Brazil would make an American loan shark blush.'[2] The number of families with a significant part of their budget being allocated to these credit card debts was also enormous: more than 50% after 2010 (when family indebtedness data began to be collected systematically) through to the present.[3]

From 2014 on, the financial gain for banks from revolving credit products began to fall[4] as unemployment rose and the population faced sharp declines in household income. It was logical that loans without collateral (as credit card debt typically is), would be the first to suffer from high rates of default. Because Brazilian banks are regularly well supported in their reserves, risking credit card defaults did not place their balance sheets under extreme pressure, but certainly triggered a decrease in profits from revolving credit.[5]

However, in a rather peculiar and unexpected manoeuvre, the resources of a Brazilian mandatory fund were released to the population. The so-called Fundo de Garantia por Tempo de Serviço (FGTS) has been in existence since 1966. It collects resources from all employers, which must regularly deposit into the fund a value proportional to staff salaries on behalf of employees. Since its creation, the fund has been directed to very restricted purposes. There are two main ways that resources can be cashed out from the fund. The first is for individuals. Until December 2016, workers could withdraw the value of deposits employers had made on their behalf in their linked accounts if they were made redundant, became ill or to purchase a home. The second is an institutional use. The FGTS managing committee chooses facilities projects to invest in to earn a return on the fund's resources. These investments can

[1]ANEFAC (2017).

[2]Horch (2014).

[3]Confederação Nacional do Comércio (2013).

[4]Larghi (2017).

[5]Marques and Oyamada (2016).

only be funded in areas legally specified by the State as a priority, typically in projects of strategic importance or that have long faced under investment: sanitation facilities, urban transport and housing construction.

On 22 December 2016, the President of Brazil approved a norm converted into Law No. 13446 of 2017 that modified the legislation of the fund to allow most of the linked accounts to be drawn down by workers. Consequently, a significant amount of money entered the Brazilian economy at a crucial point in the economic downturn. However, more than 40% of these extra funds were not spent stimulating industry, or improving commerce or the quality of life in any sense. That share of the funds (paid out to some 25.9 million workers, according to the FGTS website) was used to pay down consumer debts, in particular lines of revolving credit in the form of bank credit cards.

The fund's management has always come in for heavy criticism for yielding low returns to employees on the resources invested from employers' contributions on their behalf, sometimes lower than the national rate of inflation. In making the announcement, the government emphasised that workers would now be free to redeem the value of their linked accounts from the fund and choose the most profitable investments for whatever funds would remain as reserves for adverse situations in the future.[6] However, little attention was paid in official government statements to the fact that the release of resources reduced the offer of subsidised interest for the acquisition of housing and the formation of compulsory savings by workers for housing acquisition. There was little attention, as well, to the reduction in resources available for investment in sanitation and urban transport.

The release of FGTS funds (Medida Provisória number 763, from 2016) was announced with other microeconomic changes concerning consumer credit. Indeed, in the same announcement, the president also proclaimed an embargo against revolving credit in Brazil, as it had existed until then. From 2017 on, revolving credit can only exist for a maximum of 30 days (Brazilian Central Bank Resolution number 4549 of 26 January 2017). After the 30-day period, it becomes mandatory to establish another line of credit, with interest rates lower than those charged for the previous revolving credit. The new resolution states that interest rates should be lower, but it does not determine how much lower they should be.

The end of revolving credit was the first sign that the federal government had decided to make a policy decision to disrupt the transfer of wealth from the population to the credit card industry, which had followed the expansion of revolving credit over the previous period. The end of revolving credit and the release of FGTS funds were announced on the same day. One hypothesis to be proved by this present chapter is that this was not simply a coincidence. On the contrary, it suggests an exchange of courtesies between the two parts. The first is the federal government, which never restrained interest rates of more than 300% per year, during the more than one decade when such rates were obtaining. The second is the credit card companies, which in the end received a substantial part of the FGTS fund injection

[6] Temer (2017).

into the economy at a moment that credit card debt could well have posed a severe problem for the industry.

Regardless of the new federal government regulation, rising unemployment and falling wages would naturally see a decline in the profits of the credit card industry. Acceptance of the new regulations on revolving credit would not be a major waiver for banks: unemployment and falling incomes were the fundamental determinants of the population's lack of resources to pay interest. Not surprisingly, even after the onset of the crisis, high default rates and multiple termination of contracts coincided with continued bank profitability. In this sense, *The New York Times* put it perfectly in another article titled 'In good times or bad, Brazil banks profit'.[7]

This chapter proceeds in three parts. The first section describes how revolving credit became a business with unrivalled profitability in Brazil. The primary diagnosis points to sloppy credit selection. We consider that the determining factor for someone being approved by a credit registry was basically being employed in the context of expanding economic growth, even if this meant debt would be taken on for many years, with—at best—only partial repayment of costly revolving debts. The second part provides an overview of the economic and legal conditions of financing contracts in Brazil. The third and last part reviews some of the severe changes that the federal government has implemented in credit, especially the release of FGTS funds.

22.2 Revolving Credit and Mass Income Transfer: Profits from Defaults

The primary source of income for credit card issuers in Brazil until recently was the collection of interest from revolving credit. This is reflected in the line item 'financial income' in Table 22.1. In 2011, the financing of purchases made with credit cards 'accounted for 57% of the total, a growth of 13% compared to 2010.' Throughout the time that the analysis is done by official bodies in Brazil, financial revenues always came first as a source of profit, as shown above.

On the one hand, the primary source of profit of the card issuers comes from financial revenues. On the other, there was also an increase of delay in revolving credit over the years, what appears in Table 22.2 as 'default'.

At first, it seems illogical that profit derives from default and the use of revolving credit. Regularly, people are used to banks taking a unique model that the specialised literature calls 'risk-averse management'.[8] Under the risk-averse management 'it is assumed that managers of banks are averse to risk, so they could increase operating expenses for valuation and monitoring of loans, reducing efficiency, to compress the participation of defaults in their loan portfolios.'[9]

[7]Horch (2015).

[8]Koutsomanoli-Filippaki and Mamatzakis (2009).

[9]Tabak et al. (2010), p. 14.

Table 22.1 Changes in credit card company expenditures in Brazil (2005–2011)

Revenue	2005	2006	2007	2008	2009	2010	2011
Total	100.0	118.5	134.0	171.4	184.2	198.6	221.7
Financial income	58.9	73.4	80.7	103.8	111.7	116.4	125.3
Exchange rate income	13.3	15.8	20.1	28.1	31.0	36.4	42.1
Bearer fees	17.7	18.0	20.1	23.4	23.5	25.9	29.2
Other income	9.5	10.7	12.3	15.1	17.7	19.5	23.9
Incentive to issue	0.5	0.4	0.6	0.9	0.3	0.3	0.8
Marketing	0.1	0.1	0.2	0.1	0.0	0.0	0.4

Source (adapted): Central Bank of Brazil, Secretariat of Economic Law and Secretariat of Economic Monitoring, 2012

Table 22.2 Changes in credit card company revenue in Brazil (2005–2011)

Expenses	2005	2006	2007	2008	2009	2010	2011
Total	100.0	129.3	155.5	227.3	258.3	277.1	360.2
Default	26.8	42.6	51.2	85.2	103.1	97.2	124.0
Other expenses	45.0	54.6	63.7	92.1	103.6	129.6	118.2
Rewards programmes	–	–	–	–	–	–	30.5
Direct taxes	–	–	–	–	–	–	24.6
Data processing	15.7	16.2	18.5	22.5	20.9	19.3	23.7
Brand taxes	3.4	4.3	5.9	8.7	10.0	12.2	17.8
Risk management	2.7	2.9	4.4	6.7	9.0	8.9	12.1
Marketing and sales	6.3	8.7	11.7	12.0	11.7	9.8	9.4

Source (adapted): Central Bank of Brazil, Secretariat of Economic Law and Secretariat of Economic Monitoring, 2012
The number 100.0 represents the total amount in 2005. Other years of data show real growth because they did not consider Brazilian currency inflation

By this logic, it is presumed that management with higher risk aversion would correspond to lower rates of default. Empirical studies state that this hypothesis is confirmed:

> [T]here is the presence of risk-averse management in Brazilian banks. Initially, there seems to be a negative relationship between bank efficiency and default credits with causality occurring from efficiency to the (endogenous) risk variable, suggesting that inefficiencies in bank management could be the main cause of bank failures. As a consequence, (risk-averse) managers would tend to increase operational expenditures for loan appraisal and monitoring to control the increase in default, worsening the bank's efficiency measure, but also reducing default in the portfolios of credit.[10]

However, the inverse does not occur: 'Evidence for the inverse causal relation of defaulted credits affecting bank efficiency is disproved, since no significant coefficients are found in this aspect, in addition to the coefficients being close to zero'.[11]

[10]Tabak et al. (2010), p. 4.

[11]Tabak et al. (2010), p. 4.

Table 22.3 Redecard and Cielo profit expectations compared to other Brazilian companies (2003–2010)

	2003	2004	2005	2006	2007	2008	2009	2010
Superior limit	258.0%	48.3%	70.0%	85.2%	104.4%	17.6%	17.8%	19.4%
Fair profit	170.2%	40.5%	54.5%	63.4%	71.0%	16.7%	16.6%	17.5%
Inferior limit	82.5%	32.8%	39.0%	41.6%	37.6%	15.9%	15.4%	15.5%
Redecard	740.6%	826.2%	914.6%	1268.4%	174.4%	165.4%	193.9%	100.0%
Cielo	29.4%	44.4%	98.9%	156.6%	147.0%	876.3%	178.3%	155.1%

Source (adapted): Central Bank of Brazil, Secretariat of Economic Law and Secretariat of Economic Monitoring, 2012

This means that Brazilian banks did not become more prone to bankruptcy because they lent to those who could not pay on time, until 2014. If the logic applied to credit cards is transferred to the other forms of consumer credit, we can conclude that banks take risks and gain more from late payment charges and, astoundingly, breaches of contract.

In fact, the profits of revolving credit, based on well-designed limits (which does not mean they were a small percentage of contracts) became a business with unrivalled profitability in Brazil. It was so remarkable that the credit card industry was considered 'beyond fair profit', in the expression used by the study completed by the Brazilian Central Bank's Secretariats of Economic Law and Economic Monitoring (Table 22.3).[12]

This study developed a model for analysing the profit patterns expected by the credit card industry in comparison to other sectors of the economy. Two of the Brazilian companies showed that the credit card business had much higher profitability than might be expected from 'fair profit', that is, the expected profit from the average of other prosperous business areas.

The hegemonic discourse reproves debtors within a moral framework based on a very restricted content of obligation and responsibility. In this context, the argument that 'banks only want to lend to those customers who do not need credit' becomes fallacious. This, though, is regularly mentioned as a joke in Brazil: 'bankers would only lend an umbrella on a sunny day'. The contemporary reproduction of this discourse fractures reality to extirpate from it all the context that precedes and stimulates credit. A case in point is the granting of credit to a mass of people who undoubtedly would not be able to pay within the 30-day interest-free period.[13]

Debt by credit card is emblematic of the general posture of banks in Brazil, namely, the transfer of income in granting credit to those who, by primary experience criteria, will not pay off the entire outstanding balance during the 30-day interest-free period. Because of intentionally bad credit selection, many people in Brazil began to incur the expensive costs of revolving credit. In practice, this normal procedure undermined the economic gains of the population and transferred income

[12]Central Bank of Brazil (2012), p. 23.

[13]Hyman (2011).

Table 22.4 Brazilian National Consumer Indebtedness and Default Survey (July 2013)

Total indebtedness (% of the population)	Unable to make scheduled payments (% of those in debt)
65.2%	7.4%
Type of debt	Percentage of sample
Credit card	75.2%
Car finance	12.1%
Housing finance	5.9%

Source: National Confederation of Commerce—Confederação Nacional do Comércio—PEIC Nacional, July 2013

from individuals to the financial system. In this sense, it should be reiterated that annual interest rates on credit cards for people who were unable to pay their bills in the 30-day period were averaging 192.94% per year in 2013. After the economic crisis, as mentioned, annual interest rates on credit cards reached 500% per year between 2016 and 2017.[14]

The National Consumer Indebtedness and Default Survey (PEIC Nacional, in Portuguese) has been calculated monthly by the most important retail business association in Brazil (Confederação Nacional do Comércio) since January 2010. The data are collected in all Brazilian state capitals, surveying an average of 18,000 consumers. Table 22.4 shows data from a representative month of the period before 2014.

At a time when Brazil was experiencing its greatest expansion of wealth since the 1970s, 65.2% of the Brazilian population—estimated at 200.4 million people in 2013—was in default. There were, therefore, some 130 million people in default. Of these, 75.2% owed money to credit card companies for failing to pay the outstanding balance within 30 days. That is 97.76 million people. The most indebted households were also the most impoverished. The families that pay their debts with even longer delays are also the poorest families. It is worth noting that, with annual interest around 200%, a delay in the payment of a credit card bill for 60 days may represent an increase of one-third of the original debt.

22.3 The Rise and Fall of the Economy, Dreams of Growth, and Revolving Credit

Credit card industry earnings derived from a prosperous moment for individuals and households in Brazil. There was a decrease in poverty and an increase in income for the most deprived groups between 2003 and 2014. This movement reduced inequality.

[14]ANEFAC (2017).

A study of the period from 2006 to 2012 highlights that the decrease in inequality occurred in the range from 0% to 90% of the lowest earning households. The concentration at the peak of the income scale was quite stable, especially for the top 1%.[15] The highest earning 10% of households accounted for more than half of Brazil's income (52%). The upper 1% retained something close to a quarter (23.2%). The top 0.1% of households took about one-tenth of all income (10.6%). The top 0.05%—about 71,000 households—appropriated 8.5% of all income. Thus, although the data show an increasingly equal spread in the period, Brazil remained one of the most highly concentrated income distributions in the world.

The reasons the population's income was increasing were many. One was undoubtedly the real increase in wages. There was a real appreciation of 77% of the minimum wage between 2003 and 2016.[16] Further, poor people who received social benefits from public policies aimed at fighting poverty, (such as the famous 'Bolsa Família' and the lesser-known 'LOAS'), also had the minimum wage as the pay parameter. Besides the real increase in wages, the increase in the income of Brazilians is also based on the rise in the number of jobs. Between 2009 and 2014, there was full employment in Brazil.[17] Many other things can be mentioned to explain the income growth of the Brazilian population: (a) an increase of foreign demand for Brazilian products (especially agricultural and mining products); (b) the discovery of new oil reserves; (c) a lower level of gains in Europe and the United States for investment funds (which attracted investments to Brazil); (d) public policies aimed at fighting poverty; and (e) unprecedented levels of credit in the economy.

One of the most critical strategies used to boost Brazilian growth was giving people credit. The increase in credit did not come with an increase in local production, however. Brazilians' extra credit was used to buy imported products, mainly Chinese consumer goods. The sustainable expansion of real estate credit would have provided the opportunity to spend on something that would directly benefit Brazilians, both in the creation of jobs and in the development of companies that manufacture construction inputs. Although real estate credit grew significantly in the period, its overall volume was lower than that of credit cards.[18]

The dream of Brazilian growth began to collapse in late 2013 when Chinese instability reduced demand for Brazilian commodities. Since then, the Brazilian economy has moved well on the way to the standards of peripherical capitalism, but global consumption declined sharply. At the same time, oil reserves were shown to be more challenging to reach than was initially expected, and central economies (Europe and United States) began to find solutions to their problems. The unfortunate economic situation, associated with the discovery of corruption problems, were pointed out as crucial factors in the fall of President Dilma Rousseff in 2016.

[15]Medeiros et al. (2015), p. 982.

[16]DIEESE (2017), p. 4.

[17]IBGE (2017).

[18]Mendonça (2013).

By 2017, the unemployment rate exceeded 13%, although the official definition of unemployment refers to individuals who have declared themselves seeking employment for more than 30 days. People who have given up looking for work were categorised as 'discouraged' and are a new grade of research for the Brazilian National Institute of Geography and Statistics (Instituto Nacional de Geografia e Estatística—IBGE). The sum of those officially designated as unemployed and those who are labelled 'discouraged'—namely, the mass of the population that no longer has income to pay credit—added up to 21.1% of the workforce in January 2017.[19]

In addition to the increase in unemployment, reporting indicates that household income fell by 9.1% in 2017 compared to 2014.[20]

22.4 The Resources from the FGTS for the Extension of a Model of Income Transfer

The economic crisis profoundly changed the process of massive indebtedness that the Brazilian population was passing through. In January 2017, the new president, Michel Temer, declared that interest rates on revolving credit should be reduced (Resolution 4549 of 26 January 2017 of the National Monetary Council—Conselho Monetário Nacional, CMN). For the first time, the federal government was reacting to the transfer of income from households to the financial sector through interest rates charged by credit card companies.

At the same time, the federal authorities authorised individual workers to redeem their severance pay funds (which is one of the leading sources for the popular funding of housing finance). The government decided to release a large part of the severance pay fund (around BRL 44 billion), which historically was intended only for dismissal protection, illnesses or home purchases. The objective declared by the government was to quickly boost the economy through immediate consumption. Instead of buying homes, an activity that takes time to produce results, the draw-down of funds was expected to encourage people to buy goods immediately. However, data shows that 42% of the population did not use the funds to make purchases that would stimulate demand. Instead, they paid off debts. Among the poorest, the portion of the population that used the FGTS to pay debts was over 50%.[21] The government's intentions were therefore largely thwarted. While contrary to the stimulatory objectives of the federal government, the use of new funds to pay down debts made without guarantees (such as credit card debts) that could no longer be paid because of the severe falls in household income after 2014 was perfectly rational behaviour by households.

[19]Mota (2017).

[20]Vettorazzo and Perrin (2017).

[21]Fundação Getúlio Vargas (2017).

It is interesting to mention that FGTS funds were not the only extraordinary resource that the population used to pay down debt at that time. A survey by the National Association of Finance, Management and Accounting Executives shows how the Christmas Bonus Salary was heavily used to pay debts in Brazil. In 2014, 68% of the population planned to settle debts with the receipt of the bonuses. A further 11% intended to use the bonus to buy gifts and another 11% wanted to save for extraordinary expenses at the beginning of the year (taxes and books for children to start the school year). An additional 2% planned to renovate their homes. Some 6% had already spent the bonuses, because traditional Brazilian credit lines anticipate the rewards before December, with a considerable discount on the value that the employees will receive in that month. Only 2% intended to make savings with the bonus.[22]

FGTS resources are central to real estate financing in Brazil in two different ways. The first one is as a scheme of compulsory savings. Workers have an account that is credited with new funds every month, deposited on their behalf by their employers. The deposits correspond to a share of salaries and were, until 2016, intangible, except in cases of dismissal, severe illness or acquisition of the first owner-occupied home. The second way FGTS has been central to real estate financing is as an investment fund for subsidized credit concessions for real estate financing and sanitation facilities. The FGTS resources feed what can be translated into English as the Real Estate Finance System (in Portuguese, Sistema Financeiro da Habitação—SFH). The SFH was created in 1964 and is characterised by state regulation of real estate financing conditions, including interest rates and terms. The resources that supply this system come from two sources: a type of individual savings account and, as mentioned, the FGTS. The latter is the most important source for popular inhabiting instalments in Brazil.

Private banks may access savings accounts and the severance pay fund to broker real estate financing. Private banks have autonomy over some aspects of the contracts (especially regarding whom to offer or refuse credit), but they cannot freely determine all the clauses of these loans. In practical terms, federal government regulation makes this a subsidised type of financing, whose maximum and minimum interest rates vary, depending on household income and real estate value.[23]

It is important to note that the use of the FGTS resources for the payment of credit card debts hampered the acquisition of housing in two different ways. First, it caused the loss of people's savings. Secondly, it decreased the available resources for loans via SFH. As the popular funding system was distorted from its original vocation and drained out to pay off credit card debt, its ability to finance housing declined significantly.[24] What was left over from the FGTS severance pay fund's resources was earmarked for high-income construction. The government extended the maximum value of houses that could benefit from the subsidised popular purchasing

[22]ANEFAC (2014).

[23]Caixa (2017).

[24]UQBAR (2017).

system from BRL 950 thousand to BRL 1.5 million. Instead of providing subsidised housing for the low- or middle-income population, the resources would serve the sole mission of trying to reduce unemployment in the construction sector, even if this meant subsidising luxury real estate for high-income earners.[25]

In addition to the release of FGTS funds to pay debts, the government has also announced a new credit card system, which—as mentioned—no longer allows revolving credit to be offered indefinitely. The new rule determines that the payment of the partial invoice no longer results in a permanently renewable debt. Now, as mentioned, after 30 days of non-payment of the invoice in its total amount, revolving credit migrates to another credit line, with necessarily lower interest rates. The two measures announced jointly, seemed to be a gift and a penalty for financial institutions. The gift was reflected in the large volume of debts that were paid off with the funds from the FGTS. The penalty was demonstrated because, for the first time, the federal government outlined that the transfer of income through interest charged by credit cards had gone too far in Brazil. The gift seems, so far, to have been more efficient than the penalty. While the resources of the FGTS have already been released and debts that would probably be considered as prescribed have been paid, interest rates on the new financial products that replaced revolving credit have barely decreased.

As mentioned before, Resolution 4549 of 26 January 2017 of the National Monetary Council of Brazil stipulates that the only obligation of financial institutions is to present 'more advantageous conditions for the client'. Nothing is said about how advantageous they should be. During the year of 2017, the average interest rate was about 350% per year; quite a good advantage considering the interest rate of 500% per year obtaining the year before.[26] Nevertheless, it is astonishing when compared to comparable economies (some sources mention rates of 43% per year in Argentina, 30.5% in Colombia and 25% in Chile).[27]

22.5 Conclusions

The strategy of granting mass credit in Brazil before 2014 was structured by the returns to scale from institutionalising controlled default across the entire population. The context of economic growth and distribution of wealth was a guarantee of satisfaction for the credit card industry. The granting of credit in disproportionate amounts and at very high rates of interest for revolving credit meant a large part of the population's income was transferred to financial institutions. This model came in for criticism in terms of its social implications. But the final blow came with the perpetuation of economic gains through FGTS funds, even at the time of

[25]Eloy (2017).

[26]Central Bank of Brazil (2017).

[27]Murakawa and Marchesini (2018).

unemployment and declining incomes. The institutionalised patrimony of Brazilian workers in the FGTS was used to pay for iphones (or socks, or grocery shopping) or the other prosaic things that Brazilian lowest earning households had acquired with credit cards in the decade of booming growth and improving wealth and income distribution. The opportunity to improve home ownership opportunities, however, has slipped.

The FGTS was very much criticised for being a compulsory fund that could only be used in very few situations and offered dividends below inflation. It is also true that the original structure of the FGTS did not give workers much autonomy. However, is there any real autonomy in taking on revolving credit with interest of 500% per year? Is there genuine autonomy in allocating wealth accumulated over more than a decade to non-durable consumption to the detriment of real estate? Did the government intervention assure the best allocative outcomes? And with these results prove optimal in social terms? The intuition is that the FGTS of the old model was, for all its faults, still a better way to protect the worker's assets: a paternal model for a nation that may need a father more than ever.

References

Associação Nacional dos Executivos de Finanças, Administração e Contabilidade (ANEFAC) (2014) Pesquisa de utilização do 13 salário. https://www.anefac.com.br/uploads/arquivos/2015220162133328.pdf. Accessed 2 aug 2017

Associação Nacional dos Executivos de Finanças, Administração e Contabilidade (ANEFAC) (2017) Brazilian National Association of Executives in Finance, Administration and Accounting. Credit card interest rates data, 2013 to 2017. https://www.anefac.com.br/uploads/arquivos/2013111411130713.pdf. Accessed 14 Oct 2017

Brazil. Michel Temer, President of Brazil public announcement video. (2017) 'For Temer, the 'creative action' to release FGTS funds caused 'extraordinary outcomes''. http://www2.planalto.gov.br/acompanhe-planalto/noticias/2017/08/para-temer-acao-criativa-de-liberar-o-fgts-deu-resultado-extraordinario. Accessed 19 Feb 2018

Brazil. Medida Provisória number 763 of 2016

Brazilian Central Bank (2017) Resolution number 4549 of 26 January 2017

Brazilian Central Bank, Secretariat of Economic Law and Secretariat of Economic Monitoring (2012) Credit card industry report. Banco Central do Brasil, Secretaria de Desenvolvimento Econoômico e Secretaria de Acompanhamento Econômico. Relatório da Indústria de Cartão de Crédito. https://www.bcb.gov.br/htms/novaPaginaSPB/Relatorio_Cartoes.pdf. Accessed 26 Feb 2018

Caixa Econômica Federal (2017) Cartilha do Crédito Imobiliário. http://www.caixa.gov.br/Downloads/habitacao-documentos-gerais/Cartilha_Credito_Imobiliario.pdf. Accessed 20 Dec 2017

Confederação Nacional do Comércio (National Confederation of Commerce) (2013) PEIC Nacional–July 2013. http://cnc.org.br/central-do-conhecimento/pesquisas/pesquisa-nacional-de-endividamento-e-inadimplencia-do-consumido-25. Accessed 15 Oct 2017

DIEESE—Departamento Intersindical de Estatística e Estudos Socioeconômicos (2017) Política de valorização do salário mínimo: depois de 20 anos, reajuste fica abaixo da inflação (INPC), janeiro de 2017. https://www.dieese.org.br/notatecnica/2017/notaTecsalariominimo2017.pdf. Accessed 2 Jan 2018

Eloy C (2017) Muda o governo, seguem os equívocos. Valor. 14 de fevereiro de 2017

Fundação Getúlio Vargas (2017) Brasileiros usam FGTS inativo para pagar dívidas e consumir, aponta pesquisa. https://portal.fgv.br/noticias/brasileiros-usam-fgts-inativo-pagar-dividas-e-consumir-aponta-pesquisa
Horch D (2014) Brazil's Middle Class Finds a Lifeline at the Pawnshop. The New York Times, 3 December 2014. https://dealbook.nytimes.com/2014/12/03/for-brazilians-pawnshops-are-the-antidote-to-soaring-interest-rates/ Accessed 07 jul 2017
Horch D (2015) In good times or bad, Brazil banks profit. The New York Times. 13 August 2015. https://www.nytimes.com/2015/08/14/business/dealbook/in-good-times-or-bad-brazil-banks-profit.html. Accessed 7 Jul 2017
Hyman L (2011) Debtor Nation. Princeton University Press
IBGE—Instituto Nacional de Geografia e Estatístiica—Tabela da pesquisa mensal de emprego (2017). http://www.ibge.gov.br/home/estatistica/indicadores/trabalhoerendimento/pme_nova/. Accessed 14 Jan 2018
Koutsomanoli-Filippaki A, Mamatzakis E (2009) Performance and Merton-type default risk of listed banks in the EU: a panel VAR approach. J Bank Financ 33(11):2050–2061
Larghi N (2017) Cartões de crédito voltam a crescer com melhora da economia e concorrência. Valor. 14 December 2017
Marques F, Oyamada A (2016) Inadimplência pode atingir pico só em 2017. Valor. 30 Mar 2016
Medeiros M, Souza PHGF, Castro FA (2015) The stability of income inequality in Brasil, 2006–2012: an estimate using income tax data and household surveys. Ciência e Saúde Coletiva 20(4)
MendonÇa MJC (2013) O crédito imobiliário no Brasil e sua relação com a política monetária. Revista Brasileira de Economia. 67(4)
Mota CV (2017) Desemprego na Grande São Paulo bate recorde e atinge 16.8%. Valor Econômico. 27 Jan 2017
Murakawa F, Marchesini L (2018) Senador quer criar CPI para investigar juros de cartão de crédito. Valor. 31 Jan 2018
Tabak BM, Craveiro GL, Cajueiro DO (2010) Eficiência Bancária e Inadimplência: testes de Causalidade. Brasília: Banco Central do Brasil. Trabalhos para Discussão número 220
UQBAR (2017) Anuário 2017—Securitização e financiamento imobiliário. http://www.uqbar.com.br/institucional/oque/anuarios.jsp. Accessed 28 Jan 2018
Vettorazzo L, Perrin F (2017) População brasileira empobrece 9.1% com recessão econômica no país. Folha de São Paulo. http://www1.folha.uol.com.br/mercado/2017/03/1864296-populacao-brasileira-empobrece-91-com-recessao.shtml. Accessed 23 Jan 2017

Chapter 23
Auction Design to Procure Energy Efficiency Measures as Distributed Energy Resources

Tiago de Barros Correia, Gabriel Moreira Pinto, and Vitor Hugo da Silva Oliveira

Contents

Abstract With new technologies and business models emerging in the energy space, the use of demand-side resources to provide or substitute electricity services becomes an attainable target. However, market failures and regulatory barriers still need to be overcome. Nonetheless, policy-makers and regulators are increasing use of market-based instruments—such as white certificates, standard offers, and auctions—to foster investment in energy efficiency. This chapter discusses the Brazilian case, which is going through a transition from an initial command-and-control regulation (launched in 2001) to market-based instruments, that approach energy-

T. de Barros Correia (✉)
Florence School of Regulation, Robert Schuman Centre for Advanced Studies, European University Institute, Florence, Italy
e-mail: Tiago.DeBarros@eui.eu

G. M. Pinto · V. H. da Silva Oliveira
Brazilian Electricity Regulatory Agency (ANEEL), Brasília, Brazil

A. d. Amaral Junior et al. (eds.), *Sustainable Consumption*,
https://doi.org/10.1007/978-3-030-16985-5_23

efficiency measures as a distributed energy resource. The chapter evaluates the pilot auction proposed by the Brazilian National Electricity Agency (ANEEL) to procure energy efficiency measures in Roraima, an isolated system in the Amazon region. To evaluate the pilot, the chapter builds upon auction theory, discussing assumptions such as private value and independence, rationality, risk profile, budget constraints, symmetry, and competition. It also investigates the international experience with energy efficiency schemes from Germany, Portugal, Switzerland and two jurisdictions in the United States (PJM and Ohio), under six pillars: (i) regulation point; (ii) auction methodology; (iii) eligible technologies and measures; (iv) payment method; (v) measurement and verification rules; and (vi) funding sources.

23.1 Introduction

The world's electricity industry has undergone a period of intense technological and regulatory innovation that has enabled the emergence of new business models. Regulatory reforms began in the 1990s with the unbundling of electricity generation, transmission, distribution and commercialization activities and the creation of wholesale markets. The trigger for this disruptive process was both political and technological. The advancement of neoliberal ideas pushed for the privatization of infrastructure assets. The first wave of digitalization made the establishment of wholesale markets feasible. Moreover, improvements in thermoelectric generation with natural gas—from both aero-derivative turbines and the use of combined cycles—have produced considerable gains in energy efficiency, operational flexibility, and scale- and time-reduction in plant deployment and maturation. With this, it has been possible to introduce competitive markets at both ends of the electricity industry chain: generation and commercialization.

Since then, the process of opening electricity markets has intensified with the increase in the competitiveness of renewable generation sources, such as wind and solar, with much lower scale and implementation times than those of natural gas power plants. In the past, policymakers considered energy demand inelastic. Therefore, they used to act only on the supply side (construction of new generation plants to catch up with increased consumption). The continuous process of digitalization and the reduction of the cost of storing electricity will open the electricity market to small household customers and increase the possibilities of active use of demand response[1] as an energy resource to help control the voltage and frequency of networks and to replace or postpone the need to expand generation capacity.

[1]Demand response is a change in the level or duration of electricity consumption in response to price or other forms of financial incentives. It includes consumer actions that can change any part of the load profile, which comprise the permanent reduction in electricity consumption achieved by energy efficiency measures responding to price signals (DOE 2018).

Demand-side resources can also provide load reduction in system peak hours (demand response) and, through more efficient use of energy, achieve non-dispatchable, permanent demand reduction. Therefore, there is an increasing policy interest in opening electricity markets to demand-side resources. The procurement for energy efficiency products in capacity markets has already been in use since 2000 in ISO New England (ISO–NE), and 2012 in the PJM Interconnection (PJM) jurisdiction.[2]

In October 2012, the European Parliament and Council approved Directive 2012/27/EU acknowledging that energy efficiency is a valuable means to address the 'challenges resulting from increased dependence on energy imports and scarce energy resources, and the need to limit climate change and to overcome the economic crisis'. Furthermore, improving energy efficiency will benefit the environment, reduce greenhouse gas emissions, improve energy security, cut energy costs, and help alleviate energy poverty. Shifting to a more energy-efficient economy should also accelerate the spread of innovative technological solutions and improve the competitiveness of the industry, boost economic growth and create high-quality jobs in several sectors related to energy efficiency.

Furthermore, in November 2016, the European Commission (EC) proposed an amendment to the Directive 2012/27/EU to include a new energy efficiency target for 2030. The EC's reasoning is that the cheapest, cleanest and safest energy is that which is not used at all. Simply put, 'energy efficiency needs to be considered as a source of energy in its own right.' Energy efficiency can therefore be thought of as a demand-side distributed energy resource (DER).[3]

The use of energy efficiency as an energy resource is more natural behind the meter.[4] Energy saved through energy efficiency measures reduces the electricity bill without loss of comfort or utility. Nevertheless, energy efficiency is an energy resource at the disposal of the market. The safety and quality of electricity services depend on the existence of potential difference, which may be provided by the addition of power or by the reduction of load.

Thus, it is possible to couple the markets for energy and energy efficiency. In such a combined market, consumers could actively contribute with the balance between supply and demand for electricity, either through distributed generation in their consumer units, the management of their loads or through investments for structural reductions in their electricity consumption. The development of the electricity market based on energy efficiency as a demand-side DER has the additional advantage of reducing the environmental footprint of the electricity sector, primarily through the emission of greenhouse gases (GHG). Furthermore, investments in

[2]Liu (2017).

[3]Any resource or asset, connected to the grid at the distribution level, capable of providing or substituting electricity services, including behind-the-meter renewable and non-renewable generation, energy storage, electric vehicles, energy efficiency and other kinds of controlled loads.

[4]On the user's property rather than on the side of the electric grid/utility.

energy efficiency could have access to a much more mature and liquid market, with well-established sources of funding and financing.

However, it would be necessary to have a reliable method of measurement and verification, and a comprehensive regulatory and contractual basis for the energy-efficiency measures (EEM), converted into MWh or MW, to be traded indistinctly as electricity or capacity. The creation of commercial products that allow measurement and verification of reductions in consumers' capacity and energy requirements allow demand response actions, including consumption reductions from energy efficiency, to be a DER and substitutes for investments in generation, transmission, and distribution of electricity.

In May 2018, the Brazilian National Electricity Agency (ANEEL) initiated a process of public consultation and engagement with stakeholders for updating the current regulations to enable energy efficiency measures. The process aimed to identify and remove any existing barriers to the development of the demand response market based on energy efficiency actions, including the development of a pilot auction to meet part of the demand growth of the isolated Roraima system in the Amazon region.

Considering the successful experience with electricity auctions in Brazil over the last decades, the use of the same instrument to unlock the development of the energy efficiency market seems a natural step. The use of auctions to acquire specific products enables the regulator to mitigate existing market failures and ensure a more efficient allocation of resources. Meanwhile, the auction rules and the contract clauses must be customized considering the specificities of the procured product. Thus, complex design choices need to be made to ensure a level playing field for all participants and minimize the risk of market failure.

The objective of this work is to identify the challenges and opportunities facing the development of the market for energy efficiency as a DER and make market design recommendations for Brazil and the Roraima pilot auction.

This paper is structured as follows. Section 23.2 provides an outlook on the energy efficiency regulatory framework and discusses the existence of market failures and regulatory barriers that need to be addressed to bridge the energy efficiency gap. Section 23.3 analyzes the Brazilian Energy Efficiency Program and the proposal of the pilot auction in Roraima. Section 23.4 gives examples of auction designs to foster new markets for energy efficiency and discusses the implication of some choices over risk distribution, decision-making, and outcomes. Section 23.5 proposes the auction design for a pilot in Roraima. Section 23.6 presents the key conclusions and policy implications.

23.2 Unlocking Energy Efficiency: Issues and Tools

Ideally, energy prices should guide consumers to an optimal level of energy efficiency. However, in practice prices alone seems to be not enough to deliver (implicit) energy efficiency. According to the International Energy Agency (IEA)

Energy Efficiency 2018 report, 'it is becoming increasingly clear that energy efficiency can bring many significant economic and environmental benefits. It is also clear that huge energy efficiency potential remains untapped'. Although there are several highly cost-effective investment opportunities in energy efficiency, the global investment in energy efficiency is not on track to achieve the scale required by the Efficient World Scenario (EWS) developed by the IEA World Energy Outlook.[5] To achieve the expectations of the EWS would require average annual investment in efficient technologies to double until 2025.

Directive 2012/27/EU also highlights the need to identify and remove regulatory and non-regulatory barriers to the use of energy efficiency, underlining the assumption that there is an energy efficiency gap to be bridged. In short, there is a difference between the cost-minimizing level of energy efficiency and the level of energy efficiency realized. Both these concepts imply that society has forgone cost-effective investments in energy efficiency, although they could significantly reduce energy consumption at low cost and with attractive return rate.

23.2.1 *Issues*

This apparent failure to undertake cost-effective EEMs may reflect: (i) market failures; (ii) regulatory barriers, and/or; (iii) different perceptions about the quality of energy services. Among the possible market failures, the literature[6] usually refers to:

1. Positive externalities of energy efficiency related to climate change, energy security and social impacts, which means that private actors may not receive all the benefits of their actions, and so be less likely to take them;
2. Negative externalities of fossil fuels, including distortion in fuel prices that do not reflect the social and environmental costs associated with fuel production, distribution, and consumption;
3. Incomplete markets, as energy infrastructure requires high levels of coordination;
4. Information failures, including (a) information asymmetry between government and industry, (b) a significant level of abstraction required in the implementation of M&V, especially for the 'baseline' definition; and (c) generally poorly informed customers, since (d) there is often a lack of information on the performance of energy-efficient technologies (consumers tend not to change their behavior if little information is provided);

[5]This scenario relies on the premise that all available energy efficiency measures are implemented until 2040. All these measures, according to the IEA, use readily available and cost-effective technologies, which pays back on average by 3 years based on energy savings alone.

[6]MacGill et al. (2013), ANEEL (2018a), IEA (2017, 2018).

5. Misplaced investments and incentives–the principal-agent[7] problem and a lack of life-cycle thinking on costs and savings;
6. Infrastructure limitations. The deployment of energy efficiency technologies is highly restricted by factors such as geography, infrastructure, and human resources;
7. The scarcity of financing sources.

Besides these market failures, the energy efficiency gap may also be explained by regulatory barriers and government fiscal policies. Government policies tend to encourage energy consumption, rather than energy efficiency. For instance, government support has focused more on energy production, and the profit of electric utilities is a function of sales. Other relevant factors that seem to hold the development of the energy efficiency market are the scarcity of sources of financing. Most energy efficiency measures (EEMs) require some up-front investment that must face a discount rate to facilitate the trade-off between the initial capital investment and reduced operating costs. These EEMs also hinder investments in energy-efficiency technologies. Perceived risk of energy-efficiency investments also has a role. Consumers and businesses can be very risk-averse regarding investing in energy efficiency technologies. The uncertainties of fuel prices and the high discount rate for operating costs have both made energy-efficiency investments even more 'risky' for many decision makers.

23.2.2 Tools

On the side of the regulation, the core of the EEM is given by mandatory public policies. These include the labelling of manufactured products, setting minimum energy efficiency standards to products and new buildings, putting obligations on utilities, retail sales or end-users to invest in energy efficiency, achieving energy savings, or enforcing mandatory periodical audits of large company's energy consumption. Most of these policies are designed and defined with a command and control approach.[8]

Non-market-based approaches, however, are losing momentum. According to IEA (2018), increases in energy efficiency investment where benefits should be obvious regardless of mandatory public policy, such as buildings and industrial

[7]The agent–principal problem occurs when one the agents that will make decisions on behalf of the principal has the incentive to act in their own best interests, which are diverse to those of their principals. In particular, renters do not have incentives to invest in long-term equipment for the apartment that they do not own, and landlords lack incentives to invest in energy efficiency as they will not benefit from the energy savings (the so-called 'split incentives' problem).

[8]Except in the case of the labelling policies, which aim to reduce the asymmetry of information so that customers can make better choices, according to their preferences, and the obligations to achieve energy savings, when the regulation allows the obliged parties to procure for their EEMs.

sectors, have been minimal. Command and control approaches leave investors and customers as passive participants. Consequently, private agents see energy efficiency as a regulatory burden rather than as a business. Thus, the path to the development of energy efficiency potential may have to transition from policies based on obligations, control, and enforcement to the use of market-based instruments (MBI).

MBIs for energy efficiency establish a policy goal that market actors must achieve, without prescribing the delivery mechanisms and the specific measures. Thus, it provides market actors the freedom to choose the measures and delivery routes that work best for them to reach the policy goals.

1. the international experience with MBIs for electricity efficiency typically involves three categories: obligations for utilities to execute *standard offers* for EEM, providing electricity saving or peak consumption displacement on customers and end-users;
2. obligations on suppliers or consumers for having a given amount of *white certificates* attesting emerging saving energy; and
3. *auction and tendering programs*, where a government fund or a utility call for EEM projects that can deliver energy efficiency outcomes and the most cost-effective proposals receives long-term contracts or funding support.

Typically, in the standard offer, the obliged party must buy energy saved or peak load shifted for various pre-defined technologies (e.g., water heaters, PV rooftops, lighting, and pumps) and pays a fixed price for every MWh saved or MW displaced. It is an MBI to the extent that the government defines the price and the market defines quantities. In practice, quantity often ends-up being limited by the funds available (authorized by the regulator) to procure those demand resources. In this sense, the standard offer policy is equivalent to a Feed-in-Tariff (FiT)[9] for renewables—which typically has different prices based on eligible technologies.

A Tradable White Certificate (TWC), also called 'cap-and-trade' or 'target-and-trade' schemes, involves the settlement of a mandatory energy-saving target during a given period. Obliged parties then bear this obligation to meet individual energy-saving targets through eligible EEM. Then the white certificates are issued as evidence of realized energy savings. Obliged parties can trade the certificates and, therefore, the market clears the price of the certificate. The main argument for trading is an economic one: equalization of marginal compliance costs among obliged parties—but, parties that achieve significant energy savings are those that can do it inexpensively, so they are likely to supply the market with white certificates (Mundaca 2008).

---

[9]In most countries, a FiT or feed-in law is the primary energy policy mechanism designed to encourage the emergence and development of renewable sources of energy. Under a feed-in tariff, an obligation is imposed on regional or national electric grid utilities to buy renewable electricity from all eligible participants.

To increase the liquidity of the market, in addition to electricity market players, some authorities allow Energy Service Companies (ESCOs) that are entitled to implement measures to earn and trade TWCs. Independent organizations perform activities related to the M&V of energy savings and the coordination of trading platforms. The authorities issue TWCs, and parties that are unable to reach their target pay the penalty for non-compliance.

According to IEA, the only example of a liquid TWC market among obliged and third parties in Italy

> The white certificates programme in Italy has the longest tenure and is the only programme in Europe where a significant fraction of energy savings from accredited parties who are not also obligated entities to exist. About one-third of the white certificates in Italy have been generated by third parties since the programme began in 2005 and most white certificates in Italy are created under bilateral contracts between an energy efficiency company and an obligated energy distributor.[10]

Policymakers have also experimented with auctions as an instrument to foster more competitiveness in choosing EEM. An auction is an institution, or a set of trading rules, used to assign rights and reveal prices. A common aspect of auction-like institutions is that they elicit information, in the form of bids, from potential buyers or sellers regarding their willingness to pay or supply. The outcome—that is, who wins what and who pays how much—is determined solely based on the bid information. A second important aspect of auction-like institutions is that they are anonymous. By this, we mean that the identities of the bidders play no role in determining the winner and the price.[11]

Portugal, Switzerland, the United States, and Germany introduced competitive tenders, in 2007, 2010, 2010 and 2016, respectively. Moreover, since 2015, Brazilian regulation requires DISCOs to comply with the obligation to contract EEM through a tendering process. Genuine auction for energy efficiency, however, is still a novelty.

One reason why MBIs have not taken off so far is that public policies usually have intricate and complex motivations. Policymakers wish to see efficiency gains made in different parts of the economy and with a variety of technologies and different objectives for the outcome, such as:

1. increasing economic efficiency;
2. reducing CO_2 emissions;
3. alleviating fuel poverty;
4. reducing load at peak times;
5. reducing demand in certain locations; and,
6. encouraging innovation.

For this reason, just as renewable energy certificates programs often distinguish between photovoltaic renewable energy certificates and wind renewable energy

[10] IEA (2017).

[11] Krishna (2002).

certificates, many MBI schemes have multiple types of incentives and conditions to eligible measures and technologies. This has the side-effect of fragmenting what might be one large energy efficiency market into multiple sub-markets, which are likely to be less liquid and competitive. As seen above, this is one of the fundamental causes of the low development of the energy efficiency market.

Each of the different MBI categories has advantages and disadvantages. The standard offer has the benefit of simplicity and low transaction costs but relying on prices centrally defined by the government poses a significant risk. If the set price is too small, it will attract the interest only of a few investors and will not deliver the total potential for EEM. If the set price is too high, society may have to overpay, which would not be necessary for an efficient process.

TWC, in its turn, requires a complex set of institutions and regulations to design its market, which typically includes the creation of independent certificate verification and marketing entity, which brings relevant transaction costs. It also relies on policymakers' quantity decision. On the positive side, TWC enables the creation of an efficient market, benefitting from efficient prices and much greater engagement of obliged parties, customers and ESCO. TWS schemes are usually effective in facilitating ESCOs to assist consumers that are not highly motivated and may lack the skills to undertake EEM.

Therefore, MBIs will work better when the attribute being sold is uniform in nature and where compliance with requirements that can easily be tracked. If energy savings is traded via a simply-defined white certificate (e.g., one MWh of reduced consumption = 1 white certificate), for example, an obligation scheme could operate through a comparatively large and liquid market of buyers and sellers. Programs that reward all energy savings with the same white certificate will incentivize the take-up of least expensive measures that are the most profitable and least risky way to earn white certificates.

Finally, auctions can carry with it most of the complexity already faced in the TWC,[12] depending on the adopted scheme (including M&V transactional costs and abstraction problems in the definition of baselines).

Further, by establishing long-term commitments, even before the start of disbursement, auctions allow for a more comprehensive approach to the risks distribution among each party. They also provide financial receivables that serve as collateral for acquiring bank loans. As auctions have been used as a long-term contracting tool for renewable sources and generation capacity, their use for contracting EEM facilitates the transition to the coupling of energy and energy efficiency markets. Therefore, it is the most natural way to allow the treatment of energy efficiency as a DER.

As many of these challenges were faced during the development of the market for renewable energy, it is possible to contrast the MBI used in energy efficiency and renewable energy. It is also interesting to note that the application of the instruments

[12]Except by the need of an exclusive marketing entity.

Fig. 23.1 Correspondence between Energy Efficiency and Renewables policies (Reproduced from Maurer 2018)

also seems to follow the same pattern, suggesting that the adoption of auctions correlates with the greater maturity of the markets (See Fig. 23.1).

For countries that are familiar with Feed-in Tariffs, Standard Offers can be the most natural way to allow the treatment of energy efficiency as a DER. In turn, countries that have solid green certificates market, or even cap-and-trade markets for CO_2 emissions, may prefer white certificates. Finally, countries that make use of auctions for electricity expansion may prefer auctions for energy efficiency as well.

In any case, whatever the selected MBI scheme, its effective implementation will undergo the balance of several aspects, such as (adapted from Maurer 2018):

1. Regulation point (e.g., the obliged party);
2. Definition of eligible technologies and measures;
3. Obligation and saving targets set in energy, capacity or both;
4. Measurement and verification (M&V), including the construction of the baseline;
5. Payment method;
6. Trading rules; and
7. Funding sources for the investment on EEM.

Regarding auctions, trading rules are not relevant. On top of the six remaining ones, based on our research, we would like to add two more:

8. Auction methodology;
9. Impact on the distribution utilities.

23.3 International Experience with Tenders and Auctions for Energy Efficiency

So far, six countries have experimented with auction-like schemes for energy efficiency: Brazil, Germany, Portugal, Switzerland, the United Kingdom and the United States (multiple states). As this chapter focuses on obligations for energy savings (instead of energy capacity), it will focus on the first four (Brazil, Germany, Portugal, Switzerland) and the state of Ohio in the US.

23.3.1 Brazil

Brazil had a conjectural experience with energy efficiency MBI between June 2001 and February 2002, a period when consumers faced electricity rationing to avoid blackouts. At the time, the baseline to consumption reduction targets was set by the average consumption in May, June, and July of 2000. For low voltage customers (householders and small commercial), the goal of rationing was 20% while for high voltage customers (larger commercial and industrial), the target could range from 15 to 25%, depending on the end consumer activity. Customers who did not comply were subject to curtailment for three uninterrupted days and, in case of recurrence, 6 days.

High voltage customers, however, could trade certificates of electricity consumption reduction bilaterally.[13] The Brazilian certificates of electricity consumption reduction were traded through daily auctions. Therefore, the first Brazilian experience with energy efficiency policies was a mix between a TWC and an auction scheme with the regulation point on the high voltage customers. The difference is that, because of the urgency situation, no specific market entity was created and the certificates were negotiated in the wholesale electricity market.

However, because of an increase in hydroelectric reservoirs levels given rain, emergency construction of thermoelectric plants and massive investment in energy efficiency, the Brazilian electricity market faced an oversupply in the years immediately following 2003. Thus, the first Brazilian experience contracting was interrupted. The experience with the use of auctions as a mechanism of product selection and allocation of resources, however, was harnessed and deepened over the following years.

Auctions had great success in the introduction of renewable sources wind and solar using the representation of its technical and operational characteristics in its contractual obligations for electricity delivery. From 2004 to 2017, the installed generation capacity expanded 64,615 MW and the volume of energy traded in the wholesale market 59,820 TWh, respectively increases of 71% and 440%. The use of

[13]Silva (2003).

regulated auctions to create a market for long-term contracts reduced the sellers' exposure to the spot price, defined by the marginal cost, and remunerated the new power plants by their average cost.

In the same period, Law No. 9991/2000 created the structural Brazilian public policy for Energy Efficiency (PEE) that established the obligation for electricity distribution companies (DISCOS) to spend 0.5% of their Net Operating Income (NOI) in EEMs. Consequently, in almost 20 years of existence of the program, more than 4000 projects were carried out and around USD 1.5 billion[14] invested. However, only after 2015, the regulation established that the choice of EEMs should occur through public tenders, as a safeguard to obtain more cost-effective results. In this sense, the Brazilian PPE started as an instrument similar to standard offers with regulation point on DISCOS, in the Brazilian case. However, the mandatory target was not energy efficiency per se, but an application of minimum values in EEMs ever since it has evolved towards the auction-like system.

23.3.2 Portugal[15]

Portuguese Decree–Law No. 29/2006, obligated electricity distribution utilities (regulation point) to apply energy efficiency approved in the Plan to Promote Efficiency in Electricity Consumption (PPEC) of Portugal.[16] The PPEC has access to a government fund, originated from electricity bills (levy on tariff), that provides support up to 80% of total costs to projects approved by the program. The PPEC focuses on measures that aim at reducing the consumption of electricity or the permanent load management (only verifiable actions). Other indirect measures are also eligible, such as informational or disseminative measures that induce a more rational behavior and allows a more conscious view on the decision making of consumers. The PPEC accepts projects of two broad categories: tangible and intangible. Tangible EEMs refer to actions that contemplate the installation of equipment with energy efficiency superior to standard technology. Intangible EEMs are those that provide customers with relevant information on electricity consumption and the benefits of more efficient consumption habits, namely, actions of training, information dissemination campaigns, and energy audits.

DISCOs can submit projects, electricity traders, ESCOS, educational institutes and associations and entities that contain in their statutes the promotion and protection of customers, and business and municipal associations. Project applications

[14]R$5.7 billion.

[15]The authors thank the Portuguese Energy Regulator ERSE (Entidade Reguladora dos Serviços Energéticos) team, particularly Pedro Verdelho and Cristina Barros, for the videoconference on 20 December 2018, which has contributed to our understanding of the Portuguese experience.

[16]ERSE (2013).

should indicate the EEM market segment (e.g., household, industry and agriculture, commerce, or services), as well as:

1. the existing market barriers to the development of the EEM without the PPEC support, as well as strategies to overcome them;
2. the project budget;
3. the assessment of the EEM benefits; and,
4. the M&V plan and any other relevant information.

The PPEC will not support (i) distributed generation, (ii) tangible EEM projects with disbursements lower than 25% of the total budget in the first year of implementation, and (iii) EEM that results from legal obligations.

23.3.3 Switzerland[17]

In 2009, the Swiss Government adopted the Prokilowatt[18] program to foster the reduction of electricity consumption in about 20TWh by replacing or retrofitting existing facilities, equipment, and buildings. The Prokilowatt covers up to 30% of the costs of approved projects, and up to 100% of their administrative costs (limited to 25% of total costs). A levy on tariff funds it. The regulation point is private and public companies, including ESCOs and customers.

Projects must indicate the estimated cost-effectiveness and detail its underlying calculations. It is also necessary (i) to ensure the commitment with the reduction in electric energy consumption for at least 10 years; (ii) to prove the need for technical intervention, and; (iii) to attest that EEM performance does not rely only on the user's behavior.

The Prokilowatt does not support EEMs that:

1. are identical or similar to an EEM that is already being carried out in the same region (either from Prokilowatt or from another program);
2. promote efficiency measures in incineration and sewage treatment plants;
3. are essentially R&D or advertising;
4. promote only the substitution of conventional meters for smart meters;
5. substitute of electricity consumption by a non-renewable energy source;
6. promote the replacement of fans with a power lower than 125 W and transverse flow fans;
7. result in a reduction in production;
8. reduce the heating demand of the building by structural measures such as thermal insulation; and

[17]The authors thank the Bundesamt für Energie (BFE) team, in particular Simone Hegner, for the videoconference on 9 January 2019, which has contributed to our understanding of the Swiss experience.

[18]BMWi Bundesministerium für Wirtschaft und Energie (2017).

9. have their expected payback in fewer than 4 years, given the principle of additionality.

23.3.4 Germany[19]

In December 2014, the German Government adopted the National Action Plan for Energy Efficiency (NAPE) in which STEP UP! (*Stromeffizienzpotentiale nutzen*)[20] is one of its main pillars. The purpose of NAPE is to promote a primary use energy savings of 35 PJ. The program aims at fostering EEM that would normally be discarded because of long payback time. This way the STEP UP! covers up to 30% of the total costs of approved projects. A levy on tariff also funds STEP UP!

A call for projects in which any company based in Germany can submit projects selects the EEMs. The regulation point is private and public companies, including ESCOs and customers. Projects should indicate:

1. the EEM category;
2. a summary of the EEM's executive project;
3. the schedule of activities;
4. the financial planning; and,
5. a justification for the need for the subsidy.

23.3.5 The United States (Ohio)[21]

In the United States, energy efficiency programs have been implemented at the state level. Since December 2014, the Ohio electricity distribution company (AEP Ohio) offers the Bid4efficiency[22] program within its portfolio of energy efficiency schemes to achieve its energy savings targets. A call for projects selects the EEMs. Any non-residential AEP Ohio customers with consumption of more than 200 MWh/year or solution providers (companies that are registered with AEP Ohio to provide energy efficiency services for non-residential customers) can submit projects. Bidders should indicate a description of their EEMs and the strategy to deliver the electricity consumption reduction.

The Bid4efficiency does not support EEMs:

[19]The authors thank the VDI/VDE Innovation + Technik GmbH team, particularly Martin Richter and Felix Herziger, for the videoconference on 8 November 2018, which has contributed greatly to our understanding of the German experience.

[20]Bundesministerium für Wirtschaft und Energie (2017).

[21]The authors thank the AEP Ohio and the Overlay Consulting teams, particularly Angela Rybalt and Jon Dierking, for the videoconference on 8 November 2018, which contributed to our better understanding of the Ohio experience.

[22]American Electric Power Ohio (2018).

1. at locations that are not serviced by the AEP Ohio distribution company;
2. with a payback period lower than 1 year;
3. where electricity savings result of fuel switching (e.g., electricity to gas);
4. that promote the removal or termination of processes, companies and operations; and
5. involving distributed generation from renewable sources.

There are have been auctions schemes similar to Ohio's in the United States, such as the one executed in Missouri in 2010, as well as programs in New York and Wisconsin.

23.3.6 United States (PJM)[23]

Since 2012 PJM has allowed EEMs to directly compete against other resources in the Reliability Pricing Model (RPM).[24] The RPM is a multi-auction structure designed to procure resource commitments to satisfy the region's unforced capacity obligation through the following market mechanisms: a Base Residual Auction (BRA), three annual incremental auctions, and a Bilateral Market. The cost of the procured capacity (including EEM) is allocated to Load Serving Entities at the clearing price.[25] The majority of capacity is procured in the BRA, which is conducted 3 years in advance of the delivery year; the incremental auctions are conducted in the following years to allow adjustments in the resource commitments because of reliability requirements.

Under the capacity performance provisions for the 2018–2019 delivery year, PJM restructured the capacity products under two types: Capacity Performance and Base Capacity. Base Capacity Resources include Base Capacity Demand Resources (DR), Base Capacity Generation Resources (BG), and Base Capacity Energy Efficiency (EE) Resources, which are expected to provide permanent continuous load reduction only during the summer months.[26]

EEMs that do not involve permanent replacement of devices and installations that are user controlled or dispatchable do not qualify as an EE Resource in RPM. Examples of EE projects that do not qualify as an EE Resource include: reducing load by switching off devices, energy conservation (change of behavior) with no identified controls, load shift (e.g., user-controlled, programmable thermostats or thermal energy storage), generation (e.g., Behind-the-Meter Generation, such as

---

[23]PJM is the regional transmission organization that coordinates the movement of wholesale electricity in all or parts of Delaware, Illinois, Indiana, Kentucky, Maryland, Michigan, New Jersey, North Carolina, Ohio, Pennsylvania, Tennessee, Virginia, West Virginia and the District of Columbia. (www.pjm.com).

[24]Liu (2017).

[25]PJM (2019c).

[26]PJM (2019a).

back-up generation, solar, wind, or cogeneration); and conversion to natural gas. However, some of these projects may qualify as DR in the PJM Capacity Market.[27]

Selection Criteria The Brazilian certificates of electricity consumption reduction were traded through daily auctions. Both sellers and buyers submitted a sealed bid with price and quantity in 155 auctions. The auctioneer then constructed the supply and demand curves whose intersection defined the price that clears the market.

The Portuguese PPEC and the Brazilian PEE select EEM through a call for projects tender[28] that follows a sealed auction-like rule to receive the bids, rank the projects, select winners and clear the price (or value) that the programs will support. Winners are chosen based on a multicriteria decision-making process.

The selection criteria for PEE projects is a multi-criteria qualification based on the following conditions:

1. Cost–benefit ratio;
2. Economies of scale;
3. The share of equipment investments in total costs;
4. Direct impact on energy savings and peak load reduction;
5. Quality of the project presentation;
6. Capacity to overcome market barriers and its multiplier effect;
7. Experience in similar projects;
8. Counterpart (the percentage of total cost that the PEE will cover);
9. Diversity on end usage (as a policy to incentivize the use of multiple technologies); and
10. Educational actions (training and capacitation).

In the case of Portugal, for tangible EEMs, the auctioneer considers: (i) the ratio between costs and benefits; (ii) the size and (iii) the risk related to the scale of the measure, and; (iv) the proportion of investment in equipment in the total cost of the EEM. Intangible EEMs, in turn, are selected by: (i) the quality of the presentation of the project; (ii) the expected ability to overcome market barriers and; (iii) its multiplier effect; (iv) equity; (v) innovation and; (vi) bidder experience with similar measures. The winner of the tender must submit a semi-annual report of the measures executed to obtain his/her due payments.

The Swiss Prokilowatt and the German STEP UP! also select EEMs through calls for projects with a sealed auction-like rule to receive the proposals and the winners are chosen based on cost-effectiveness. In Prokilowatt, projects should demonstrate the estimated cost-effectiveness (Rp./kWh) and justify the underlying calculations. It is also necessary to ensure the commitment with the reduction in electric energy consumption with a term greater than 10 years, to prove the need for technical intervention, and to attest that EEM performance is independent of the user's

[27]PJM (2019b).

[28]We define a tender as a bidding process where the winners are chosen by multicriteria analysis that may include the nature and features of the project.

behavior. In the STEP UP! scheme, the winners are chosen based on the ratio cost effectiveness. For bids with the same effectiveness and payback of more than 3 years, the winner will be the project with the lowest cost.

The American Bid4efficiency uses a reverse English auction[29] rule to receive bids. The auction starts with a maximum price, and the bidders submit bids of lower prices until the total offers of EEM meet the energy savings target. The total payment received by the seller, however, will be limited to 50% of total project costs. Conversely, the winner will receive the minimum between the amount of energy saved times the winning bid (price) and 50% of project costs.

Measurement and Verification (M&V) In the Brazilian certificates auction, the energy saving was measured directly in the electricity wholesale market. In the Brazilian PEE and the Portugal and USA programs, the winner must inform its measuring plan, compatible with the characteristics of its EEM that will be verified by the DISCO,[30] in the Brazilian case, or by an independent auditor, in the other ones. The German and Swiss programs also require an auditor to confirm the energy savings, but these two provide the measuring methodology. Both programs provide the step-by-step to calculate the energy savings to facilitate the work for both parties.

Risk and Policy Impact on the Distribution Utility In the Brazilian certificates auction, there were no implementation or performance risks associated with the agents, the government or the regulator. The real issue was to establish a mechanism to minimize the social cost of the rationing program. In this sense, the program aimed at avoiding the risk of a blackout. On this side, the program was successful. Under the Brazilian PEE, the DISCO must comply with a minimum of CBR of 0.8 for energy efficiency measures. If the project underdelivers, the company will get only the respective share of total project costs. If the regulator does not approve yearly expenses adding up to 0.5% of the company NOI within 2 years, the company is subject to a fine.

In the cases of Portugal, Switzerland, Germany, and Ohio, the winners bear the risk related to the costs and performance of the EEM. If the project does not comply with the requirements and standards of the regulation, the amount supported by the program will be reduced or even suspended. However, there is no additional regulatory or contractual penalty for failure or underperformance on delivering the committed EEM. Therefore, the society bears the risk (opportunity costs) that the winners quit trying to deliver the initial project and settle for an alternative that is good enough to assure some support from the program, but that would not be selected in the bidding process. Further, it is important to remark that, in Germany, the proof of fraud subject the perpetrators to prison penalty, which helps to mitigate the moral hazard risk.

---

[29]We define an auction as a bidding process that selects winners solely based on bids of price and/or quantity.

[30]Subject to ANEEL's oversight.

Furthermore, the decision to not support projects with quick payback times, present in the Swiss, German and American schemes, may distort the decision-making process of private agents selecting a new investment. They may decide to implement projects with lower energy and economic efficiency preferable than other options which will have no access to the program funding—the agent–principal problem, resulting in a loss of social welfare.

Concerning the impact on the distribution services, Brazilian DISCOs are directly affected by the market loss from its EEMs until the next regulatory review, which takes place on average every 5 years. In Portugal, because of decoupling, the DISCO is not affected. In the USA, the DISCO loses the market, but it also receives remuneration from its efficiency services.[31]

23.4 The Brazilian Energy Efficiency Program and the Pilot Auction Proposal in Roraima

The analysis of projects carried out under the Brazilian Energy Efficiency Program (PEE) shows that approximately 54% of the resources were applied in EEM projects for low-income residential consumer units. Additionally, only 4.5% of the resources were invested in industrial consumer units and 2.8% of the resources in consumer units classified as commercial, segments where the scale would allow for more effective energy saving and peak demand reduction.[32]

The regulatory impact assessment by ANEEL indicates that the Brazilian strategy to promote energy efficiency presents a structural contradiction.[33] On the one hand, Law 9991 assigned the DISCOs the obligation to collect resources from customers and invest 0.5% of their NOI in energy efficiency projects. On the other hand, the distribution tariff is defined *ex ante* (price-cap model), and the DISCO bears the risk of market realization (MWh consumption). In this way, the DISCO is the protagonist in the execution of a public policy that compromises its revenue—the more efficient its implementation, the lower the market realization and the DISCO's revenues.

Another aspect that discourages the DISCO from developing energy efficiency projects is the transaction cost of the PEE—the company must implement and monitor an extensive portfolio of small projects, outside of its main business, that ANEEL must approve one by one with risks of costs not being recognized. An indication of the DISCOS's lack of interest is the unapplied resources of about USD 435 million.[34] If there were a friendly regulatory framework for the expenditure of these resources, the DISCO would do so more assiduously.

---

[31]No specific information was found about the possible impact on distribution services for Germany and Switzerland.

[32]ANEEL (2018a).

[33]Ibid.

[34]R$1.7 billion.

It is true that, in specific cases, the DISCO can benefit from the energy efficiency resulting from EEMs aimed at low income or budget constrained consumers (as may be the case of local governments and public organizations) that could reduce the level of payments defaults. EEM in areas with overload also can delay the need for investments and reduce electric losses, and efficiency campaigns generate value in the image of the distributor. However, in general, the loss of revenue is relevant and discourages DISCOS from truly engaging with PEE.

Regarding the cost of the project, the Cost–Benefit Ratio index (established in the PEE regulation) only requires energy efficiency projects to cost less than the avoided generation, but ANEEL acknowledges that there is no guarantee that agents are executing EEMs at the lowest possible cost. Thus, the problem is an inefficient regulatory design to achieve the purposes of the public policy of encouraging energy efficiency.[35]

To overcome the identified regulatory failures, ANEEL began to study the possibility of strengthening the use of MBIs to foster EEMs with two central assumptions. First, the program must rely on an economic foundation, using auctions to select the lowest cost EEM for the same energetic output. In this way, regulation should be technologically neutral and friendly to the creation of new business models, to allow creativity and innovation by the investors. Likewise, participating companies should have the incentives aligned with ANEEL, to increase energy efficiency with a consequent reduction of consumption.

Second, the regulation should enable EEMs that ensure reductions in electricity consumption (MWh) to be a DER and thus a substitute for electricity generation assets. Thus, instead of creating a market for energy efficiency, the regulator would allow the insertion of the EEMs in the electricity market, which has much more liquidity, sources of funding and financing and instruments for risk management.

Differently, from most regulators and policymakers who implement new regulations nationwide, ANEEL decided to apply some key concepts of the 'design thinking' process—in particular, its Stage 4, Prototyping.[36] According to designers, there are eight ways of prototyping. ANEEL started with 'sketches and diagrams' and 'paper interfaces'[37] to illustrate the relationships among all players, a risk map, and the program, step-by-step. Based on these quick products, developed in a few weeks, ANEEL invited key executives from the energy efficiency industry to provide their feedback. Given that the market feedback was very positive, ANEEL spent a couple of months developing a second prototype: financial models simulating the impacts of the auction under multiple perspectives (consumers, entrepreneurs, government, suppliers). By doing so, it identified several problems and risks

[35]ANEEL (2018a).

[36]https://www.interaction-design.org/literature/article/stage-4-in-the-design-thinking-process-prototype.

[37]https://www.interaction-design.org/literature/article/prototyping-learn-eight-common-methods-and-best-practices.

and develop strategies to mitigate it or to present the involved trade-off to the stakeholders and decision makers.

ANEEL then proposed a first pilot project in Roraima, the perfect place to do so as it is the biggest isolated load from the national system, allowing for much better monitoring of the impacts of the first energy efficiency auction on multiple stakeholders. The pilot auction has been known as a *reverse generation auction* or a *negative energy auction*. The pilot scheme proposes a reduction of 35 GWh per year, capped at about 5 million EUROS/year.[38]

ANEEL made public all documentation (including sketches, diagrams, and financial models) from the start of its public consultation process, enabling the debate with market players. Following that, the team carried out a field mission, to collect feedback on the project from local consumers, government authorities, the DISCOs, and suppliers. ANEEL then organized in its headquarters multiple workshops to share these insights with development banks, private investors, academic institutions, multilateral development banks, and foreign regulators (Portugal, Germany, Switzerland, and the US). The next form of prototyping will be the release for a public discussion of all contracts and legal documents. Finally, ANEEL plans to organize a mock auction on its headquarters, so much that all interested companies can simulate their bids. By the mid of 2019, it will execute its first pilot project in Roraima and choose two other locations as the next pilots.

Any person or company can participate in the auction, except the local government (state and municipal), the local distribution company and its related parties, as they would face a conflict of interest. Moreover, any measure that reduces electricity consumption would be eligible, including intangible actions (such as behavioral interventions) and small renewable distributed generation (microgeneration with local consumption, including homes, small commerce and condominiums, systems limited to 75 kV). The winners will be denominated Consumption Reduction Agents (CRA).

The CRA will commit to the annual reduction of electricity consumption of its own or a portfolio of aggregated customers under his responsibility for five years, known as Electricity Savings Commitments (ESC). To avoid the risk of having the entire outcome of the pilot depending on the performance of a single winner, the auctioneer will procure for small ESC contracts, and each participant will be able to sign at most two ESC, ensuring that the pilot project will have between four and eight winners.

The pilot will exempt previous technical qualification, focusing entirely on results regarding electricity (MWh) saved, with contractual penalties for annual under delivery. To avoid inexperienced participants, the bidders must present financial guarantees or insurances for their bids and for the compliance with the contract, as well as to prove the availability of enough resources to a minimum equity investment on proportion to the total expected investments.

---

[38]Cap of R$600/MWh $\times$ 35,040 MWh/year = R$21,024,000.

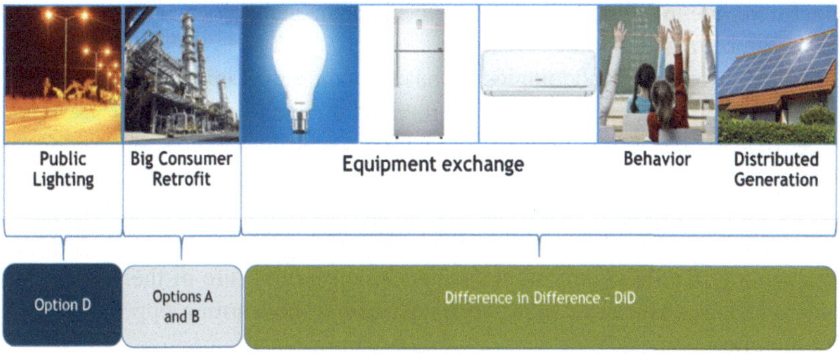

Fig. 23.2 M&V methodology applied to each type of EEM in the Brazilian pilot (Source: ANEEL 2018b)

The rules of the auction will already define the Measurement & Verification (M&V) methodology applied to each type of EEM. For public lighting and retrofit of large consumers EEMs, consumption reduction would be certified based on the International Performance M&V Protocol (options A, B, and D).[39] For EEMs aimed for the small and medium-sized consumer, there would be a need for *ex ante* certification to trace a baseline. The external parameters would be measured and verified based on the comparison of the performance of the consumers subjected to EMMs (treatment group) with the other consumers (control group), with the Difference-in-Difference (DiD) methodology (Fig. 23.2).

Concerning risks to the winner, the remuneration is variable, and the regulator will not see the real costs of the project. In this way, the CRA can either make a big profit or lose a lot. The first risk, therefore, is inefficient pricing in the auction. Especially because it will make a firm commitment—namely, a winner not delivering results, in addition to not receiving, will still pay a fine of 20% for the revenue equivalent to each unit of energy not delivered. The pilot project mapped eight main risks for the CRA:

1. **Financial**: high, as it is a new business model that banks are not familiar with;
2. **Political**: low, as the federal, state and city governments are on board;
3. **Commercial**: medium, because of the PEE, many Brazilian companies are experienced with selling/donating energy efficiency equipment to the population, including low income;
4. **Operational**: low, as most EEMs available in Roraima are simple;
5. **Environmental**: low, as the only issue is to recycle the replaced goods;

[39] **Option A: Partially Measured Retrofit Isolation** involves isolation of the energy use of the equipment affected by an EEM from the energy use of the rest of the facility and uses partial measurement and stipulation to determine the energy saving. **Option B: Retrofit Isolation** has savings determination techniques identical to those of Option but requires full measurement. **Option D: Calibrated Simulation**, savings are determined through simulation of the energy use of components or the whole facility.

6. **Performance**: medium/high, because besides the technology risk, the CRA also bears the behavior risk for 5 years;
7. **M&V**: low, as the methodologies are based on the actual invoice from the DISCO and the criteria are parameterized on the auction documents;
8. **Credit**: low, because a pool of large DISCOs from all over the country will finance the auction.

From the regulator's point of view, the main risks are:

1. the winning EEMs are not implemented because of failure of the bid to foresee and assess its costs and risks properly—wasting the learning opportunity aimed by the pilot;
2. collusion distorted the price cleared by auction;
3. the program resources are earmarked for energy efficiency measures that would be adopted without the need for public support (the principle of additionality);
4. the division of costs and benefits among society increases inequality; and
5. The consumption reduction is undertaken in socially undesirable ways, e.g., with loss of comfort for the consumers or even reduction of productive activity by the trade or industry (Table 23.1).

23.5 Auction Design Proposal for the Roraima Pilot

The schemes described in Section 22.3 adopted variations of two auction designs. The Brazilian certificates auction was designed as pay-as-clear[40] sealed bid auction. The Brazilian PEE, and the Portuguese, Swiss, and German schemes use the pay as bid sealed bid auction design. The Ohio scheme used the open auction design that is equivalent to reverse English auction (pay-as-clear open bid auction).

In the case of the Brazilian certificates auction, the option for a sealed bid design does not seem to have been successful. Of the 155 auctions held, only 77 ended with some energy transaction. In other cases, the auction ended without any agreements that satisfied the price and quantity conditions presented in both buyers' and sellers' bids. An open auction could allow adjustment of expectations.

The preference for the pay-as-bid[41] sealed bid auction in the other studied schemes is justified by the fact that it is a modality of easy implementation by the auctioneer and that allows a long window for projects submission. Further, issues related to the complexity of strategy development in first-price auctions are mitigated by the fact that the compensation of the EEMs is a function of the project cost rather than the bidder's price. That approach simplifies participants' decision-making process but adds a significant transaction cost to the schemes because it requires a

[40] Also known as a second price auction. For more detail, see Correia et al. (2009).

[41] Also known as a first price auction. For more detail, see Correia et al. (2009).

Table 23.1 Comparative table of the different schemes of EEM auctions (compiled by the authors)

Program	Roraima Pilot	Certificates Auction	PEE	PPEC	Prokilowatt	STEP UP!	RPM	Bid4 efficiency
Non-eligibleEEM	None	None	None	First year budget <25% or 20%	Smart metering	Energy storage	Reducing load by switching off devices, change of behavior with no identified controls, and load shift	Change of fuel source
Distributed generation	Yes	Yes	Yes	No	No	No	No	No
Remuneration	Cleared price	Winning bid	Up to 100% of project costs	Up to 80% of project costs	Up to 30% of project costs	Up to 30% of project costs	Clear price	Winning bid, up to 50% of project costs + small performance bonus
Selection criteria	Cost-effectiveness (R$/MWh)	Cost-effectiveness (R$/MWh)	Multi-criteria	Multi-criteria	Cost-effectiveness (Rp./kWh)	Cost-effectiveness (€/kWh)	Cost-effectiveness (U$/MW. day)	Cost-effectiveness (U$/MWh)
Measurement	Difference-in-difference, board data, IPMVP	Own method	IPMVP	Own method	Method pre-defined by the regulator	Method pre-defined by the regulator	EE resource provider must submit an M&V Plan to PJM	Own method

(continued)

Table 23.1 (continued)

Program	Roraima Pilot	Certificates Auction	PEE	PPEC	Prokilowatt	STEP UP!	RPM	Bid4 efficiency
	options A, B and D							
Verification	DISCO	Wholesale market entity	DISCO	Independent auditor	Independent auditor or BFE	Independent auditor	Independent auditor	Independent auditor
Penalty	20% of what is not delivered	None	None	None	None	Prison time or cash fine (just for fraud)	Non-performance charge rate	None
Minimum classification criteria	Warranty, minimum net equity, performance bond	None	Benefit-cost ratio <0.8	Positive 'Social Test' (NPV > 0)	Payback >4 yearsCER < 8 Rp./kWh	Payback >3 yearsCER < 0.1 €/kWh	The period of an EE installation determines whether an installation is eligible to be a capacity resource for a delivery year	Warranty + region compatibility

high degree of monitoring and auditing of the costs estimated in the projects and executed by the winners. Thus, although they are competitive processes, it would be more accurate to define them as tenders. At first glance, the auction name looks more appropriate for the American scheme. Bid4efficency allows the auction to clear the price through a more dynamic process and the agents are remunerated by the cleared price. However, it is also capped at 50% of the project cost. Therefore, up to now, there is no pure auction for energy efficiency in place.

In the case of the Roraima Pilot, the choice of the auction modality involves the evaluation of the structure and maturity of the energy efficiency market in Brazil and the existence of sunk costs and other elements that may affect the agents' risk willingness during the auction. The arguments will be detailed below, but there is a clear recommendation for adopting an open and pay-as-clear auction design. Such an approach increases the attractiveness of the auction by simplifying the process of developing the bid strategy and reduces the risk of the so-called *winners' curse.*[42] Issues related to the structure and maturity of the market and the risk of price distortion by collusion or cartel are addressed by the inclusion of additional safeguards in the systematic auction.

First, in a scenario of an immature market, bidders will probably have some degree of value interdependence, as each bidder may learn and therefore be influenced by the valuations of the other bidders. It means that the announcement of the victory causes a decrease in the initial estimated value and this may represent the conclusion that the value had been overestimated. A failure to foresee and consider this effect fully when formulating bidding will result in the *winners' curse.*

In the pilot project financial model, it can be seen that each lot of 4.4 GWh/year delivers a Net Present Value (NPV) from R$ 21.1 to R$ 32.6 million in avoided subsidies throughout 10 years to Brazilian customers, depending on the winning bid. In the base-case scenario, the expected clearing bid is EU 0.10/avoided kWh, yielding a NPV of R$ 24.9 million to Brazilian customers in avoided subsidy. For every R$ 1 reduced at the bidding price in the range of R$ 0/MWh and R$ 600/MWh (auction cap), the marginal expected benefit of the bid is only R$ 19,215. Therefore, the downside of not implementing the project is of at least R$ 21.1 million, whereas the upside of fighting for a higher bid is at most R$ 12.5 million—in terms of expected value, every R$ lower that the bid may clear is only worthy if the probability of winner's curse is lower than 0.08%.[43]

The design of the auction, however, can reduce the risk of the *winners' curse.* In his seminal paper, Vickrey[44] (1961) demonstrated that the expected payment function of a bidder does not depend on the particular auction form and the expected revenue in any auction is the same if the following assumptions are ensured:

---

[42]Defined by Krishna (2002) as the possibility that the winner overbids and pays more than the true value.

[43]Analysis executed based on the financial models published by ANEEL at the Public Consultation n. 007/2018.

[44]Vickrey (1961).

(a) *Private value & independence*—each bidder knows the value of the object to himself at the moment of the bid, and the bidders' private information values are not correlated;
(b) *Rationality*—all bidders seek to maximize their expected profits;
(c) *Risk neutrality*—Bidders are indifferent between a sure chance of gain or loss and an unsure chance with the same expected amount of gain or loss;
(d) *The absence of budget constraints*—all bidders have enough budget to pay up to their respective values;
(e) *Symmetry*—all bidders draw their value according to the same distribution function;
(f) *Competition*—the absence of collusion among bidders.

Intuitively, such a result occurs because of a compensatory balance between the dominant bidding strategy and the price definition rule. In a pay-as-clear auction, for example, each bidder knows that his bid is the sole criterion to determine his winning chances but has no effect over the expected price in a victory scenario. Thereby, the dominant strategy is to bid his true reservation value.

In contrast, in a pay-as-bid auction equilibrium behavior could not be to bid an amount equal to the true reservation value because it would guarantee only a zero payoff. Thus, bidders face a simple trade-off: an increase in the bid will increase the probability of winning while, at the same time, will reduce the gains from winning. Consequently, the dominant strategy is to 'shade' his reservation value and to bid an amount that equals the expected bid of the second higher valuation bidder.[45] Therefore, in the presence of interdependent values, the bidder will have to shade his bid a little beyond the expected second reservation value in a pay-as-bid auction, in other to avoid winners' curse.

Moreover, if the values are also affiliated, which means that the bidders' information interdependence is positively correlated, bidders will be able to have better value estimation in an open auction. In fact, in an English (or open ascending) auction the various bidders can follow the bids of the others directly and know the prices at which bidders quit and the number of the remaining active bidders at each stage. It allows the remaining active bidders to make inferences about the information that the bidders that had left the auction had to make this decision and then update their reservation value. A sealed-bid, by its very nature, makes no such exchange of information possible.

According to Krishna,[46] the English auction should outperform the pay-as-clear sealed auction, which in turn, out-performs the pay-as-bid sealed auction. Such revenue ranking is a consequence of the different strategies to mitigate the winners' curse risk available in each auction methodology. In the same way, the release of

---

[45]Beyond formal differences among the various types of auctions, we can state that: (a) Dutch auctions are strategically equivalent to FPSB auctions and that (b) under certain hypotheses, English auctions are strategically equivalent to SPSB auctions.

[46]Krishna (2002).

public information in any of the four designs should lead to higher revenues and a lower difference between the expected revenues.

Second, as the auction will procure energy savings, bidders will have to spend considerable time and money only to prepare the projects. This means sunk cost, and it can influence bidders' risk tolerance. The higher the sunk cost, the greater the risk aversion of losing the auction and the greater the aggressiveness of the participants. When bidders are averse to the risk of losing the auction, the marginal increment in the payoff associated with a successful, slightly higher bid weight less than the possibility of losing the auction. A risk-averse bidder thinks that it is more important for him not to risk losing the auction than the small gain that he can make by slightly increasing his bid. This leads risk-averse bidders to shade their bids less than risk-neutral bidders. Consequently, the auctioneer may expect a greater revenue under the pay-as-bid sealed auction than under a pay-as-clear auction, but the trade-off will be a higher risk of the winner's curse in this scenario, especially with interdependence among the bidder's valuation (Riley and Samuelson 1981). However, since it is a pilot whose main purpose is to generate information for later scaling the program nationwide, ANEEL should privilege the auctioning modality that reduces the bidders' risks.

Third, regarding the risk of collusion, the Brazilian Competition Authorities advice that every public bidding process must pay special attention to the presence of conditions that facilitate the formation of a cartel.[47] In the case of the Brazilian energy efficiency market, there are few firms with technical and financial capacity to design and execute EEMs. Such concentration of the market facilitates the existence of tacit collusion.[48] Moreover, for reasons of auditing and transparency, public tenders reveal a large body of information about the participants, which facilitates the structuring of new cartels (Bajari and Yeo 2009).

Robinson (1985) shows that assuming a one-time stage game; cartels are stable (i.e., incentive-compatible) in open ascending-bid (English) auctions, but not in sealed bid auctions. The dominant strategy in the English auction is to keep bidding until reaching his valuation. That combined with the assumption that cartels fully share information and the possibility to identify the bidders and retaliate non-compliers (given by open or oral auctions designs), is sufficient to ensure that no other bidder will make positive profits betraying the cartel. However, in the sealed-bid, there is an incentive to cheat on the cartel and, therefore, no cartel strategy is a Nash equilibrium in a sealed auction.

An important condition of the Robinson (1985) theorems is that no private information remains inside the cartel. If bidders have private information, they will frequently have positive expected profits, even if they do not cooperate. To have a stable cartel, every member must prefer both participation and compliance. For the

[47]SDE (2008).

[48]Circumstance where the bidders' dominant strategy results in the appropriation of a larger portion of the consumer's surplus than would be possible in a perfect competition environment. It does not require any kind of exchange of information or coordination.

independent-values case, there is no advantage to private information in the oral auction. Division rules are in place, which gives bidders an incentive to participate in the cartel, so long as valuations are credibly revealed to cartel members. In the common-value case, if bidders are symmetric and if the value estimates are credibly revealed to the other cartel members, a cartel solution guaranteeing every member an equal share of either the *ex-ante* or the *ex-post* cartel profit will be a Nash equilibrium in the English auction.

Additionally, the auction to procure for EEMs is not a singular event. A sequence of auctions turns it into a repeated game, where the reputation of the participants becomes relevant in the formulation of dominant bid strategies. In this sense, repeated auctions are more susceptible to cartel activity, by allowing retaliation of participants in subsequent auctions. The difference between the open and the closed auction, therefore, become less relevant.

However, considering the possibility of collusion, it is important to include safeguards against a cartel Nash equilibrium in the auction design. Bajari and Yeo (2009) discuss how the US Federal Communications Commission (FCC) introduced rule changes including click box bidding, the presence of large minimum opening bids and anonymous bidding that potentially make collusion more difficult. Click box bidding makes it harder for bidders to signal each other directly. Large minimum opening bids give bidders fewer rounds for communication via bids to work out a split of the licenses before prices become too high. Anonymous bidding disguises the identity of bidders during the auction, making it difficult, and perhaps impossible, for the cartel to use the bids to monitor and enforce collusive agreements.

The Brazilian experience with electricity procurement auctions also include the use of safeguard rules against collusion, containing a click-biding and an anonymous bidding instruments similar to that of the US FCC and the holding of hybrid auctions with a sealed bidding phase to encourage betrayal, and an open bidding phase, to reduce the risk of winner's curse. Another common practice in Brazilian auctions is to perform the formal qualification of the participants only after the presentation of the proposal and selection of the winners. This avoids the bureaucracy of analysing the documentation of all participants and reduces the risk of administrative and judicial disputes over the disqualification of competitors who did not win. Further, about the risk of collusion, most of the information concerning market participants remain with reserved access. In short, the risk of collusion or cartel because of the Brazilian energy efficiency market structure is not enough to justify a sealed bid auction. An open auction, with tailor-made safeguards, is the recommended option.

Finally, the auction design and contractual rules may also address other issues like the additionality principle and the concern with distributive impacts. The additionality principle means that the resources used in a given public policy must bring real benefits to society. Otherwise, there would only be the redistribution of income. Therefore, it has relation with the market failure of positive externalities in the sense that, as private actors do not receive all the benefits of their actions and are less likely to take them, the public policy assures additionality, defined as the impact beyond the business as usual.

This concern, however, has less relevance when contracting an energy resource to reduce subsidies. Costumers will pay the EEM a market cleared price for an effectively accomplished service, that is a substitute for another energy source. In the specific case of the Roraima pilot, EEM will substitute diesel-fired power plants with the most expensive generation cost (six times the national average, about R $1200/MWh) in Brazil and that receive a federal subsidy,[49] funded by all the Brazilian electricity customers to be affordable by the local customers. The reduction of the subsidy burden also helps to mitigate the regulator's concern with the risk of the division of costs and benefits among society increases inequality. The development of the Energy Efficiency market can be an important instrument to reduce the cost of other public policies. In the case of Brazil, in addition to the subsidy to fossil fuels generation in isolated regions, it is also possible to reduce the cost of the subsidy for irrigated agriculture and low-income consumers (Table 23.2).[50]

23.6 Conclusions and Policy Implications

There is significant space for the development of the energy efficiency market. Investments in EEMs have important positive externalities and can be instruments to help social development and facilitate the transition to a low carbon economy. Nonetheless, the monitoring of investment opportunities in energy efficiency indicates the existence of a gap, in the sense that private agents are not adopting EEMs with economic attractiveness.

The public energy efficiency policies commonly adopted by the countries, while recognizing the existence of market failures and regulatory barriers, have focused on command and control solutions. Besides, little attention has been paid to the economic impacts of the energy efficiency support schemes on distribution and other consumers. In many countries, where distribution utilities bear the market risk, their shareholders' interests are directly opposite to efficiency. Nevertheless, often regulators count on them as the main player to implement energy efficiency policies. The lack of alignment between policy objectives and business rationale plays an important role in the energy efficiency gap. Customers and firms perceive energy efficiency standards more as a regulatory burden than as a business opportunity. More attention must be given on the regulation point decision (distribution company, ESCOs or customer) and electricity tariff design.

[49] In accordance with Brazilian Law No. 12111/2009, the Fuel Consumption Account subsidizes the generation of electricity in isolated systems. The subsidy equals the difference between the local generation cost and the average generation cost in the rest of the country. Its resources are collected through a levy on distribution and transmission tariffs.

[50] In accordance with the law n. 10.438, this subsidy aims to increase the agricultural productivity of the central region of Brazil and reduce the pressure to expand the agricultural frontier north towards the rain forest.

Table 23.2 Comparative table of the different auction designs (compiled by the authors)

Program	Roraima Pilot	Certificates Auction	PEE	PPEC	Prokilowatt	STEP UP!	RPM	Bid4 efficiency
Design	Auction	Auction	Tender—Call for Projects	Tender—Call for Projects	Tender—Call for Projects	Tender—Call for Projects	Auction	Auction
Bid	Open	Sealed	Sealed	Sealed	Sealed	Sealed	Sealed	Open
Price	Pay-as-clear	Pay-as-clear	Pay-as-bid	Pay-as-bid	Pay-as-bid	Pay-as-bid	Pay-as-clear	Pay-as-clear
Approaches the addionality principle	No	No	No	Yes	Yes	Yes	No	Yes
Approaches the winner's risk	Yes	No	No	No	No	No	Yes	No
Approaches collusion risk	Yes	No	No	No	No	No	Yes	No

In the same context, regulators should also look at the impact of energy efficiency on remaining customers. In countries where the tariff design is volumetric, energy efficiency has distributive implications similar to those verified in distributed generation—as some customers (typically the wealthier ones) become more efficient and reduce their consumption, they will pay less for the network. Network fixed costs will have to be covered by other customers in the same concession area (often the less wealthy). Therefore, energy efficiency incentives must be thought holistically, especially about the implementation of multi-part tariffs. On the flip side, this factor encourages energy efficiency projects on subsidized customers (such as low-income citizens in some countries) and on the public sector (i.e., government buildings, schools, and hospitals), which will result in lower energy tariffs and taxes in the future. So far, experiences with MBIs suffer from great fragmentation and the adoption of regulatory solutions that solve specific problems but cause perverse side-effects. The existence of several small energy efficiency markets increases the cost of transactions and reduces liquidity and the availability of financing. Failure to provide a comprehensive regulatory framework based on a more in-depth regulatory impact assessment can lead to unexpected market distortions with counterproductive results.

The Swiss, German and American schemes approach to address the additionality principle is a fundamental example. If the schemes award a subsidy to every unit of energy saved, the amount of private investment in energy efficiency will increase, but this will mean that EEM that do not need public support has also received resources from the program. Further, by attempting to mitigate this risk through the analysis of projects paybacks, the amount of private investment in energy efficiency may remain unchanged, and society will pay to subsidize less effective EEM without achieving any benefit.

Moreover, the concern with the principle of additionality seems exaggerated in the actual context of the energy efficiency gap. Regulators and policymakers should first remove the barriers to market development of EEMs that are economically viable before putting regulatory safeguards for such EEMs to receive public support.

An alternative to overcome these issues would be the coupling of energy efficiency and electricity markets. Thus, instead of attempting to provide subsidies and financial incentives for a change in consumer attitudes, the EEMs would be remunerated for electricity services. The idea that energy efficiency is an energy resource associated with the demand response was already present in Directive 2012/27/EU and reinforced by the amendment presented in 2016. Regulators and policymakers, however, does not appear to have assimilated its entire disruptive potential.

As seen above, any resource connected to the grid at the distribution level capable of providing or substituting electricity services is a DER. It concept comprises supply and demand-side resources. The safety and quality of electricity services depend on the existence of potential difference, which may be provided by the addition of power or by the reduction of the load.

Usually, demand response is more associated with controllable measures for change or reduction of consumption in peak time. Resources with these characteristics offer great operational flexibility. Nevertheless, it is important to note that, however desirable, operational flexibility and control are not attributes present on all supply-side energy resources and it would be unfair to require them to allow EEMs to enter the electricity market. Energy efficiency meets all requirements to be a demand-side DER, as it is a change in the level of electricity consumption in response to price or other forms of financial incentives.

Additionally, it is necessary to recognize that the EEMs have special characteristics that hinder their immediate coupling with all the markets of electricity, among them the challenges regarding M&V. Moreover, EEMs share some important features of other DERs such as the predominance of fixed costs relative to variables and the need for upfront investments that makes project viability very dependent on loans.

The issue of M&V will probably require the revision of the International Protocol to introduce new approaches, such as the one developed by ANEEL to determine the amount of energy savings (differences-in-differences) or other possibilities made possible by the strong process of digitization of meters and network elements of the electricity industry (such as randomized controlled trials, regression discontinuity design or matching).[51]

Regarding the dependence on financing, the international experience with auctions for contracting renewable sources can help. Several countries are using procurement auctions that offer contracts or long-term receivables to facilitate credit access through the modality of project finance. Brazil, in particular, has extensive experience in this regard (Araujo et al. 2008; Moreno et al. 2010). Still, as shown above, the definition of the design of an auction and the contractual obligations is not trivial. Regulators need to be alert to the structure and maturity of the market to make weighted choices about the risks of the winner's curse or collusion, for example.

The similarities between EEM and DER also suggest that the first step towards the coupling of electricity and energy efficiency markets would be the integration of energy efficiencies with the other DER. However, except for the novel auction design in Brazil, the schemes presented in section 22.4 do not allow competition or integration, albeit with restrictions, between EEM and distributed generation.

Regulators and policymakers typically have multiple objectives and concerns, such as funding sources, cost and benefit sharing, and the principle of additionality whose harmonization is not simple either. ANEEL's strategy of conducting an in-depth regulatory impact assessment together with a broad engagement process with stakeholders is essential to considering alternatives and trade-offs. However, because it is an innovative approach, it is necessary to highlight the relevance of adopting the prototyping method and pilots, which accompanies the design thinking approach adopted by ANEEL and allows early identification of risks and failures from bad regulatory design.

[51]For a comprehensive and accessible introduction to impact evaluation, see the Inter-American Development Bank and World Bank's 2016 publication: 'Impact Evaluation in Practice, Second Edition', available at: https://openknowledge.worldbank.org/bitstream/handle/10986/25030/9781464807794.pdf?sequence=2&isAllowed=y.

Finally, in relation to the auction design that ANEEL will adopt in the case of the pilot of Roraima, considering the evaluation of the Brazilian energy efficiency market structure and maturity, the expected bidders' risk tolerance, and the possibility of winners curse, there is a clear recommendation for adopting an open and pay-as-clear auction. Such a strategy increases the attractiveness of the auction by simplifying the process of developing a bid strategy and reduces the risk of the winners' curse. Additional safeguards in the systematic auction and contractual clauses is enough to address the risk of price distortion by collusion.

References

American Electric Power Ohio (2018) Participant Program Guidelines; Frequently Asked Questions. Available at: https://aepohio.com/save/business/programs/EnergyEfficiencyAuction/

ANEEL (2018a) Technical Report n. 158, 5th of June of 2018. Brazilian Electricity Regulatory Agency, Brasília, Brazil. Available at: www.aneel.gov.br

ANEEL (2018b) Public Consultation n. 7/2018. Obter subsídios sobre o conceito de Leilão de Eficiência Energética e o conjunto de metodologias e premissas utilizado na Análise de Impacto Regulatório de projeto piloto a ser realizado em Roraima. Available at: www.aneel.gov.br

Araujo J, Costa A, Correia T, Melo E (2008) Reforms of the reforms in Brazil: problems and solutions. In: Sioshansi (ed) Competitive electricity markets. Design, implementation, performance. Elsevier, pp 543–572

Bajari P, Yeo J (2009) Auction design and tacit collusion in FCC spectrum auctions. Inf Econ Policy 21(2):90–100

BMWi Bundesministerium für Wirtschaft und Energie (2017) Guideline for the Promotion of Electricity Savings in Competitive Tenders: Using Power Efficiency Potentials—STEP up! Federal Ministry of Economics and Energy. Available at: https://www.stepupenergieeffizienz.de/teilnehmen/archiv

Correia T, Masili G, Resende J, Costa A (2009) Using information disclosure to design optimal electricity auctions under imperfect competition. In: 32nd Conference of the International Association for Energy Economics, San Francisco – USA

Directive 2012/27/EU of the European Parliament and of the Council of 25 October 2012 on energy efficiency. Available at: https://eur-lex.europa.eu/legal-content/EN/TXT/?qid=1399375464230&uri=CELEX:32012L0027

DOE (2018) Department of Energy. Available at: https://www.energy.gov/oe/activities/technology-development/grid-modernization-and-smart-grid/demand-response

ERSE (2013) Rules of the plan to promote efficiency in electricity consumption approved under the tariff regulation. Entity of regulatory energetic services. Available at: http://www.erse.pt/pt/planodepromocaodaeficiencianoconsumoppec/edicoesPPEC/ppec1314/Paginas/default.aspx

Gertler PJ, Martinez S, Premand P, Rawlings LB, Vermeersch CMJ (2016) Impact evaluation in practice, 2nd edn. Inter-American Development Bank and World Bank, Washington, DC. Available at: https://openknowledge.worldbank.org/handle/10986/25030 Licence: CC BY 3.0 IGO

IEA (2017) Market-based instruments for energy efficiency: policy choice and design. International Energy Agency

IEA (2018) Energy efficiency 2018: analysis and outlooks to 2040. International Energy Agency

Krishna V (2002) Auction theory. Elsevier

Liu Y (2017) Demand response and energy efficiency in the capacity resource procurement: case studies of forward capacity markets in ISO New England, PJM and Great Britain. Energy Policy 100:271–282

MacGill I, Healy S, Passey R (2013) Trading in energy efficiency – a market-based solution to market failure, or just yet another market failure? In: Sioshansi (ed) Energy efficiency: towards the end of demand growth. Elsevier, pp 563–590

Maurer L (2018) Energy efficiency auctions: a market based mechanism. Unpublished Presentation in Brasília – Brazil on July 16th, 2018

Moreno R, Barroso LA, Rudnick H, Mocarquer S, Bezerra B (2010) Auction approaches of long-term contracts to ensure generation investment in electricity markets: lessons from the Brazilian and Chilean experiences. Energy Policy 38:5758–5769

Mundaca L (2008) Markets for energy efficiency: exploring the implications of an EU-wide 'tradable white certificate' scheme. In: Energy energy economics, vol 30. Elsevier, pp 3016–3043

PJM (2019a) 2018/2019 RPM base residual auction results. Available at: https://www.pjm.com/~/media/markets-ops/rpm/rpm-auction-info/2018-2019-base-residual-auction-report.ashx

PJM (2019b) RPM energy efficiency (EE) FAQs. Available at: https://www.pjm.com/-/media/markets-ops/rpm/rpm-auction-info/rpm-energy-efficiency-faqs.ashx?la=en

PJM (2019c) Manual 18: PJM capacity market. Revision: 41. Effective Date: January 1, 2019

Proposal Amendment to the Directive 2012/27/EU of the European Parliament and the Council. Available at: https://eur-lex.europa.eu/legal-content/EN/TXT/?uri=CELEX:52016PC0767R%2801%29

Riley J, Samuelson W (1981) Optimal auctions. Am Econ Rev 71(3):381–392

Robinson M (1985) Collusion and the choice of auction. J Econ 16(1):141–145

SDE (2008) Combate a Cartéis em Licitações. Um guia prático para pregoeiros e membros de comissões de licitação. Brazililian Ministry of Justice Available at: file:///C:/Users/tdebarro/Downloads/cartilha_licitacao.pdf

Silva A (2003). Leilões de certificados de energia elétrica: máximo excedente versus máxima quantidade negociada. Doctoral Dissertation, Available at: http://repositorio.unicamp.br/bitstream/REPOSIP/264537/1/Silva_AdrianoJeronimoda_M.pdf

Vickrey W (1961) Counterspeculation, auctions and competitive sealed tenders. J Financ 16:8–37

Chapter 24
Consumer Law, Sustainable Energy Consumption and Mini- and Microgrid Decentralized Generation in Brazil

Claudia Lima Marques and Matheus Linck Bassani

Contents

Abstract This chapter analyses the legal framework for mini- and microgrid decentralized generation specified in Brazilian National Agency of Electricity (ANEEL) Normative Resolution (REN) No. 482/2012 and verifies the advantages and disadvantages therein. In this context, the chapter addresses the question of why there are so few mini- and microgrid energy connections in Brazil considering the vast potential of the country's solar and wind resources. Global frameworks, such as the as UN Sustainable Development Goals (SDGs)—in particular, SDG 7 (Ensure access to affordable, reliable, sustainable and modern energy for all), SDG 12 (Ensure sustainable consumption and production patterns), and SDG 17 (Revitalize the global partnership for sustainable development)—and the UNCTAD Guidelines for Consumer Protection, provide a basis for analysis. The Brazilian Consumer Protection Code (BCPC) has been targeted by proposals to mandate sustainable consumption as a market principle in Brazil, and the chapter will analyse this. It will then assess whether the legal framework in Brazil is fit for promoting growth in sustainable energy generation. The chapter contends that the lack of positive

C. L. Marques · M. L. Bassani (✉)
Faculty of Law, Federal University of Rio Grande do Sul (UFRGS), Porto Alegre, Brazil

A. d. Amaral Junior et al. (eds.), *Sustainable Consumption*,
https://doi.org/10.1007/978-3-030-16985-5_24

incentives established by the legal framework is an obstacle to increase the adoption of the decentralized system by consumers.

24.1 Introduction

The present chapter seeks to analyse the legal framework for mini- and microgrid decentralized generation of electricity specified in Brazilian National Agency of Electricity (ANEEL) Normative Resolution (REN) No. 482/2012 and to verify the advantages and disadvantages therein. The link between energy production and consumption and the environment is well known.[1] Electricity has been considered the basis for subsistence and a driver of world development since the early twentieth century,[2] and its role may become even more relevant in the future because of population growth, expanding consumption and the increased focus on production and well-being in postmodern[3] society.

Since the Stockholm United Nations Climate Conference in 1972, sustainable development has been evolving through multiple academic and professional domains, including economics, law, engineering, biology and others.[4] The concept was based on mutual respect and equilibrium across social, environmental and economic domains, which are considered essential to provide minimum levels of opportunity, preservation and income growth.[5]

Production and consumption are at the heart of the world trade system, as indicated in the foundation of the World Trade Organization in 1994. Nonetheless, the need to protect the environment is an urgent matter for human beings and biodiversity more generally. Sustainability is part of a cultural movement to ensure the world remains 'alive' for future generations. International organisations like the UN have been launching initiatives—including the Sustainable Development Goals (SDGs)—to incentivise members to comply with several broader policies, such as the reduction of poverty, health and education, energy and water access, gender equality and others.

It is necessary to highlight that words 'decentralized' and 'distributed' are in this chapter considered similars in the sense of fragmented generation or production of electricity. The word 'energy' is used in a broad sense, considering primary and secondary energy sources with relevance for any final consumer or customer, such as industry or households.[6] More precisely, primary sources are natural resources present on the Earth, such as water, wind, sun, crude oil and coal, all of which are

[1] See Lyster and Bradbrook (2006), p. xxi.
[2] See Kapseu et al. (2012), p. 17.
[3] Lyotard (1989), and Bauman (1992).
[4] See Sarlet and Fensterseifer (2014), p. 102.
[5] See Marques (2015).
[6] See Fatouros (2008), pp. 367–368.

used in multiples ways. Gasoline and electricity are secondary sources because they are processed or refined forms based on primary energy sources.[7]

Innovations such as decentralized energy systems are considered one way to provide sustainable energy consumption on a small scale, and are expected to help reducing emissions and saving grid capacity, while also providing opportunities for renewable energy. They could thus form a constituent part of a more sustainable future. Broad implementation of distributed generation, however, would imply thoroughgoing structural change, as well as a surge in innovation.[8]

In this context, the chapter addresses the question of why there are so few mini- and microgrid energy connections in Brazil considering the vast potential of the country's solar and wind sources for power generation and consumption. The chapter proceeds as follows. First, global initiatives such as the UN Guidelines for Consumer Protection (UNGCP), the SDGs—particularly SDG 7 (Ensure access to affordable, reliable, sustainable and modern energy for all), SDG 12 (Ensure sustainable consumption and production patterns) and SDG 17 (Revitalize the global partnership for sustainable development)—will be identified as they provide the basis for the analysis. The chapter then discusses the Brazilian legal framework and assesses how it is failing to provide sustainable energy generation growth in Brazil. Consumer law will be analysed to then suggest possible adjustments as including the concept of sustainable consumption. Through Senate Law Project (PLS) No. 281/2012,[9] the Brazilian Consumer Protection Code (BCPC) has been targeted by proposals to include sustainable consumption as a guideline for market actors, which could provide a new ground for future analysis. Finally, REN No. 482/2012 of ANEEL, which is the legal framework for mini- and microgrid decentralized generation in the country will be assessed in terms of its capacity to contribute to sustainable consumption. The lack of positive incentives within the legal framework could influence consumer decision-making in adopting decentralized generation of electricity. The final section will summarise the arguments and present some concluding thoughts.

24.2 Global Initiatives Signifying the Relevance of Energy to Sustainable Development

The concept of sustainable development incorporates economic progress, environmental protection and social development and embodies at least four major legal principles: intergenerational equity, sustainable use, equitable use and integration. Since the Brundtland Report, the concept has evolved from a political position to a

[7]See International Energy Agency (2005), p. 18.

[8]Pehnt (2006), pp. 1–18.

[9]Brazil. Senate Law No. 281/2012.

legal concept and is now used as a tool for balancing and harmonising legal decisions, applied as an interstitial 'meta-norm' or as a concept that, despite not prescribing precise results, allows reducing tensions between trade and environmental objectives.[10] Sustainable energy consumption and production can be considered as one of the numerous—and specific—areas that the general concept of sustainable development operates in. Considering that energy is an essential asset to provide a decent life, the topic should have long been at the heart of UN discussions as a basic right of human beings.

The 1985 UNGCP, revised and updated in 1999 and 2015, clearly provide the link between consumption and energy. The UN Guidelines were first adopted by the General Assembly in Resolution 39/248 of 16 April 1985. In 1999, the UNGCP were expanded significantly by ECOSOC to include a new section on sustainable consumption.[11] The specific areas section of the Guidelines was recently revised by the General Assembly in Resolution 70/186 of 22 December 2015, including a guideline on energy (Guideline 76)[12] and public utilities (Guideline 77).[13]

In 2011, Sustainable Energy for All (SE4All) was launched to catalyse action by 2030.[14] In September 2015, the SDGs introduced a range of objectives to be pursued by Member States. The present research focuses on two direct goals—SDG 7 (Guarantee for all to affordable, reliable, sustainable and modern access to energy) and SDG 12 (Production and consumption patterns). It also touches on SDG 17, which concatenates institutions to make possible the application of all the other goals.[15] SDG 12 has many relevant targets to be achieved by 2030, emphasising the efficient use[16] of natural resources and the removal of subsidies for the inefficient use of fossil fuels affecting the industry, commerce and services.

[10] Morosini and Marceau (2013), pp. 59–92.

[11] See UNCTAD (2015), Guidelines 59 and 60.

[12] See UNCTAD (2015), Guideline 76 on energy: 'Member States should promote universal access to clean energy and formulate, maintain or strengthen national policies to improve the supply, distribution and quality of affordable energy to consumers according to their economic circumstances.'

[13] See UNCTAD (2015), Guideline 77 on public utilities: 'Member States should promote universal access to public utilities and formulate, maintain or strengthen national policies to improve rules and statutes dealing with provision of service, consumer information, security deposits and advance payment for service, late payment fees, termination and restoration of service, establishment of payment plans and dispute resolution between consumers and utility service providers, taking into account the needs of vulnerable and disadvantaged consumers.'

[14] See United Nations (2016).

[15] See United Nations (2015).

[16] Energy intensity is measured by the amount of energy required for an activity, whereby the lower the energy used to produce a product, the lower the intensity. Intensity is distinct from energy efficiency, which improves when a certain level of service is provided with reduced quantities of energy inputs or services, in relation to the amount of energy previously foreseen. If there is an improvement in energy efficiency, the intensity will be reduced. But other factors among the four large final consumers (industry, transportation, households and trade) may change the intensity, such as a change in the mode of production of a heavy industry, or a reduction in the quantity

It is necessary to emphasise the relevance of the word *access* in the goals referred to above. In SDG 12, access to energy is implicit because there can be no energy consumption without access (through transmission and distribution grids), and the same applies to sustainable energy consumption generated through renewable sources. In SDG 7, access to energy is the guarantee of a society with more opportunities, dignity and well-being. By way of illustration, it is estimated that 1.3 billion people have inadequate access to energy.[17] An alternative is off-grid generation, allowing the use of power in remote areas without being connected to the grid. Such goals could be achieved through actor cooperation and institutions, provided by SDG 17: *Revitalize the global partnership for sustainable development.*[18]

Given the context presented, there is a need for greater effort to speed the transition to a sustainable energy system, based on the renewability of energy sources without affecting the possible use of environmental goods by future generations.[19] Development is not possible without energy, and full sustainable development is not possible without sustainable energy. The challenge is to develop a service that is financially viable and environmentally sustainable and that remains within the limits of consumer accessibility, through an intergenerational contract. According to Gawel:

> In the course of the energy transition, today's generations are investing in a sustainable system of energy supply. The energy transition is consequently an intergenerational contract. The temporally limited (additional) costs being borne today are supposed to lead to permanently lower costs and risks tomorrow. This goal makes the energy transition into an outstanding sustainability project.[20]

The energy transition is based on the analysis of a future projection of energy demand, related to population growth and economic activity. In the event of reduction or maintenance of the current level of global consumption, it would be enough to increase the generation of power by renewable sources to reduce the use of conventional sources.

---

produced, which could reduce the intensity without, however, changing efficiency. US Department of Energy (2017).

[17]International Energy Agency (2010).

[18]See United Nations (2017).

[19]Renewable resources can adjust faster by renewing their inventories over a given period, but only if they are economically relevant. However, intervention in these stocks over a period can generate effects (positive or negative) on inventories in subsequent periods, such as fish stocks, other fauna and water reservoirs. Once exhaustible natural resources are consumed, they cannot be renewed. Further, in the event that renewal is so slow that the stock has no economic relevance, these resources are also considered expendable. In this sense, Sweeney (1993), pp. 759–761 has developed the concept of exhaustion based on the following assumptions: '(1) the stock decreases over time, whenever the resource is being used, (2) the stock never increases over time, (3) the rate of stock decrease is a monotonically increasing function of the rate of resource use [i.e., the more you use the lower the inventory] and (4) no use is possible without a positive stock.'

[20]Gawel et al. (2014), p. 6.

However, assuming an increase in the level of consumption, as confirmed by the projection of the International Energy Outlook 2017, more effort will be required. An increase in power generation through renewable sources on a higher scale than the growth of the consumption level will be needed if the use of fossil resources is to be controlled. According to the forecast, in 2040 renewables will still provide the same share of world electricity generation as coal (31%).[21]

24.3 A Specific Analysis of Sustainable Energy Consumption

Producers and consumers have always been linked and dependent on each other. Historically, a productivist bias was dominant until the end of the Second World War. A consumption perspective developed subsequently, notably by Jean Baudrillard[22] in his book *Consumer Society*. On this point, it is interesting to note Ritzer's observation regarding the discussion by Zygmunt Bauman on the question:

> Bauman argued that the society of concern to classical theorists such as Marx was work-based. It was one in which its members were engaged 'primarily as producers'. In contrast, a 'later-modern, or second-modern or post-modern stage engages its members primarily in their capacity as consumers'. More extremely, Bauman discusses the 'passage from producer to consumer society'.[23]

It is also necessary to highlight that the distributed generation of electricity is a means to make both the generation (i.e., production) and the consumption sustainable simultaneously when the consumer himself generates the energy he consumes. The analysis of the production or consumption point of view, therefore, has no significant implications in this chapter.

David Goldblatt notes that sustainable consumption can be considered as a broader concept in contrast with sustainable energy consumption:

> [E]nergy consumption is more often treated independent of other types of consumption, takes its place in the broad spectrum of consumption types. This is appropriate since energy use responds to many of the same general driving forces, is shaped by many of the same social practices and norms, and can be viewed through some of the same lenses that analysts use to study general material or service consumption.[24]

[21]International Energy Outlook (2017), p. 80.

[22]Baudrillard (2011), pp. 13–120.

[23]Bauman (1992), cited in Ritzer (2014), pp. 9–10.

[24]Goldblatt (2005), p. 3.

Nevertheless, the decision to alter the mode of production and consumption takes time even on vital sectors of the economy. Precisely on the energy field, Goldblatt states that 'some part of a person's domestic energy use is more highly discretionary, and this part may sometimes be measured through economic or psychological devices'.[25] It depends on whether individuals are convinced and able to invest in new, energy efficient gadgets or energy generators such as solar panels. The legal framework is an essential key to provide incentives for this conduct.[26]

It is interesting to note that significant investments in renewable energy infrastructures are made mostly when legal certainty is established through a legal framework, generally linked with a clear role for government. Private households also need legal certainty to decide whether it is beneficial to invest to consume sustainably.

Positive incentives coming from public policies[27] can provide a legal framework enabling private parties to invest in energy distributed generation infrastructures, as done in Germany through the *Energiewende* policy. Nonetheless, the individual's discretion to choose (behaviour[28]) to consume based on a sustainable manner has its limits, once social, legal, technological and cultural elements (plural factors) might be obstacles for ideal implementation.

Because the implementation of a decentralized generation system is mainly based on a private decision, positive incentives based on sound regulation become a necessity to ensure long-term stability. Some movements have been done in Brazil fostering sustainable energy consumption, which will be analysed below.

[25]Ibid, p. 18.

[26]A more in-depth analysis of the discretion (the possibility to choose) of the consumer to pursue sustainable consumption can be verified in the figure in Goldblatt (2005), p. 19 entitled: 'A depiction of society's current "discretion continuum" for resource or energy consumption, with future alternative trajectories'.

[27]UNCTAD (2015): '58. Member States are encouraged to create or strengthen effective regulatory mechanisms for the protection of consumers, including aspects of sustainable consumption.'

[28]UNCTAD (2015): '62. Member States and other relevant organizations should promote research on consumer behaviour related to environmental damage in order to identify ways to make consumption patterns more sustainable.'

24.4 Fostering Sustainable Energy Consumption in Brazil

24.4.1 The Proposal to Include Sustainability in the Brazilian Consumer Protection Code (BCPC)

Briefly, the Brazilian CPC (Federal Law No. 8078/1990) has traditionally been applied to consumption contract of electricity between the consumer and the public service utilities, as defined in Article 3 of the Code.[29] Essential goods such as electricity, water, telecommunications and gas, are objects of public service provision by a legal entity, public or private, comprising, therefore, the concessionaires/utilities.

The application of the CPC is corroborated by a reading of its Article 4, VII, whose National Policy on Consumer Relations establishes the rationalisation and improvement of public services (Section VII) as a principle to meet the needs of consumers with respect to their dignity, health and safety, protection of their economic interests, improvement of their quality of life, as well as the transparency and harmony of consumer relations. Still, Article 6 ensures that the adequate and effective public services provisions in general (Section X) are basic consumer rights.[30]

Under Article 175 of the Brazilian Federal Constitution,[31] the competence to provide public services[32] rests with the public authority and can be performed by the State itself or by third parties after bidding, through concession or permission. Between utilities and clients, a long-term[33] consumer contract relationship can be verified with dependence,[34] captivity or subordination of the consumer.[35] That the

---

[29]The supplier/producer definition is provided by Article 3 of the CPC, which states: ‘The supplier is any natural or legal person, public or private, national or foreign, as well as depersonalized entities, which carry out production, assembly, creation, construction, transformation, import, export, distribution or commercialization of goods or service's provision’ (Brazil. Federal Law No. 8078/1990).

[30]Miragem (2013), p. 171.

[31]Brazil. Federal Constitution (1988), Article 175: ‘The Public Power, under the law, directly or under concession or permission, always through public bidding, is responsible for the provision of public services’ (authors' translation).

[32]The concept of public service has been defined by Mello (2014) as follows: ‘Public service is all activity offering material usefulness or comfort to the satisfaction of the community in general, but uniquely fruitful to the administered, which the State assumes as pertinent to its duties and lends itself or even by those who do it, under a regime of public law—therefore, consecrator of prerogatives of supremacy and special restrictions—established in favour of the interests defined as public in the normative system’ (p. 695, authors' translation).

[33]See Shunck (2016), pp. 27–75.

[34]Pachauri (2011).

[35]On long-term ‘captivity’ contracts, see Marques (2016a), pp. 97–115.

service is essential demonstrates the end users' vulnerability[36] and the need for consumer/user protection.[37]

Bruno Miragem states that:

> The regulation of the electric sector, as well as other public services, due to the provisions established in specific legislation, or even as a result of the application of art. 22 of the BCPC, must ensure the protection of consumer rights, not only about the conduct of the supplier in the direct relationship, but also in the formulation and application of administrative regulations.[38]

To promote the legal basis for sustainable consumption, the UNGCP include sustainable consumption lines, following the suggestions of the United Nations Commission on Sustainable Development.[39] In addition to consumer protection, the promotion of sustainable consumption is a goal (Clause I, h), and achieving sustainable consumption patterns is a general principle of the Guidelines (Clause III, 5, i). Furthermore, Clause V, H, is in line with the concept of general sustainable development established and provides that:

> **H. Promotion of sustainable consumption**
>
> 49. Sustainable consumption includes meeting the needs of present and future generations for goods and services in ways that are economically, socially and environmentally sustainable.[40]

The Guidelines predict energy and environmental protection provisions in Clause V, which states:

> V. Guidelines
>
> (...)
>
> G. Education and information programmes
>
> (...) 44. Consumer education and information programmes should cover such important aspects of consumer protection as the following:
>
> (...)
>
> *(f) Environmental protection; (...);*
>
> *(i) Efficient use of materials, energy and water.*
>
> K. Measures relating to specific areas
>
> (...). 76. Energy. Member States should promote universal access to clean energy and formulate, maintain or strengthen national policies to improve the supply, distribution and quality of affordable energy to consumers according to their economic circumstances.

---

[36]Marques and Miragem (2014), p. 116.

[37]Miragem (2013), p. 172.

[38]Miragem (2004), p. 69.

[39]The Guidelines were first adopted by the General Assembly in Resolution 39/248 of 16 April 1985, later expanded by the Economic and Social Council in Resolution 1999/7 of 26 July 1999 and revised and adopted by the General Assembly in Resolution 70/186 of 22 December 2015.

[40]UNCTAD (2015).

> Consideration should be given to the choice of appropriate levels of service, quality and technology, *regulatory oversight*, the need for awareness-raising programmes and the importance of community participation (authors' emphasis).

Cooperation and regulation in strategic areas are instruments for implementing the sustainability of consumption patterns using the best practices, including by encouraging the removal of subsidies in sectors that are no longer in line with sustainable production and consumption patterns, combating programmed obsolescence and promoting recyclable products, as well as seeking social needs and avoiding potential adverse effects on market access. It aims to reduce the use of natural resources through new technologies, always considering the safety and quality of products.[41]

Faced with the established affirmative guidelines, civil society promoted the drafting of a legislative amendment by including normative text in Senate Bill (PLS) No. 281/2012.[42] The Senate report agreed with the content of the bill. The text of Article 4, e, XI of the proposed law reads as follows:

> **Art. 4** The National Policy on Consumer Relations aims to meet the needs of consumers, respect for their dignity, health and safety, protection of their economic interests, improvement of their quality of life, protection of the environment, and such as the transparency and harmony of consumer relations, according to the following principles:
>
> (...) II–[governmental action to effectively protect the consumer:].
>
> (e) by encouraging sustainable production and consumption patterns.
>
> (...) IX–promotion of sustainable production and consumption patterns, to meet the needs of the current generations, to allow better living conditions, promoting economic development and social inclusion, without compromising environmental quality and service needs of future generations (free translation).

According to Marques, the doctrine states that this inclusion is justified:

> To clarify that the sustainable consumption patterns imposed by the UN Consumer Protection Guidelines consolidates the position of the consumers to be considered users and being protected when the consumers use public services. Thus, the sentence clarifies that the BCPC applies to public services remunerated individually (*uti singuli*) by the user or consumer. Public services provided to consumers is a sensitive and vital issue for the Brazilian population and in this sense, Parliament has a chance here to complete the work already initiated by Art. 22 of the [Consumer Protection Code].[43]

If this text is approved, the National Policy on Consumer Relations, in addition to meeting the needs of consumers, will aim at respect for their dignity, health and safety, protection of their economic interests, better living conditions and the protection of the environment, imposing a permanent dialogue between these two rights.[44]

---

[41] UNCTAD (2015). Clause V, H, 50–62.

[42] Brazil. Senate Bill No. 281/2012.

[43] Marques (2016b), pp. 419–425.

[44] Ibid, pp. 419–425.

With this standard, the notion of sustainable consumption would be included in the BCPC, and it becomes clear that Section IX is a principle of the Code. Likewise, it seems very important that the new rule must clarify that government action in the consumer market should encourage sustainable consumption and production patterns. Article 4 of the BCPC is a narrative norm of significant impact and the most cited article of the BCPC,[45] hence the relevance in introducing there the bases of sustainable consumption in the country, including the energy sector in that scope.

Finally, it should be pointed out that Clause V, K, 77, of the UNGCP establishes the duties of Member States and that the user of remunerated public services is a consumer and must be protected by the State, especially if she is hyper-vulnerable or has aggravated vulnerability.[46]

The updated BCPC proposals are fundamental to the evolution of a mature dialogue between consumption and environmental protection, in order to expressly demonstrate the need for goods and services to be adapted to the new required standards to achieve sustainability in Brazil, including, logically, the sector of the distributed generation of electricity that will be specifically dealt with below.

24.4.2 Decentralized Generation As a Means of Promoting Sustainable Energy Consumption

Micro and mini-generation are salient in the context of the so-called regulated captive market, where the consumer is obliged to purchase energy from a particular energy supplier. Under the Resolution (REN) No. 482/2012[47] issued by the Brazilian National Agency for Electricity (ANEEL), micro-generators are those with installed power less than or equal to 75 kilowatts (kW), and mini-generators are those whose installed power is between 75 kW and 5 megawatts (MW).[48]

Relevant advantages emerge from this system, such as the energy matrix diversification with fossil fuel dependence mitigation, consumption near the generation site avoiding losses and postponing investments in the transmission and distribution sectors, and increased consumer autonomy because of the reduction in supply disruption.[49]

Recently, the term 'prosumer' has been given to any consumer of electricity that starts to generate power for her own consumption or to send it to the grid,

[45]Marques et al. (2013), p. 78.

[46]UNCTAD (2015), p. 24.

[47]Agência Nacional de Energia Elétrica (2012).

[48]Agência Nacional de Energia Elétrica (2012). Article 2, I and II and REN No. 482/2012, respectively.

[49]Gesel (2018), pp. 19–20; 25–26.

accumulating both functions (producing and consuming).[50] The active role of consumers provides therefore new criteria to be considered in the prosumer concept:

> [T]oday many consumers are playing a more active role in energy markets. These hybrid energy consumers and producers are most accurately described as 'prosumers'. (. . .) However, 'prosumer' is more useful as a general term because the modern energy consumers described above do more than generate electricity. They also provide essential grid services, such as storage, regulation, and demand response in both retail and wholesale markets.[51]

Although, in the Brazilian case, prosumers cannot sell the electricity generated to the grid operator (i.e., via feed-in-tariffs as occurs in many countries[52]), but are only allowed to consume for themselves according to an explicit long-term contract in which future consumption is rebated as a credit for any electricity generated.[53] Both the utility and the prosumer depend on grid access; the former to distribute energy in the traditional design and the latter to feed the surplus power generated to receive a credit that can be applied to future consumption.[54] This is limited to a term of 60 months[55] for the same consumer unit (this is also called net metering).

The Brazilian government developed this compensation system to foster distributed energy generation, a commendable initiative that introduced this new modality into the Brazilian power sector. Updated by ANEEL Resolution (REN) No. 687/2015, REN No. 482/2012 unequivocally has the advantages of: (i) decentralising generation and promoting energy security; (ii) reducing greenhouse gases emissions—and thus a positive incentive to implement the COP 21 agreement, SDGs 7 and 12 and the UN Guidelines and; (iii) increasing access to energy[56] in interconnected grid systems, thus ensuring freedoms and enabling sustainable development.[57] Being able to consume electricity can be considered as source of well-being and may even facilitate other socioeconomic opportunities.[58]

---

[50]Briefly, the term 'prosumer' combines the notion of producer and consumer in one individual. See Toffler (1980) and, for business perspectives, see Ritzer (2014).

[51]Jacobs (2017), p. 524.

[52]Jansen (2011), pp. 329–331.

[53]Noronha (2013), pp. 96–97.

[54]Agência Nacional de Energia Elétrica (2012). Article 2, III, REN No. 482/2012.

[55]Agência Nacional de Energia Elétrica (2012). Article 6, I, REN No. 482/2012.

[56]On energy access in rural or isolated areas (such as the Amazon), see Bassani and Ferreira (2017).

[57]To deepen the discussion, the illustrious Nobel Prize-winning economist Amartya Sen (2010), pp. 10–25 has discussed at length criteria for deciding whether a given society can be considered developed, based not only on income but also on the extent to which deprivations of liberty that limit the choices and opportunities of persons to exercise their condition as agent are eliminated. Sen emphasizes five types of freedom, which he considers instrumental: (i) political freedoms; (ii) economic opportunity; (iii) social mobility; (iv) guarantees of transparency and; (v) personal security. These rights and opportunities help to promote a person's overall capacities and can complement each other. Instrumental liberties interrelate, connecting with one another, and contribute to the increase of human freedom in general.

[58]Costa (2009).

However, the regulatory framework was insufficient to popularise the deployment of the system as designed. ANEEL's Technical Note (ATN) No. 56/2017-SRD[59] re-evaluated the projections of mini- and microgrid connections. The 2015 ANEEL projection for 2017 was 40,909 grid connections. When the ATN was released in May 2017, there were 11,780 connections (10,453 using solar PV). The revised 2017 ANEEL projection for 2017 is now 26,834 grid connections.

However, even the 2017 forecast was also an overestimate. There were 14,277 grid connexions in September 2017 (14,144 through solar PV) and the total connections were 20,663 (20,498 through solar PV) as of 21 January 2018. In 12 June 2018, there were 30,604 connections to the grid (30,410 through solar PV); in 29 May 2019, due to equipment cost reduction and power tariff increase, there were 78,337 connections (78,041 through solar PV),[60] below ANEEL's forecast. Despite recognising the positive initiative of the legal regime governing distributed energy generation in Brazil, it is clear that there are many obstacles to be overcome if it is to be adopted widely.

Private investment needs incentives to encourage adoption of this decentralized type of energy generation. From the point of view of the active consumer (the potential prosumer), the nature of the compensation system could be an obstacle because the long investment payback (generally 4–8 years), depending on factors such as the cost of installation, the region, solar incidence and radiance, PV-panel efficiency, etc. From a strictly financial point of view, the remuneration system tends to be more attractive in the short term because the payback to the prosumer is immediately. The compensation system also assumes the consumer has the capital available (or access to such funds) to install the generation system in her home, which is not common. From the broader point of view of the electricity system, this transition logically depends on the regulatory framework that is in operation. The German example, based on a feed-in tariff, is a case of success, but questions have nevertheless arisen about investment limits and the degree of government intervention required to reduce dependence on nuclear power and improve energy efficiency.[61]

Another obstacle is the requirement that generation and compensation be in the same concession area, which creates a clear territorial limitation. While REN No. 482/2012 was a crucial initial milestone, it could have been more flexible in abolishing the territorial restriction, for example by allowing energy to be generated in more desirable places and consumed in nearby urban centres.[62] Issues regarding how such a proposal would operate in practice would require ongoing adaptation as the regulatory framework evolves, but it is considered feasible in the sense that electricity is a fungible and contractible good.

Another issue is the freedom of choice for the captive consumer. According to the president of Abraceel, the Brazilian Association of Energy Marketers,

---

[59] Agência Nacional de Energia Elétrica (2017).

[60] Agência Nacional de Energia Elétrica (2018).

[61] Gawel et al. (2014).

[62] On the right to self-generation, see: Weissman and Wellinghoff (2015).

> the consumer would have an invoice with two services. One would be the spinning service, which should be paid to the current utility owner of the grid connected to the distributed generator power system. The other would be the power that could be bought.[63]

The consumer could change the supplier in a system similar to the one already implemented in the telecommunication sector, which is also considered as a natural monopoly based on network structures.[64] Certainly, the power network should be adapted to this possibility, which would require long-term effort.

Another relevant point is taxation. In Brazil, ICMS (value-added tax) and PIS/COFINS (a social tax) are applied on the total power generated by the prosumer. However, according to ANEEL's own suggestion, taxation should only focus on the difference, if positive, between final consumption and the power surplus fed into the grid. If the difference between the power consumed and generated is lower than the minimum consumption rate, the tax base should be applied only on the cost of availability.[65]

Through Article 8 of Federal Law No. 13169/2015, the PIS/COFINS rate was reduced to zero. By discretionary decision of the federal states (required in the case of ICMS), there was an adhesion to the Agreement of the National Council of Finance Policy—Confaz No. 16/2015—authorising an exemption in the internal operations related to the circulation of electricity, subject to billing under the energy compensation system established by REN No. 482/2012.[66] An issue that may emerge is the conflict between Section 1 of the first clause of the Confaz No. 16/2015—which limits the application of the exemption to systems with loads of up to 1 MW—and REN 482/2012, which states a limit of 5 MW. It should also be noted that the government could reverse the decisions to reduce the PIS/COFINS rate to zero and to exempt the ICMS requirement at any time, which is obviously more likely in times of financial crisis or indebtedness. Given the few connections of distributed generation in the country, reversing these tax incentive measures would be a setback and would not have much impact on the public accounts.

Some benefits from decentralized generation of electricity are clear, including limiting the need for new investments in expensive transmission and distribution systems, low environmental impact, and minimisation of losses and diversification of the energy matrix.[67] One issue to be addressed is the difficulty in understanding technical and complex information by consumers in general about the adoption of the system, leading to the need to hire specialised advisers to verify equipment and installation costs and, where possible, verify financing offers. The costs for adequacy of the metering system, necessary to implement the energy compensation system, are the responsibility of the interested party under the Article 8 of the REN No. 482/2012. To foster more sustainable energy generation, more accurate

[63] Brazil. Câmara Legislativa.

[64] See: Nester (2006).

[65] Agência Nacional de Energia Elétrica (2016).

[66] Confaz (2015).

[67] Gesel (2018).

information on how to achieve this adequacy and its risks should be widely disseminated based on the duty to provide information by the supplier.

The implementation of micro or mini-generation is complex and depends on the ability of the consumer to analyse the cost–benefit ratio, the choice of type of energy source (solar panels, wind turbines, biomass generators, etc.), equipment technology, local installation (rural or urban), the value of the tariff to which the consumer unit is subject, conditions of project payment/financing and existence of other consumer units that can enjoy the credits of the energy compensation system.[68] This is an even more relevant obstacle considering the informational vulnerability of the consumer in a society where information is increasingly relevant and capable of transformation.[69]

The positive updated (through REN No. 687/2015) made possible the power generation by multiple consumer units such as condominiums, whose modality allows the distribution of energy generated among the owners in percentages defined by themselves. Another innovation was the shared generation, allowing stakeholders to form a consortium or cooperative to install a distributed micro or mini-generation to reduce the bills of the consortium or cooperative.

Regarding costs, there is a trend on equipment's price reducing[70] in the medium and long term because of the production dissemination and maturity (for example, higher volume and higher PV efficiency) with greater competitiveness of renewables in relation to the energy generation by traditional sources. Evidence proving the reduced value of the MW was the recent New Energy Auction promoted in Brazil at the end of 2017.[71]

More precisely, these data demonstrate that fixed costs are lower, as the generation of energy by renewable sources (sun and wind) have a variable cost close to zero—unlike other power plants using coal and oil with a high variable cost from the purchase of fossil fuels.[72]

It should be noted that the need to expand the decentralized generation is also necessary from the point of view of the National Electric System (SIN). When generating energy in a decentralised way, there is a trend to reduce the drawdown of hydropower reservoirs and the load in the integrated system because

> the electricity is consumed in the place where it is generated. It avoids and displaces the generation of a source that would be dispatched to meet the load if there were no such generation (Denholm et al., 2014). The plant that will stop generating due to the diffusion of the distributed photovoltaic generation will be the one with the highest variable cost being dispatched, considering that the generation is defined by order of merit.[73]

[68]Agência Nacional de Energia Elétrica (2016).

[69]Marques (2016a), pp. 332–342.

[70]National Renewable Energy Laboratory (2017), pp. 23, 31, 40.

[71]Empresa de Pesquisa Energética (2017).

[72]Gesel (2018), p. 5.

[73]Ibid, p. 12 (authors' translation).

The challenge of financing the up-front purchasing of equipment is a great obstacle because of the high rates of interest charged by public and private banks. Interest rates vary according to criteria such as locality (rural/urban) and borrower profile. Long-term investment is impracticable for most consumers.[74] The Brazilian Social and Development Bank (BNDES) launched an initiative in June 2018 by to provide credit for renewable energy equipment acquisitions, which might be a solution in the short term.[75]

Finally, positive incentives can be developed on several fronts, such as Senate Bill (PLS) No. 204/2014, which amends Federal Law No. 9991/2000 to include distributed micro-generation in the mandatory minimum percentage of resources applied in energy efficiency projects by concessionaires and permission holders (utilities) of power distribution. This may cover up to 80% of the resources of its efficiency programs for consumer units benefitting from the Social Electricity Tariff—whether in low-income communities, rural communities, or (if the bill becomes law) in decentralized micro-generation projects.[76]

24.5 Conclusion

Global initiatives form a baseline of public policies that Member States must foster and implement. SDGs 7 and 12—on access and efficiency—provide the basis of the SE4All program. They also present an interweaving of the objectives for the energy sector in the three lines of action: SE4All, the SDGs and COP 21. The difference is that, in the latter, climate change is the guiding thread whereas, in the others, sustainable development takes pole position.

Decentralized generation of electricity covers both production and consumption through renewable sources, a matter that deserves to be highlighted when acting on two fronts with only one positive action, fulfilling the requirements of any public policy to be fostered. The increase of this type of generation tends to be a catalyst for development and for an energy transition, radiating its effects in several areas, such as reducing inequalities, increasing income, creating more equitable opportunities, objectives that are pursued in international agreements.

Sustainable energy consumption in Brazil has evolved over the last 20 years, mainly through new energy auctions, which encompass more installed power and are intended to diversify the energy matrix, which today depends on hydropower. Regarding distributed generation of electricity, the number of current connections is below government forecasts, which prove the incentives are still insufficient to provide positives decisions to invest in the installation of the necessary equipment.

[74]Banco do Nordeste (2017); Banco Nacional de Desenvolvimento Econômico e Social (2017); Agência de Desenvolvimento Paulista (2017).

[75]Banco Nacional de Desenvolvimento Econômico e Social (2018).

[76]Brazil. Senate Bill No. 204/2014.

In its update through REN No. 687/2015, REN No. 482/2012 could have been more incisive to encourage the promotion of renewable energy generation, which since its inception has been limited to creating an initial regulatory framework to lay the foundations for the compensation system, probably because of the significant difficulties of adaptation to a traditional electrical system. Therefore, investment incentives for this form of generation are not yet sufficiently attractive for widespread adoption of distributed energy generation systems, despite the immense potential. Like all novelties, adapting to a new system requires overcoming obstacles.

A first challenge is the hard work of making consumers aware that owning a renewable energy generation system is considered as a local action to mitigate climate change. The proposed BCPC amendments are a way to promote the need for sustainable production and consumption and to fortify the right of the technically vulnerable prosumer to make choices to mitigate the impacts on the environment. After awareness, the next most challenging obstacle is to positively boost consumer discretion in acquiring the distributed generation system. The positive decision to invest often depends on clear information and understanding of a complex system, usually requiring a third-party expert. Another relevant constraint is capacity to pay, mainly because of the few viable financing alternatives available to most Brazilians. The long term payback (because Brazil adopted the compensation system-net metering) also affects the decision. If equipment costs fall, the investment and the payback period will also decline, which may increase the connections to the grid.

Predictability and legal certainty are important factors as well. The risk that the state governments will increase the PIS/COFINS tax rate and revoke the ICMS tax exemption cannot be ruled out. The prosumer who invests in generation equipment expecting a full return in 4–8 years, can be overwhelmed if governments choose to raise tax rates, which is always possible. It would then be left to the courts to decide how prosumers might be compensated. Other issues involving the suspension of the compensation by the utility or delays in connection procedures may also hamper the spread of decentralized generation of electricity.

Certainly, advances in the implementation of this new modality of generation and consumption of electricity have occurred. The greatest payoff is that this modality is entirely based on non-polluting sources renewed according to the laws of nature, avoiding the natural resources depletion. It has been proven in this study, therefore, that (i) decentralized generation approach enhances sustainability in the generation and consumption of electricity; (ii) that connections to the grid below initial forecasts demonstrate the need (ii.i) to improve legal provisions of sustainable patterns, and (ii.ii) to foster positive incentives including legal protection to enable consumers to become more active, i.e., prosumers.

References

Agência de Desenvolvimento Paulista (2017) Linha Economia Verde. http://www.desenvolvesp.com.br/empresas/opcoes-de-credito/projetos-sustentaveis/linha-economia-verde/. Accessed 15 Jan 2017

Agência Nacional de Energia Elétrica (2012) REN No. 482/2012. http://www2.aneel.gov.br/cedoc/ren2012482.pdf. Accessed 25 Aug 2017

Agência Nacional de Energia Elétrica (2016) Micro e minigeração distribuída: sistema de compensação de energia elétrica, 2nd edn. ANEEL, Brasília. http://www.aneel.gov.br/documents/656877/14913578/Caderno+tematico+Micro+e+Minigera%C3%A7%C3%A3o+Distribuida+-+2+edicao/716e8bb2-83b8-48e9-b4c8-a66d7f655161. Accessed 30 Aug 2017

Agência Nacional de Energia Elétrica (2017) ANEEL's Technical Note (ATN) No. 0056/2017-SRD/ANEEL, 24 May 2017. Processo No. 48500.004924/2010-51. Assunto: Atualização das projeções de consumidores residenciais e comerciais com microgeração solar fotovoltaicos no horizonte 2017–2024. http://www.aneel.gov.br/documents/656827/15234696/Nota+T%C3%A9cnica_0056_PROJE%C3%87%C3%95ES+GD+2017/38cad9ae-71f6-8788-0429-d097409a0ba9. Accessed 10 Aug 2017

Agência Nacional de Energia Elétrica (2018) Unidades consumidoras com geração distribuída. http://www2.aneel.gov.br/scg/gd/GD_Fonte.asp. Accessed 12 June 2018

Banco do Nordeste (2017) Programa FNE-SOL. https://www.bnb.gov.br/programas_fne/programa-de-financiamento-a-micro-e-a-minigeracao-distribuida-de-energia-eletrica-fne-sol. Accessed 15 Jan 2018

Banco Nacional de Desenvolvimento Econômico e Social (2017) Pronaf Eco. https://www.bndes.gov.br/wps/portal/site/home/financiamento/produto/pronaf-eco. Accessed 15 Jan 2018

Banco Nacional de Desenvolvimento Econômico e Social (2018) BNDES muda regra e pessoas físicas podem investir em energia solar. https://www.bndes.gov.br/wps/portal/site/home/imprensa/noticias/conteudo/bndes-muda-regra-e-pessoas-fisicas-podem-investir-em-energia-solar/!ut/p/z0/pY7NTgMxDISfhUOOUayytHBcQSXUH_UCUsmlcnfdrUvXTpNsgbcnrDhz4TQeaebzOO-2zgteucPMKngu_s1Pd6uH5fy52sAK7l4qqJ9uZ9XrbA7L-4lbOP93oBD4dLn42vlGJdNndtu9tJR2LClzHprxk4Gj9mSA-xBJEhoQzdwwJgNjb2jVwFi0_dCijdRFtGQDpaSY7IETN0WDttRblisVerTlJqHYMdqkZ4w_eyZx_bjunA-YjyV50N9F_wOHd7__-qhvvgERUgT_/. Accessed 11 June 2018

Bassani ML, Ferreira LV (2017) The energy access in rural or isolated areas in Brazil: a viability review. In: Brant LNC et al (eds) Sustainable development and energy matrix in Latin America: the universal clean energy accessibility. Konrad Adenauer Stiftung/Cedin, Belo Horizonte, pp 353–379

Baudrillard J (2011) A sociedade de consumo (Transl. Artur Morão). Lisboa, Edições 70

Bauman Z (1992) Intimations of postmodernity. Routledge, London

Brazil. Federal Law No. 8.078/1990–Consumer Protection Code

Brazil. Federal Law No. 12.965/ 2014

Brazil. Federal Law No. 13.169/2015

Brazil. Civil Code

Brazil. Senate Law Bill No. 431/2014

Brazil. Senate Law Bill No. 281/2012

Brazil. Senate Law Bill No. 204/2014

Brazil. Brazilian Federal Constitution

Brazil. Câmara Legislativa. http://www2.camara.leg.br/camaranoticias/radio/materias/ULTIMAS-NOTICIAS/475906-CAMPANHA-DA-ABRACEEL-QUER-INCENTIVAR-CONSUMIDOR-AO-MERCADO-LIVRE-DE-ENERGIA.html. Accessed 15 Oct 2014

Confaz (2015) Convênio No. 16/2015. https://www.confaz.fazenda.gov.br/legislacao/convenios/2015/cv016_15. Accessed 15 Jan 2018

Costa MA (2009) O direito de acesso à energia: meio e pré-condição para o exercício do direito ao desenvolvimento e dos direitos humanos. Ph.D. Thesis, Universidade de São Paulo

Empresa de Pesquisa Energética (2017) Leilões de Energia Nova. http://www.epe.gov.br/sites-pt/publicacoes-dados-abertos/publicacoes/PublicacoesArquivos/publicacao-268/Informe%20Leil%C3%B5es%202017%20-%20Portugues.pdf. Accessed 15 Jan 2017

Fatouros AA (2008) An international legal framework for energy. In: Collected courses of the Hague Academy of International Law, vol 332. Martinus Nijhoff Publishers, The Hague

Gawel E et al (2014) The future of the energy transition in Germany. Energy Sustain Soc 4(1):1–9

Gesel (2018) Grupo de Estudos do Setor Elétrico/UFRJ. Impactos Sistêmicos da Micro e Minigeração Distribuída. TDSE Texto de Discussão do Setor Elétrico, n° 79, janeiro de 2018, Rio de Janeiro. http://www.gesel.ie.ufrj.br/app/webroot/files/publications/46_tdse79.pdf. Accessed 19 Jan 2018.

Goldblatt DL (2005) Sustainable energy consumption: personal, technological, or social change? Springer Netherlands, Dordrecht

International Energy Agency (2005) Energy statistics manual. http://www.iea.org/publications/freepublications/publication/statistics_manual.pdf. Accessed 8 Mar 2018

International Energy Agency (2010) World Energy Outlook–Energy Access. http://www.worldenergyoutlook.org/resources/energydevelopment/globalstatusofmodernenergyaccess/#d.en.8609. Accessed Oct 2015

International Energy Outlook (2017). https://www.eia.gov/outlooks/ieo/pdf/0484(2017).pdf. Accessed 15 Jan 2018

Jacobs SB (2017) The energy prosumer. Ecol Law Q 43(1):519–580

Jansen B (2011) Community wind power: making more americans energy producers through feed-in tariffs. Kansas J Law Public Policy 20:329–349

Kapseu C et al (2012) Énergie renouvelables en Afrique Subsaharienne. Harmattan, Paris

Lyotard JF (1989) A condição pós-moderna. Gradiva, Lisboa

Lyster R, Bradbrook A (2006) Energy law and the environment. Cambridge University Press, Melbourne

Marques CL (2015) Atualização do Código de Defesa do Consumidor e o diálogo entre o direito do consumidor e o direito ambiental: estudo em homenagem à Eládio Lecey. In: Benjamin AH, Leite JM (Eds) 20° Congresso Brasileiro de Dir. Ambiental–Ambiente, Sociedade e Consumo Sustentável

Marques CL (2016a) Contratos no Código de Defesa do Consumidor, 8th edn. São Paulo, Revista dos Tribunais

Marques CL (2016b) 25 anos de Código de Defesa do Consumidor e as sugestões traçadas pela Revisão de 2015 das Diretrizes da ONU de proteção dos consumidores para a atualização. In: Miragem B, Marques CL, Oliveira AF (eds) 25 anos de Código de Defesa do Consumidor: trajetória e perspectivas. Revista dos Tribunais, São Paulo, pp 395–438

Marques CL, Miragem B (2014) O novo direito privado e a proteção dos vulneráveis, 2nd edn. Revista dos Tribunais, São Paulo. http://www.planetaverde.org/arquivos/biblioteca/arquivo_20150602200928_5210.pdf. Accessed 10 Feb 2018

Marques CL, Benjamin AH, Miragem B (2013) Comentários ao Código de Defesa do Consumidor, 4th edn. Revista dos Tribunais, São Paulo

Mello CAB (2014) Curso de Direito Administrativo, 32nd edn. Malheiros, São Paulo

Miragem B (2004) A regulação do serviço público de energia elétrica e o direito do consumidor. Revista de Direito do Consumidor 51:68–100

Miragem B (2013) Curso de Direito do Consumidor, 4 edn. rev. atual. e ampl. Revista dos Tribunais, São Paulo

Morosini FC, Marceau G (2013) The status of sustainable development in the law of the World Trade Organization. In: Celli U Jr, Basso M, Amaral A Jr (eds) Arbitragem e comércio internacional: estudos em homenagem a Luiz Olavo Baptista. Quartier Latin, São Paulo, pp 59–92

National Renewable Energy Laboratory (2017) U.S. Solar Photovoltaic System Cost Benchmark: Q1 2017. https://www.nrel.gov/docs/fy17osti/68925.pdf. Accessed 20 Dec 2017
Nester AW (2006) Regulação e Concorrência: compartilhamento de infraestruturas e rede. Dialética, São Paulo
Noronha F (2013) Direito das obrigações, 4 edn. rev. atual. e ampl. Saraiva, São Paulo
Pachauri S (2011) Reaching an international consensus on defining modern energy access. Curr Opin Environ Sustain 3(4):235–240
Pehnt M (2006) Micro cogeneration technology. In: Micro cogeneration: towards decentralized energy systems. Springer, Berlin, pp 1–18
Ritzer G (2014) Prosumption: evolution, revolution, or eternal return of the same? J Consum Cult 14(1):3–24
Sarlet IW, Fensterseifer T (2014) Princípios do Direito Ambiental. Saraiva, São Paulo
Sen A (2010) Desenvolvimento como liberdade. Companhia das Letras, São Paulo
Shunck GB (2016) Contratos de longo prazo e dever de cooperação. Almedina, São Paulo
Sweeney JL (1993) Economic theory of depletable resources: an introduction. In: Sweeney JL, Kneese AV (eds) Handbook of natural resource and energy economics. Elsevier B.V, Amsterdam
Toffler A (1980) The third wave. Bantam Books, New York
UNCTAD (2015) United Nations Conference on Trade and Development. United Nations Guidelines for Consumer Protection. http://unctad.org/en/PublicationsLibrary/ditccplpmisc2016d1_en.pdf. Accessed 15 Dec 2017
United Nations (1972) Stockholm Declaration. http://www.onu.org.br/rio20/img/2012/01/estocolmo1972.pdf. Accessed 12 Oct 2017
United Nations (2015) Transforming our world: the 2030 agenda for sustainable development. https://sustainabledevelopment.un.org/post2015/transformingourworld. Accessed 26 Apr 2016
United Nations (2016) Sustainable energy for all strategic framework for results 2016–2021. June 2016. Available at http://seforall.org/sites/default/files/2016_EUSEW.pdf. Accessed 13 Jan 2018
United Nations (2017) SDG 17. http://www.un.org/sustainabledevelopment/globalpartnerships/. Accessed 23 Dec 2017
US Department of Energy (2017) Energy Intensity Indicators: Efficiency vs. Intensity. https://www.energy.gov/eere/analysis/energy-intensity-indicators-efficiency-vs-intensity Accessed 20 Aug 2017
Weissman S, Wellinghoff J (2015) The right to self-generate as a grid-connected customer. Energy Law J 36:305–326

Chapter 25
Planned Obsolescence Resulting from Electrical and Electronic Equipment: Waste Rights and Brazil's National Solid Waste Policy

Cláudio José Franzolin

Contents

Abstract Contemporary concerns about the sustainability of consumption do not derive solely from the fact that the products available in the market incorporate new technologies with as yet unknown risks. The deliberate market strategy of planned obsolescence—the rapid and continuous release of newer versions of products—is also pertinent. While this strategy to manufacture products with shorter life-cycles to be quickly replaced by newer versions moves the market, it also means products are discarded faster, resulting in residues and waste. This is particularly true of electrical and electronical equipment (EEE), which, once discarded, become electronic waste (e-waste). This chapter analyses planned obsolescence of EEE within the waste rights, more specifically, within Brazil's National Solid Waste Policy (NSWP), (Brazilian Law No. 12305/2010). Having in mind sustainable consumption, a systematic interpretation of the consumer rights was carried out.

C. J. Franzolin (✉)
Pontifical Catholic University of Campinas (PUC-Campinas), Center for Postgraduate Program in Law, Campinas, Brazil

A. d. Amaral Junior et al. (eds.), *Sustainable Consumption*,
https://doi.org/10.1007/978-3-030-16985-5_25

25.1 Introduction

Not all countries express concern regarding the destination of electronic products in the post-consumption phase and not all of them have an effective waste management system in place. As much was said recently in a publication on technical standards published by the International Telecommunication Union (ITU),[1] written by authors belonging to a wide variety of international bodies, such the United Nations University and the World Health Organization, among others. As with many studies involving waste, our analysis focuses on electrical and electronic equipment. As the ITU report states, GDP growth[2] is an indication of the acceleration of the consumption of such equipment on a global scale and, consequently, the impact in the post-consumption stage, when discarded, when they become electronic waste, technological waste or e-waste.[3]

Electronic devices have made a significant contribution to improving people's lives over the decades. We have, however, become increasingly dependent on them. In addition to that, suppliers increasingly encourage consumers not only to acquire new devices, but also newer versions of existing ones, either because of technological innovations by adding new functions, or because a new designs or style emerges, or through deliberate reduction in the life-cycles of such products.

The ITU report notes that, in the global context, we will have reached almost 50 million metric tonnes of e-waste at the end of 2017; of this total, 5 million was produced in Latin America. The report also states that in the Western Hemisphere, the United States, Brazil and Mexico are the largest generators of electrical and electronic waste.[4] In another report, prepared by Greenpeace, it is estimated that by 2017, the volume of technological waste may be 67 million metric tonnes.[5]

We see now an abundance of electronic devices—such as cell phones, tablets, microcomputers, notebooks, monitors, smart televisions and batteries, among other electronics, with shorter life-cycles. Moreover, consumers are co-opted and driven by advertising and planned obsolescence, among various strategies of suppliers—retailers, manufacturers, etc—to exchange their products for newer versions, discarding previous ones, which culminates in the emergence of e-waste. In the face of the continuous launch of new electronic products in the market, the

---

[1]See Sustainable Management of Waste Electrical and Electronic Equipment in Latin America (2016), p. 1.

[2]See Sustainable Management of Waste Electrical and Electronic Equipment in Latin America (2016), p. 1.

[3]The terms electrical and electronic waste, technological waste and e-waste will be used synonymously throughout the chapter.

[4]See Sustainable Management of Waste Electrical and Electronic Equipment in Latin America (2016), p. 1.

[5]See Cook and Jardim (2017), p. 4.

accelerating discarding of previous versions by consumers and the generation of technological waste[6] there must be more of a commitment to sustainable consumption.[7]

For this reason, we focus our study on the consumption of electronic equipment, in the light of Brazil's National Solid Waste Policy (NSWP), enshrined in Law No. 12305/2010, which emphasises the need to 'achieve sustainable production and consumption patterns of goods and services in order to meet the needs of the current generations and to enable better living conditions without compromising environmental quality and meeting the needs of future generations'.[8]

Thus, a dialogic approach between Consumer Law and waste rights[9]—a new field under construction—is adopted through a systematic interpretation that allows a 'search for solutions that are respectful of the organization, in its ... axiological dimension and in its non-linear historical character, understood as a holistic project, potentially coherent and permeable to evolutionary mutations'.[10] After all, the legal system gravitates around a constitutional project', which allows a more renewed meaning,[11] given the new concrete problems that emerge from consumption relations which demand a rationality based on environmental[12] and constitutional values.

The chapter is divided in two. The first section details the concept of planned obsolescence and its impacts in the field of electrical and electronic equipment (EEE). Secondly, it analyses Brazil's (NSWP), highlighting the importance of consumer rights and its relation to laws governing waste disposal.

25.2 Planned Obsolescence: From Equipment to Electronic Residues

According to Bauman,[13] society is no longer one of producers, but one of consumers, a shift further accentuated by the advent of the post-industrial society[14] where the so-called consumer society emerges. Furthermore, we note that the production process is no longer either standardised or homogenised; it becomes

[6]See Sustainable Management of Waste Electrical and Electronic Equipment in Latin America (2016), p. 6.

[7]Magalini et al. (2015), p. 5.

[8]See Article 3, XIII.

[9]See Aragão (2003), pp. 55–60.

[10]Aragão (2003), p. 8.

[11]See: Konder (2015), p. 44.

[12]Leff (2012), p. 23.

[13]See: Bauman (2008).

[14]Bell (1973), p. 148.

part of the fragmentation and ephemerality of production, 'having a direct impact on the sphere of consumption'.[15] Thus, according to Baumann,[16] we have moved from consumption to consumerism because—in Bauman's view the latter foregrounds the recycling of wills, desires and anxieties, the ephemeral and the routine. From this perspective, consumption—which was once based on the demand for reliable, regular, transparent, durable, resistant and safe ways to meet basic human needs—must now be instantaneous and fast; if the product does not meet the consumer's need, he discards it and it becomes waste.[17] As summarised by Baudrillard: 'What is produced today is not produced according to the respective use value or the possible duration, but, depending on its death',[18] leading to a product life-cycle that is ephemeral; it is a form of 'technological sabotage'.[19]

In this vein, Magera states that electronic devices 'no longer belong to the category of utility tools',[20] because consumers want to be the first to buy the latest versions which contain modern technologies.[21] Thus, electronic equipment purchased today, is already outdated tomorrow[22] and is discarded. It is in this scenario that the planned obsolescence[23] and e-waste emerge.

25.2.1 *Defining the Concept of Planned Obsolescence*

Magera states that the electronic product disposal takes place because of both planned obsolescence and general media. However, this study will focus on planned obsolescence only.[24]

As Franzolin has noted:

> Planned obsolescence is a strategy adopted by suppliers to provide a flow and cycle of continuous movement of products available in the market, making them, after a reduced life-cycle, old-fashioned and uninteresting. The result is inevitable disposal and replacement by newer versions.[25]

Vance Packard (following the industrial designer Brooks Stevens)[26] observes that the modern economy is based on planned obsolescence because sound products are

[15]Retondar (2008), p. 140.
[16]Bauman (2008), p. 44.
[17]Ibid.
[18]Baudrillard (2014), p. 44.
[19]Ibid, p. 45.
[20]Magera (2013), p. 99.
[21]Ibid.
[22]Loubet (2007), p. 246.
[23]Magera (2013), p. 15.
[24]Ibid.
[25]Franzolin (2017), p. 45; see, also: Moraes (2015).
[26]Packard (1965), p. 50.

made, and people are persuaded to buy them, but then something is added in the latest version of the product, and the old product becomes outdated and obsolete. From there, Packard[27] delineates three types of obsolescence. Functional obsolescence is when 'an existing product becomes old-fashioned when a product that performs the function better is introduced'.[28] In this obsolescence, Packard recognises the advantage of the improvement, however, the author himself acknowledges the possibility of manipulation[29] by the manufacturer. Then there is quality obsolescence, where 'a product breaks or degrades over time, usually not too long'.[30] Here a fixed life-cycle for the product is developed by the supplier, typically to promote replacement sales. Finally, there is psychological obsolescence of a strategy consists of 'spending the product in the mind of the owner'[31]; thus, the product does not need to improve functionally, enough, change it in its format and presentation.

Planned obsolescence reveals itself in almost all everyday electronic devices, so much so that Christoph Behrends[32] states that the twentieth century was marked by obsolescence. Behrends cites the planned obsolescence applied to video equipment in the 1990s, claiming that the DVD technology rereleased to consumers to replace VHS and BETACAM had already been superseded by technology although it was marketed to consumers as cutting edge. The next generation of consumer recording and playback technology (i.e., DivX often known in the market as Blu-ray DVD) *had already been invented* but the main manufacturers decided to leverage the potential of first-generation DVD technology in the market first. Conversely, manufacturers had already developed Blu-ray but delayed its launch in the market strategically to boost profits. Consumers, having shelled out for inferior technology ended up replacing it, shortly after, by the most up-to-date version (Blu-ray).

Another product subject to elaborate programmed obsolescence is the printer. Magera notes how printers are designed to fail after a certain number of pages printed, depending on the model and manufacturer.[33] Printer cartridges are also notoriously designed to deplete more quickly than they should do. Indeed, obsolescence is common in the computer market. Magera explains that the hardware (equipment) does not align with software (programs and applications), because 'when the market launches new programs, it becomes necessary to acquire new machinery with higher availability of memory',[34] which will make the old machinery obsolete.[35] These are typical examples involving quality obsolescence in the field of electronic equipment, to reduce the product life-cycle.

---

[27]Ibid.

[28]Ibid, p. 51.

[29]Ibid.

[30]Ibid.

[31]Ibid, p. 63.

[32]Behrends (2004), p. 3.

[33]Magera (2013), p. 15.

[34]Ibid.

[35]Ibid.

The market for EEE is also subject to the so-called 'incompatibility obsolescence'.[36] Here, a product that has already been marketed and sold becomes incompatible with newer versions and even with competing brands. Cell phones are a case in point.[37] Current-generation smart phones offer a wide range of functions, allowing users to make calls but also to connect to social media, access e-mail, listen to music, take photos, receive offers, download useful applications, and so on.[38] In the context of these mobile devices, there is an obsolescence of incompatibility[39] and style. In terms of style, the consumer replaces the device not because it offers new functions, but because of newer designs[40] and sizes. Another type of observed obsolescence that is said to feature in the smart phone market is 'standard functional obsolescence',[41] namely, when one component fails (e.g. the battery) that renders the device less useful or causes it to fail entirely.

In sum, there are several ways of artificially reducing the product life-cycle of EEE to further stimulate consumption via frantic and rapid replacement. What results is the discarding of vast quantities of material (often with environmentally damaging components) and large increases in electronic waste. Excessive generation of waste, however, is a source of danger to life, health and the balance of ecosystem.[42] After all, it causes disarrangements and new problems, which demand the need for new solutions.

25.2.2 *Defining Electrical and Electronic Waste*

Patrícia Faga Iglecias and João Múcio note that, traditionally, a product life-cycle was understood to have three phases: production, distribution and consumption.[43] This, naturally, occluded the fourth phase: post-consumption. After all, discarded products must go somewhere and, indeed, almost all ends up as waste. All of this serves to confirm Paulo Machado's pertinent observation that 'there is an undeniable connection between generation of waste and consumption'.[44]

The UN has declared clearly that the proper disposal of waste is not only a domestic concern for nation-states, but also an international problem.[45] Therefore, when dealing with the impacts of waste, the UN has stated:

[36]See CEC-ZEV.

[37]See CEC-ZEV.

[38]See Souza et al. (2007).

[39]See CEC-ZEV.

[40]See CEC-ZEV.

[41]See CEC-ZEV.

[42]Aragão (2003), p. 9.

[43]Lemos and Mendes (2013), p. 43.

[44]Machado (2016), p. 682.

[45]UN Economic and social development (n.d.).

21.38 By the end of the century, over 2.0 billion people will be without access to basic sanitation, and an estimated half of the urban population in developing countries will be without adequate solid waste disposal services. As many as 5.2 million people, including 4 million children under five years of age, die each year from waste-related diseases. The health impacts are particularly severe for the urban poor. The health and environmental impacts of inadequate waste management, however, go beyond the unserved settlements themselves and result in water, land and air contamination and pollution over a wider area. Extending and improving waste collection and safe disposal services are crucial to gaining control over this form of pollution.

21.39 The overall objective of this programme is to provide health-protecting, environmentally safe waste collection and disposal services to all people. Governments, according to their capacities and available resources and with the cooperation of the United Nations and other relevant organizations, as appropriate, should: a. By the year 2000, have the necessary technical, financial and human resource capacity to provide waste collection services commensurate with needs; b. By the year 2025, provide all urban populations with adequate waste services; c. By the year 2025, ensure that full urban waste service coverage is maintained and sanitation coverage achieved in all rural areas.

European law also has spoken on the question. European Economic Community Directive 75/442/EEC was adopted on 25 July 1975 was the first to provide a regulation on solid waste. Later, Directive 2006/12/EC (5/4/2006) was enacted and was amended by Directive 2008/98/EC of 19 November 2008 by the European Parliament. In Article 3, paragraph 1 the of the latter, waste is defined as: 'any substance or object which the holder discards or intends to or is required to discard'. Note that the text opted for a more extended conceptualisation of waste, offering wide scope for the issue to be addressed within the domestic legislation of each country, while it nevertheless defined and classified some types of waste.

Regarding electronic equipment and waste, Directive 2012/19/EU of the European Parliament and of the Council of 4 July 2012, expressly ruled on the subject:

Article 3, I

(a) 'electrical and electronic equipment' or 'EEE' means equipment which is dependent on electric currents or electromagnetic fields in order to work properly and equipment for the generation, transfer and measurement of such currents and fields and designed for use with a voltage rating not exceeding 1 000 volts for alternating current and 1 500 volts for direct current; (...); (e) 'waste electrical and electronic equipment' or 'WEEE' means electrical or electronic equipment which is waste within the meaning of Article 3(1) of Directive 2008/98/EC, including all components, sub-assemblies and consumables which are part of the product at the time of discarding.

Further, the reading of this directive presents different ways of classifying EEE. For instance, in Annex I of the Directive, a classification of electrical and electronic products to be followed up to 2018 is provided under Article 2, I (b). The period after 2018 is covered by one of the categories in Annex III. It is worth noticing that

EEE products have been categorized into various groupings by pieces of national legislation considering their original purpose, size, composition and/or weight. Internationally, the suggested classification has been used as a reference by the 2002 European Directive reflected in Annex II Indicative list of EEE which falls within the categories of Annex I (53)

In domestic Brazilian law, waste is defined in Article 3, XVI of Law No. 12305/2010, but no specific conceptual delimitation for EEE waste is provided. Although there is no legal definition of EEE waste, the NSWP[46] has established that manufacturers, importers, distributors:

1. have shared responsibility for handling waste (Article 30) and must commit to effective 'reverse logistics' systems (Article 33, VI, § 3);
2. must provide information on these actions (Article 33, § 8); and
3. should develop products with lower impacts on human health and environmental quality through the product life-cycle (Article 42, II), among other aspects, imposing duties that directly or indirectly relate to suppliers of electrical and electronic products.

For this reason, it is even more necessary to recognise how important the effectiveness of NSPW is, because waste is: (a) inseparable from social relations; (b) a source of pollution; and (c) produces environmental damage, the effects which are not even known, and may take years until they manifest in human health and the environment.

In this regard, problems related to post-consumption electric/electronic waste arise. This problem is not found in Brazil, but in many countries. In Brazil, this type of waste has to be analysed considering the PNRS (L. 12.305/2010-LPNRS).[47]

25.3 Brazil's National Solid Waste Policy

According to Ricardo Abramovay, economic growth—based on the ever-expanding marketing of goods and services on terms largely framed by the 'business world' itself—cannot be evaluated solely in terms of jobs created and production generated.[48] It should also assess the side effects on nature, the ecosystem and the climate, and therefore the impact on overall well-being and human development. Finally, it is necessary for the market to develop new paradigms,[49] incorporating ecological conditions of sustainability. Electric and electronic products containing heavy elements (lead, cadmium, etc.), if discarded as waste without adequate planning,[50] affects not only the consumer, but also the collective and future generations in a diffuse way.

For this reason, what is called for is a greater attention to what are known as waste rights. Here, integrating consumption and waste means (re-)constructing consumer relations from a perspective based on new bases that allows assimilating new ways

[46]See Lemos and Mendes (2013), p. 47.

[47]Leff (2012), p. 25.

[48]Abramovay (2012), p. 34.

[49]See: Candemil (2012), pp. 13–44.

[50]See: Gouveia and Prado (2010), pp. 859–866.

of acting and thinking about the agents involved in the market. This is so much so that the NSWP itself highlights, among its principles, 'the systemic vision, in the management of solid waste, that considers the environmental, social, cultural, economic, technological and public health variables'.[51] This imposes a more holistic and expanded perspective regarding the confrontation of the socio-environmental impacts resulting from the generation of electrical and electronic waste.

In general, waste—regardless of its category or kind—is concerning because it is produced in abundance, which only shows that today there is a clear inefficiency in the articulation between production and consumption, causing a 'residual conflict'[52] related to an 'environmental and social conflict'.[53] For that matter, Maria Alexandra Aragão,[54] presents the following reasons to study waste rights:

1. the inherent lack of interest of those who produce the waste in how it is disposed of. After all, 'waste is a substance the production of which was not the main objective of the business activity, which is either extraction, manufacture, distribution or commercialization';[55]
2. the heavy disposal costs;
3. the enigmatic nature of the material itself, because the physicochemical composition of the waste is often unknown to those who deal closely with it;
4. waste is the genesis of the most varied types of pollution;
5. the difficult territorial management, particularly as the adequate location for the receipt of waste; and
6. waste, especially hazardous waste, is a major health hazard and a threat to the balance of ecosystems.

It is possible to recognise the need to establish a set of principles for waste rights, helping to strengthen the growing autonomy of this new field in greater interconnection with others, such as environmental rights and consumer rights, to:

1. raising awareness of more sustainable production;
2. allocation of new responsibilities;
3. the definition of new concerns in waste management; and
4. concern about the product life-cycle.

Regarding the principles of waste rights, Aragão mentions those of precaution, prevention, polluter-pays, appreciation and self-sufficiency, among others. The precaution, prevention and polluter-payer principles are today expressly recognised in Article 3, I and II of the NSWP. As for the valorisation principle, for Aragão this means the intention to value raw materials and fuels and thus avoid the impacts that

[51] Article 6, III, of Law 12305/2010.

[52] Aragão (2006), p. 93.

[53] Aragão (2003), p. 8.

[54] Ibid.

[55] Ibid.

they cause on the environment.[56] Under item 15 from Article 3 of Directive 2008/98/EC of the European Parliament and of the Council of 19 of November 2008:

> Valorisation means any operation for which the main result is the processing of the waste in order to serve a useful purpose by replacing other materials which would otherwise have been used for a specific purpose or the preparation of the waste for that purpose in the plant or in the whole of the economy. Annex II contains a non-exhaustive list of valorisation operations.

Notwithstanding, the principle of self-sufficiency is contained in the directive, when it establishes that waste generators themselves can dispose of their waste:

> It is necessary, in order to enable the Community as a whole to become self-sufficient in waste disposal and in the recovery of mixed municipal waste collected from private households and to enable the Member States to move towards that aim individually, to make provision for a network of cooperation as regards disposal installations and installations for the recovery of mixed municipal waste collected from private households, taking into account geographical circumstances and the need for specialised installations for certain types of waste.

In the context of Brazilian domestic law, it is necessary to strengthen the dialogue between Brazil's Consumer Protection Code and the NSWP, enshrined in Law No. 12305/2010, which is now the basic and fundamental legal framework on waste. The NSPW includes the following core elements:

1. a definition of basic concepts (Article 3,), among them, waste;
2. specific principles (Article 6);
3. fundamental objectives (Article 7);
4. instruments for waste management (Article 8) to operationalise those objectives;
5. legal principles, such as shared responsibility for the product life-cycle (Article 30); and
6. coordinated management of solid waste, with the participation of both government and private enterprise (Articles 31, § 1 and 34).

Thus, the NSWP seeks to establish a minimum legal framework to introduce instruments, principles and duties in the national legal system, through standards of open content. In this way, the interpretation of the law can contribute to the new debates about planned obsolescence, about the reduced life cycle of products and about the challenges to build more sustainable consumption.

After all, waste is undoubtedly a source of new and complex conflicts involving multiple interests.[57] That is to say, it is only through a cooperation between the various state governments, combined with internal policies and legislation, that an awakening to a new social order can be promoted, in a perspective that demands more sensitivity to man in the various concrete manifestations, as to the production habits consumption, considering that they will have an impact on future generations and on the environment. Moreover, this production of waste and the resulting

---

[56] Aragão (2003), p. 8.

[57] Aragão (2003), p. 7.

impacts are very much in line with the analysis we have presented through the chapter concerning EEE, which, when it turns into waste, undoubtedly damages the environment and, therefore, the health of every consumer.

25.3.1 The National Solid Waste Policy and the Disposal of Electrical and Electronic Equipment

Portilho notes how the Rio-92 Conference has come to be seen as the origin-point for the debate on consumption and the environment.[58] Agenda 21 followed Rio-92, and it has highlighted the conditions and habits that stimulate debates about lifestyles based on consumption and the resulting responsibilities. The 1992 United Nations Conference, recognising the need for sustainable consumption, culminated in a document titled *The Future We Want*.

In relation to consumer rights, Claudia Lima Marques[59] reminds us that, at various times, the Brazilian CPC has been attentive to environmental concerns. The author highlights paragraph 2 of Article 37 of the CPC, deeming abusive advertisement and contract clauses which do not follow environmental norms. I Marques goes on to say that the NWSP is closely tied to post-consumer responsibility to establish new parameters and duties for sustainable consumption.

These points corroborate Patrícia Faga's statement recognising the need for a more expanded analysis of consumption, as 'consumption ends up having repercussions that exceed the person of the consumer'[60] precisely because after the product life-cycle is exhausted (irrespective of whether this has been deliberately engineered by suppliers)[61] there is inevitable waste. Without proper disposal, pollution[62] and negative impacts on the environment occur.[63] Therefore, 'it is necessary to think of macro solutions that involve education, awareness, change of position',[64] whether by the supplier, the consumer or the civil society.

This view, although still incipient, is increasingly prevalent, according to a study published in the Greenpeace Guide,[65] which states that some companies in the electronics sector are beginning to pay genuine attention to sustainability by recognising concerns related to the return and recycling of their products in the post-consumption stage.[66] However, such an initiative is insufficient. It is simply a

[58]Portilho (2005), p. 51.
[59]Marques (2015), p. 129.
[60]Lemos (2012), p. 25.
[61]Ibid.
[62]Baudrillard (2014), p. 33.
[63]Tonani (2011), p. 76.
[64]Lemos (2012), p. 34.
[65]Cook and Jardim (2017), p. 6.
[66]Ibid.

self-justification for producers to produce evermore EEE for profit. After all, as Patrícia Iglecias has warned, post-consumption residues are an environmental problem 'of whose solution and routing depends our survival on Earth'.[67] In this context, Brazil's NSWP, in highlighting the importance of solid waste management, establishes that the supplier must observe an order of priority, namely, not generation, reduction, reuse, recycling, treatment of the solid wastes and the final environmentally adequate disposition of the wastes.[68]

Furthermore, and as mentioned above, Article 33 of the NSWP establishes shared responsibility and demands suppliers work to enhance reverse logistics. Specifically, items I to VI of Article 33 identify products and packaging whose manufacturers, importers, distributors and traders are obliged to implement reverse logistics for, after the use and disposal of the product by the consumer. Reverse logistics mandates had already existed in some resolutions of the Brazilian National Council of the Environment, for example, in relation to agrochemical packaging (the return of these packages for reuse or recycling). What makes the NSWP so innovative is its enshrining of the shared responsibility of the producers and consumers in the entire product life-cycle, including post-consumption.

The NSPW, therefore, increases and expands the responsibility of producers once they are to adopt inverse logistics. Simply put, reverse logistics is a list of processes implemented by the producers to reuse products which were discarded after having been used by consumers. Therefore, if producers adopt planned obsolescence strategy to sell more and more, they will also have an increase in cost, related to the inverse logistics.[69]

Defined in Article 3 of the NSPW, reverse logistics was conceived as an instrument of the PNRS (Solid Residues National Policy Law); Decree No. 7.404/2010 may be firmed by certain instruments (Articles 15, I, II and III of L. 12.305/2010–NSWP), either by sectoral agreements, regulated by the public authorities, under terms of commitment. Reverse logistics emerges, thus, as the post-consumer responsibility of suppliers in general which must be applied to certain products, among them, electronic and electrical equipment. In short, it is not enough to think of sustainable consumption in an abstract way. It is necessary to implement effective mandates and regulations in the practice of consumer relations, mainly for EEE, considering the risks of the materials used in production. It is in this sense that the concern with EEE waste should guide its destination through the most varied instruments of the NPSW, whose implications are not only national, but also cross-border, once the issue of waste, including, but not limited to electric or electronic waste, has cause a new phenomenon, that is the transfer of waste from rich countries to poor countries. In an attempt to face the transboundary movements, the Basel Convention's Control of Transboundary Movement of Hazardous Wastes and their Disposal entered into force, and it is valid to this day. Thus, each country

[67]Lemos (2012), p. 85.

[68]Article 9 of Law No. 12305/2010.

[69]Machado (2016), p. 647.

should adopt the internal necessary measures to manage and eliminate its own waste generated in its own territory—and not to transfer the waste to other countries. However, when it comes to electronic waste, such convention requires further studies, which is not the scope of this paper.

25.4 Conclusion

Whatever the type of waste—construction, domestic, medical, among others—this study has focused on electrical and electronic waste, considering that it contains a large quantity of heavy metals and is generally subject to the strategic management of the product life-cycle via planned obsolescence. But this is a strategy that encourages the generation of waste and is fundamentally unsustainable. For waste generation, as Aragão notes, 'is at the origin of the most varied types of pollution, causing environmental disturbances in the hydrosphere, atmosphere, lithosphere and biosphere [constituting] a source of hazards to life and health, as well as for the balance of ecosystems'.[70] It is no wonder then that Vance Packard has claimed that consumers have become 'garbage manufacturers'.[71]

Electronic and electrical equipment, when discarded without proper or effective planning, becomes electrical and electronic waste and can produce severe socio-environmental impacts. This situation is greatly exacerbated by planned obsolescence, which is standard industry practice. This chapter has focused the analysis on suppliers of electrical and electronic equipment: manufacturer, sellers, and importers that, directly or indirectly, participate in or integrate the market for these products. It has argued that suppliers must assume new duties related to the post-consumption stage in the product life-cycle, namely, the moment the product is discarded becomes electronic waste. In so doing, the chapter has also pointed to the importance of 'waste rights' as a means to strengthen dialogue between waste law and Consumer Law. Finally, we have discussed how the analysis relates to current regulatory practice, drawing on the case of Brazil's NSWP, which has emerged as the basic regulatory framework for waste across the country and could be a model for other countries to draw upon. While the framework does not give specific treatment to electrical and electronic waste as a distinct category, it certainly covers that kind of waste in its regulatory framework.

If our goal is to expand the remit of sustainable consumption, the question of management of e-waste must take on a sharper focus for consumers, producers and regulators alike. Planned obsolescence has long been the industry standard for the marketing of consumer electric and electronic devices as a means to boost sales, expand profits and ensure a constant flow of new consumers for the firms that market these products. Moreover, while such an approach has the benefit of dazzling

[70] Aragão (2003), p. 9.
[71] Packard (1965), p. 8.

consumers with ever more sophisticated features and options, the downside is the ever-expanding volumes of harmful e-waste that threaten human life, health and the balance of the ecosystem. The choice is stark and the conclusion clear: strategic obsolescence is unsustainable in the long term.

References

Abramovay R (2012) Muito além da economia verde. Abril, São Paulo

Aragão MAS (2003) O direito dos resíduos [Cadernos CEDOA—Centro de Estudos de Direito do ordenamento, do urbanismo e do ambiente]. Almedina, Coimbra

Aragão MAS (2006) O princípio do nível elevado de proteção e a renovação ecológica do direito do ambiente e dos resíduos [Colecção teses]. Almedina, Coimbra

Baudrillard J (2014) A sociedade de consumo (trans: Artur Morão), Edições 79. Lisboa

Bauman Z (2008) Vida para o consumo: a transformação das pessoas em mercadorias. Zahar, Rio de Janeiro

Behrends C (2004) Consequences of planned obsolescence for consumer culture and the promotional self. Grin Verlag, Munich

Bell D (1973) O advento da sociedade pós-industrial: uma tentativa de previsão social (trans: Heloysa de Lima Dantas). Cultrix, São Paulo

Candemil R (2012) Mudanças de paradigmas para uma sociedade sustentável: um novo desafio para o direito brasileiro? Revista de Direito Ambiental 68:13–44

CEC-ZEV Centre Européе de la consommation-Zentren Frür Europaischen Verbraucherschutz E.V. L'obsolescence programmée ou les dérives de la société de consommation'. http://www.europe-consommateurs.eu/fileadmin/user_upload/eu-consommateurs/PDFs/publications/etudes_et_rapports/Etude-Obsolescence-Web.pdf. Accessed 10 Oct 2017

Cook G, Jardim E (2017) Guide to Greener Electronics Company. Greenpeace, Washington, p 4. http://www.greenpeace.org/usa/wp-content/uploads/2017/10/Guide-to-Greener-Electronics-2017.pdf. Accessed 25 Oct 2017

DIRECTIVe 2008/98/EC/CE European Parliament and of the Council of 4th July 2012. http://eur-lex.europa.eu/legal-content/PT/TXT/PDF/?uri=CELEX:32008L0098&from=ES. Accessed 10 Oct 2017

DIRECTIVe 2012/19/EU of the European Parliament and of the Council of 4th July 2012. file:///F:/ARTIGOS%20ASSUNTOS%202017/consumo%20sustent%C3%A1vel/DIRETIVA%202012%2019%20UE%20DO%20PARLAMENTO%20EUROPEU%20E%20DO%20CONSELHO.pdf. Accessed 19 Oct 2017

European Commission. Waste electrical & electronic equipment (WEE). http://ec.europa.eu/environment/waste/weee/index_en.htm. Accessed 20 Oct 2017

Feniss-Fundacion (2016) Los 10 artículos por excelencia afectados por la obsolescencia programada. http://feniss.org/los-10-articulos-por-excelencia-afectados-por-la-obsolescencia-programada/. Accessed 15 Oct 2017

Franzolin CJ (2017) Obsolescência planejada e pós-consumo e a tutela do consumidor. Revista de direito do consumidor 109:39–75

Goergescu-Roegen N (2012) O decrescimento: entropia, ecologia, economia. [Trans: Maria José Perillo Isaac]. Senac, São Paulo

Gouveia N, Prado RR (2010) Riscos à saúde em áreas próximas a aterros de resíduos sólidos urbanos. Rev Saúde Pública 44(5):859–866. http://www.producao.usp.br/bitstream/handle/BDPI/9412/art_GOUVEIA_Riscos_a_saude_em_areas_proximas_a_2010.pdf?sequence=1. Accessed 1 Nov 2017

Konder CN (2015) Distinções hermenêuticas da constitucionalização do direito civil. In: Konder CN (ed) Direito civil constitucional. Gen-Atlas, São Paulo, pp 25–45

Leff E (2012) Aventuras da epistemologia ambiental: da articulação das ciências ao diálogo de saberes (trans: Silvana Cabucci Leite). Cortez, São Paulo

Lemos PFI (2012) Resíduos sólidos e responsabilidade civil pós-consumo, vol 2. Revista dos Tribunais, São Paulo

Lemos PFI, Mendes JMA (2013) Resíduos eletroeletrônicos e seu panorama jurídico no Brasil: desafios regulatórios e oportunidades de implementação de sistemas de logística reversa. Revista de Direito Ambiental 72:39–63

Loubet LF (2007) Contornos jurídicos da responsabilidade pós-consumo. In: Freitas WP (ed) Direito ambiental em evolução, vol 5. Juruá, Curitiba, pp 245–269

Machado PAL (2016) Direito ambiental brasileiro, vol 24. Malheiros, São Paulo

Magalini F, Kher R, Baldé P (2015) E-waste in Latin American. Statistical analysis and policy recommendations. GSMA Latin America; UNU-IAS; UNU-IAS SCYCLE, p 5. https://www.gsma.com/latinamerica/wp-content/uploads/2015/11/gsma-unu-ewaste2015-eng.pdf. Accessed 28 Nov 2017

Magera M (2013) Os caminhos do lixo: da obsolescência programada à logística reversa. Átomo, Campinas

Marques CM (2015) Atualização do Código de Defesa do Consumidor e o diálogo entre o direito do consumidor e o direito ambiental: estudo em homenagem à Eládio Lecey [Reform of the Brazilian Consumer Code and the dialogue between Consumer Law and Environmental Law]. In Benjamin AH, Leite JRM (eds) Congresso Brasileiro de Direito Ambiental: Ambiente, sociedade e consumo sustentável [recurso eletrônico] 20. Congresso Brasileiro de Direito Ambiental, 10. Congresso de Direito Ambiental dos Países de Língua Portuguesa e Espanhola, 10. Congresso de Estudantes de Direito Ambiental. Instituto O Direito por um Planeta Verde, São Paulo, pp 126–146

Moraes KG (2015) Obsolescência planejada e direito: (in) sustentabilidade de consumo à produção de resíduos. Livraria do Advogado, Porto Alegre

Packard V (1965) Estratégia do desperdício [Trans: Aydano Arruda]. IBRASA-Instituto Brasileiro de difusão cultural, São Paulo

Política Nacional De Resíduos Sólidos. http://www.planalto.gov.br/ccivil_03/_ato2007-2010/2010/lei/l12305.htm. Accessed 1 Oct 2017

Portilho F (2005) Sustentabilidade ambiental, consumo e cidadania. Cortez, São Paulo

Retondar AM (2008) A (re) construção do indivíduo: a sociedade de consumo como 'contexto social' de produção de subjetividades. Sociedade e Estado 23:137–160. http://www.scielo.br/pdf/se/v23n1/a06v23n1.pdf. Accessed 25 Feb 2018

Souza AS, Silva CP et al (2007) A obsolescência planejada: uma reflexão frente aos problemas socioambientais do planeta. In XXVII Encontro Nacional de Engenharia de Produção. Foz do Iguaçu, 9–11 October 2007. http://repositorio.furg.br/bitstream/handle/1/669/A%20obsolesc%C3%AAncia%20planejada%20uma%20reflex%C3%A3o%20frente%20aos%20problemas%20socioambientais%20brasileiros.pdf?sequence=1. Accessed 16 Oct 2017

Sustainable Management of Waste Electrical and Electronic Equipment in Latin America (2016) ITU, Secretariat of the Basel Convention, UNESCO, UNIDO, UNU, WIPO, BCRC-South America, ECLAC

Tonani P (2011) Responsabilidade decorrente da poluição por resíduos sólidos: de acordo com a L. 12.305/2010: Institui a Política Nacional de Resíduos Sólidos, vol 2. Gen Editora, São Paulo

UN Economic and Social develepment (n.d.) https://sustenabledevelopment.un.org/content/dsd/agenda21/res_agenda21_21.shtml

Chapter 26
Potential Legal Avenues for Managing the Environmental Risks of Nanotechnology

Wilson Engelmann, Haide Maria Hupffer, and Raquel Von Hohendorff

Contents

Abstract We are on the verge of a nanotech revolution but to date almost no State has legislated effectively to manage the potential environmental risks that might be generated by nanoparticles. Instead, self-regulation and the creation of norms by other social actors have emerged to fill the gap. The temporality of these new forms of regulation—and the ability to deal with future risks—represent challenges for the legal order, which must intervene to grant legal effect to new regulatory production. By analysing publications from ISO TC229 (on nanotechnology) and the comprehensive principles regarding nanotechnologies and materials advanced by NanoAction (a project of the International Centre for Technology Assessment), this chapter proposes a new self-regulatory model to manage nanotechnological risks and serve as legal guidance for researchers, laboratories, research centres and nanoscale industries. Considering a range of social and environmental factors, the model systematically links the NanoAction principles to existing rules and principles in Brazilian law.

W. Engelmann (✉) · R. Von Hohendorff
Universidade do Vale do Rio dos Sinos (UNISINOS), São Leopoldo, Brazil
e-mail: wengelmann@unisinos.br

H. M. Hupffer
Universidade Feevale, Novo Hamburgo, Brazil
e-mail: haide@feevale.br

A. d. Amaral Junior et al. (eds.), *Sustainable Consumption*,
https://doi.org/10.1007/978-3-030-16985-5_26

26.1 Introduction

The twenty-first century has been characterised by the emergence of an unprecedented technical–scientific revolution, driven by advances in new products and devices with nanotechnologies. The number of nanotechnology-based products is growing rapidly in the consumer market: according to the Nanotechnology Products Database,[1] as of December 2017, there were 7286 products on the market with at least one nanotechnology-based component, produced by 1438 companies located in 52 countries. The main sectors in which nanotechnology is being employed in production are food, cosmetics, electronics, household appliances, oil, renewable energy, textiles, and medicine. Nanotech is also strong in the environmental and health sectors—including in groundwater remediation and water purification, wastewater treatment and soil purification—and sporting goods and fitness. There is a great versatility in the application of nanoscale technology, which guarantees success and rapid growth and in a short time.

As stated by Schwab,[2] nanotechnologies are at the centre of the structure known as the fourth industrial revolution. To understand the implications, effects and possibilities that nanotechnology may bring, it will be necessary to develop 'systemic thinking' (i.e., to integrate and promote interaction between different areas of knowledge, especially through interdisciplinarity). Conversely, in working out how to work systemic thinking demands an answer to the following question: 'Why must scientists work together to save the world?'[3]

This is the main purpose of this chapter, namely to study the basis for the integration of law in research and advances in nanoscale that are being developed by the Hard Sciences. In terms of nanoscience, a movement has been observed, in recent years, from 'concepts' to 'application'. The law seeks to integrate the advances of the nanotechnology revolution, helping to build 'systemic thinking'. This way of thinking will require the breaking down of disciplinary barriers to shed light on the challenges that are generated from the nanoscale to the global scale, and their impacts and consequences.[4]

The functionalist method will be used, drawing on the systemic-constructivist perspective of Niklas Luhmann, the risk studies approach developed by Ulrich Beck, and the logic of content analysis of Laurence Bardin. Methodologically, we draw heavily on the functionalist approach, proposed by Luhmann, that 'uses the process of relating in order to understand the existing and contingent, and distinguish them as comparable.' That is, 'the relationship between problem and solution of the problem here is not regarded as an end in itself; but serves as a guide to the question of other possibilities, such as threads in the search for functional equivalence.'[5] It is the

[1]StatNano (2017).

[2]Schwab (2016).

[3]Interdisciplinarity (2015).

[4]Ulijn and Riedo (2016).

[5]Luhmann (1990).

systemic-functionalist perspective that seeks to establish this link between the problem and a solution to be built on constructivist grounds, notably by observing the regulatory frameworks that can cope with the challenges posed by nanotechnology.[6]

This is what might be called a nanotechnological fact, a phenomenon that is being experienced by society in the present historical moment. Its effects are invisible and also largely undetermined, when considered in the set of techniques, materials and uses of nanotechnologies. In this regard, we draw on Beck's *Theory of Risk Society*[7] to understand the context of scientific uncertainty and the dimensions of risk and danger. In a public lecture given on 15 February 2006, at the London School of Economics, Beck said that in his first publication in 1986 he had described 'risk society' as 'an inescapable structural condition of advanced industrialization' and criticised the 'mathematicized morality' of expert thinking and public discourse on 'risk profiling'. While policy-oriented risk assessment posited the manageability of risks, he pointed out that 'even the most restrained and moderate-objectivist account of risk implications involves a hidden politics, ethics and morality'.[8]

Risk, Beck continued in his lecture, 'is not reducible to the product of probability of occurrence multiplied with the intensity and scope of potential harm'. Rather, it is a socially constructed phenomenon, in which some people have a greater capacity to define risks than others. Not all actors really benefit from the reflexivity of risk only those with real scope to define their own risks. Risk exposure is replacing class as the principal inequality of modern society, because of how risk is reflexively defined in the specialised literature: 'In risk society relations of definition are to be conceived analogous to Marx's relations of production'. The inequalities of definition enable powerful actors to maximise risks for 'others' and minimise risks for 'themselves'. Risk definition, essentially, is a power game. This is especially true for world risk society where Western governments or powerful economic actors define risks for others. As Beck noted: 'Risks presuppose human decisions. They are partly positive, partly negative, and are the consequences of human decisions and interventions'.[9]

The decision on more specific regulation or not, or even self-regulation, will generate risks, which will be added to the risks that could be generated by the manipulation of the nanoscale itself. Herein lies the main question that we seek to address in the chapter: what would be the framework of a safe alternative to dealing with the unknown part of the nanoworld? Conversely, what structural elements are likely to produce a framework that can offer guidelines for research and product development, with consequent commercialisation based on nanotechnology, to a scenario with an uncertain risk assessment, no legislative regulation by the State and the presence of regulatory types developed on an ad hoc basis by public and private actors?

[6]Ibid.

[7]Beck (1992).

[8]Beck (2006).

[9]Ibid, p. 333.

In response, we propose the following hypothesis. The principles formulated by NanoAction[10] can be used as elements for structuring such a framework in view of the existence of scattered research on potential nano risks and damage that may be generated to human beings and living organisms, as well as the environment in general, taking into account the current corpus of regulatory texts produced by public and private actors, but without legislative participation by the State.

Some toxicological tests carried out with specific materials (such as nano silver, nano carbon tubes and others), have pointed to the existence of hazards. These results, for now, are only samples of a world of possibilities offered by nanotechnology, but they are an indication that the debate on the issue is necessary and that this is indeed a legal issue to be studied and communicated to the consumer public and non-specialist experts alike. This method is suitable for interdisciplinary research development, notably from the recently discussions published in the journal *Nature*.[11]

A comprehensive list of keywords was used to structure data collection in 80 documents published by the Organisation for Economic Co-operation and Development (OECD 2017) on the question of nanotech: 'risk', 'environmental safety' or 'environment', 'human health' and 'manufactured nanomaterial'. These same general keywords serve to structure searches in the abstracts of publications from the Web of Science database in the period 2010–2016. For this latest research, the group of keywords was extended, searching for publications with specifications and details of the searched words of the OECD documents. This was again undertaken through a search using the keywords listed above of Web of Science with full text available on the Journals Portal CAPES/Brazil (an institution linked to the Ministry of Education of Brazil).

Moreover, the present research includes comprehensive content analysis,[12] carried out in three stages: (a) pre-analysis; (b) exploration of the material; and (c) treatment of results, including inference and interpretation.

26.2 Nanotechnologies: Some Preliminary Considerations

To begin, we must address conceptual issues, in particular the question of the terms 'nanotechnology', and 'nanomaterial'. There is no uniform definition of these two terms in the literature.[13] Some, like Maynard, argue that one should not work with definitive designations, for 'basing regulations on a term with no scientific justification will do more harm than good'.[14] The International Organization for

[10] NanoAction (2007).

[11] Interdisciplinarity (2015).

[12] Bardin (2013).

[13] Stone et al. (2017).

[14] Maynard (2011).

Standardization (ISO) through its Technical Committee 229 on Nanotechnologies issued a definition of nanotechnology which contains the same elements as those that have applied generally in recent decades, namely that it involves 'understanding and control of matter and processes at the nanoscale, typically, but not exclusively, below 100 nanometers in one or more dimensions where the onset of size-dependent phenomena usually enables novel applications.'[15] To this, the ISO further that nanotech also utilises the 'properties of nanoscale materials which differ from the properties of individual atoms, molecules, and bulk matter, to create improved materials, devices, and systems that exploit these new properties.'[16]

The first feature relates to particle size and this may generate risks, considering the physicochemical changes that are created in this size range. In this scenario, a recent editorial in *Nature Nanotechnology* stated: '[D]espite these fundamental advances, nanotechnology is confronted with a critical bottleneck. We are still struggling to translate the fundamental advances reported in the scientific literature into tangible technological applications that can be appreciated at the layman's level.'[17] The editors went on to detail two key implications of the problem:

> First, the properties of matter change when scaled up, just like they change when scaled down to the nanoscale; in particular, the level of control can be exerted at the nanoscale or at the single-object level tends to wane at the meso and macroscales or when dealing with a large number of objects. And second, industry is reluctant to invest money in developing large-scale new processes for nanomaterial fabrication unless they are guaranteed a sizeable profitable return.[18]

This is also a concern for Warheit, who notes that the evaluation of environmental and health risks of nanomaterials is difficult work that involves multi-discipline knowledge and requires global views.[19] Therefore, arguments over—and even misunderstandings about—nanotoxicity are common. Much attention should thus be paid to the future design of environmentally friendly and biocompatible nanomaterials.

This opens a space of still unknown consequences. For the law, there is clearly a need to regulate the risks and minimise potential future damage. This, however, is a great challenge because the law always assesses facts in hindsight, as it were, giving them legal effects that are supposed to condition future behaviours, at least in theory. In the case of nanotechnology, the effects—scratches and damages, so to speak—that may arise in the present and in the future, remain unknown. This also poses a problem for the State in its legislative regulation on the matter: there is still no sufficient scientific information to inform formal regulation. In this chapter we therefore seek to signal some regulatory possibilities for this risk scenario, much about which—as mentioned—remains unknown, to predict the eventuality of future

[15] ISO/TC 229 (2017).

[16] Ibid.

[17] Nature Nanotechnology (2016), p. 733.

[18] Ibid.

[19] Warheit (2010).

damage. Furthermore, some recent studies have indicated a disconnect between academic research and its application on the industrial scale, which opens new possibilities for risk.

The research must also deal with the important distinction between *danger* and *risk*. In the environmental scenario, danger refers to the intrinsic characteristics of a product or process that make it harmful or damaging; risk refers to the degree or likelihood of exposure to such danger.[20] When examining the life cycle of nanomaterial, it is possible to observe that the danger may be in the following phases or processes: the raw material; the production process; the final product (and packaging); its application or use; its reuse, recycling and final disposal (i.e., the product's 'end of life').[21] Risk, then, is reflected in the exposure of workers, consumers or the environment in each of these phases.[22] Here, a question arises: will the last, 'end of life' stage effectively be the final part, especially considering the possibilities for reuse and recycling? This is another issue that requires further study.

The life cycle of nanotechnology applied to food, for example, has many uncertainties regarding the characteristics of risks. This is because it can affect the bioavailability and nutritional value of food. Nanofoods improve food security, extend life, enhance flavour and nutrients, allow for the detection of pathogens/ toxins/pesticides, and the serving of functional foods.[23] Another example is remediation of groundwater, for which '[v]arious nanomaterials have been developed'.[24] These two examples show the paradoxical perspective of nanotechnologies: they bring benefits, while bearing problems and difficulties that are as yet unknown, or only partially understood. Therefore, according to Warheit, 'the assessments of nanomaterial-related health risks must be accurate and verifiable'.[25] It is especially important to focus on 'nanoparticle physicochemical characterization, as well as consideration of potential routes of exposure, justification of nanoparticle doses, and inclusion of benchmark controls.'[26] These, then, are aspects that demand further research and analysis, if the creation of more precise regulatory frameworks is to be made feasible.

[20]Foladori and Invernizzi (2016).

[21]Vaseashta (2015).

[22]Shatkin and Kim (2015).

[23]He and Hwang (2016).

[24]Santhosh et al. (2016), p. 1116.

[25]Warheit (2010), p. 4777.

[26]Ibid, p. 4777.

26.3 What Is the Content of the Scientific Publication: Nanotechnology or Nanotoxicology?

Toxicological effects and impacts of nanomaterials on human health and the environment have not received the same attention from researchers if compared to the positive aspects of products with nanotechnology widely touted in the media and in scientific journals. Compared with nanotoxicology, nanosafety has received wide consideration. However, nanosafety is not nanotoxicology. Nanotoxicology focuses on the physiology, pathology and biomolecular mechanisms of nanomaterials. Nanosafety focuses on the evaluation of nanomaterial risks in natural environments and biology. The main results and conclusions of nanotoxicology are from the lab, while the main results and conclusions of nanosafety should not match the current environment and real organisms, where field studies are important. Studies on toxicology still need to be improved and deepened. Moreover, laboratory studies are important, but fieldwork is also required.

Fieldwork involves the investigation of nanomaterial risks by analysing the samples from the natural environment, and potential workers exposed to nanomaterials. The laboratory data integrating the results from fieldwork would present more real, nonmaterial risks than laboratory experiments alone.[27] At this point there is a great empty space to be filled. The limited volume of research that has been published on the topic of nanosafety speaks for itself, compared to the almost exponential rise over the past 15 years or so in the number of articles on nanotoxicology. Although only a couple of hundred papers had appeared on the topic of 'nanomaterials' and 'environmental and health effects' before the year 2000 this number has exploded to over 10,000 since 2001. Most of those studies, however, do not offer any kind of clear statement on the safety of nanomaterials. On the contrary, most of them are either self-contradictory or arrive at completely erroneous conclusions.[28] In a data base research from the Web of Science, between the years of 2010 and 2017, from the key words presented in the abstract of the articles and evidenced below, it has been verified data in Table 26.1.

It can be observed that the publications on 'nanotechnology' are much greater than the concern with issues relating to 'toxicology' or the relationship between 'nanotechnology and toxicology'. The ratio of the publications is very unbalanced, leading to the conclusion that manipulation at the nanoscale is safe. Questions dealing with the impacts, there including human and environmental risks also have publications in low amounts, when compared with the total number of published articles on nanotechnology. Although many articles reviewed for this publication mention concern about the risks, it appears that at the junction of 'nanotechnology' and 'risks' is found in only 1219 articles that had these words on their abstracts. The result indicates that the interest in the publications continues to be in nanotechnology

[27]Xiangang et al. (2016).

[28]Krug (2014).

Table 26.1 Publications in Web of Science (2000–2017)

Number of items	Combinations of words
18,870	Nanotechnology
5106	Nanotechnology and materials
1258	Nanotechnology and environmental
1219	Nanotechnology and risks
885	Nanotechnology and benefits
473	Nanotechnology and regulation
437	Nanotechnology and physics
404	Nanotechnology and risk assessment
377	Nanotechnology and consumer
360	Nanotechnology and environmental risks
320	Nanotechnology and environmental impacts
312	Nanotechnology and waste
264	Nanotechnology and nanotoxicology
223	Nanotechnology and law
200	Nanotechnology and toxicology
175	Nanotechnology and consumer risks
87	Nanotechnology and social impacts

itself and its applications without effectively concerning about research and publication of the effects that working with the nanoscale may cause.

Herein lies the 'conundrum' studied by Owen, in his book of the same title, where he questions whether 'scientific innovation, increased efficiency, and good intentions can make our energy and climate problems worse', his underlying point being 'that, even when we act with what we believe to be the best of intentions, our efforts are often at cross-purposes with our goals. That's the conundrum.'[29] Engineered nanoparticles are found in the following sources: consumer products, industrial products and in the manufacturing process. So, these are the potential environmental elements to be exposed: air, water and soil. Herein lies the greatest possibility of risks, and the 'conundrum': even when conduct follows what is believed to be the best of intentions (water remediation, healthier food, drugs with fewer adverse effects, clothing that needs to be washed less often, among other possibilities), our efforts are constantly at cross-purposes with our goals. In short, despite the great opportunities nanotechnologies represent, they also open possibilities for the generation of unknown risks.

Another purpose of this study, based on the 80 documents of the OECD 'Series on the Safety of Manufactured Nanomaterials', is to analyse the perception of risks and the way that the assign legal effects of the unknown future damages that might be generated from the manipulation of the nanoscale, especially in relation to human health and the preservation of the environment. Comprehensive keywords were used to enable data collection from the 80 documents: 'risk', 'environmental safety' or

[29]Owen (2011), p. 5.

'environment', 'human health' and 'manufactured nanomaterial'. The word 'nanomaterial manufactured' was repeated 4934 times, followed by 'risk' (4214), 'environment' (2204) and 'human health' (1478). This shows extreme concern about the risks that nanomaterials could pose to the environment and human health. The observed keywords show that manufactured nanomaterials may generate risks for the environment and human health, with little concern for environmental safety, which has only 48 repetitions. Thus, it is important to structure the framework of the legal risk management tool for nanotechnology companies to gather information and deal with future uncertainties and damage, through the evaluation of the stages in the nanomaterial life cycle. A comprehensive evaluation of the ethical and social impacts in the structuring of self-regulation is relevant in this scenario also.

Comparing the data shown in Table 26.1 with the findings in the 80 OECD documents, we found that in both data the concern is with nanotechnology and nanomaterials. The fear of the risks and impacts that nanomaterials may cause in relation to human health and the environment is sketchy. Therefore, despite the preoccupation in announcing risks and impacts, in fact, the main focus is still on the benefits and possibilities that nanoscale can provide. These data show that there is still much to be done. The risks should also be studied with the same emphasis given to applications. There is also an obstacle to legislative regulation by the State: while the risks are unknown, creating traditional formal regulation is very complicated.

Nanoparticles access the human body in different ways, including inhalation, skin contact and oral ingestion (food, water, drugs, etc.), which can generate numerous reactions. If research studying these reactions is scarce, what can we say about effective and aligned actions to monitor 'nanopathologies'?[30] It can be said that the point of view of human health research development to analyse the toxicity of nanoparticles and consequences of associations with nanoparticles that accumulate in the human body and the environment is still a new field and it needs much investment in science to advance.

26.4 Regulatory Challenges: Are We on the Right Track?

A fairly comprehensive survey of existing regulations in the US, the European Union, some Asian countries (such as Japan, South Korea and China), in addition to the Latin American countries, especially Brazil, can be found in Wackera, Proykova and Santos.[31] The regulatory mechanisms already developed will not be addressed, as they are already sufficiently well studied. What we do intend to show are some regulatory initiatives that have emerged in recent years that are not related to legislative output by the State, as shown in Snir and Ravid:

[30] Gatti and Montanari (2008).

[31] Wackera et al. (2016).

> A closer examination of the key actors in the networks further shows the interrelation between public and private actors in regulatory innovation, and the role of intermediary organizations. On the other hand, most of the global market has no distinct regulations regarding consumer products [made up] of nanomaterials.[32]

Considering the nanosafety of humans and the environment, are highlighted the following international organisations which edited structures with regulatory characteristics: the international organisations (the OECD, REACH–Registration, Evaluation, Authorization and Restriction of Chemicals, the ISO, etc.) and the developed countries (USA, EU and Japan, etc.) are trying to create guidelines and standardisation for toxic evaluation and regulation plans or research projects for nanotechnology. Those organisations also indicate regulatory perspectives for intergovernmental organisations such as the World Health Organization (WHO). As Park and Yeo have noted: 'The analysis may also suggest that by going through private international decision-making venues, national regulators can strategically diffuse their regulatory approaches to other countries.'[33] Furthermore, as Rodine-Hardy has observed:

> Two of the largest economic actors, the EU and the US, have made very different regulatory decisions toward nanotechnology. The EU introduced an official definition of nanotechnology and created several nano-specific new regulations in recent years, whereas the United States has followed more of a 'wait and see' policy.[34]

Risk and damage may be found at different levels in the same life cycle of a nanomaterial. A guideline for a proposed Life Cycle Assessment is presented as follows: the development of 'nanoecotoxicology'. As Kahru and Ivask note, this new subdiscipline of ecotoxicology faces two important and challenging problems: 'the analysis of the safety of nanotechnologies in the natural environment and the promotion of sustainable development while mitigating the potential pitfalls of innovative nanotechnologies'.[35] What this means is that the growth of nanotechnology applications should be guided by sustainability, focused on preservation of life on Earth for present and future generations of living beings, human or otherwise. In a recent article, Hischier and Walser identified 17 studies with different forms of structuration of life-cycle assessments.[36] According to the researchers, a Life Cycle Assessment 'is a comprehensive framework that quantifies ecological and human health impacts or the product or system over its complete life cycle'.[37]

Risk assessment forms the basis for regulatory actions with respect to nanomaterials. The question of contextualisation, categorisation and analysis of risk that may arise from nanoparticles in the various steps that are part of the life cycle is of fundamental importance. The testing of current and potential

---

[32]Snir and Ravid (2015), p. 21.

[33]Park and Yeo (2016), pp. 225–226.

[34]Rodine-Hardy (2016), p. 89.

[35]Kahru and Ivask (2013), p. 823.

[36]Hischier and Walser (2012).

[37]Nadoveza et al. (2013), p. 410, cited in Hischier and Walser (2012).

nanomaterials is very challenging: first, because of the large number of nanomaterials; and second, because of small changes in the nanomaterial, such as its physical structure or surface coatings, what can significantly change its interactions with biological and environmental systems, especially at the molecular and cellular level. Consequently, it is critical that in assessing risk to health and safety (i.e., unwanted and negative interactions), one should be able to identify the mechanism by which these negative interactions operate. The scope of such interactions should be determined in scientific laboratories and during stages of the life cycle.[38]

We propose the following practical approach to developing a regulation model, to be followed by application by industry, as a way to mitigate any future damage that may arise from the use of nanoscale materials in relation to workers and consumers' health, in addition to environmental issues.

26.5 Structuring the Framework as a Regulatory Alternative

The framework that is proposed is inspired by the following key elements: (a) a definition of nanotechnology and nanomaterials as new substances, according to the ISO and the OECD; (b) thresholds whereby nanomaterials should be recognised as much more reactive than their bulk counterparts, thereby increasing the risk of harmful impact by nanomaterials compared to an equivalent mass of bulk materials; (c) testing protocols; (d) market surveillance; and (e) transparency of consumer information and worker protections.[39] Furthermore, to deal with the risk of future damage, in a context of scientific uncertainty about its potential extent, the framework must be guided by the precautionary principle. In this context, consideration must be given to precautionary decision-making as follows: (a) comprehensive treatment of information and knowledge; (b) integration of multiple values in decision-making; (c) a more democratic decision-making process; (d) defining the range of solutions and (e) adopting a common procedural framework. The required regulatory framework must be coherent, proportionate and effective, and suited to the nature of the potential dangers, with common procedures that organise research, expertise, public information and debate.[40]

Based on this outline, the framework should be able to answer the following questions, seeking to guide decision-making around nanotech applications:

1. Does any modification introduced actually affect design properties (chemical composition, crystallinity, surface chemistry/charge, primary size, particle size distribution and its evolution in testing and in life-cycle media)?

[38]Materials on the Nanoscale (2016), p. 25.

[39]Azoulay and Buonsante (2014).

[40]Gonçalves (2013).

2. If yes, does the modification affect risk determinant properties (structural alerts) that can be considered relevant for estimating potential hazardous?
3. If yes, do the changes in risk determinant properties reduce in vitro or in vivo toxicity according to the established biomarkers of mode of action? Evidence collected can be used to read-across risk for human health and establishing new exposure limits.
4. If yes, different scenarios can occur:

   a. Tested toxicology endpoints do not show a coherent response (e.g. some results show the reduction of toxicity potential, while others demonstrate an increase). In this case, the further mechanistic investigation is necessary to better correlate changes in design and risk determinant properties and toxicology end points.
   b. Tested toxicology endpoints show a coherent response (e.g. all the results for the selected modification show an increase or decrease of potential toxicity); in this case, the cost/effectiveness evaluation should not be considered before discharging or validating the design solution proposed.[41]

These questions might be taken as a starting point for one or more regulatory texts of non-state-owned legislative origin, which are summarised in Wackera, Proykova and Santos[42] and in Stone et al.[43] In all steps of the framework it is recommended to follow the following guiding principles: (a) precautionary foundation; (b) health and safety of the public and workers; (c) environmental protection; (d) transparency; (e) public participation; (f) inclusion of broader impacts and; (g) manufacturer liability (Fig. 26.1).[44]

In the various steps of this framework, a central theme emerges: the identification of risks. They will have different intensity and impact depending on the type of the nanoparticle, its size and the environment where they will be exposed. It should also be considered that exposure may occur along different pathways: direct exposure (occupational, consumer and/or environment) or indirect (general population exposure through the environment), each one with its own mitigation measures.[45] Therefore, the bigger the threshold of the 'impact of risks or estimating potential hazards' at the junction with the 'likelihood of damage', much more attention must be given to the various stages of the life cycle. The records of events and reactions in each step should be noted, generating useful information. Moreover, the last step is fundamental, wherein the following question must be addressed: Is this the end of nanomaterial life or just the beginning of a new cycle of that life? This will be especially important when preparing for recycling.

---

[41] Stone et al. (2017).

[42] Wackera et al. (2016).

[43] Stone et al. (2017).

[44] NanoAction (2007).

[45] Stone et al. (2016).

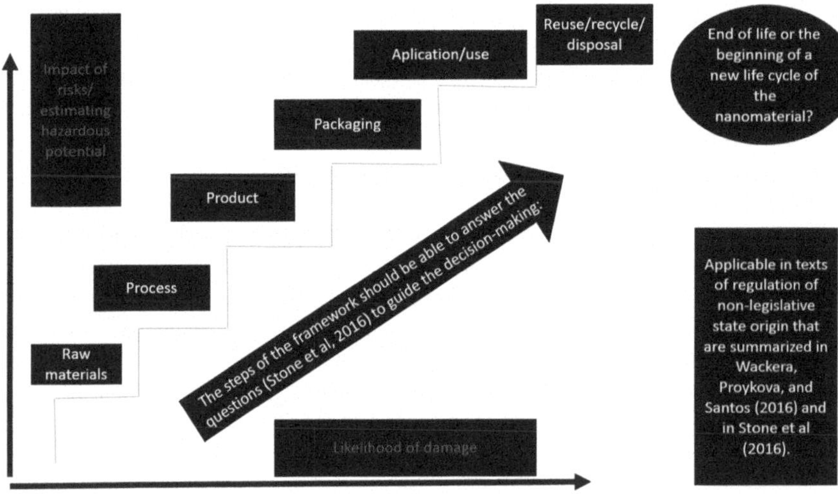

Fig. 26.1 Framework of the life cycle assessment. Source: Prepared by the authors

26.6 Conclusion

The construction and use of frameworks, with the application of a small adjustment to existing nanotechnologies, and the other legal texts in force in each country, although not nano-specific, do have application at the nanoscale, and thus comprise the alternative that must be adopted at this moment. Throughout the life cycle of a nanomaterial, the explicit coordinates in the framework should be observed, within a framework of 'shared responsibility'. Under Article 3, XVII of Brazilian Law No. 12305 of 2 August 2010, this involves[46]:

> [A set] of individual and chained assignments of manufacturers, importers, distributors and traders, consumers and owners of public services for urban cleaning and solid waste management to minimize the volume of solid waste and generated waste, as well as to reduce impacts on human health and environmental quality of the resulting product life-cycle management, under this Act.[47]

In the various steps of the framework information on nanomaterials should be created, aimed to fill any gaps that currently exist. This is the industrial side of the production of nanotechnologies. At the same time, research on nanotoxicological effects—both *in vivo* as *in vitro*—should increase. The current body of research and data is still very small.

In the various steps examined, the law must be present, seeking to legally structure the information that will be collected. From this information, it is possible to review each step by adding new non-state legislative regulatory texts that might

[46]Brazil (2010).

[47]Brazil (2010).

arise. In the long term, and with more accurate information, it will be possible to plan the structure of regulatory texts that meet the specific requirements of different countries, through international treaties. Because of globalisation and the characteristics of the 'risk society', localised regulatory texts are not the answer here but rather the extension of a general order beyond the territory of each country. The precautionary principle will ground the link between the *nanotechnologies we have* and *the nanotechnologies we want.*

The panorama presented throughout this present research, seeking to respond the problem formulated in the introductory section, points to the confirmation of the hypothesis that structured this chapter: despite the importance of the study of the risk highlighted in the extracted publications from the Web of Science database, there are still few scientific publications on the implications of nanoscale materials or nanotechnology concerning human health and environmental preservation for present and future generations. The absence of legislative regulation by the State increases the importance of regulatory texts already published by public and private actors, as well as their effective usage throughout the various steps of the life cycle and how they can be integrated within the framework proposed here, where the social, environmental and legal impacts generated by nanotechnologies can be evaluated. The possibilities offered by the framework would go a long way to help regulate the various impacts that nanotechnologies can generate, especially in the context of sustainable consumption; meaning, both in the manufacture of products based on nanomaterials and in the disposal of waste that may carry nanoparticles.

Acknowledgements The authors would like to thank the following people. We are grateful to the various interviewees for the time they took to talk to us about their work, the anonymous referees for their very helpful suggestions, and various people at conferences, seminars, and so on, over the last few years for their comments on previous versions of this chapter. We are particularly grateful to the students and researchers of the JUSNANO Research Group (Brazil) and the Centro de Investigação e Desenvolvimento sobre Direito e Sociedade (CEDIS), the Faculdade de Direito da Universidade Nova de Lisboa (Portugal) and the Instituto Jurídico Portucalense (IJP), and the Universidade Portucalense, Porto (Portugal).

Funding This research has been conducted by the authors in the context of the following research projects:

1. 'Nanotechnologies as an example of innovation: in search of structuring elements to evaluate the benefits and risks produced from the nanoscale in the scenario of Responsible Research and Innovation (RRI) and of the Ethical, Legal and Social Impacts–ELSI', *Project Support Research/Call CNPq/MCTI (Brazil) n. 25/2015 Humanities, Social and Applied Social Sciences.*
2. 'Observatory on the Legal Impacts of Nanotechnologies: structuring essential elements for the development of dialogue between the Sources of Law from regulatory indicators to research and industrial production based on the nano scale', *Support for Research Projects/MCTI/CNPq/Universal 14/2014* (Brazil).

References

Azoulay D, Buonsante V (2014) Regulation of nanomaterials in the US: proposed measures to fill in the gap. Eur J Risk Regul 2:228–235. https://doi.org/10.1017/S1867299X00003652

Bardin L (2013) L'analyse De Contenu, 2nd edn. PUF, Paris

Beck U (1992) Risk society: towards a new modernity (published in association with theory, culture & society). SAGE, London

Beck U (2006) Living in the world risk society. A Hobhouse Memorial Public Lecture given on Wednesday 15 February 2006 at the London School of economics. Econ Soc 35(3):329–345. https://doi.org/10.1080/03085140600844902

Brazil (2010) Law No. 12305 of 2 August 2010. http://www.planalto.gov.br/ccivil_03/_ato2007-2010/2010/lei/l12305.htm. Accessed 18 Feb 2018

Foladori G, Invernizzi N (2016) La Regulación de las nanotechnologies: una mirada desde las diferencias EUA-UE. Vigilância Sanitária em Debate 4(2):8–20. https://doi.org/10.3395/2317-269x.00726

Gatti AM, Montanari S (2008) Nanopathology. The health impact of nanoparticles. Pan Stanford, Singapore

Gonçalves V (2013) Critical approach of the use of economic models in precautionary risk management. Eur J Risk Regul 3:335–345. http://ejrr.lexxion.eu/article/EJRR/2013/3/286. Accessed 18 Feb 2018

He X, Hwang HM (2016) Nanotechnology in food science: functionality, applicability, and safety assessment. J Food Drug Anal 24(4):671–681. http://www.jfda-online.com/article/S1021-9498(16)30075-8/pdf. Accessed 18 Feb 2018

Hischier R, Walser T (2012) Life cycle assessment of engineered nanomaterials: state of the art and strategies to overcome existing gaps. Sci Total Environ 425:271–282. https://doi.org/10.1016/j.scitotenv.2012.03.001

Interdisciplinarity (2015) Nature 525:325. http://www.nature.com/news/interdisciplinarity-1.18295. Accessed 18 Feb 2018

ISO/TC 229 (2017). http://www.iso.org/iso/iso_technical_committee?commid=381983. Accessed 18 Feb 2018

Kahru A, Ivask A (2013) Mapping the dawn of nanoecotoxicological research. ACC Chem Res 46(3):823–833. https://doi.org/10.1021/ar3000212

Krug HF (2014) Nanosafety research-are we on the right track? Angew Chem Int Ed 53:12304–12319. https://doi.org/10.1002/anie.201403367

Luhmann N (1990) Sociedad y sistema: la ambición de la teoría. Paidós, Barcelona

Materials on the Nanoscale—uniform description system version 2.0 (2016) Chem Int 38(6):25–25. https://doi.org/10.1515/ci-2016-0620

Maynard AD (2011) Do not define nanomaterials. Nature 475:31. http://www.nature.com/nature/journal/v475/n7354/full/475031a.html. Accessed 18 Feb 2018

Nadoveza D, Koukias A, Karakoyun F, Kiritsis D (2013) A concept for graph-based LCA analysis tool. In: Conference: advances in production management systems (APMS), Sep 2013, State College, PA, United States, vol 415. Springer, IFIP Advances in Information and Communication Technology, AICT, Berlin, pp 410–417. https://doi.org/10.1007/978-3-642-41263-9_51

NanoAction (2007) Principles for the oversight of nanotechnologies and nanomaterials. http://www.centerforfoodsafety.org/files/final-pdf-principles-for-oversight-of-nanotechnologies_80684.pdf. Accessed 18 Feb 2018

Nature Nanotechnology (2016) A matter of scale. v.II, set. 2006 (editorial). https://www.nature.com/articles/nnano.2016.180. Accessed 24 Feb 2018

OECD (2017) Working party on nanotechnology (WPN): vision statement. https://www.oecd.org/sti/nano/oecdworkingpartyonnanotechnologywpnvisionstatement.htm. Retrieved 18 May 2019

Owen D (2011) The conundrum: how scientific innovation, increased efficiency, and good intentions can make our energy and climate problems worse. Penguin Group, New York

Park HG, Yeo MK (2016) Nanomaterial regulatory policy for human health and environment. Mol Toxicol 12:223–236. https://doi.org/10.1007/s13273-016-0027-9
Rodine-Hardy K (2016) Nanotechnology and global environmental politics: transatlantic divergence. Global Environ Polit 16(3):89–105. https://doi.org/10.1162/GLEP_a_00367
Santhosh C et al (2016) Role of nanomaterials in water treatment applications: a review. Chem Eng J 306:1116–1137. https://doi.org/10.1016/j.cej.2016.08.053
Schwab K (2016) The fourth industrial revolution. Edipro, São Paulo
Shatkin JA, Kim B (2015) Cellulose nanomaterials: life cycle risk assessment, and environmental health and safety roadmap. Environ Sci Nano 2:477–499. https://doi.org/10.1039/C5EN00059A
Snir R, Ravid G (2015) Global nanotechnology regulatory governance from a network analysis perspective. Forthcoming in Regulation & Governance, pp 1–31. https://papers.ssrn.com/sol3/papers.cfm?abstract_id=2626808. Accessed 26 Feb 2018
StatNano (2017) Nanotechnology products database 2017. http://product.statnano.com/. Accessed 18 Feb 2018
Stone V et al (2016) Research priorities relevant to development or updating of nano-relevant regulations and guidelines. http://www.safenano.org/news/news-articles/eu-nanosafety-cluster-publishes-research-regulatory-roadmap. Retrieved 18 May 2019
Stone V et al (2017) Research priorities relevant to development or updating of nano-relevant regulations and guidelines. http://www.safenano.org/news/news-articles/eu-nanosafety-cluster-publishes-research-regulatory-roadmap. Accessed 18 Feb 2018
Ulijn RV, Riedo E (2016) Learning to think systems. Nat Nanotech 11:824. http://www.nature.com/nnano/journal/v11/n9/full/nnano.2016.161.html. Accessed 18 Feb 2018
Vaseashta A (2015) Life cycle analysis of nanoparticles: risk, assessment, and sustainability. DEStech Publications, Lancaster
Wackera MG, Proykova A, Santos GML (2016) Dealing with nanosafety around the globe-regulation vs. innovation. Int J Pharm 509:95–106. https://doi.org/10.1016/j.ijpharm.2016.05.015
Warheit DB (2010) Debunking some misconceptions about nanotoxicology. Nano Lett 10 (12):4777–4782. https://doi.org/10.1021/nl103432w
Xiangang H, Dandan L, Gao Y, Zhou Q (2016) Knowledge gaps between nanotoxicological research and nanomaterial safety. Environ Int 94:8–23. https://doi.org/10.1016/j.envint.2016.05.001

Chapter 27
Looking Back to Look Forward: A Future Research Agenda for Sustainable Consumption, Law and Development

Alberto do Amaral Junior, Lucila de Almeida, and Luciane Klein Vieira

Contents

Abstract This chapter aims to look forward to a future research agenda for sustainable consumption, law and development by way of conclusion. While it is for the readers to retrieve the lessons addressed by each contribution in this edited book, for us it is compelling to close this journey by looking forward. Therefore, we suggest and seek for a future research agenda that aligns the concept of sustainable consumption with the interdisciplinary debate of law and development.

This book has sought to develop the topic of sustainable consumption and the law using a variety of research methods, and from the perspective of multiple legal disciplines. With that purpose in mind, we invited a diverse group of legal scholars from more than ten different nationalities and with a range of expertise to identify and analyse the issue of sustainable consumption within their fields of research.

The result is an original edited volume resembling a handbook for ongoing reference work. The collection of chapters covers the subject of sustainable consumption and the law comprehensively. It considers the issue from the perspective of traditional legal disciplines—such as Consumer Law, Constitutional Law, Tort Law, International Public and Economic Law—but not only. The structure of the volume

A. d. Amaral Junior
University of São Paulo (USP), São Paulo, Brazil

L. de Almeida (✉)
Florence School of Regulation, Robert Schuman Centre for Advanced Studies, European University Institute, Florence, Italy
e-mail: lucila.dealmeida@eui.eu

L. Klein Vieira
Universidade do Vale do Rio dos Sinos (UNISINOS), São Leopoldo, Brazil

The University of Buenos Aires (UBA), São Leopoldo, Brazil
e-mail: lucianevieira@unisinos.br

A. d. Amaral Junior et al. (eds.), *Sustainable Consumption*,
https://doi.org/10.1007/978-3-030-16985-5_27

also encompasses more modern systematisation of law derived from an understanding of fragmentation[1] and pluralism[2] in law. There are chapters that shed much-needed light on sector-specific engagements with sustainable consumption in national, transnational, and international legal orders. Gathering a set of legal scholars with such diverse backgrounds has required an enormous effort. We are pleased about the efforts and commitment of all the contributors. The collection of chapters here addressed crucial questions whatever the standpoint.

Notwithstanding the size of the task, this edited book is neither exhaustive concerning issues of sustainable consumption and the law at this moment and into the future, nor was it our intention for it to be so. Quite the opposite. The primary objective of this edited volume is, and remains, to inspire legal scholars, legal pratitioners, and policymakers to take sustainable consumption seriously in the domain of legal argumentation in theory and practice, analytical and normative propositions, here and after. Technological innovations will continue to puzzle lawyers to solve issues like planned obsolescence. Cyclical economic crises will continue raising concerns of over indebtedness. Opposition to climate change's policies, although endorsed by minorities yet, will continue demanding our surveillance to ensure the maintenance of our rights to a healthy environment.

By way of conclusion, rather than look back to the 25 substantive chapters and summarise the content of the aforementioned pages, this final chapter aims to look forward to a future research agenda. While it is for the readers to retrieve the lessons addressed by each contribution in this edited book, for us it is compelling to close this journey by looking forward.

Therefore, we suggest and seek for a future research agenda that aligns the concept of sustainable consumption with the interdisciplinary debate of law and development.[3] We do so for two reasons. Firstly, we aim to anticipate the answer to critics based on the misleading assumption that sustainable consumption is incompatible with the goal of sustainable development, which was not developed in depth in this edited book. Secondly, adding the concept of development to sustainable consumption and the law brings to the debate the element of interdisciplinarity that was not our main endeavour into this project.

As discussed throughout this edited book, sustainable consumption has a broader meaning than environmental friendly consumption. Sustainable consumption and the law, therefore, refers to legal arrangements that uphold individuals and institutions to consume *better* (e.g., eco-labels through contracts in food supply chains), *less* (e.g., waste management to reduce the consumption of raw material), and even 'prosuming' (e.g., solar panels in households).

Although the meaning of sustainable consumption alludes to less consumption, the former should not be read as incompatible with sustainable development. Scholars who read the effects of sustainable consumption in a short-run could

[1]Teubner and Fischer-Lescano (2004) and Koskenniemi (2006).

[2]Teubner (1997), Michaels (2005), Berman (2007) and Krisch (2010).

[3]Kennedy (2006), Trubek (2012) and Trubek et al. (2012).

contest its goals as incompatible with sustainable development mostly by measuring the latter through economic growth, i.e., GDP.

We claim, instead, that sustainable consumption is not incompatible sustainable development. On the contrary, an in-depth understanding of sustainable consumption renders impossible to separate sustainable consumption from sustainable development. To understand the intertwined relationship of those two policy goals, the narrative needs to go back and revive the first endorsement to sustainable development by the United Nations, which happened concomitantly with the support to environmental sustainability.

Sustainable development and environmental sustainability entered into the policy agenda of international organisations concomitantly in the 1970s. In 1972, the first UN Conference on the Human Environment in Stockholm introduced two major results: the consolidation of principles on the environmental protection[4]; and the UN Program for the Environment.[5] In the same year, the Club of Rome released the report entitled *The Limits of Growth*, which warned about dire consequences stemming from unbridgeable exploitation of natural resources. It is worth notice that sustainable development, despite mentioned in the Stockholm Conference, gained strength among government and non-governmental organisations only in 1980 with the release of the 'World Conservation Strategy: Living Resource Conservation for Sustainable Development'. The report explicitly stressed its meaning and, in doing so, did not hesitate to build a strong connection between economic development and sustainable consumption[6]:

> [H]uman beings, in the quest for economic development and enjoyment of the riches of nature, must come to terms with the reality of resource limitation and the carrying capacity of ecosystems, and must take account of the needs of future generations.

Following the 1983 World Commission on Environmental and Development, the report of the United Nations Commission on Environment and Development, whose chairperson was Gro Harlem Brundtland, became the phrase sustainable development currently used across the world. The report timely defined[7]:

> Sustainable development is development that meets the needs of the present without compromising the ability of future generations to meet their own needs

Under the auspices of the UN, the 1992 Rio de Janeiro Conference on Environment and Development, also called Rio Earth Summit, adopted the UN Framework Convention on Climate Change (UNFCCC) and the Convention on Biological Diversity (CBD), two agreements of utmost importance. Furthermore, in the wake of the Rio Summit, the UN adopted the Convention on Combat Desertification in 1994.

---

[4]United Nations General Assembly (1972).

[5]Meadows (1972).

[6]International Union for Conservation of Nature and Natural Resources, United Nations Environment Programme, World Wildlife Fund, Food and Agriculture Organization of the United Nations, and UNESCO (1980).

[7]World Commission on Environment and Development (1987), p. 41; Sachs (2015), p. 5.

Since the Rio Conference, sustainable development influenced international instruments and meeting to debate environmental themes. Besides the equity within the same generation, sustainable development embraces intergenerational ethics in a way that we cannot pass on the coming generations a planet worse than we received from our ancestors. On the occasion of World Summit on Sustainable Development, held in Johannesburg in 2002, the WSSD Plan of Implementation considered being interdependent and mutually reinforcing pillars economic development, social development and environmental protection. Identical commitment to sustainable development echoed on the Rio+20, in 2012, Summit, convened to celebrate the 1992 Rio Conference on Environment and Development.

With the intention to carefully examine measures carried out at different levels, governmental representatives met in Rio in 2012 to participate in the UN Conference on Sustainable Development, the so-called Rio+20. In the document 'The World we want', all participants observed[8]:

> We commit ourselves to re-invigorating the global partnership for sustainable development that we launched in Rio in 1992. We recognize the need to impart new momentum to our cooperative pursuit of sustainable development and commit to work together with Major Groups and other stakeholders in addressing the implementation gaps.

Lastly, the UN decided to push forward global environmental agenda by means of a set of targets designed to mobilise governments, business, NGOs, and civil society to be achieved from 2016 to 2030. Sustainable Development Goals, SDGs, the label usually known, was adopted at a special summit of the UN General Assembly in September 2015. Ten SDGs with specific orientations seek to guide policies at all levels of world policies. SDG 2, especially, pursues[9]:

> Achieve economic development within planetary boundaries. This goal means all countries have a right to economic development as long as that development respects planetary boundaries, ensures sustainable production and consumption patterns, and help to stabilize the global population by midcentury.

In the same vain, the United Nations Guidelines for Consumption Protection of 2016 foresaw[10]:

> 6. Unsustainable patterns of production and consumption, particularly in industrialized countries, are the major cause of the continued deterioration of the global environment. All Member States should strive to promote sustainable consumption patterns; developed countries should take the lead in achieving sustainable consumption patterns; developing countries should seek to achieve sustainable consumption patterns in their development process, having due regard for the principle of common but differentiated responsibilities. The special situation and needs of developing countries in this regard should be fully taken into account.

States certainly look for continued economic growth, notably in developing countries, but only growth that is environmentally sustainable within the planetary

---

[8]United Nations Development Group (2012).

[9]United Nations General Assembly (2015).

[10]United Nations Conference on Trade and Development (2016).

boundaries can be supported. Among the topics at stake are the ways to use and produce energy, food, design and, build cities.

In line with the political and legal discourse always supported by international organisations, especially United Nations, Jeffrey Sachs distinguishes an analytical and a normative concept of sustainable development. The first dimension involves four complex interacting systems dealing with a global economy: the social interactions of trust, ethics, inequality, and social support networks in communities. The analytical dimension, therefore, analyses the changes to complex Earth systems, such as climate and ecosystems, and studies the problems of governments and business.[11] The scientific study of complex systems interactions is decisive under this view.

Nevertheless, the normative concept of sustainable development defines and guides the future course of economic and social development of the planet. Thus, sustainable development in an ethical sense, preaches the widespread of economic progress, eradication of extreme poverty, encouragement of social trust through policies devoted to strengthening the community and the environmental protection where human-induced degradation is prevented. Lacking good governance, none of those ambitious intentions can be achieved.

Good governance encompasses not only governments but also multinational companies, which have become fundamental actors over the years. Such companies directly affect the human well-being whether they obey the law, respect the natural environment and help the communities to eliminate extreme poverty. To Sachs the concept of sustainable development envisions four basic goals: economic prosperity, social inclusion and cohesion; environmental sustainability and good governance by major social actors, including governments and business.

References above provide an adequate framework to grasp sustainable consumption from a new perspective. Concomitantly, this analytical framework shows how to approach sustainable consumption pursuant to risks that the likelihood of an environmental disaster poses.

References

Berman PS (2007) Global legal pluralism. South Calif Rev 80:1155

International Union for Conservation of Nature and Natural Resources et al (1980) World conservation strategy: living resource conservation for sustainable development. IUCN, Gland

Kennedy D (2006) Three globalizations of law and legal thought: 1850–2000. In: Trubek D, Santos A (eds) The new law and economics development: a critical appraisal. Cambridge University Press, Cambridge

Koskenniemi M (2006) Fragmentation of international law: difficulties arising from the diversification and expansion of international law: report of the study group of the international law commission. http://dag.un.org/handle/11176/260027. Accessed 2 Jan 2019

[11] Sachs (2015), p. 485.

Krisch N (2010) Beyond constitutionalism: the pluralist structure of postnational law. Oxford University Press, Oxford

Meadows DH (1972) The limits to growth: a report for the club of Rome's project on the predicament of mankind. Universe Books, New York

Michaels R (2005) The re-state-ment of non-state law: the state, choice of law, and challenge from global legal pluralism. Wayne Law Rev 51:1209

Sachs JD (2015) The age of sustainable development. Columbia University Press, New York

Teubner G (1997) Global Bukowina: legal pluralism in the world society. In: Teubner G (ed) Global Law Without a State. Dartmout

Teubner G, Fischer-Lescano A (2004) Regime-collisions: the vain search for legal unity in the fragmentation of global law. Mich J Int Law 25:999

Trubek DM (2012) Law and development 50 years on. Univ. of Wisconsin legal studies research paper no. 1212. https://papers.ssrn.com/sol3/papers.cfm?abstract_id=2161899. Accessed 2 Jan 2018

Trubek DM, Coutinho D, Schpiro MG (2012) Towards a new law and development: new state activism in Brazil and the challenge for legal institutions. World Bank Legal Rev:291–314

United Nations Conference on Trade and Development (2016) United Nations guidelines for consumer protection. https://unctad.org/en/PublicationsLibrary/ditccplpmisc2016d1_en.pdf. Accessed 15 Dec 2018

United Nations Development Group (2012) A million voices: the world we want. https://undg.org/wp-content/uploads/2016/12/The-World-we-Want.pdf. Accessed 14 Dec 2018

United Nations General Assembly (1972) United Nations conference on the human environment. https://www.refworld.org/docid/3b00f1c840.html. Accessed 14 Dec 2018

United Nations General Assembly (2015) Tranforming the world: the 2030 agenda for sustainable development, A/RES/20/1. https://www.refworld.org/docid/57b6e3e44.html. Accessed 14 Dec 2018

World Commission on Environment and Development (1987) Our common future. Oxford University Press, Oxford

The manufacturer's authorised representative in the EU is Springer
Nature Customer Service Centre GmbH, Europaplatz 3, 69115 Heidelberg,
Germany. If you have any concerns regarding our products, please
contact ProductSafety@springernature.com

Printed and bound by CPI Group (UK) Ltd, Croydon, CR0 4YY
30/09/2026
02225434-0018